7th RILEM International Conference on Cracking in Pavements

RILEM Bookseries

Volume 4

RILEM, The International Union of Laboratories and Experts in Construction Materials, Systems and Structures, founded in 1947, is a non-governmental scientific association whose goal is to contribute to progress in the construction sciences, techniques and industries, essentially by means of the communication it fosters between research and practice. RILEM's focus is on construction materials and their use in building and civil engineering structures, covering all phases of the building process from manufacture to use and recycling of materials. More information on RILEM and its previous publications can be found on www.RILEM.net.

For further volumes:
http://www.springer.com/series/8781

A. Scarpas · N. Kringos · I. Al-Qadi · A. Loizos
Editors

7th RILEM International Conference on Cracking in Pavements

Mechanisms, Modeling, Testing, Detection, Prevention and Case Histories
Volume 2

Editors

A.(Tom) Scarpas
Delft University of Technology
Delft
The Netherlands

Imad L. Al-Qadi
University of Illinois at Urbana-Champaign
Urbana-Champaign
USA

Niki Kringos
KTH Royal Institute of Technology
Stockholm
Sweden

Andreas Loizos
National Technical University of Athens
Athens
Greece

ISSN 2211-0844
ISBN 978-94-007-4565-0
Printed in 2 Volumes
DOI 10.1007/978-94-007-4566-7
Springer Dordrecht Heidelberg New York London

e-ISSN 2211-0852
e-ISBN 978-94-007-4566-7

Library of Congress Control Number: 2012937232

© RILEM 2012

This work is subject to copyright. All rights are reserved by the Publisher, whether the whole or part of the material is concerned, specifically the rights of translation, reprinting, reuse of illustrations, recitation, broadcasting, reproduction on microfilms or in any other physical way, and transmission or information storage and retrieval, electronic adaptation, computer software, or by similar or dissimilar methodology now known or hereafter developed. Exempted from this legal reservation are brief excerpts in connection with reviews or scholarly analysis or material supplied specifically for the purpose of being entered and executed on a computer system, for exclusive use by the purchaser of the work. Duplication of this publication or parts thereof is permitted only under the provisions of the Copyright Law of the Publisher's location, in its current version, and permission for use must always be obtained from Springer. Permissions for use may be obtained through RightsLink at the Copyright Clearance Center. Violations are liable to prosecution under the respective Copyright Law.
The use of general descriptive names, registered names, trademarks, service marks, etc. in this publication does not imply, even in the absence of a specific statement, that such names are exempt from the relevant protective laws and regulations and therefore free for general use.
While the advice and information in this book are believed to be true and accurate at the date of publication, neither the authors nor the editors nor the publisher can accept any legal responsibility for any errors or omissions that may be made. The publisher makes no warranty, express or implied, with respect to the material contained herein.

Printed on acid-free paper

Springer is part of Springer Science+Business Media (www.springer.com)

Preface

Because of vehicular and environmental loading, pavement systems have been deteriorating at a rapid rate. A series of six earlier RILEM Conferences on Cracking in Pavements in Liege (1989) (1993), Maastricht (1996), Ottawa (2000), Limoges (2004) and Chicago (2008) have clearly demonstrated that cracking constitutes one of the most detrimental, frequent and costly pavement deterioration modes.

Unfortunately, despite intense international efforts, there is still a strong need for methodologies that enable the construction and rehabilitation of crack resisting and/or tolerant pavements which at the same time are smooth, quiet, efficient, cost effective and environmentally friendly.

In the resent past, new materials, laboratory and in-situ testing methods and construction techniques have been introduced. In addition, modem computational techniques such as the finite element method enable the utilization of sophisticated constitutive models for realistic model-based predictions of the response of pavements. The **7th RILEM International Conference on Cracking in Pavements** aims to provide an international forum for the exchange of ideas, information and knowledge amongst experts involved in computational analysis, material production, experimental characterization, design and construction of pavements.

All submitted contributions were subjected to an exhaustive refereed peer review procedure by the Scientific Committee, the Editors and a large group of international experts on the topic. On the basis of their recommendations, 129 contributions which best suited the goals and the objectives of the Conference were chosen for presentation and inclusion in the Proceedings.

The strong message that emanates from the accepted contributions is that, by accounting for the idiosyncrasies of the response of pavement engineering materials, modern sophisticated constitutive models in combination with new experimental material characterization and construction techniques provide a powerful arsenal for understanding and designing against the mechanisms and the processes causing cracking and pavement response deterioration. As such they enable the adoption of truly "mechanistic" design methodologies.

The Editors would like to thank the Scientific Committee and the pavement engineering research community who took the responsibility of reviewing the manuscripts and ensuring the excellent quality of the accepted papers and the members of the Organizing Committee for their contribution to the management of the Conference affairs.

We hope that the Conference will contribute to the establishment of a new generation of asphalt and concrete pavement engineering design methodologies based on rational mechanics principles and in which computational techniques, advanced constitutive models and material characterisation techniques shall constitute the backbone of the design process.

Delft, March 2012

The Editors

A.(Tom) Scarpas
Delft University of Technology, The Netherlands

Niki Kringos
KTH Royal Institute of Technology, Sweden

Imad L. Al-Qadi
University of Illinois at Urbana-Champaign, USA

Andreas Loizos
National Technical University of Athens, Greece

Contents

Volume 1

Laboratory Evaluation of Asphalt Concrete Cracking Potential

Characterization of Asphalt Mixture's Fracture Resistance Using the Semi-Circular Bending (SCB) Test 1
L.N. Mohammad, M. Kim, M. Elseifi

The Flexural Strength of Asphalt Mixtures Using the Bending Beam Rheometer 11
M.I. Turos, A.C. Falchetto, G. Tebaldi, M.O. Marasteanu

Experimental Study of the Precracking 21
R. Mitiche_Kettab, A. Boulanouar, A. Bali

Comparison between 2PB and 4PB Methodologies Based on the Dissipated Energy Approach 31
M. Pettinari, C. Sangiorgi, F. Petretto, F. Picariello

Evaluation of Thermal Stresses in Asphalt Layers Incomparison with TSRST Test Results 41
M. Pszczoła, J. Judycki

A Four-Point Bending Test for the Bonding Evaluation of Composite Pavement 51
M. Hun, A. Chabot, F. Hammoum

Assessment of Cracking Resistance of Bituminous Mixtures by Means of Fenix Test 61
R. Miró, A. Martínez, F. Pérez-Jiménez, R. Botella, G. Valdés

Development of Dynamic Asphalt Stripping Machine for Better Prediction of Moisture Damage on Porous Asphalt in the Field 71
M.O. Hamzah, M.R.M. Hasan, M.F.C. van de Ven, J.L.M. Voskuilen

Effect of Wheel Track Sample Geometry on Results 83
P.M. Muraya, C. Thodesen

Performance of 'SAMI'S in Simulative Testing 93
O.M. Ogundipe, N.H. Thom, A.C. Collop, J. Richardson

Towards a New Experimental and Numerical Protocol for Determining Mastic Viscosity 103
E. Hesami, D. Jelagin, B. Birgisson, N. Kringos

Interference Factors on Tests of Asphalt Biding Agents Destinated to Paving Works Using a Statistic Study 115
E.F. Amorim, A.C. de Lara Fortes, L.F.M. Ribeiro

Development of an Accelerated Weathering and Reflective Crack Propagation Test Methodology 125
K. Grzybowski, G.M. Rowe, S. Prince

Pavement Cracking Detection

The Use of Ground Penetrating Radar, Thermal Camera and Laser Scanner Technology in Asphalt Crack Detection and Diagnostics 137
T. Saarenketo, A. Matintupa, P. Varin

Asphalt Thermal Cracking Analyser (ATCA) 147
H. Bahia, H. Tabatabaee, R. Velasquez

Using 3D Laser Profiling Sensors for the Automated Measurement of Road Surface Conditions 157
J. Laurent, J.F. Hébert, D. Lefebvre, Y. Savard

Pavement Crack Detection Using High-Resolution 3D Line Laser Imaging Technology 169
Y. (James) Tsai, C. Jiang, Z. Wang

Detecting Unbounded Interface with Non Destructive Techniques 179
J.-M. Simonin, C. Fauchard, P. Hornych, V. Guilbert, J.-P. Kerzrého, S. Trichet

New Field Testing Procedure to Measure Surface Stresses in Plain Concrete Pavements and Structures 191
D.I. Castaneda, D.A. Lange

**Strain Measurement in Pavements with a Fibre Optics Sensor
Enabled Geotextile** 201
O. Artières, M. Bacchi, P. Bianchini, P. Hornych, G. Dortland

Field Investigation of Pavement Cracking

**Evaluating the Low Temperature Resistance of the Asphalt Pavement
under the Climatic Conditions of Kazakhstan** 211
B. Teltayev, E. Kaganovich

**Millau Viaduct Response under Static and Moving Loads Considering
Viscous Bituminous Wearing Course Materials** 223
S. Pouget, C. Sauzéat, H. Di Benedetto, F. Olard

**Material Property Testing of Asphalt Binders Related to Thermal
Cracking in a Comparative Site Pavement Performance Study** 233
A.T. Pauli, M.J. Farrar, P.M. Harnsberger

**Influence of Differential Displacements of Airport Pavements on
Aircraft Fuelling Systems** 245
*A.L. Rolim, L.A.C.M. Veloso, H.N.C. Souza, P.L. de O. Filho,
L.V. de A. Monteiro*

**Rehabilitation of Cracking in Epoxy Asphalt Pavement on Steel
Bridge Decks** 255
L. Chen, Z. Qian

Long-Term Pavement Performance Evaluation 267
L. Petho, C. Toth

Structural Assessment of Cracked Flexible Pavement 277
L.W. Cheung, P.K. Kong, G.L.M. Leung, W.G. Wong

**Comparison between Optimum Tack Coat Application Rates as
Obtained from Tension- and Torsional Shear-Type Tests** 287
S. Hakimzadeh, N.A. Kebede, W.G. Buttlar

**Using Life Cycle Assessment to Optimize Pavement
Crack-Mitigation** 299
A.A. Butt, D. Jelagin, B. Birgisson, N. Kringos

**Preliminary Analysis of Quality-Related Specification Approach for
Cracking on Low Volume Hot Mix Asphalt Roads** 307
D.J. Mensching, L.M. McCarthy, J.R. Albert

**Evaluating Root Resistance of Asphaltic Pavement Focusing on
Woody Plants' Root Growth** 317
S. Ishihara, K. Tanaka, Y. Shinohara

20 Years of Research on Asphalt Reinforcement – Achievements and Future Needs 327
A.H. De Bondt

Concrete Pavement Strength Investigations at the FAA National Airport Pavement Test Facility 337
E.H. Guo, D.R. Brill, H. Yin

Pavement Cracking Modeling Response, Crack Analysis and Damage Prediction

The Effects Non-uniform Contact Pressure Distribution Has on Surface Distress of Flexible Pavements Using a Finite Element Method 347
D.B. Casey, A.C. Collop, G.D. Airey, J.R. Grenfell

Finite Element Analysis of a New Test Specimen for Investigating Mixed Mode Cracks in Asphalt Overlays 359
M.R.M. Aliha, M. Ameri, A. Mansourian, M.R. Ayatollahi

Modelling of the Initiation and Development of Transverse Cracks in Jointed Plain Concrete Pavements for Dutch Conditions 369
M. Pradena, L. Houben

Pavement Response Excited by Road Unevennesses Using the Boundary Element Method 379
A. Almeida, L.P. Santos

Discrete Particle Element Analysis of Aggregate Interaction in Granular Mixes for Asphalt: Combined DEM and Experimental Study 389
G. Dondi, A. Simone, V. Vignali, G. Manganelli

Recent Developments and Applications of Pavement Analysis Using Nonlinear Damage (PANDA) Model 399
E. Masad, R.A. Al-Rub, D.N. Little

Laboratory and Computational Evaluation of Compact Tension Fracture Test and Texas Overlay Tester for Asphalt Concrete 409
E.V. Dave, S. Ahmed, W.G. Buttlar

Crack Fundamental Element (CFE) for Multi-scale Crack Classification 419
Y. Huang, Y. (James) Tsai

Cracking Models for Use in Pavement Maintenance Management 429
A. Ferreira, R. Micaelo, R. Souza

Contents

Multi-cracks Modeling in Reflective Cracking 441
J. Pais, M. Minhoto, S. Shatnawi

Using Black Space Diagrams to Predict Age-Induced Cracking 453
G. King, M. Anderson, D. Hanson, P. Blankenship

Top-Down Cracking Prediction Tool for Hot Mix Asphalt Pavements ... 465
C. Baek, S. Thirunavukkarasu, B.S. Underwood, M.N. Guddati, Y.R. Kim

A Theoretical Investigation into the 4 Point Bending Test 475
M. Huurman, R. Gelpke, M.M.J. Jacobs

Multiscale Micromechanical Lattice Modeling of Cracking in Asphalt Concrete 487
A.D. Banadaki, M.N. Guddati, Y.R. Kim, D.N. Little

Accelerated Pavement Performance Modeling Using Layered Viscoelastic Analysis 497
M. Eslaminia, S. Thirunavukkarasu, M.N. Guddati, Y.R. Kim

Numerical Investigations on the Deformation Behavior of Concrete Pavements 507
V. Malárics, H.S. Müller

Fatigue Behaviour Modelling in the Mechanistic Empirical Pavement Design 517
M.F. Saleh

Theoretical Analysis of Overlay Resisting Crack Propagation in Old Cement Concrete Pavement 527
Y. Zhong, Y. Gao, M. Li

Calibration of Asphalt Concrete Cracking Models for California Mechanistic-Empirical Design (CalME) 537
R. Wu, J. Harvey

Performance of Concrete Pavements and White Toppings

Shear Failure in Plain Concrete as Applied to Concrete Pavement Overlays 549
Y. Xu, J.N. Karadelis

Influence of Residual Stress on PCC Pavement Potential Cracking 561
X. Li, D. Feng, J. Chen

Plain Concrete Cyclic Crack Resistance Curves under Constant and Variable Amplitude Loading 571
N.A. Brake, K. Chatti

Influence of External Alkali Supply on Cracking in Concrete Pavements 581
C. Sievering, R. Breitenbücher

Plastic Shrinkage Cracking Risk of Concrete – Evaluation of Test Methods 591
P. Fontana, S. Pirskawetz, P. Lura

Compatibility between Base Concrete Made with Different Chemical Admixtures and Surface Hardener 601
M.T. Pinheiro-Alves, A.R. Sequeira, M.J. Marques, A.B. Ribeiro

Compatibility between a Quartz Surface Hardener and Different Base Concrete Mixtures 607
M.T. Pinheiro-Alves, A. Fernandes, M.J. Marques, A.B. Ribeiro

Suitable Restrained Shrinkage Test for Fibre Reinforced Concrete: A Critical Discussion 615
A. Reggia, F. Minelli, G.A. Plizzari

Influence of Chemical Admixtures and Environmental Conditions on Initial Hydration of Concrete 625
A.B. Ribeiro, V.A. Medina, A.M. Gomes

Application of Different Fibers to Reduce Plastic Shrinkage Cracking of Concrete 635
T. Rahmani, B. Kiani, M. Bakhshi, M. Shekarchizadeh

Volume 2

Fatigue Cracking and Damage Characterization of Asphalt Concrete

Evaluation of Fatigue Life Using Dissipated Energy Methods 643
C. Maggiore, J. Grenfell, G. Airey, A.C. Collop

Measurement and Prediction Model of the Fatigue Behavior of Glass Fiber Reinforced Bituminous Mixture 653
I.M. Arsenie, C. Chazallon, A. Themeli, J.L. Duchez, D. Doligez

Fatigue Cracking in Bituminous Mixture Using Four Point Bending Test 665
Q.T. Nguyen, H. Di Benedetto, C. Sauzéat

Top-Down and Bottom-Up Fatigue Cracking of Bituminous Pavements Subjected to Tangential Moving Loads 675
Z. Ambassa, F. Allou, C. Petit, R.M. Eko

Contents XIII

Fatigue Performance of Highly Modified Asphalt Mixtures in
Laboratory and Field Environment 687
R. Kluttz, J.R. Willis, A.A.A. Molenaar, T. Scarpas, E. Scholten

A Multi-linear Fatigue Life Model of Flexible Pavements under
Multiple Axle Loadings 697
F. Homsi, D. Bodin, D. Breysse, S. Yotte, J.M. Balay

Aggregate Base/Granular Subbase Quality Affecting Fatigue
Cracking of Conventional Flexible Pavements in Minnesota 707
Y. Xiao, E. Tutumluer, J. Siekmeier

Fatigue Performance of Asphalt Concretes with RAP Aggregates and
Steel Slags 719
M. Pasetto, N. Baldo

Fatigue Characterization of Asphalt Rubber Mixtures with Steel
Slags 729
M. Pasetto, N. Baldo

Fatigue Cracking of Gravel Asphalt Concrete: Cumulative Damage
Determination 739
F.P. Pramesti, A.A.A. Molenaar, M.F.C. van de Ven

Fatigue Resistance and Crack Propagation Evaluation of a
Rubber-Modified Gap Graded Mixture in Sweden 751
*W. Zeiada, M. Souliman, J. Stempihar, K.P. Biligiri, K. Kaloush, S. Said,
H. Hakim*

On the Fatigue Criterion for Calculating the Thickness of Asphalt
Layers.. 761
M. Livneh

Acoustic Techniques for Fatigue Cracking Mechanisms
Characterization in Hot Mix Asphalt (HMA) 771
M. Diakhaté, N. Larcher, M. Takarli, N. Angellier, C. Petit

Fatigue Characteristics of Sulphur Modified Asphalt Mixtures 783
A. Cocurullo, J. Grenfell, N.I.M. Yusoff, G. Airey

Effect of Moisture Conditioning on Fatigue Properties of Sulphur
Modified Asphalt Mixtures 793
A. Cocurullo, J. Grenfell, N.I.M. Yusoff, G. Airey

Fatigue Investigation of Mastics and Bitumens Using Annular Shear
Rheometer Prototype Equipped with Wave Propagation System 805
M. Buannic, H. Di Benedetto, C. Ruot, T. Gallet, C. Sauzéat

Effect of Steel Fibre Content on the Fatigue Behaviour of Steel Fibre
Reinforced Concrete 815
M.F. Saleh, T. Yeow, G. MacRae, A. Scott

Effect of Specimen Size on Fatigue Behavior of Asphalt Mixture in
Laboratory Fatigue Tests 827
N. Li, A.A.A. Molenaar, A.C. Pronk, M.F.C. van de Ven, S. Wu

Evaluation of the Effectiveness of Asphalt Concrete Modification

Long-Life Overlays by Use of Highly Modified Bituminous Mixtures ... 837
D. Simard, F. Olard

Investigation into Tensile Properties of Polymer Modified Bitumen
(PMB) and Mixture Performance 849
E.T. Hagos, M.F.C. van de Ven, G.M. Merine

Effect of Polymer Dispersion on the Rheology and Morphology of
Polymer Modified Bituminous Blend 859
I. Kamaruddin, N.Z. Habib, I.M. Tan, M. Komiyama, M. Napiah

Effect of Organoclay Modified Binders on Fatigue Performance 869
N. Tabatabaee, M.H. Shafiee

Effects of Polymer Modified Asphalt Emulsion (PMAE) on Pavement
Reflective Cracking Performance 879
Y. Chen, G. Tebaldi, R. Roque, G. Lopp

Characterization of Long Term Field Aging of Polymer Modified
Bitumen in Porous Asphalt 889
D. van Vliet, S. Erkens, G.A. Leegwater

Bending Beam Rheological Evaluation of Wax Modified Asphalt
Binders 901
G.L. Baumgardner, G.M. Rowe, G.H. Reinke

Reducing Asphalt's Low Temperature Cracking by Disturbing Its
Crystallization 911
E.H. Fini, M.J. Buehler

Mechanistic Evaluation of Lime-Modified Asphalt Concrete
Mixtures 921
A.H. Albayati

Crack Growth Parameters and Mechanisms

Determination of Crack Growth Parameters of Asphalt Mixtures 941
M.M.J. Jacobs, A.H. De Bondt, P.C. Hopman, R. Khedoe

Differential Thermal Contraction of Asphalt Components 953
I. Artamendi, B. Allen, C. Ward, P. Phillips

**Mechanistic Pavement Design Considering Bottom-Up and
Top-Down-Cracking** 963
A. Walther, M. Wistuba

Strength and Fracture Properties of Aggregates 975
I. Artamendi, C. Ward, B. Allen, P. Phillips

**Cracks Characteristics and Damage Mechanism of Asphalt Pavement
with Semi-rigid Base** 985
A. Sha, S. Tu

**Comparing the Slope of Load/Displacement Fracture Curves of
Asphalt Concrete** 997
A.F. Braham, C.J. Mudford

**Cracking Behaviour of Bitumen Stabilised Materials (BSMs):
Is There Such a Thing?** 1007
K. Jenkins

**Experimental and Theoretical Investigation of Three Dimensional
Strain Occurring Near the Surface in Asphalt Concrete Layers** 1017
D. Grellet, G. Doré, J.P. Kerzrého, J.M. Piau, A. Chabot, P. Hornych

**Reasons of Premature Cracking Pavement Deterioration –
A Case Study** 1029
D. Sybilski, W. Bańkowski, J. Sudyka, L. Krysiński

**Effect of Thickness of a Sandwiched Layer of Bitumen between Two
Aggregates on the Bond Strength: An Experimental Study** 1039
S. Mondal, A. Das, A. Ghatak

**Hypothesis of Existence Semicircular Shaped Cracks on Asphalt
Pavements** 1049
D. Hribar

**Quantifying the Relationship between Mechanisms of Failure and the
Deterioration of CRCP under APT: Cointegration of Non-stationary
Time Series** 1059
E.R. de Vos

**Influence of Horizontal Traction on Top-Down Cracking in Asphalt
Pavements** 1069
C.S. Gideon, J.M. Krishnan

Evaluation, Quantification and Modeling of Asphalt Healing Properties

Predicting the Performance of the Induction Healing Porous Asphalt Test Section 1081
Q. Liu, E. Schlangen, M.F.C. van de Ven, G. van Bochove, J. van Montfort

Determining the Healing Potential of Asphalt Concrete Mixtures – A Pragmatic Approach 1091
S. Erkens, D. van Vliet, A. van Dommelen, G.A. Leegwater

Asphalt Durability and Self-healing Modelling with Discrete Particles Approach 1103
V. Magnanimo, H.L. ter Huerne, S. Luding

Quantifying Healing Based on Viscoelastic Continuum Damage Theory in Fine Aggregate Asphalt Specimen 1115
S. Palvadi, A. Bhasin, A. Motamed, D.N. Little

Evaluation of WMA Healing Properties Using Atomic Force Microscopy 1125
M. Nazzal, S. Kaya, L. Abu-Qtaish

Cracking and Healing Modelling of Asphalt Mixtures 1135
J. Qiu, M.F.C. van de Ven, E. Schlangen, S. Wu, A.A.A. Molenaar

Reinforcement and Interlayer Systems for Crack Mitigation

Effects of Glass Fiber/Grid Reinforcement on the Crack Growth Rate of an Asphalt Mix 1145
C.C. Zheng, A. Najd

Asphalt Rubber Interlayer Benefits in Minimizing Reflective Cracking of Overlays over Rigid Pavements 1157
S. Shatnawi, J. Pais, M. Minhoto

Performance of Anti-cracking Interface Systems on Overlaid Cement Concrete Slabs – Development of Laboratory Test to Simulate Slab Rocking 1169
K. Denolf, J. De Visscher, A. Vanelstraete

The Use of Bituminous Membranes and Geosynthetics in the Pavement Construction 1181
P. Hyzl, M. Varaus, D. Stehlik

Stress Relief Asphalt Layer and Reinforcing Polyester Grid as Anti-Reflective Cracking Composite Interlayer System in Pavement Rehabilitation 1189
G. Montestruque, L. Bernucci, M. Fritzen, L.G. da Motta

Characterizing the Effects of Geosynthetics in Asphalt Pavements 1199
S. Vismara, A.A.A. Molenaar, M. Crispino, M.R. Poot

Geogrid Interlayer Performance in Pavements: Tensile-Bending Test for Crack Propagation 1209
A. Millien, M.L. Dragomir, L. Wendling, C. Petit, M. Iliescu

Theoretical and Computational Analysis of Airport Flexible Pavements Reinforced with Geogrids 1219
M. Buonsanti, G. Leonardi, F. Scopelliti

Optimization of Geocomposites for Double-Layered Bituminous Systems 1229
F. Canestrari, E. Pasquini, L. Belogi

Sand Mix Interlayer Retarding Reflective Cracking in Asphalt Concrete Overlay 1241
J. Baek, I.L. Al-Qadi

Full Scale Tests on Grid Reinforced Flexible Pavements on the French Fatigue Carrousel 1251
P. Hornych, J.P. Kerzrého, J. Sohm, A. Chabot, S. Trichet, J.L. Joutang, N. Bastard

Thermal and Low Temperature Cracking of Pavements

Low-Temperature Cracking of Recycled Asphalt Mixtures 1261
N. Tapsoba, C. Sauzéat, H. Di Benedetto, H. Baaj, M. Ech

Thermal Cracking Potential in Asphalt Mixtures with High RAP Contents 1271
Q. Aurangzeb, I.L. Al-Qadi, W.J. Pine, J.S. Trepanier, I.M. Abuawad

Micro-mechanical Investigation of Low Temperature Fatigue Cracking Behaviour of Bitumen 1281
P.K. Das, D. Jelagin, B. Birgisson, N. Kringos

The Study on Evaluation Methods of Asphalt Mixture Low Temperature Performance 1291
T. Yiqiu, Z. Lei, S. Liyan, J. Lun

Cracking Propensity of WMA and Recycled Asphalts

Permanent Deformations of WMAs Related to the Bituminous Binder Temperature Susceptibility 1301
F. Petretto, M. Pettinari, C. Sangiorgi, A. Simone

Cracking Resistance of Recycled Asphalt Mixtures in Relation to Blending of RA and Virgin Binder 1311
M. Mohajeri, A.A.A. Molenaar, M.F.C. Van de Ven

Warm Mix Asphalt Performance Modeling Using the Mechanistic-Empirical Pavement Design Guide 1323
A. Buss, R.C. Williams

Shrinkage and Creep Performance of Recycled Aggregate Concrete 1333
J. Henschen, A. Teramoto, D.A. Lange

Effect of Reheating Plant Warm SMA on Its Fracture Potential 1341
Z. Leng, I.L. Al-Qadi, J. Baek, M. Doyen, H. Wang, S. Gillen

Fatigue Cracking Characteristics of Cold In-Place Recycled Pavements 1351
A. Loizos, V. Papavasiliou, C. Plati

Author Index 1361

RILEM Publications 1367

RILEM Publications Published by Springer 1377

Evaluation of Fatigue Life Using Dissipated Energy Methods

Cinzia Maggiore[1], James Grenfell[1], Gordon Airey[1], and Andrew C. Collop[2]

[1] University of Nottingham, UK
[2] De Montfort University, Leicester, UK

Abstract. Flexural fatigue is a process of cumulative damage. It is also one of the main failure modes in asphalt mixtures and flexible pavement structures. This means good prediction of a pavement's fatigue life will help to develop and improve pavement design procedures.

Different approaches have been used to characterise the fatigue behaviour of asphalt including phenomenological-based models, fracture mechanics and dissipated energy methods.

This paper evaluates fatigue, paying particular attention to the dissipated energy criteria developed so far for asphalt materials. Two methods have been chosen to measure dissipated energy in tension-compression fatigue tests. Fatigue life obtained from the two different methods is compared with the traditional fatigue failure criterion Nf_{50}, identifing both their advantages and disadvantages.

An asphalt mixture has been chosen for the laboratory activity and different tension-compression fatigue tests have been undertaken at 15 Hz, at 20°C at different stress levels, in order to better understand fatigue behaviour of asphalt mixtures.

1 Introduction

Flexural fatigue is one of the main failure modes in asphalt mixtures. Different approaches are usually used to characterise fatigue behaviour of HMA: the phenomenological approach; the fracture mechanics approach; and the energy and dissipated energy approach. This paper focuses on the latter of the three methods.

1.1 Dissipated Energy Concept

When a constant load is applied in a viscoelastic material as asphalt, the deformation and therefore strain increases over time and, when the load is removed, it tends recover some of the deformation but maintains some residual deformation [1].

Viscoelastic materials are characterised by a hysteresis loop because the unloaded material traces a different path to that when loaded (phase lag is recorded between the applied stress and the measured strain), as shown in Figures 1; in this case the energy is dissipated in the form of mechanical work, heat generation, or damage.

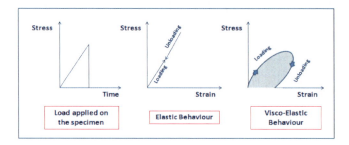

Fig. 1. Elastic and Visco-Eastic behaviour

The area of the hysteresis loop represents the dissipated energy in a load cycle and the following equation can be used to calculate its value in a linear viscoelastic material:

$$W_i = \pi \sigma_i \varepsilon_i \sin\varphi_i \qquad (1)$$

Where: W_i is the dissipated energy in cycle i, σ_i is the stress level in cycle i, ε_i is the strain level in cycle i, and φ_i is the phase angle in cycle i.

During a fatigue test, where repeated stresses are applied to a sample below the failure stress, the stiffness reduces and microcracks are induced in the material; therefore the dissipated energy, W, varies per loading cycle and it, usually, increases for controlled stress tests and decreases for controlled strain tests, as shown in the Figure 2 [2].

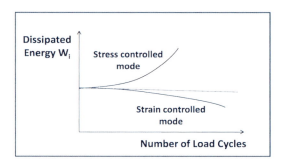

Fig. 2. Dissipated Energy versus Load cycle for different loading modes [2]

The hysteresis loop and the stress and strain sinusoidal waveforms start to change; in the beginning the waveform path is well defined, then it starts to deform (initial failure: cracking) and in the end it is flat (fatigue failure) [3].

Different energy methods have been developed to evaluate the fatigue life in asphalt material [4-7].

Van Dijk was one of the earliest researchers to apply the dissipated energy concepts to study fatigue in asphalt mixtures and he determined an equation that relates the cumulative dissipated energy (CDE) to the number of cycles to failure [8-9].

Some other researchers believe that in order to have damage in the material there should be a change in the hysteresis loop, and thus a change in dissipated energy [5,10,11]. Since the dissipated energy is history dependent, it is a parameter well related to the accumulated damage in a specimen.

The same researchers have suggested the Ratio of Dissipated Energy Change, RDEC as parameter to describe fatigue in asphalt materials. They believe that the RDEC is a true indicator of damage because it is able to eliminate the other forms of dissipated energy due to mechanical work or heat generation; so it can be considered a good parameter to describe the fatigue process in asphalt

2 Laboratory Testing

2.1 Materials

A 10 mm Dense Bitumen Macadam (DBM) or asphalt concrete was chosen for the experimental work. This type of mixture is the most commonly used in UK; it is a continuously graded mixture relying on aggregate interlock for its mechanical properties. A 100 Penetration grade bitumen was chosen for the mixture. The aggregates type selected was a crushed limestone.

2.2 Specimen Preparation

For the experimental work, a dogbone shaped specimen was choosen to undertake tension-compression tests. The specimen is 150 mm high with a 50 mm square central cross section. (see Figure 4). After placing the asphalt mixture in a characteristic mould, a kango vibrating hammer was used to compact the material.

Before starting fatigue tests, each specimen was glued between two plates and to ensure parallel alignment of the plates with specimen, a right angle jig was used.

2.3 Tension-Compression Test Procedure

During a fatigue test the stress is applied using a continuous pulsating load through the actuator in a uniaxial manner to the dogbone specimen as shown in Figure 3. Two Linear Variable Differential Transducers (LVDTs) are glued vertically to the specimen at diametrically opposite positions. They record the axial deformation in the specimen during the test; the average of the two measurements is considered. Two strain transducers were used to measure the horizontal deformation as illustrated in Figure 4.

 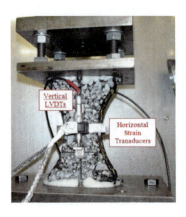

Fig. 3. Tension-Compression Test **Fig. 4.** Strain transducers

3 Results and Discussion

For the preliminary laboratory activity, fatigue tests were conducted at 20 °C and 15 Hz, considering different load levels from 1.5 kN to 2.1 kN, equivalent from 600 KPa to 840 KPa. Figure 5 shows a schematic of the load chosen for the experimental work.

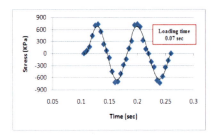

Fig. 5. Test performed with and without rest periods

The cumulative Dissipated Energy (CDE) approach was the first method considered. CDE is the total energy dissipated by the material during the fatigue test; in particular it is the sum of all areas within the stress-strain hysteresis loop for every cycle until failure, relating the fatigue behaviour to both initial and final cycles. (see Figure 6).

Van Dijk determined an equation that relates the CDE to the number of cycles to failure as follows:

$$W_f = A(N_f)^z \tag{2}$$

Where: W_f is the cumulative dissipated energy to failure, N_f is the number of load cycle to failure, and A, z are the mixture dependent constants (experimentally determined).

Figure 11 shows that the correlation between the increasing of the CDE and the decreasing of stiffness modulus could exist; but CDE behaviour changes depend on the mode of loading [4], (see Figure 2); also, this parameter does not distinguish the amount of DE due to damage rather than viscoelasticity. For all those reasons, CDE method was not taken in consideration for the final analysis.

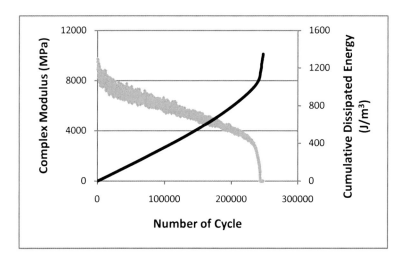

Fig. 6. Stiffness and CDE behaviour

According to Hopman et al [14, 2] cumulative dissipated energy is well correlated to the crack initiation N1, by mean of the Energy Ratio R_σ. The energy ratio is defined as follows:

$$Energy\ Ratio = \frac{nW_0}{W_i} = \frac{n(\pi\sigma_0\varepsilon_0 sin\varphi_0)}{\pi\sigma_i\varepsilon_i sin\varphi_i} \qquad (3)$$

Where W_0 is the energy dissipated in the first cycle, W_i is the energy dissipated at i^{th}-cycle. Figure 7 shows the plot of the energy ratio and the complex modulus. Researchers [4,16] believed that the peak value of the energy ratio plot represent the number of fatigue cycles in which the crack initiates (N1). Researchers said that this phenomenon usually occurs in a range between 40-50% reduction of the initial complex modulus value. During the experimental work, it was found that this value is bigger than the traditional number of fatigue to failure Nf_{50} (when the initial stiffness reduces of 50%).

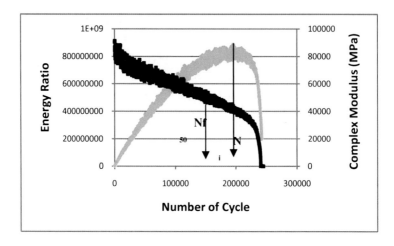

Fig. 7. Energy Ratio and complex modulus behaviour

As mentioned before, dissipated energy changes during a fatigue test, due to the beginning of microcracking during the fatigue process; the evolution of dissipated energy is shown in Figure 8.

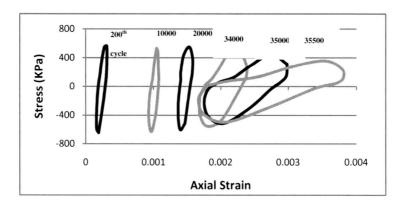

Fig. 8. Evolution of Dissipated Energy in a fatigue test

As can be seen, in the first stage of the test, the hysteresis loop is well defined and the area is small. After this the loop changes shape. It starts to rotate and the area becomes bigger. At the end of the tests, dissipated energy is usually characterised by a flat and irregular hysteresis loop. Thus, change in dissipated energy should be a good parameter to describe fatigue phenomenon in asphaltic material and it is considered the starting point for the development of the fatigue model.

Some researchers [5, 10, 15, 18] have suggested the Ratio of Dissipated Energy Change, RDEC as a parameter to describe fatigue in asphalt materials. The same researchers believed that the RDEC is a true indicator of damage because it is able to eliminate the other forms of dissipated energy due to mechanical work or heat generation; so it can be considered a good parameter to describe the fatigue process in asphalt, and is calculated with the following expression:

$$\text{RDEC} = \frac{DE_{n+1} - DE_n}{DE_n} \qquad (4)$$

Where: RDEC is ratio of the dissipated energy change per load cycle, DE_n is dissipated energy produced in load cycle n, and DE_{n+1} is dissipated energy produced in load cycle n+1.

Figure 9 shows the variation of the RDEC plotted against the Number of load cycles.

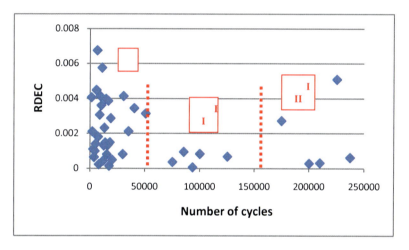

Fig. 9. The variation of the RDEC as a function of Load Cycles

It is possible to notice three main phases during a fatigue test. The RDEC, after a rapid decrease (I stage), reaches a plateau stage in which a plateau value (PV) can be obtained; this represents an energy plateau where an almost constant rate of energy input is being turned into damage. It is verified that PV is uniquely related to fatigue life. After the RDEC increases rapidly until true fatigue failure (III stage). The same researchers correlated the plateau value with a number of fatigue to failure cycles by means of statistical approach, by means of the following equation [11]:

$$\text{PV} = cN_f^d = 0.4428 N_f^{-1.1102} \qquad (5)$$

Also the stiffness follows a three stage evolution process (see Figure 10):
after a rapid evolution of stiffness (phase I), stiffness decrease seems more regular

(phase II); fracture occurs in the final stage (phase III) and it is characterised by an acceleration of stiffness drop. Therefore another good point is a correlation between the three-stage evolution of stiffness with the three-stage evolution of dissipated energy.

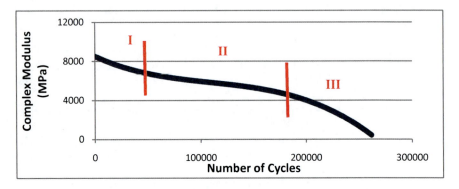

Fig. 10. Variation of stiffness as a funciotn of Load Cycles

Many researchers [6, 7] have defined failure as the point at which the specimen's flexural stiffness is reduced to 50% of the initial flexural stiffness in fatigue testing. Usually, this initial stiffness is defined as the specimen's flexural stiffness measured at the 50th load cycle. This traditional failure criterion was considered to make the comparison between the dissipated energy approaches. It was found that cumulative dissipated energy approach tends to overestimate fatigue life in asphalt mixture and the RDEC approach tends to underestimate it, because the plateau stage is often in a range of the number of cycles to failure lower than the N_{f50}. (see Figure 11).

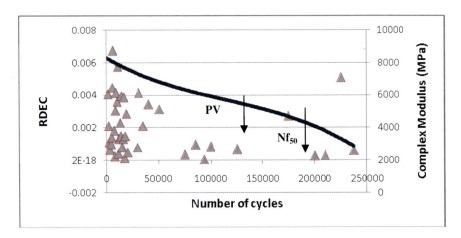

Fig. 11. Nf_{50} and PV

Table 1 shows the different range of cycles to failure for a fatigue test undertaken at 15 Hz, at a temperature of 20°C, applying a load of 1.5 KN (600KPa).

Table 1. Comparison of fatigue life of a test carried put at 1.5 KN.

Failure criterion	Nf_{50}	Energy Ratio	RDEC
Cycles to failure	165000-170000	180000-190000	125000-135000

4 Conclusions

Different approaches have been used to characterise fatigue behaviour of a particular asphalt mixture (10 mm dense bitumen macadam) including phenomenological-based models, fracture mechanics and dissipated energy methods. This work focuses on the latter approach.

Comparing fatigue lives based on these different approaches is the main point of this paper.

The following conclusions can be made:

- Two methods were compared with the traditional failure criterion Nf_{50}: energy ratio and the RDEC approach. It was found that the energy ratio tends to overestimate fatigue life for the asphalt mixture considered.
- The RDEC principles are very interesting because they focus the attention to the change of DE, but were found to underestimate fatigue life for the asphalt mixture considered.
- The change in dissipated energy is a good parameter to describe the fatigue phenomenon in asphaltic material, because it takes into account only the damage effect in the material (not contribution due to mechanical work or viscoelastic properties of the material). For this reason it is considered the starting point for the future work. Laboratory testing such as 2 Point Bending and 4 Point Bending have already been planned to undertake more fatigue tests under different loading conditions.

The fatigue data used in this paper are mainly based on tension-compression tests for a 10 mm DBM. Further investigation should be carried out to verify if the results found are still valid for other asphalt mixtures, mode of loading (i.e. strain controlled mode) and testing configurations (i.e. bending tests).

References

[1] Hamed, F.: Evaluation of Fatigue Resistance for Modified Asphalt Concrete Mixtures Based on Dissipated Energy Concepts. Department of Civil engineering and Geodesy, Technische Universitat Darmstadt, Germany (2010)

[2] Rowe, G.: Application of the dissipated energy concepts to fatigue cracking in asphalt pavements. Departement of Civil Engineering. University of Nottingham, Nottingham, UK (1996)
[3] Al-Khateen, G., Shenoy, A.: A distinctive fatigue failure. Journal of the Association of Asphalt Paving Technologists (2004)
[4] Rowe, G.: Performance of Asphalt Mixtures in the Trapezoidal Fatigue Test. In: Proceedings of the Association of Asphalt Paving Technologist (1993)
[5] Shen, S., Carpenter, S.: Dissipate Energy Concepts for HMA Performance: Fatigue and Healing, Departement of Civil and Environmental Engineering. University of Illinois at Urbana-Champaign: Urbana, Illinois (2007)
[6] SHRPA-404, Fatigue Response of Asphalt-Aggregate Mixes, Asphalt Research Program Institute of Transportation Studies University of California, Berkeley. Strategic Highway Research Program National Research Council, Washington, DC (1994)
[7] Baburamani, P.: Asphalt fatigue life prediction models: a literature review, Research Report ARR 334. ARRB Transport Research (1993)
[8] Van Dijk, W.: Pratical Fatigue Characterization of Bituminous Mixes. In: Proceedings of the Association of Asphalt Paving Technologist (1975)
[9] Van Dijk, W., Visser, W.: The Energy Approach to Fatigue for Pavement Design. In: Proceedings of the Association of Asphalt Paving Technologist (1977)
[10] Ghuzlan, K., Carpenter, S.: Energy-Derived, damage-based failure criterion for fatigue testing. Transportation Research Record - TRR, 1723 (2000)
[11] Ghuzlan, K., Carpenter, S.: Fatigue damage analysis in asphalt concrete mixtures using the dissipated energy approach. Canadian Journal of Civil Engineering (2001)
[12] Kim, R., Little, D., Benson, F.: Chemical and mechanical evaluation on healing mechanism of asphalt concrete. Association of Asphalt Paving Technologists (1990)
[13] Kim, Y., Little, D., Burghardt, R.: SEM Analysis on Fracture and Healing of Sand-Asphalt Mixtures. Journal of Materials in Civil Engineering, 140–153 (1991)
[14] Hopman, Kunst, Pronk: A renewed interpretation method for fatigue measurement, verification of Miner's rule. In: 4th Eurobitume Symposium Madrid, pp. 557–561 (1989)
[15] Carpenter, S., Ghuzlan, K., Shen, S.: A fatigue endurance limit for higway and airport pavement. Journal of Transportation Research Record - TRR, 1832 (2003)
[16] Walubita, L.F.: Comparison of fatigue analysis approaches for two hot mix asphatl concrete mixtures. Ph.D Thesis, Texas A&M University, Texas Transportation Institute, Austin, Texas (2005)

Measurement and Prediction Model of the Fatigue Behavior of Glass Fiber Reinforced Bituminous Mixture

I.M. Arsenie[1,2], C. Chazallon[1], A. Themeli[1], J.L. Duchez[2], and D. Doligez[3]

[1] Laboratory of Design Engineering (LGeCo), INSA de Strasbourg,
ioanamaria.arsenie@insa-strasbourg.fr
[2] Epsilon Ingénierie, Parc de Ruissel – Avenue de Lossburg, 69480 Anse, France
[3] 6D Solutions, 17 Place Xavier Ricard, 69110 Sainte Foy les Lyon, France

Abstract. Geocomposite materials such as fiber glass grids are frequently employed in asphalt pavement design as reinforcement interlayers, having the role to delay the occurrence and propagation of cracking. This paper studies the fatigue behavior of a standard bituminous mixture (EB 10 wearing course 35/50 EN 13108-1) and of a composite made of the same bituminous mixture and glass fiber grid. The aim of this study is to characterize the fatigue behavior of the two structures, to compare them and to quantify the increase in fatigue life due to the use of the glass fiber grid. The fatigue behavior is described using four point bending (4PB) laboratory tests results and using a finite elements modeling. The fatigue tests are carried out with two specimens: a simple bituminous beam and another one reinforced with a glass fiber grid placed at its bottom. The testing conditions are sinusoidal excitation with haversine waveform at 25 Hz and 10°C, using controlled strain mode. The reinforcement role of the fiber glass grid is evidenced by the 4PB tests results. Along with the tests, a finite element modeling is used to make a prediction of the fatigue behavior. The asphalt damage prediction model of Bodin, implemented in the finite element software CAST3M, is tested on the bituminous mixture and on the glass fiber reinforced bituminous mixture. The behavior obtained with the experiment is compared to the computational models' prediction.

1 Introduction

The service life of an asphalt pavement depends on its performance under the action of traffic loads and thermal stresses. The repeated loads induced by traffic produce the phenomenon known as "fatigue cracking", which is one of the major causes of pavement deterioration. Among the different methods used to delay this phenomenon, the method using coated glass fiber grid called "geogrid" has gained acceptance since the 90's. This method has applications in: treatment of cracking produced by thermal changes in rigid and semi-rigid pavements, cracking of flexible pavements and treatment of longitudinal cracks in the case of road extensions. In practice the complex made of bituminous mixture and geogrid has a very good behavior with time under repeated loads. Nowadays the grid is employed as reinforcement of bituminous layers to delay the fatigue crack initiation and propagation. In this paper, we study the fatigue behavior of the composite material, in order to compare the test results with a prediction based on the Bodin model.

2 Fatigue Four Point Bending Test Configuration and Bending Device

2.1 Choice of the Test Configuration

The composite material behavior under cyclic load was firstly studied in laboratory and secondly was modeled with a non-local damage approach. After the review of the European Standards [1] dealing with fatigue tests, and taking into account the sizes of the composite material, the choice was made to use the 4PB test. The test is well recognized by other authors [7] and is favored because failure happens in an area of uniform stress corresponding to the central part of the beam, between the two load lines. The equations of the beam theory can be applied if the hypothesis of an elastic homogeneous isotropic material is taken into account. In the reinforced specimen we consider that the geogrid is perfectly incorporated within the bituminous mix with a layer of residual bitumen emulsion. Neither the shear phenomenon nor the possible slips of the grid inside the specimen's structure are taken into consideration.

2.2 Four Point Bending Device and Procedures

The standard 4PB device is adapted to the sizes of the materials, according to the standard [1]. The standard suggests a 450 x 50 x 50 mm bituminous beam. The minimum dimension of the beam (width or height) respects the condition related to the granular size: this dimension must be superior to three times the maximum granular size of the bituminous mixture. Because there is no condition concerning the dimensions of reinforced bituminous specimens, we consider that the same condition applies to the non-reinforced bituminous specimens. For reinforced beam, this condition leads to a minimum of three coated fiber glass yarns placed in the width of the beam. We obtain a minimum width of 100 mm. From the beam conditions, the length has to be 6 times the maximum value (width or height), we obtain the beam dimensions 620 x 100 x 90 mm. In figure 1 the test configuration is represented. Figure 2 shows the special bending device used for testing.

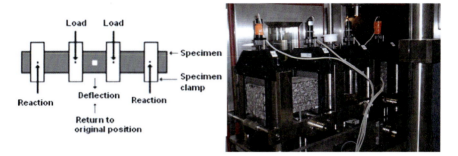

Fig. 1. (left) Four point bending configuration [5] **Fig. 2.** (right) Four point bending device ZWICK at EPSILON

The first type of 4PB test available in [1] is performed with a haversine waveform. In this test, the central part is bent from the initial position to the double amplitude of deflexion. The rotation is free in the reactions and the load lines. The vertical deflexion in the center of the beam is measured at the bottom side with a linear variable differential transducer (LVDT). The material response and the phase angle are measured all along the test. The laboratory fatigue tests results presented in this paper were obtained with this fatigue test configuration. The second standard type of bending test available can be performed with a sinusoidal waveform, creating a symmetrical bending around the original position (from a positive to a negative deflexion). This fatigue test configuration presented in section 6 was employed to model damage evolution.

3 Materials

The bituminous mixture presented in table 1 is a standard EB 10 class III according to the European classification [3], with an elastic modulus of 9 GPa, obtained in 4PB rigidity tests performed at 15°C and 10 Hz and in indirect tensile rigidity tests (ITT) performed at 15°C with 124 ms loading time. The geogrid used as reinforcement is the coated glass fiber grid CIDEX 100 SB of 6D Solutions, represented in table 2. The product is an elastic composite made of warp and filling yarns and a nonwoven part of polyester fiber (mesh: 40 x 40 mm^2). Both yarn types are made of continuous glass fiber glass and resin. In an asphalt pavement structure, the geogrid is placed at the interface between surface course and base course with an emulsion made of residual bitumen. The geogrid is employed in pavement rehabilitation as a reinforcement material for the surface course. Its main role is to delay cracking phenomenon.

Table 1. Formula and Granulometric curve of the bituminous mixture

FORMULA   EB 10 class III
45%   0/3
15%   3/8
40%   8/15
5,97% BITUMEN 35/50
Compactness 93 – 95%

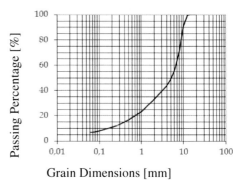

Table 2. Properties and image of the geogrid

GRILLE CIDEX 100 SB

Grid: Fiber glass + resin (type SB) : 383 g/m²
Fiber of polyester: 17 g/m²
Mechanical resistance at failure: 100 KN/m
Mechanical resistance at 1% deformation: 35 KN/m
Residual bitumen for embedding: 600 g/m²

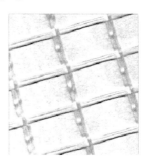

4 Sample Preparation

Two types of samples are made and subjected to 4PB test in the laboratory EPSILON Engineering.

There are three steps in the fabrication process of the asphalt slabs illustrated in figure 3, respectively:

  A. the compaction of the first bituminous layer,
  B. the insertion of the geogrid with an emulsion applied at ambient temperature (22-23°C)
  C. the compaction of the second bituminous layer.

In order to obtain the beam, the last operation is sawing the 620 x 400 x 150 mm slabs to the dimensions 620 x 100 x 90 mm. The beam vertical section from the bottom to the top is: 29 mm bituminous layer, 1mm geogrid with bitumen emulsion and 60 mm bituminous layer. In the case of non-reinforced specimens the fabrication remains the same, adding only an emulsion layer between the two bituminous layers.

Four specimens are tested: two bituminous beams "B1" and "B2" and two reinforced bituminous beams "RB1" and "RB2". There is a difference in the number of warp yarns between the beams RB1 and RB2: two warp yarns in the case of RB1 and three for RB2.

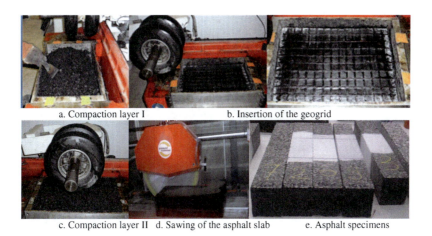

a. Compaction layer I b. Insertion of the geogrid

c. Compaction layer II d. Sawing of the asphalt slab e. Asphalt specimens

Fig. 3. Fabrication process in laboratory (a, b, c, d, e)

5 Fatigue Tests and Results

5.1 Fatigue Tests

The fatigue tests consist of repeated bending. They are performed with a controlled strain, which is kept constant during the test. The deflexion is constant and the response of the material is the force to keep a constant deflexion. The force decreases with the number of cycles. The tests are performed at 10° C and 25 Hz, these values have to be considered in the French Pavement Design Method [2]. The "fatigue resistance" of bituminous mixtures, considered in the design of a road pavement, represents its ability to withstand repeated bending without any fracture [8].

The strain value chosen by the authors is equal to 200 µm/m, in order to obtain a fatigue life of the non-reinforced bituminous mixture around 100 000 cycles. The "fatigue life" is defined by the European standard [1] as the number of cycles corresponding to a material stiffness decrease to half of its initial value. In the case of a controlled strain fatigue test, the force also decreases to half of its initial value. We consider this definition as the criterion I in the result interpretation. Because there are no existing fatigue criteria concerning the composite bituminous mixtures, another criterion has been introduced by the authors. The criterion II considers the fatigue life as the number of cycles corresponding to a 80% decrease of the initial material stiffness, respectively force. The criterion II supposes that the geogrid reinforcement role remains efficient after the 50% material stiffness decrease. Both criteria have been used in the result interpretation.

5.2 Fatigue Test Results

Figure 4[4] illustrates the evolution of the ratio between the measured force and the initial force, with the number of load cycles. Three cases are compared: the non reinforced samples B1 and B2, the reinforced sample RB1 and the reinforced sample RB2. We compare the fatigue life obtained on B1, B2, RB1 and RB2, for each criterion mentioned in tables 3 and 4.

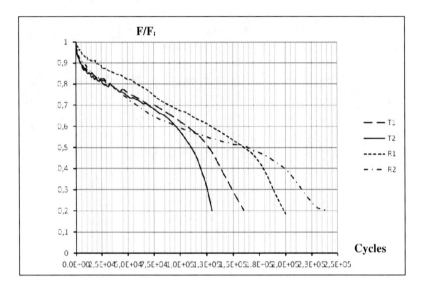

Fig. 4. Evolution of the force during 4 fatigue tests

According to criterion I (table 3), the reinforcement used in RB1, respectively in RB2, improves on average the fatigue life of the non reinforced samples of 1.37 times, respectively of 1.46 times. According to criterion II (Table 4), the reinforcement used in RB1, respectively in RB2, improves on average the fatigue life of the non reinforced samples of 1.37 times, respectively of 1.68 times. As expected, the results show that the fatigue life of the sample increases with the number of fiber glass yarns. The fatigue life increase is 10% between RB1 and RB2 in the case of criterion I, respectively 31% in the case of criterion II. The fatigue life remains the same using both criteria for RB1. On the contrary, the increase in fatigue life is significant for RB2 using criterion II.

The results are presented in the tables below. The corresponding abbreviations represent:

Nf: the number of cycles corresponding to the fatigue life defined for each criterion, N_{fB_i}: the fatigue life of B_i with i= (1; 2), N_{fRB1}: the fatigue life of RB1,

N_{fRB2}: the fatigue life of RB2, N_{f1}: improvement in fatigue life due to the use of the geogrid in RB1 as respect with B1 and B2, N_{f2}: improvement in fatigue life due to the use of the geogrid in RB2 as respect with B1 and B2.

Table 3. Fatigue life according to criterion I

Specimen	N_f	$N_{f1} =$ N_{fRB1}/N_{fBi} Average = 1.37	$N_{f2} =$ N_{fRB2}/N_{fBi} Average = 1.46	$N_{f1} - N_{f2}$ Average = 0.09
B1	125 000	1.28	1.36	0.08
B2	109 000	1.47	1.56	0.09
RB1	160 000	-	-	-
RB2	170 000	-	-	-

Table 4. Fatigue life according to criterion II

Specimen	N_f	$N_{f1} =$ N_{fRB1}/N_{fBi} Average = 1.37	$N_{f2} =$ N_{fRB2}/N_{fBi} Average = 1.68	$N_{f1} - N_{f2}$ Average = 0.31
B1	160 000	1.23	1.50	0.28
B2	130 000	1.51	1.85	0.34
RB1	196 000	-	-	-
RB2	240 000	-	-	-

6 Damage Evolution Prediction Using the Finite Elements Method

6.1 Finite Element Model

Two 2D models of the fatigue behavior under 4PB loading have been performed using the finite element software Cast3M: one for the non-reinforced bituminous beam called "B" and one for the reinforced bituminous beam called "RB". The damage evolution is predicted by the Bodin model [6] based on a "three regimes" law.

In Bodin's model, the damage variable is introduced as a scalar as presented in the Eqn. (1):

$$\sigma_{ij} = (1-D) \cdot C^O_{ijkl} \cdot \varepsilon_{kl} \quad (1)$$

in which σ_{ij} is the stress tensor,

ε_{kl} is the strain tensor,

C^O_{ijkl} is the elasticity matrix,

D is the damage scalar variable with $0 \leq D \leq 1$.

During the test the damage evolves in three phases. This evolution is given by Eqn. (2):

$$\dot{D} = f(D) \cdot \bar{\varepsilon}^{\beta} \cdot \langle \dot{\bar{\varepsilon}} \rangle \tag{2}$$

in which $f(D)$ is expressed in the Eqn. (3):

$$f(D) = \frac{\alpha_2}{\alpha_1 \cdot \alpha_3} \cdot \left(\frac{D}{\alpha_2}\right)^{1-\alpha_3} \cdot \exp\left(\frac{D}{\alpha_2}\right)^{\alpha_3} \tag{3}$$

and $\bar{\varepsilon}$ is calculated in the Eqn. (4):

$$\bar{\varepsilon}(x) = \frac{1}{V_r(x)} \cdot \int_{\Omega} \psi(x-s) \cdot \tilde{\varepsilon}(s) \cdot ds \tag{4}$$

in which $V_r(x)$ is calculated in the Eqn. (5):

$$V_r(x) = \int_{\Omega} \psi(x-s) \cdot ds \tag{5}$$

and $\tilde{\varepsilon}$ is calculated in the Eqn. (6):

$$\tilde{\varepsilon} = \sqrt{\sum_{i=1}^{3}\left(\frac{\langle \sigma_i \rangle}{E \cdot (1-D)}\right)^2} \tag{6}$$

where: $\bar{\varepsilon}$ is the average of the equivalent strain,

$\langle \dot{\bar{\varepsilon}} \rangle$ is the strain rate. The Macauley brackets are used to account only the positives values,

β is a parameter related to the slope p of the fatigue curve in log – log coordinates $p = -(\beta+1)$,

α_1, α_2, α_3 are damage evolution parameters,

$V_r(x)$ is a representative volume in a point of coordinate x,

Ω is the studied volume,

$\psi(x-s)$ is the weight function,

s is a relative coordinate from the point of coordinate x,

$\tilde{\varepsilon}$ is the equivalent strain,

$\langle \sigma_i \rangle$ is the principal traction stresses,

E is the complex modulus (10° C, 25 Hz).

The hypothesis of plane stresses is used to model one half of the asphalt beam thanks to the symmetry of the problem. Figure 5 represents the finite element model, its geometry and the boundary conditions. The applied fatigue solicitation is constant strain.

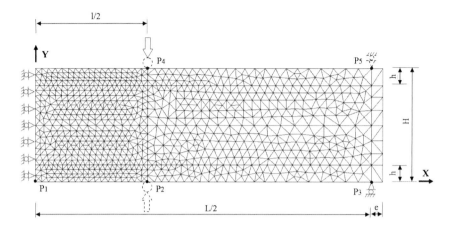

Fig. 5. Finite Element Model

The 4PB fatigue test modeled with the Bodin model, takes into account a sinusoidal waveform, creating a symmetrical bending around the original position. As a consequence, a symmetric material is chosen, resulting in a double reinforcement of the beam. Two positions of the double reinforcement are tested: h = 0 mm and h = 15 mm, where "h" represents the height as indicated in Figure 5.

According to fatigue tests carried out only on the geogrid, not presented in this paper, the CIDEX 100 SB grid presents a fatigue limit larger than $7.3 \cdot 10^{-3} m/m$, which is larger than the constant strain used in the fatigue test ($\varepsilon = 140 \cdot 10^{-6} m/m$ in tests presented in Figure 6). Since the solicitation level is smaller than the fatigue limit, the geogrid is not damaged during the tests. Considering this fact, the geogrid is modeled as a non-damageable material with a Young's modulus "E" of 43 GPa and a Poisson's coefficient "ν "equal to 0.35.

The elements used for modeling the reinforcement are bare type linear elements. Two sections of glass fiber are considered for the 2D models. The first section called "S" is equal to 0.1163 mm^2, which is the equivalent section of the three warp yarns which are in the width of the tridimensional (3D) real beam. The second section called "2S" is calculated as the sum between S and the equivalent section of the filing yarns which are in the longitudinal section of the real beam.

There is a difference between the tests and the modeling consisting in the waveform which has been used. The haversine waveform is used for tests and the sinusoidal waveform is used in the Bodin model. Considering this fact, the model couldn't be used to predict the laboratory results. However, the damage prediction is made taking into account the mechanical characteristics of a bituminous mixture studied by Bodin [6] in the case of sinusoidal waveform: E = 12 GPa, ν = 0.35 and the fatigue evolution parameters. The fatigue evolution parameters of the Bodin 'three regimes' law are: β = 4.0, $\alpha_1 = 5.58 \cdot 10^{-15}$, $\alpha_2 = 0.42$ and $\alpha_3 = 3.0$ – see Eqn. (3). The bituminous mixture is modeled by triangular plan elements.

6.2 Damage Modeling Results

Figure 6 represents the damage evolution of the beams B and RB in the 4PB fatigue test under a constant strain load $\varepsilon = 140 \cdot 10^{-6} m/m$. In this figure we can observe that the third phase of the damage evolution is influenced by the presence of the reinforcement i.e. its section (S or 2S) and its position (h) in the asphalt beam.

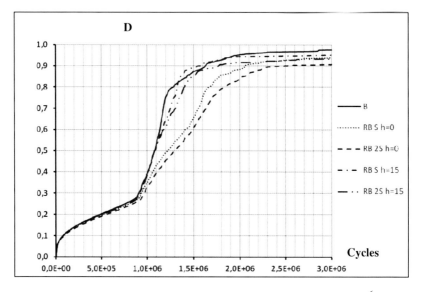

Fig. 6. Evolution of damage under constant strain fatigue test $\varepsilon = 140 \cdot 10^{-6} m/m$

It is notable that the glass fiber increases considerably the fatigue life of the reinforced specimen in comparison with the non-reinforced one. The reinforcement is more efficient when the geogrid is placed at the beam extremities (h = 0 mm). The section increase has an influence over the fatigue life, which is more significant when criterion II is considered. These results suggest the fact that the geogrid representative section in the 2D model is a value between S and 2S. This value is calculated by the sum of S and the reduced filing yarns section, contributing effectively to the bending efforts. The second term is obtained by the multiplication of the filling yarns section with a subunit coefficient. The coefficient is going to be estimated when a significant number of experimental results will be available.

The following load levels: $\varepsilon = 130 \cdot 10^{-6} m/m$, $\varepsilon = 140 \cdot 10^{-6} m/m$, $\varepsilon = 150 \cdot 10^{-6} m/m$, $\varepsilon = 160 \cdot 10^{-6} m/m$, $\varepsilon = 170 \cdot 10^{-6} m/m$ and $\varepsilon = 180 \cdot 10^{-6} m/m$ are tested with the Bodin damage model, having the reinforcement placed at the beam extremities. The results are presented in table 5. Three millions of cycles are simulated in the modeling. We observe that the gain in fatigue life for the lowest and the highest load levels is inferior to the gain obtained in the other load cases. This influence on the modeling results, currently under study, may be influenced by the calculated number of points per cycle. Considering the mean result over the 6 strain levels and criterion II, it can be estimated that the reinforcement increases the beam fatigue life time from 33.5% to 45.5%.

Table 5. Damage modeling results - CAST3M. The percentages represent the gain in fatigue life of RB S and RB 2S reported to B.

Epsilon $\varepsilon \cdot 10^{-6} m/m$	Damage modeling results - CAST3M					
	B		RB S		RB 2S	
	D=0,5	D=0,8	D=0,5	D=0,8	D=0,5	D=0,8
130	1 612 537	2 030 382	1 777 726	2 638 630	1 885 806	2 616 694
	-	-	10.24%	29.96%	14.49%	28.88%
140	1 082 367	1 277 898	1 223 180	1 672 255	1 295 969	1 830 000
	-	-	13.01%	30.86%	16.48%	43.20%
150	761 152	853 810	857 929	1 158 389	910 689	1 342 299
	-	-	12.71%	35.67%	16.42%	57.21%
160	527 891	556 783	607 529	821 284	657 247	919 216
	-	-	15.09%	47.51%	19.68%	65.09%
170	386 607	418 870	436 898	570 437	471 179	627 465
	-	-	13.01%	36.18%	17.95%	49.80%
180	270 109	283 891	307 603	342 882	308 263	366 138
	-	-	13.88%	20.78%	12.38%	28.97%
Average Gain ε from 130 to 180			12.99%	33.49%	16.23%	45.53%

7 Conclusions and Perspectives

The objective of the 4PB tests was to show the improvement in fatigue life of the bituminous mixture due to the use of the coated glass fiber grid. The number of specimens is not sufficient to give a general value of this improvement. However, the fatigue behavior of the composite material confirmed that the reinforcement role should be tested on reinforced bituminous beams with three warp yarns (RB2) and analyzed with criterion II (beyond the standard definition of fatigue life). In this case, we obtain a gain of 68% in fatigue life.

The purpose of the Cast3M simulations was to test the Bodin damage evolution model of bituminous mixtures with a reinforced bituminous mixture. The results confirm the positive effect of the grid observed in laboratory and the fact that criterion II is more adapted than criterion I (the standard definition of fatigue life) for the reinforcement role evaluation. In this case, the reinforcement role of the geogrid leads to a gain in fatigue life between 33.5% and 45.5%, depending on the fiber glass section.

Both laboratory and modeling results point up the gain in fatigue life due to the geogrid. Current work consists in performing more tests with both haversine and sinusoidal waveform, in order to confirm the reinforcement role. A second target is the adjustment of damage evolution model to reinforced bituminous mixtures in order to obtain an accurate prediction model of damage.

References

[1] EN 12697-24, Méthodes d'essai pour mélange hydrocarboné à chaud. Résistance à la fatigue, French version (2007)
[2] Technical Guide, Technique, Conception et dimensionnement des structures de chaussées, SETRA-LCPC (1994)
[3] EN 13 108-1, Specification des matériaux Partie1. Enrobés bitumineux, French version (2006)
[4] Arsenie, I.M., Chazallon, C., Duchez, J., Doligez, D., Themeli, A.: Fatigue behavior of a glass fiber reinforced asphalt mix in 4 Point bending test and damage evolution modeling. In: Gerdes, A., Kottmeier, C., Wagner, A. (eds.) Proceedings of the International Conference on Climate and Constructions, pp. 275–286. Karlsruhe Institute of Technology, Germany (2011)
[5] Bacchi, M.: Analysis of the variation in fatigue life through four-point bending test. In: Pais (ed.)Proceedings of the 2nd Workshop on Four point Bending: From Theory to Practice, , pp. 205–215. University of Minho, Portugal (2009) ISBN 978-972-8692-42-1
[6] Bodin, D., Pijaudier-Cabot, G., De La Roche, C., Piau, J.M., Chabot, A.: Continuum damage approach to asphalt concrete fatigue modeling. Journal of Engineering Mechanics 130(6), 700–708 (2002)
[7] Huurman, P., Pronk, A.C.: Theoretical Analysis of the 4Point Bending Test. In: Scarpas, Al-Quadi (eds.) Advanced Testing and Characterization of Bituminous Materials, Loizos, Partl, p. 978. Taylor and Françis Group, London (2009) ISBN 978-0-415-55854-9
[8] Pais, J.C., Minhoto, M.J.C., Kumar, D.S.N.V.A., Silva, B.T.A.: Analysis of the variation in fatigue life through four-point bending test. In: Pais (ed.) Proceedings of the 2nd Workshop on Four point Bending: From Theory to Practice, pp. 287–291. University of Minho, Portugal (2009) ISBN 978-972-8692-42-1

Fatigue Cracking in Bituminous Mixture Using Four Point Bending Test

Q.T. Nguyen, H. Di Benedetto, and C. Sauzéat

University of Lyon/Ecole Nationale de Travaux Publics de l'Etat Département Génie
Civil et Bâtiment (DGCB) (CNRS 3237),
Rue Maurice Audin, 69518 Vaulx en Velin, France
{quang-tuan.nguyen,herve.dibenedetto,cedric.sauzeat}@entpe.fr

Abstract. This paper describes investigation into cracking in bituminous mixture using the four point bending notched fracture (FPBNF) test, which has been developed at the University of Lyon/ Ecole Nationale de Travaux Publics de l'Etat (ENTPE). A special loading path is applied on the notched beam specimen at a constant temperature of - 4.5°C. A monotonic loading was first applied until the peak load and after unloading, many loading/unloading cycles at small amplitude were carried out until the final failure of specimen. Deflection of the beam and crack mouth opening displacement (CMOD) are measured. Crack length is determined experimentally using crack propagation gauges. It is also obtained with an improved method, called Displacement Ratio Crack length (DRCL) method, developed at ENTPE laboratory, which allows back calculating the crack length. This method is based on the relation between two experimental displacement measurements: the crack mouth opening displacement (CMOD) and the deflection of the beam. The results obtained from this method are discussed and compared with the crack length measured with crack propagation gauges. During the test, the fracture behaviour is investigated. The crack propagation is studied as a function of loading/unloading cycle number. The stress intensity factor is evaluated. Two different domains of crack evolution are distinguished: the first domain where pre-existing crack progressively re-opens, the second domain where crack propagates. The Paris fatigue law could be applied in the domain where crack propagates.

1 Introduction

Cracking is one of the major distresses in asphalt pavements. The main causes of cracking include the road traffic (fatigue cracking…) and the climatic conditions (low temperature and temperature cycling). In this paper, the fatigue cracking of bituminous mixtures is investigated in mode one using a four point bending notched fracture (FPBNF) test. In the literature, few works are reported considering this kind of test for bituminous mixtures. Some results obtained on FPBNF test from our team were published recently [1-3]. Meanwhile, the fracture characteristics of bituminous mixtures have been widely investigated in mode one with other tests such as: the single edge notched beam (SENB) test [4-6], the semi-circular bending (SCB) fracture tests [7-9] and the disk-shaped compact tension test [10]... One of the interests of the FPBNF test is that a constant moment field is created in the middle part of the beam.

In this paper, the description of the FPBNF test with the different measurements made on the specimen are first detailed. Then, experimental campaign is presented. It includes tested material and considered special loading path. Lastly, the analysis of the test result is proposed. The validity or not of the Paris law [11] is checked in the two observed different domains: the first domain where existing crack progressively re-opens, the second domain where crack propagates.

2 Presentation of FPBNF Test

The FPBNF test was designed at ENTPE laboratory to investigate the cracking of bituminous mixtures (Figure 1) in mode one. The tested specimens are prenotched prismatic beams (55 cm long, 7 cm high and 6.5 cm wide). A 2 cm high initial notch (a_0) is made in the middle of the beam. During the test, the specimen is placed in a temperature-controlled chamber. A temperature sensor (PT100 type), which is fixed on the specimen, gives the temperature at the surface of the sample. The test is performed using a hydraulic press (INSTRON 1273). The chosen load cell has a capacity of 10kN. This system makes it possible to control either in load or displacement mode. During our experimental campaign, monitoring is made from the displacement of the piston.

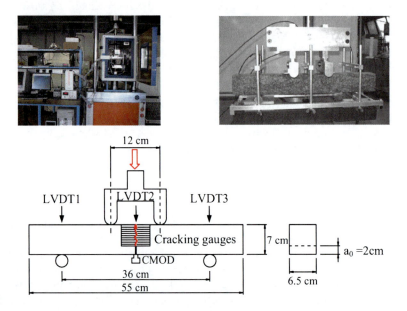

Fig. 1. Four point bending notched fracture (FPBNF) test

Fig. 2. Cracking gauge glued on the specimen (one gauge on each side)

Three linear variable differential transducers (LVDT) measure vertical displacements on top of the beam: in the centre of the beam (LVDT2), and above the two lower supports (LVDT1 and LVDT3). Taking into account the punching effect at the supports, the "true" deflection of the beam f is obtained by Eqn. (1).

$$f = LVDT2 - \frac{LVDT1 + LVDT3}{2} \qquad (1)$$

An extensometer, placed under the beam, is used to measure the crack mouth opening displacement (CMOD) of the notch. In order to follow the crack propagation, cracking gauges are also used (Figure 2). These gauges are constituted of 21 parallel wires separated each other by 2.5mm. These wires when breaking indicate the crack position. The used gauges are 8cm long and about 5cm high. One gauge is glued on each lateral faces of the specimen, so that the first wire is just over the initial notch.

3 Experimental Campaign

3.1 Material and Specimens

The tested bituminous mixture is a "BBC" ("Béton Bitumineux Clouté" in French) according to the French classification (NF P 98-133). Aggregate is a 0/6 mm grading from "La Noubleau" quarry. Aggregates fraction content is given in Table 1. Grading curve is presented in Figure 3. Bituminous mixture contents 6.85% (aggregate weight) of pure bitumen (35/50 pen grade). The specimens 55 x 6.5 x 7 cm (Figure 1) were sawn from plate (600 x 400 x 11 cm) made using LPC plate compactor. The tested specimen has 3.5% (total volume) of air void. The pre-notch a_0 (2 cm) is performed in two steps. Firstly, a circular saw gives a 1.5cm deep and 5mm wide notch. Then the notch is ended with care using a hacksaw. Thus, at the top of the notch width is about 1mm.

Table 1. Aggregates size of tested bituminous mixture

Aggregate type	La Noubleau 4/6	La Noubleau 2/4	La Noubleau 0/2
Percent in weight (%)	33.5	5.5	61

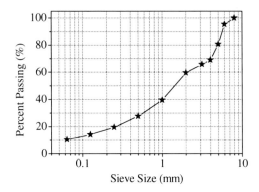

Fig. 3. Aggregates grading curve of tested bituminous mixture

3.2 Description of the Test

The four point bending test is performed with constant imposed displacement rate of the piston and at a constant temperature in the thermal chamber. The presented test is performed at -4.5°C with 1mm/min displacement rate. A special loading path is considered as presented in Figure 4.

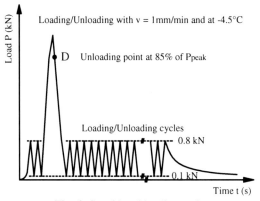

Fig. 4. Considered loading path

First, two small cycles of loading/unloading are carried out to allow setting of specimen inside the loading frame. Then, a monotonic loading is applied. Just after the peak (point D figure 4), the specimen is unloaded down to 0.1 kN. Then small loading/unloading cycles are carried out until the final failure of specimen. All loadings are applied at the same constant imposed displacement rate (1mm/min). The small cycles are between $P_{cycle\_max} = 0.8$kN and $P_{cycle\_min} = 0.1$kN.

4 Analysis of the Results

Figure 5 presents the result of the test in the axes load (P) vs. deflection (f) (Eqn. (1)). For legibility of the figure, not all cycles are presented. Excepted indicated cycle number only one cycle every 190 cycles is plotted. The role of the first large monotonic loading at the beginning of the test is to create an initial crack in the beam that may heal when unloading. The healing phenomenon has been studied by several authors [12-14] but clear identification and modelling of this phenomenon remain to be improved.

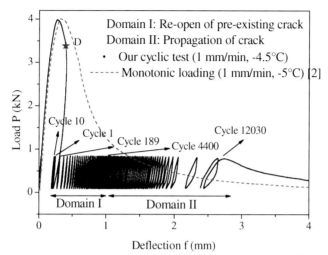

Fig. 5. Loading vs. deflection, otherwise mentioned only one cycle on 190 is plotted. Two domains of crack evolution are observed.

During the first 10 cycles, the maximum deflection of the beam is decreasing. Moreover, 189 small cycles are needed before the maximum deflection level reaches the deflection just after the first large unloading. Under the effect of small loading/unloading cycles, the crack propagates progressively. After cycle 10, two main different domains can be distinguished: the first domain where deflection difference between two cycles decreases (this may be a phase where pre-existing crack, created by the first large loading, progressively re-opens), the second domain where deflection difference between two cycles increases up to total failure (this may be explained by crack propagation of the pre-existing crack). These observations are Visible in Figure 5. The cycle 4400, where deflection difference between two consecutive cycles is minimum, is the boundary between the two domains I and II. As a comparison, a monotonic loading test performed at -5°C with 1mm/min displacement rate on a specimen of the same material and the same dimension, which is realized by our team [2], is also presented in the Figure 5.

4.1 DRCL Method for Back Calculation of Crack Length

The DRCL (Displacement Ratio method for predicting Crack Length) method is a new method, which makes it possible to back calculate the crack length during the four point bending crack propagation test [1, 2]. This method has been developed at laboratory of University of Lyon/ENTPE. It is based on the relation between two measured displacements: the crack mouth opening displacement (CMOD) and the deflection of the beam (f). In the axes CMOD-f, the slope of the linear part at the beginning of each loading/reloading cycle is named r_d. Thanks to FEM calculation considering the hypothesis of isotropic linear elastic or viscoelastic behaviour, the DRCL method gives a relation between r_d and the crack length "a" for a beam whose dimensions have been defined. Therefore, using this method, the crack length "a" can be calculated from the experimental slope values "r_d" measured in the test. Due to the limited space in this paper, the details of this method are not presented here. Reader can consult reference [1, 2] for more explanation. The great interest of the method is to be valid for linear viscoelastic materials.

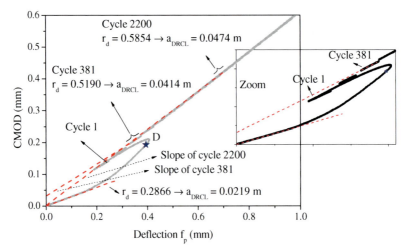

Fig. 6. Some values of crack lengths "a_{DRCL}" (DRCL method) obtained from slope r_d determined from the CMOD-f curve

Figure 6 presents some calculated crack lengths "a_{DRCL}" values using DRCL method for some chosen cycles presented in axes CMOD vs. deflection. From the slope values of the linear part at the beginning of each loading/unloading cycle, the crack lengths "a_{DRCL}" could be back calculated. The calculated value a_{DRCL} for the first loading at the beginning of the test is 2.19cm. One can observed that it is very close to the targeted initial pre-notch value (a_0 = 2cm), which confirms the accuracy of the method and gives a validation of the approach. The results show that from cycle 2200, the curve of the loading part of the cycles is on a line passing though the origin (as a first approximation), which is not the case for the previous cycles. This remark also applied for other FPBNF tests realized at ENTPE.

4.2 Experimental Crack Length Obtained by Cracking Gauges and Comparison with DRCL method

Direct measurement of crack length is obtained from cracking gauges glued on each side of the specimen. The obtained crack length values are noted a_{gauge1} and a_{gauge2}. Results given by the two cracking gauges, for the first large loading/unloading cycle, are plotted in Figure 7. It can be seen that the crack length still increases for a while after the peak, even if load decreases. During the large unloading period, crack length given by the gauges decreases. This indicates that the crack mouth is closing but it does not give any information about healing of the two lips as no information on how the two lips are linked can be obtained at this step.

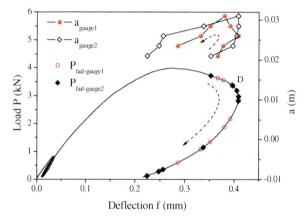

Fig. 7. Load and crack length given by the gauges during the first large loading/unloading cycle

Figure 8 presents a comparison between the crack lengths obtained from the two cracking gauges (a_{gauge}) and from DRCL method (a_{DRCL}). The measured crack length is slightly different on each side. The crack may also have a different evolution inside the beam. The observed difference between a_{gauge} and a_{DRCL} can be explained by the strength deformation of the gauge wires. Only macro-cracks having a minimal width can be detected. Therefore, the detection of crack is delayed comparing with a_{DRCL}. Moreover, before macro-crack propagation, an initiation phase exists where damage occurs and only micro-cracks develop [15] in front of the crack tip. The crack length obtained by DRCL method should be considered as the sum of the macro-crack and a fictitious crack representing the damage zone at the crack tip [1, 2]. The value of a_{DRCL} takes into account the effect of the damaged zone in front of the crack. A fitting curve is proposed in Figure 8 to approximate the evolution of a_{DRCL} as a function of the number of cycles.

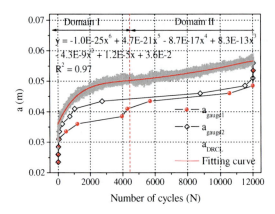

Fig. 8. Crack length during the small loading/unloading cycles

4.3 Validity of the Paris Law

The crack propagation fatigue laws relate the stress intensity factor range (ΔK) to crack propagation rate. The Paris law [11] is the most popular fatigue crack model. The expression of the Paris law in mode I is given by Eqn. (2).

$$\frac{da}{dN} = c(\Delta K_I)^m \qquad (2)$$

Where N is the number of cycles, da/dN is crack length increase per cycle, c and m are constants. $\Delta K_I = K_{I\ Pmax} - K_{I\ Pmin}$ is the difference between the stress intensity factor at maximum and minimum loads. Considering a linear elastic behaviour, the stress intensity factor in mode I is given in case of the four point bending fracture test by Eqn. (3) [16]:

$$K_I = \frac{3}{2}\frac{P(L-l)}{BW^2}Y(x)\sqrt{a} \qquad (3)$$

Where Y(x) is a geometry factor; x is the relative crack length (a/W); L is the lower span of the beam; l is the upper span of the beam; B is the width of the beam; W is the height of the beam. Y(x) is calculated using FEM calcuation [2, 16].

The variation of a in Eqns. (2) and (3) are given by the fitting curve of a_{DRCL} presented in figure 8. Figure 9 shows a plot of da_{DRCL}/dN versus ΔK_I in logarithmic axes for the two identified domains. It can be seen that slope of the curve is opposed in the two domains. In domain I, da_{DRCL}/dN decreases with the increase of ΔK_I. It is the same tendency as observed during initiation phase. In domain II, da_{DRCL}/dN increases with the increase of ΔK_I. The results show that the Paris law can only be applied in the domain II, where crack propagates (see fitting curve in figure 9). The two obtained constants for Paris law (equation 2) are, c=7.11E-7 and m = 0.83. It should be underlined that these constants are very sensitive to the chosen fitting curve for "a" (Figure 8).

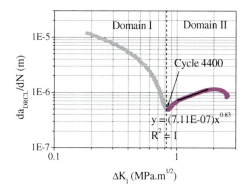

Fig. 9. da_{DRCL}/dN versus ΔK_I for the tested material

5 Conclusions

The aim of the research presented in this paper is to characterize fatigue cracking of bituminous mixtures. A four point bending notched fracture (FPBNF) test designed at ENTPE laboratory is used. From the obtained results, the following conclusions can be drawn:

- The crack length is measured experimentally during the test with two cracking gauges. These values are compared with the one obtained from the DRCL (Displacement Ratio method for predicting Crack Length) back analysis method developed by our team. The results show the coherence between the two values and confirm the pertinence of DRCL method.
- In the presented test, two different domains of crack evolution are distinguished: the first domain where pre-existing crack progressively re-opens, the second domain where crack propagates.
- The Paris fatigue law could only be applied for the tested material in the domain where crack propagates.

References

[1] Nguyen, M.L., Di Benedetto, H., Sauzéat, C., Wendling, L.: Investigation of cracking in bituminous mixtures with a four point bending test. In: 6th RILEM International Conference on Cracking in Pavements, Chicago Illinois, pp. 283–293 (2008)

[2] Nguyen, M.L.: Etude de la fissuration et de la fatigue des enrobés bitumineux, PhD. ENTPE-INSA Lyon. Mecanique, Energétique, Acoustique et Génie Civil, p. 276 (2009) (in French)

[3] Di Benedetto, H., De la Roche, C., Baaj, H., Pronk, A., Lundstrom, R.: Fatigue of bituminous Mixtures. Materials and Structures 37, 202–216 (2004)

[4] Kim, K.W., El Hussein, M.: Variation of fracture toughness of asphalt concrete under low temperatures. Construction and Building Materials 11(7-8), 403–411 (1997)

[5] Artamendi, I., Khalid, H.A.: Fracture characteristics of crumb rubber modified asphalt mixtures. In: Third Eurasphalt & Eurobitume Congress, Vienna, pp. 1317–1326 (2004)
[6] Wendling, L., Xolin, E., Gimenez, D., Reynaud, P., De la Roche, C., Chevalier, J., Fantozzi, G.: Characterisation of crack propagation in butuminous mixtures. In: Fifth International RILEM Conference on Cracking in Pavements, pp. 639–646. RILEM Publications S.A.R.L, France (2004)
[7] Li, X., Marasteanu, M.O.: Investigation of Low Temperature Cracking in Asphalt Mixtures by Acoustic Emission. International Journal of Road Materials and Pavement Design 7(4), 491–512 (2006)
[8] Marasteanu, M.O., Dai, S., Labuz, J.F., Li, X.: Determining the Low-Temperature Fracture Toughness of Asphalt Mixtures. Transport Research Record (1789), 191–199 (2002)
[9] Molenaar, J.M.M., Molenaar, A.A.A.: Fracture toughness of asphalt in the semi-circular bend test. In: Proceedings of the 2nd Eurasphalt & Eurobitume Congress, Barcelona, pp. 509–517 (2000)
[10] Wagoner, M.P., Buttlar, W.G., Paulino, G.H.: Disk-shaped Compact Tension Test for Asphalt Concrete Fracture. Experimental Mechanics 45(3), 270–277 (2005)
[11] Paris, P.C., Erdogan, F.: A critical analysis of crack propagation laws. Journal of Basic Engineering 85(4), 528–534 (1963)
[12] Bodin, D., Soenen, H., De la Roche, C.: Temperature Effetcs in Binder Fatigue and Healing Tests. In: 3rd Euraspahlt & Eurobitume Congress, Vienna, pp. 1996–2004(2004)
[13] Kim, Y.-R., Little, D.N., Lytton, R.L.: Fatigue and healing characterization of asphalt mixtures. Journal of Material in Civil Engineering 15(1), 75–83 (2003)
[14] Planche, J.-P., Anderson, D.A., Gauthier, G., Le Hir, Y.M., Martin, D.: Evaluation of fatigue properties of bituminous binders. Materials and Structures 37, 356–359 (2004)
[15] Di Benedetto, H., Corté, J.-F.: Matériaux routiers bitumineux, vol. 2, p. 283. Lavoisier Publisher (2005) (in French)
[16] Fantozzi, G., Orange, G., R'Mili, M.: Rupture des matériaux, GEMPPM, INSA de Lyon (1988) (in French)

# Top-Down and Bottom-Up Fatigue Cracking of Bituminous Pavements Subjected to Tangential Moving Loads

Zoa Ambassa[1,2], Fatima Allou[1], Christophe Petit[1], and Robert Medjo Eko[2]

[1] Groupe d'Étude des Matériaux Hétérogènes – Equipe Génie Civil et Durabilité, Université de Limoges, boulevard Jacques Derche, 19300 Egletons, France
[2] University of Yaoundé I, P.O. 812 Yaoundé, Republic of Cameroon

Abstract. A model allowing for the determination of bituminous pavement degradation on traffic circle is presented. The development work has relied on the viscoelastic modeling of bituminous pavements subjected to multiple-axle traffic loads, using the following variables: pavement structure, load speed (or frequency), load configuration, and bituminous materials temperature. The method derived has successfully simulated the phenomenon under investigation. Results obtained indicate that the design concept based on bending fatigue in bituminous layers is not sufficient to realistically predict the degradation of a bituminous pavement structure. The phenomenon of shear at bituminous interface must also be taken into account, as revealed by the simulation results for the degradation of two pavement structures from the French design code [10] (i.e. flexible pavement and thick asphalt pavement).

1 Introduction

Trucks driving over pavements across the world feature multiple axle configurations, ranging from a single axle up to 8 axles [1-2]. Motor vehicles with such configurations also cause pavement degradations, the extent of which has yet to be sufficiently assessed. To date, the relative intensity of multiple-axle loads has been determined through the application of Miner's law. From a material mechanics standpoint, the fatigue life expectancy of a bituminous layer is evaluated in the laboratory using a two-point flexural strength test (NF EN 12697-24), consisting of inserting into the small base of a trapezoid a tube embedded in its larger base. With this position, the set-up emits a continuous sinusoidal signal that helps determine the number of load cycles before failure occurs [3]. It is observed that the load signal shape affects the fatigue life expectancy of asphalt pavements [1-5].

Pavement structure materials are made to be subjected upon every truck crossing to both fast and short loadings. The accumulation of damage generated in these materials, which is reflected in a loss of stiffness, leads to fatigue cracking. The objective of this paper is to develop a model which will allow determining the pavement degradations which occur on traffic circle. The work performed has been based on the following variables: type of pavement structure, load speed (or frequency), load configuration, and bituminous material temperature. Various moving loads configurations (single, dual standard (130-kN), tandem and tridem-axle loads) will be considered. The effects of tire loading on pavement degradation on traffic circle will be assessed by taking into account not only the vertical component, but also the centrifugal strengths (transverse component) and the effect of braking (longitudinal component). The viscoelastic law of the Generalized Kelvin-Voigt model, which has been incorporated into the Cast3M FE code [6], will be used to estimate the mechanical behavior of bituminous layers.

2 Bituminous Materials Behavior

The pavement structure is composed of two identical viscoelastic bituminous layers (BB: asphalt concrete and GB: asphalt gravel). Their complex modulus was measured in the LCPC Laboratory [7]. The complex modulus responses of the mixes measured in the laboratory were first replicated using the 2S2P1D model, which is a generalization of the Huet-Sayegh model [8]. In reference to the 2S2P1D model, the Kelvin-Voigt body values were fixed. Table 1 lists the Kelvin-Voigt parameters used for bituminous materials (BB and GB) [9].

Table 1. The Generalized Kelvin-Voigt model parameters at 20° and 30°C.

n°	BB			GB		
	E_i (MPa)	η_i(20°C) (MPa.s)	η_i(30°C) (MPa.s)	E_i (MPa)	η_i(20°C) (MPa.s)	η_i(30°C) (MPa.s)
	33500			31000		
1	1479000	9,15E-06	5,34E-07	1522500	1,04E-05	8,95E-07
2	1397400	8,65E-05	5,04E-06	1438500	9,79E-05	8,46E-06
3	1091400	6,76E-04	3,94E-05	1123500	7,65E-04	6,61E-05
4	705840	4,37E-03	2,55E-04	726600	4,95E-03	4,27E-04
5	433500	2,68E-02	1,57E-03	446250	3,04E-02	2,62E-03
6	265200	1,64E-01	9,57E-03	273000	1,86E-01	1,61E-02
7	159120	9,85E-01	5,74E-02	163800	1,12E+00	9,63E-02
8	90576	5,61E+00	3,27E-01	93240	6,35E+00	5,48E-01
9	46716	2,89E+01	1,69E+00	48090	3,27E+01	2,83E+00
10	20196	1,25E+02	7,29E+00	20790	1,42E+02	1,22E+01
11	7007	4,34E+02	2,53E+01	7214	4,91E+02	4,24E+01
12	1999	1,24E+03	7,22E+01	2058	1,40E+03	1,21E+02
13	495	3,06E+03	1,79E+02	509	3,47E+03	2,99E+02
14	99	6,12E+03	3,57E+02	102	6,93E+03	5,98E+02
15	22	1,33E+04	7,77E+02	22	1,51E+04	1,30E+03
16	216	1,34E+06	7,81E+04	223	1,52E+06	1,31E+05
17	733	4,54E+07	2,65E+06	755	5,14E+07	4,44E+06
18	19278	1,19E+10	6,96E+08	19845	1,35E+10	1,17E+09
19	504900	3,13E+12	1,82E+11	519750	3,54E+12	3,06E+11

3 The Pavement Model Considered for the Present Analysis

The French design code for pavement structures [10-11] proposes a set of sizes (materials and layers thicknesses) for pavement structures. This set is based on the class of trucks traffic (TCi), the class of pavement foundation performance (PFi) and the road category. The mechanics associated with the material properties and layer thicknesses of the selected pavement structures are demonstrated in Figure 1.

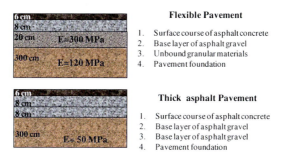

Fig. 1. The selected Asphalt pavements samples

3.1 Model Geometry and Mesh

Different types of loads can be considered in a pavement design. These are related to single, dual, tandem or tridem-axle loads. To take into account the effects of different configurations of loading, the French design method consists of modeling single or dual loads effects on an elastic pavement. The vertical contact stresses in this study equal 662 kPa (for the single, dual standard and tridem axle) and 535 kPa (for the tandem axle), centrifugal stress is 472 kPa and braking stress is 347 kPa.

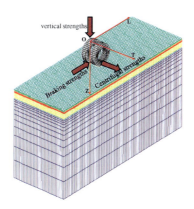

Fig. 2. The 3D FE model geometry for a dual standard

Effects of moving tire loading were estimated on traffic circle by taking into account not only the vertical component, but also the transverse component and the effect of braking (longitudinal component). The structures were modeled in 3D by means of the Cast3M FE code [6]. The FE mesh employed and the modeled pavement geometry are both shown in Figure 2 (for a thick asphalt pavement). The eight-node hexahedral "brick" elements were used for the FE analysis. When using the dual axle configuration at 42 km/h in the thick asphalt pavement structure, the mesh contains 90,752 elements and a total of 98,338 nodes. The interfaces between pavement layers were assumed to be perfectly bonded.

3.2 Moving Load Analysis and Boundary Conditions

The considered pavement section was composed of homogeneous and isotropic elastic material layers (pavement foundation and unbound granular materials) as well as bituminous layers with linear viscoelastic behavior. The Generalized Kelvin-Voigt model was implemented for this purpose (its parameters were listed in Table 1). For each layer, Poisson's ratio of 0.35 has been used. A mechanical calculation of this pavement structure has been performed in using small strains under a moving load at constant speed, over same time period as the OL axis (longitudinal axis) (Fig. 2). The pavement response due to a moving tire load has been estimated at 42 km/hr. The load duration depends on both the vehicle speed S and length l of the tire contact area. It is reasonable to assume that the load exerts practically no effect when it is located a distance of $6l$ from the point under consideration. In this paper, the tire contact pressure is uniformly distributed over a rectangular surface area 270*184 mm^2, with the inter-axle distance equal to 1.35 m for multi-axle configurations. The vertical, transverse and longitudinal displacements of the bottom plane of the model (Fig. 2) are fixed. The nodes of both sides of the L-Z plane, which transversely limit the model, are constrained relative to the T axis (transverse axis), and the movements of both transverse T-Z planes are constrained with respect to the L axis (longitudinal axis).

4 Pavement Response and Damage Analysis

In this section, the effects of loading a moving tire on the pavement will be discussed. These results have been derived from the simulation run on Cast3M [6]. The numerical calculation resulting from the simulation of single, dual, and multiple-axle loading conditions will then be given. For the elastic calculus, viscoelastic effects due to asphalt materials are taken into account only through an equivalent elastic modulus, which is determined from complex modulus test. The values of frequency of request are computed from the signals of longitudinal strain at the bottom of the Bituminous Gravel (GB) in the single wheel path. The period T is measured between both peaks in contraction of the longitudinal signal [12] (see figure 3). The frequency determined by this method is equal to 8 Hz [9]. It corresponds to a vehicle speed of 42 km/h at a reference temperature. This frequency was determined in the thick asphalt pavement at a depth of 22 cm from the surface of pavement. The equivalent elastic modulus obtained for the reference temperatures (20 and 30 °C) and a frequency of 8 Hz are the followings:

- BB at a frequency of 8 Hz: E=6000 MPa at 20 °C; E=1300 MPa at 30 °C.
- GB at a frequency of 8 Hz: E=6000 MPa at 20 °C; E=2000 MPa at 30 °C.

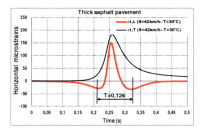

Fig. 3. Frequency measured between both peaks in contraction of the longitudinal signal at the bottom of the GB layer

4.1 Numerical Calculation Results

Figure 4 presents the horizontal strain signals at the bottom of the GB3 layer for tridem axle configuration. These results are simultaneously obtained thanks to a viscoelastic and elastic calculus on Cast3M. The signals of the viscoelastic calculus are expressed according to the speed (S) and to the temperature (T) whereas those of the equivalent elastic calculus are expressed according to the frequency (f) and the temperature. Furthermore, the slowing in the recovery of transverse strain cannot be predicted by the elastic model. This delay is ascribable to viscoelasticity, as illustrated by the experimental results, which clearly indicate that bituminous material viscoelasticity needs to be taken into account in order to generate a more realistic simulation of strains produced by loads moving at low speed on flexible pavements.

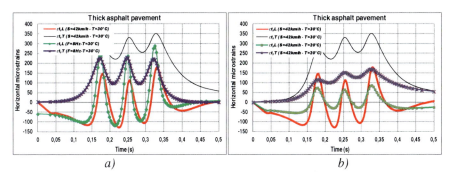

Fig. 4. Computed longitudinal ($\varepsilon_{t,L}$) and transverse ($\varepsilon_{t,T}$) strains at the bottom of the GB layer (for a thick asphalt pavement) for the tridem axle configuration

The main modifications brought about by the viscoelastic modeling with respect to the elastic modeling (Fig. 4-a) are the following: the maximal amplitudes of transverse strain signals are higher than the longitudinal (thick asphalt pavement), the viscoelastic signals are dissymmetrical, the amplitudes of maximum strain obtained by the viscoelastic modeling are obtained after the middle of the load crosses the point of measure, this occurs as later as temperature is increased. Along the length direction, the area of strain distribution increases as the temperature drops (Fig. 4-b). The loading time gradually decreases as the number of axles rises. Figure 5 displays computed longitudinal shear stresses of the BB/GB interface layer for the tridem axle configuration. The latest fatigue test studies bituminous material behavior in either bending or tension-compression. At the present time, the design concept based solely on bending fatigue is insufficient to realistically predict the fatigue life expectancy of a pavement structure; the phenomenon of shear at the BB/GB interface must also be taken into account. At this interface, the maximum shear stress develops in the bituminous pavement structure [13]. To date, identifying the fatigue behavior of pavement structures by the bending of bituminous layers has been preferred, as well as by the shear at the BB/GB interface. These two approaches complement each other in deriving a global prediction of the fatigue life expectancy of pavement structures.

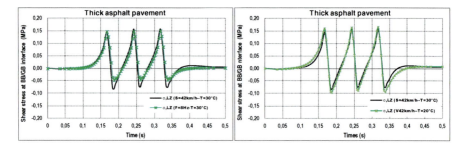

Fig. 5. Computed longitudinal (τ_{LZ}) shear stresses at the BB/GB interface (thick asphalt pavement) for the tridem axle configuration

The results presented in this section reveal that both the quantification and qualification of the strains and shear stress signal depend on the loading history experienced by the pavement structure. The developed method indicates that the intrinsic magnitudes helping to characterize strains and shear stress signals are not neccessarily constant, but instead functions of the following main parameters:

- pavement structure (material behavior, thickness and stiffness of pavement layers, temperature, type of interface between layers, etc.);
- load configuration (intensity, type of wheels, axles, etc.);
- load speed (or frequency).

4.2 Damage Analysis

Characterizing the bituminous pavement damage caused by multiple-axle loads requires both the quantification and summation of pavement responses. The Load Equivalent Factor (LEF) is defined as the damage of an axle group relative to the damage of a dual standard axle (130-kN). In this paper, the LEF has been calculated according to the strain area mathod. This method has been devoloped by Chatti and Hajek [1-2,14] (Equat. 1 and 2).

$$LEF = \frac{Damage_{Axle.j}}{Damage_{s\tan dard..axle.s}} = \frac{\int_0^t |\varepsilon_j^{ni}| dt}{\int_0^t |\varepsilon_s^{ni}| dt} \quad (1) \qquad LEF_\tau = \frac{\int_0^t |\tau_j^{ni}| dt}{\int_0^t |\tau_s^{ni}| dt} \quad (2)$$

$\varepsilon_{s,j}$ is the horizontal strain acting upon the dual standard axle s (respectively axle j); t is the time, if this strain is expressed in the time domain, or the distance, if the strain is expressed in the space domain; $\tau_{s,j}$ is the shear stress at the BB/GB interface subjected to the dual standard axle s (respectively axle j); and n_i is the integration method exponent (n=1 in this paper).

Table 2 presents the LEF of loading of a moving tire on the bituminous pavement structure on traffic circle. It is determined by using the relationship 1 and 2.

Table 2. The LEF on the GB3 layer and LEF_τ at the BB/GB interface

Axle	Single	Dual standand (130 kN)	Tandem	Tridem
LEF	0.86	1.00	1.61	2.36
LEF_τ	0.85	1.00	1.32	1.80

4.3 Towards a Proposition of a New Fatigue Law of Multi-axle Configurations

The current law, which verifies the fatigue of bituminous layers in bending, as defined by the relationship in (Equat. 3), does not strictly allow taking into account all parameters influencing pavement structure behavior, as presented in the Section 4.1.

$$\log(N_f) = a.\log(\varepsilon_t) + b \quad (3)$$

A new fatigue law has therefore been adopted. Such a law overcomes the gaps in the current fatigue law for multi-axle configurations; it uses both the LEF (Equat. 1) and strain amplitude as inputs. For these special configuration, the critical strain will represent the maximum peak in tension. The expression of this new fatigue law is given by the relationship (4). Similarly, a new fatigue law to identify cracking occurring at the interface is given by the relationship (5).

$$N_f = \frac{1}{LEF} \cdot \left(\frac{\varepsilon_t}{a}\right)^{-\frac{1}{b}} \quad (4) \qquad\qquad N_{f.i} = \frac{1}{LEF_\tau} \cdot \left(\frac{\tau_{max}}{a}\right)^{-\frac{1}{b}} \quad (5)$$

4.4 Predictions of Pavement Lives

The French design method [10] uses the following fatigue criterion on bituminous layer:

$$\varepsilon_t = \varepsilon_6 (10°C, 25Hz) \left(\frac{E(10°C)}{E(T_{eq})}\right)^{0,5} \cdot \left(\frac{N_f}{10^6}\right)^b \cdot k_c \cdot k_r \cdot k_s \quad (6)$$

Similar to the fatigue cracking model, the interface fatigue model can be written as follows [15-17]:

$$\tau = \tau_6 \left(\frac{K_s(10°C)}{K_s(20°C)}\right)^{-1.697} \cdot \left(\frac{N_{f.i}}{10^6}\right)^{-0.223} \quad (7)$$

with: $\begin{cases}(10°C): \tau_6 = 0,36 MPa; K_s = 104 MPa/mm \\ (20°C): \tau_6 = 0,13 MPa; K_s = 57 MPa/mm\end{cases}$

The damage is computed as the inverse of the fatigue life expectancy as shown in equation 8.

$$D = \frac{1}{N_f} \quad (8)$$

Tables 3, 4 and 5 present the results of fatigue life and damage on pavement structure analyzed.

Table 3. Fatigue life and damage of the GB layer in the thick asphalt pavement

Axle configuration	Temperature 20°C		Temperature 30°C		Temperature 20°C		Temperature 30°C	
	Viscoelastic calculus (vehicle speed:42km/h)				Elastic calculus (Frequecy: 8Hz)			
	N_f	$(1/N_f)$	N_f	$(1/N_f)$	N_f	$(1/N_f)$	N_f	$(1/N_f)$
Single	2.06E+5	4.85E-6	8.61E+4	1.16E-5	5.34E+4	1.87E-5	5.56E+4	1.80E-5
Dual standand (130kN)	2.65E+4	3.78E-5	1.56E+4	6.42E-5	1.31E+4	7.65E-5	3.60E+3	2.78E-4
Tandem	2.25E+4	4.44E-5	8.15E+3	1.23E-4	4.28E+4	2.34E-5	9.00E+3	1.11E-4
Tridem	1.65E+4	6.07E-5	6.84E+3	1.46E-4	1.08E+5	9.25E-6	1.80E+4	5.55E-5

Table 4. Fatigue life and damage of the GB layer in the flexible pavement

Axle configuration	Temperature 20°C		Temperature 30°C		Temperature 20°C		Temperature 30°C	
	Viscoelastic calculus (vehicle speed:42km/h)				Elastic calculus (Frequecy: 8Hz)			
	N_f	$(1/N_f)$	N_f	$(1/N_f)$	N_f	$(1/N_f)$	N_f	$(1/N_f)$
Single	5.14E+5	1.95E-6	8.30E+5	1.20E-6	2.96E+5	3.38E-6	5.11E+5	1.96E-6
Dual standand (130kN)	7.41E+4	1.35E-5	3.05E+5	3.28E-6	2.27E+4	4.40E-5	4.28E+4	2.34E-5
Tandem	1.92E+5	5.20E-6	3.90E+5	2.57E-6	4.45E+4	2.25E-5	6.00E+4	1.67E-5
Tridem	1.98E+5	5.06E-6	5.51E+5	1.81E-6	8.61E+4	1.16E-5	1.40E+5	7.15E-6

Table 5. Fatigue life and damage at the BB/GB interface in the bituminous pavement structures

Axle configuration	Viscoelastic: 42km/h		Elastic: 8Hz		Viscoelastic: 42km/h		Elastic: 8Hz	
	Thick asphalt pavement (temperature 20°C)				Flexible pavement (temperature 20°C)			
	N_f	$(1/N_f)$	N_f	$(1/N_f)$	N_f	$(1/N_f)$	N_f	$(1/N_f)$
Single	2.75E+5	3.63E-6	3.06E+5	3.27E-6	1.15E+5	8.66E-6	9.96E+4	1.00E-5
Dual standand (130kN)	3.68E+4	2.72E-5	3.62E+4	2.76E-5	1.75E+4	5.73E-5	1.46E+4	6.85E-5
Tandem	6.95E+4	1.44E-5	8.20E+4	1.22E-5	2.98E+4	3.36E-5	2.83E+4	3.53E-5
Tridem	1.89E+5	5.28E-6	2.17E+5	4.60E-6	7.32E+4	1.37E-5	7.16E+4	1.40E-5

Figure 6 clearly shows the level of global damage of the pavement structures on traffic circle. For the thick asphalt pavement structure, the mode of main damage is the "bottom-up cracking" type; whereas the "top-down cracking" damage is the type displayed for the flexible pavement. To date, identifying the fatigue behavior of pavement structures by the bending of bituminous layers has been preferred, as well as by the shear at the BB/GB interface. These two approaches complement each other in deriving a global prediction of the fatigue life expectancy of pavement structures.

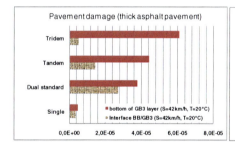
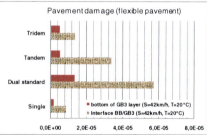

Fig. 6. Pavement damage

5 Conclusion

The objective of this paper was to develop a model allowing the determination of damage of bituminous pavement on traffic circle. The parameters input to simulate the above fatigue were: pavement structure, load speed (or frequency), load configuration, and bituminous material temperature. The method developed in this paper has successfully replicated the observed phenomenon. This paper has also shown that the design concept based solely on the bending fatigue of asphalt layers may not be sufficient to predict damage. Moreover, shear fatigue phenomena at the

BB/GB interface must be taken into account. These two approaches complement each other for a global prediction of the fatigue life expectancy of the bituminous pavement structures.

References

[1] Chatti, K., Salama, H.K.: Evaluation of fatigue and rut damage prediction methods for asphalt concrete pavements subjected to multiple axle loads. International Journal of Pavement Engineering 12(1), 25–36 (2011)
[2] Chatti, K., Manik, A., Salama, H.K., Chadi, M., Lee, S.: Effect of Michigan multi-axle trucks on pavement distress, Final Report MDOT, p. 312 (2009)
[3] Homsi, F.: Endommagement des chaussées bitumineuses sous chargements multi-essieux, Ph.D. Thesis in french, Ecole Centrale de Nantes, France, p. 203 (2011)
[4] Kogo, K., Himeno, K.: The effect of different waveforms and rest period in cyclic loading on the fatigue behavior of the asphalt mixture. In: Al Qadi, Scarpas, Loizos (eds.) Pavement Cracking (2008)
[5] Bodin, D., Merbouh, M., Balay, J.-M., Breysse, D., Moriceau, L.: Experimental study of the waveform shape effect on asphalt mixes fatigue. In: Proceeding of the 7th International RILEM Symposium on Advanced Testing and Characterization of Bituminous Materials (ATCBM 2009), Rhodes, May 26-28, vol. 2, pp. 725–734 (2009)
[6] Cast3M, Cast3M is a research FEM environment; its development is sponsored by the French Atomic Energy Commission (2010), http://www-cast3m.cea.fr/cast3m
[7] LCPC, Vérification du comportement mécanique des matériaux du manège (2003)
[8] Olard, F., Di Benedetto, H.: General "2S2P1D" Model and Relation Between the Linear Viscoelastic Behaviours of Bituminous Binders and Mixes. Road Materials and Pavement Design 4(2), 185–224 (2003)
[9] Zoa, A., Allou, F., Petit, C., Medjo, R.: Modélisation viscoélastique de l'endommagement des chaussées bitumineuses sous chargement multi-essieux. In: Actes des 29e Rencontres Universitaires de Génie Civil, Tlemcen, Algérie, Mai 29-31, vol. 3, pp. 120–129 (2011)
[10] LCPC–SETRA, Conception et dimensionnement des structures de chaussée, Guide technique, Paris (1994)
[11] LCPC-SETRA, Catalogue des structures types de chaussées neuves, Ministère de l'équipement, des transports et du logement (1998)
[12] Domec, V.: Endommagement par fatigue des enrobés bitumineux en condition de trafic simulé et de température, Ph.D. Thesis in french, Université de Bordeaux 1, France, p. 277 (2005)
[13] Zoa, A., Allou, F., Petit, C., Medjo, R.: Importance de la modélisation des interfaces dans la conception rationnelle des chaussées. In: Actes des 28e Rencontres Universitaires de Génie Civil, La bourboule, France, Juin 02-04, pp. 1112–1121 (2010)
[14] Hajek, J.J., Agarwal, A.C.: Influence of Axle Group Spacing on Pavement Damage. Transportation Research Record (1286), 138–149 (1990)
[15] Diakhaté, M.: Fatigue et comportement des couches d'accrochages dans les structures de chaussée, Ph.D. Thesis in french, Université de limoges, France, p. 241(2007)

[16] Petit, C., Diakhaté, M., Millien, A., Phelipot-Mardelé, A., Pouteau, B.: Pavement Design for Curved Road Sections - Fatigue Performances of Interfaces and Longitudinal top-down Cracking in Multilayered Pavements. Road Materials and Pavement Design 10(3), 609–624 (2009)

[17] Diakhaté, M., Millien, A., Petit, C., Phelipot-Mardelé, A., Pouteau, B.: Experimental investigation of tack coat fatigue performance: Towards an improved lifetime assessment of pavement structure interfaces. Construction and Building Materials 25, 1123–1133 (2011)

Fatigue Performance of Highly Modified Asphalt Mixtures in Laboratory and Field Environment

Robert Kluttz[1], J Richard Willis[2], André A.A. Molenaar[3], Tom Scarpas[3], and Erik Scholten[1]

[1] Kraton Polymers
[2] Auburn University
[3] Delft University of Technology

Abstract. High levels of SBS polymer modification lead to a bituminous binder with improved resistance to both rutting and fatigue cracking. Beam fatigue and modeling predict that, using this binder, pavement thickness can be reduced and still achieve equal or superior long term performance. To test this, a section at the National Center for Asphalt Technology (NCAT) was paved using a binder with a nominal grade of PG 88-22 for all three lifts. This pavement was constructed in August 2009 at a thickness of 145 mm compared to a standard thickess of 180 mm for the control and other pooled-fund study sections. In August 2010 an unrelated section that experienced failure was rehabilitated with a similar structure. This paper reports beam fatigue analysis and modelling and compares the results with observed field performance over both structurally sound and weak subgrades. The beam fatigue data predicts a very high endurance limit for the mixtures. Although conclusions are premature, to date neither structure shows surface distress.

1 Introduction

As traffic loadings increase and budgets for construction and maintenance shrink, agencies and researchers look for innovative ways to design and build asphalt pavements that have lower initial construction costs and longer lifetimes to achieve overall reduction in life cycle costs. Polymer-modified asphalt (PMA) is a well-established product for improving the effectiveness of asphalt pavements. In particular, styrene-butadiene-styrene (SBS) polymers are commonly used to improve permanent deformation resistance and durability in wearing courses. [1, 2] Use of SBS in intermediate and base courses has been limited due partly to the perception that base courses, with narrower temperature spans than surface courses, do not need modification. However, the ability of SBS polymers to resist fatigue cracking could, in theory, be used to reduce the overall cross-section of a flexible pavement. This is of particular importance for perpetual pavements that often feature high-modulus intermediate asphalt layers and fatigue-resistant bottom layers. There is a need for materials that have enhanced fatigue characteristics and can carry load more efficiently through a reduced cross-section.

At loadings of 2½ to 4%, SBS polymers give a significant increase in permanent deformation resistance which is well-recognized. However, the increase in toughness and durability is more modest. However, at higher loading, 7 to 8% SBS, the bitumen-polymer blend undergoes a phase inversion so that the polymer becomes a fully continuous phase. This paper outlines an experimental program to test, model, and field validate the performance of highly modified bitumen for both permanent deformation and cracking resistance.

2 Early Work

The initial work in this program was carried out by Road and Railway Engineering section and the Structural Mechanics section at the Delft University of Technology. Material testing comprised two parts: four-point bending beam fatigue studies, and monotonic, uniaxial tensile and compression tests to develop model parameters. [3-5]

2.1 Material Testing

An overview of beam fatigue test results is presented in Figure 1. All mix designs were the same as was the base bitumen, the only variable being the exact type and loading of SBS polymer.

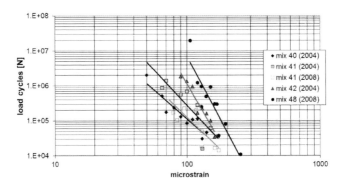

Fig. 1. Beam Fatigue Test Results

The impact of polymer loading is quite significant. Mix 40 is unmodified. Mixes 41 and 42 contain 6% SBS, while mix 48 contains 7.5%. Based on these data alone, one might expect one to two orders of magnitude in fatigue life going from unmodified to 7 to 8% loading.

2.2 Finite Element Modelling

Concurrently, modelling work was conducted using the designs shown in Figure 2. The designs compare the performance of monolithic bound layers over structurally sound subbase and subgrade. The thickness of the unmodified control was 250 mm while the highly modified section was 150 mm for a 40% reduction in thickness.

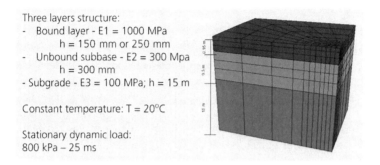

Three layers structure:
- Bound layer - E1 = 1000 MPa
  h = 150 mm or 250 mm
- Unbound subbase - E2 = 300 Mpa
  h = 300 mm
- Subgrade - E3 = 100 MPa; h = 15 m

Constant temperature: T = 20°C

Stationary dynamic load:
800 kPa – 25 ms

Fig. 2. Model Pavement Structures and Test Parameters

Modelling was conducted with the Delft Asphalt Concrete Response (ACRe) 3-D finite element continuum damage model. Initial data was fitted to a Schapery model using tensile strength, fracture energy and compliance slope. Some of the results are shown in Table 1.

Table 1. Calculated Fatigue Model Parameters – m based on E* vs. frequency

Modification Type	A-ratio Improvement (m)
48- experimental SBS	68 (0.38)
42 – standard SBS	8 (0.38)
41 – standard SBS	2 (0.41)

The fatigue theory was well in line with the bending beam material test results, with one to two orders of magnitude improvement.

A series of advanced modelling studies were conducted utilizing single point loading simulation, then full wheel track loading simulation. Some results from the wheel track loading are shown in Table 2. Both deformation and cracking damage are reduced, on average, by about a factor of two, despite the fact that the highly modified pavement was 40% thinner than the unmodified pavement.

Table 2. Comparative Damage Calculated from Wheel Track Model

Distress	250 mm unmodified	150 mm highly modified
Shear deformation	2.05E-2	0.78E-2
Compressive deformation	1.27E-2	0.70E-2
Longitudinal cracking	1.31E-3	0.02E-3
Vertical cracking	7.72E-4	4.41E-4
Transverse cracking	8.65E-4	0.79E-4

3 NCAT Sections

In 2009, two test sections (Figure 3) were constructed as part of a field experiment to validate using high polymer mixtures. [6] Each section was built using three asphalt layers. The primary differences between the two test sections were the amount of polymer used in the mixtures and the overall pavement thickness. One section was built 145 mm thick using two High Polymer Mixes (HPM) while the second test section was built 178 mm thick using conventional asphalt concrete.

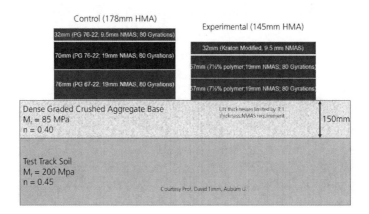

Fig. 3. Pavement Structures for NCAT Control and Experimental Section N7

Also in 2009, a third section was rehabilitated. [7] This section was part of a 2006 experiment to determine perpetual pavement thickness over a weak subgrade using conventional PG 67 and PG 76 binders. The 356 mm pavement is still performing well, but the 254 mm pavement had failed at the end of the 2006 cycle. In 2009 it was rehabilitated with a 127 mm mill and inlay, but ten months later the inlay was cracked through and had 30-35 mm ruts extending into the subgrade.

To stabilize the section through the rest of the 2009 cycle, NCAT and Oklahoma elected to try a highly modified structure. This provided an excellent head-to-head comparison with adjacent highly modified sections, one over a strong base structure and one over a weak base. In the latter case, there was discussion over the optimum design hinging on the most important distress. For the remaining life of the section, was it more important to mitigate cracking or rutting? In the end, Oklahoma opted for a finer, richer mix (Figure 4) for the bottom lift to mitigate reflective cracking at the (hopefully minor!) cost of slightly reduced modulus.

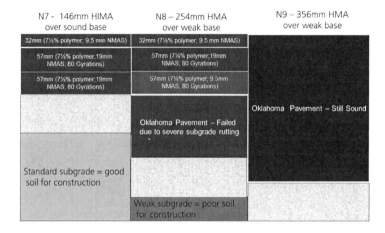

Fig. 4. Pavement Structures for Experimental Rehabilitation Section N8

3.1 Material Testing

Extensive material testing has been conducted on the asphalt mixtures in this study. [8] The HPM binder in N7 and N8 graded at approximately PG 88-22. Bending beam fatigue testing was conducted on the two base mixtures constructed in the control and the HPM test sections in accordance with AASHTO T 321-07. Nine beams were fabricated for fatigue testing to 7 ± 1 per cent air voids. Three beams of each mixture were tested at both 400 and 800 microstrain, but due to the fatigue performance of the HPM, the remaining three HPM beams were tested at 600 microstrains while the control beams were tested at 200 microstrains.

AASHTO T 321-07 was used to define beam failure as a 50% reduction in beam stiffness in terms of number of cycles until failure. Normally, the test would run to approximately 40% of initial stiffness, but as a factor of safety and to ensure a complete data set, the beams for this project were allowed to run until the beam stiffness was reduced to 25% of the initial stiffness. Upon finding the cycles to failure at three different strain magnitudes, the fatigue endurance limit was calculated for each 19 mm mix design using the NCHRP 9-38 procedure. [9] Test results are shown in Figure 5.

The difference between the average fatigue life of the control mixture to that of the HPM at two strain levels tested in this study was determined using the failure criterion (50% reduction in beam stiffness). This information helps evaluate important aspects of the material behavior shown in Figure 5 as follows:

- At the highest strain magnitude, the HPM was able to withstand almost four times more loading cycles than the control mixture.
- At 400 με, the average fatigue life of the HPM was greater than the control mixture. The average cycles until failure for the control mixture was 186,000 and while the HPM averaged 6,000,000 loading cycles.

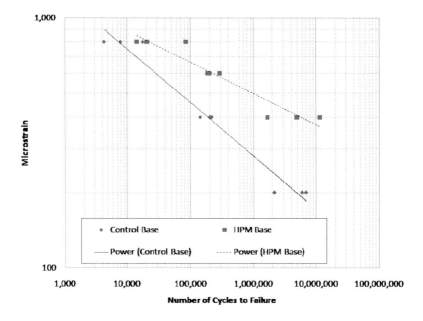

Fig. 5. Beam Fatigue Test Results [8]

The 95 percent one-sided lower prediction of endurance limit for each of the two mixes tested in this study based on the number of cycles to failure was determined in accordance with AASHTO T 321-07. The procedure for estimating the endur-ance limit was developed under NCHRP 9-38. The calculated fatigue endurance limit for the control base mixtures was 77 microstrain while the endurance limit for the HPM base mixtures was 231. Thus, the HPM base mixture could theoretically withstand three times the strain magnitudes without accruing damage when compared to the control base mixture.

3.2 Conventional and S-VECD Modelling

Advanced modelling and beam fatigue endurance limit determinations both indicate very good fatigue performance for these mixes. However, both procedures are costly and time consuming and so are difficult to implement for routine pavement analysis and design. A straightforward analysis using the Mechanistic Empirical Pavement Design Guide (MEPDG) methodology [10] is shown in Figure 6.

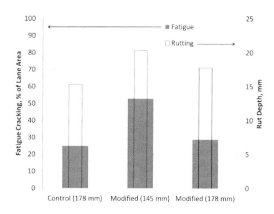

Fig. 6. MEPDG Modelling Using Conventional Calibration Factors

In its current form, the MEPDG relies on a layered elastic model for distress development. The dynamic modulus mastercurve is the only material input data so fatigue response is assumed. Without adjusting calibration factors to account for the improved fatigue performance, the MEPDG actually predicts poorer performance for the HPM pavement than the control. The calibration factors may readily be adjusted, but that reverts to advanced testing/modelling to determine appropriate values.

Several researchers have been working to develop a happy medium, models that will recognize and correctly attribute a broader range of material properties yet will require simpler input and execution. In particular, the research group at North Carolina State University has developed a Simplified Viscoelastic Continuum Damage (S-VECD) model [11-13] which requires only straightforward input data and is readily executable.

Material test data from several of the NCAT 2009 pooled fund sections was developed by NCAT and by the FHWA Turner-Fairbank Highway Research Center laboratory. The results of the S-VECD analysis are shown in Figures 7 and 8. [14]

The S-VECD model also predicts a higher endurance limit, roughly double the average value for the various conventional structures. ACRe, beam fatigue and S-VECD all predict improved performance to varying degrees. It is now up to field performance to determine the relative accuracies.

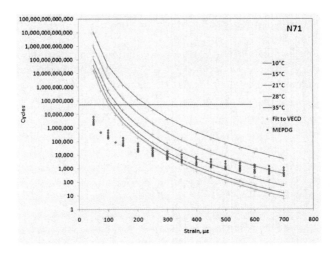

Fig. 7. Calculated Fatigue Performance with S-VECD Model [14]

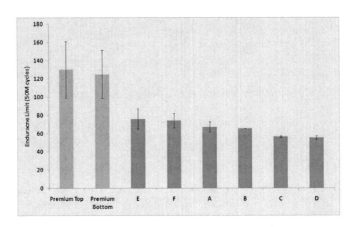

Fig. 8. Endurance Limit (50MM cycles) from Range of Temperatures [14]

3.3 Field Performance to Date

To date, both Sections N7 and N8 are performing well. Both sections exhibit minimal rutting, < 2 mm, about one third of the control. Section N7 has no cracks, but none of the sections in the pooled fund group has cracked. At this point we can simply state that this thinner, highly modified section is showing superior rutting resistance. There is not yet any discrimination on cracking. Section N8 recently developed its first crack (Figure 9) at 5.27 MM ESALS. The 2009 rehabilitation first developed a crack at 2.7 MM ESALS. It is early to draw conclusions, but if deterioration of Section N8 proceeds at the same sort of extended rate as initial crack formation, then the lifetime of the HPM rehabilitation will exceed the lifetime of conventional high performance rehabilitation by a factor of two.

Fig. 9. Photo of N8 Highly Modified Rehabilitation at 5.27 MM ESALS

4 Conclusions

- Highly SBS-modified asphalt mixes demonstrate improved fatigue performance in modelling, material testing and, by early data, field performance, which may allow thickness reduction in pavement design.
- An advanced 3-D finite element model, ACRe, and a simplified continuum damage model, S-VECD, both predict improved fatigue performance in line with beam fatigue predictions.
- While the field test is not complete, the preliminary observation of first crack appearance at about twice the time for the conventional structure is in line with the modelling and beam fatigue results.

References

[1] Anderson, R.: Asphalt Modification and Additives. In: The Asphalt Handbook MS-4, 7th edn., pp. 86–89. Asphalt Institute, Lexington (2007)
[2] von Quintus, H., Mallela, J., Buncher, M.: Quantification of Effect of Polymer-Modified Asphalt on Flexible Pavement Performance. In: Transp. Res. Rec. No. 2001, pp. 141–154. Transportation Research Board of the National Academies, Washington, DC (2007)
[3] van de Ven, M., Poot, M., Medani, T.: Advanced Mechanical Testing of Polymer Modified Asphalt Mixtures. Report 7-06-135-3, Road and Rail Engineering, Delft University of Technology, The Netherlands (2007)
[4] Molenaar, A., van de Ven, M., Liu, X., Scarpas, A., Medani, T., Scholten, E.: Advanced Mechanical Testing of Polymer Modified Base Course Mixes. In: Proc., Asphalt – Road for Life, Copenhagen, pp. 842–853 (2008)
[5] Kluttz, R., Molenaar, A., van de Ven, M., Poot, M., Liu, X., Scarpas, A., Scholten, E.: Modified Base Courses for Reduced Pavement Thickness and Improved Longevity. In: Proc. Intl. Conf. Perpetual Pavement, Columbus, OH (2009)

[6] Timm, D., Robbins, M., Kluttz, R.: Full-Scale Structural Characterization of a Highly Polymer-Modified Asphalt Pavement. In: Proc. 90th Annual Transp. Res. Board, Washington, DC (2011)
[7] Timm, D., Powell, R., Willis, J., Kluttz, R.: Pavement Rehabilitation Using High Polymer Asphalt Mix. In: submitted for the Proc. 91st Annual Transp. Res. Board, Washington, DC (2012)
[8] Willis, J., Timm, D., Kluttz, R., Taylor, A., Tran, N.: Laboratory Evaluation of a High Polymer Plant-Produced Mixture. In: submitted for the Assoc. In: Asphalt Paving Technol. Annual Meeting, Austin, TX (2012)
[9] Prowell, B., Brown, E., Daniel, J., Bhattacharjee, S., von Quintus, H., Carpenter, S., Shen, S., Anderson, M., Swamy, A., Maghsoodloo., S.: In: Validating the Fatigue Endurance Limit for Hot Mix Asphalt, NCHRP Report 646, Transportation Research Board, National Academies of Sciences (2010)
[10] Applied Research Associates, Arizona State University, Mechanistic-Empirical Pavement Design Guide Version 1.100 (2009)
[11] Hou, T., Underwood, B., Kim, Y.: Fatigue Performance Prediction of North Carolina Mixtures Using the Simplified Viscoelastic Continuum Damage Model. J. Assoc. Asphalt Paving Technol. 79, 35–80 (2010)
[12] Underwood, B., Kim, Y., Guddati, M.: Improved Calculation Method of Damage Parameter in Viscoelastic Continuum Damage Model. Intl. J. Pavement Eng. 11(6), 459–476 (2010)
[13] Kim, Y.R., Guddati, M., Underwood, B., Yun, T., Subramanian, V., Savadatti, S., Thirunavukkarasu, S.: Development of a Multiaxial VEPCD-FEP++, Final report to the Federal Highway Administration, Project Number DTFH61-05-RA-00108 (2008)
[14] Gibson, N., Kim, Y.: presented at the spring Federal Highway Administration Expert Task Group Meetings, Phoenix, AZ (2011)

Publication Disclaimer. We believe the information set forth above to be true and accurate, but any recommendations or suggestions that may be made in the foregoing text are without any warranty or guarantee whatsoever, and shall establish no legal duty or responsibility on the part of the authors or their employer. Furthermore, nothing set forth above shall be construed as a recommendation to use any product in conflict with any existing patent rights. Kraton Polymers expressly disclaims any and all liability for any damages or injuries arising out of any activities relating in any way to this publication.

A Multi-linear Fatigue Life Model of Flexible Pavements under Multiple Axle Loadings

Farah Homsi[1,2], Didier Bodin[3], Denys Breysse[2], Sylvie Yotte[2], and Jean Maurice Balay[1]

[1] Université Nantes Angers Le Mans, IFSTTAR, Route de Bouaye, CS4, 44341 Bouguenais cedex, France
[2] Université Bordeaux 1, I2M UMR CNRS 5295, Département Génie Civil et Environnement (GCE), av des facultés, F-33405 Talence cedex, France
[3] ARRB Group Ltd, 500 Burwood Highway, Vermont South VIC 3133, Australia

Abstract. The fatigue damage of a pavement under repeated traffic loadings is a key issue for pavement design. At the material scale, the fatigue performance of asphalt mixtures can be assessed with laboratory tests. The standard fatigue test consists of the application of a continuous sinusoidal signal on the specimens and enables to write the fatigue life as a function of the strain level. Real loadings are more complex. Additional parameters may therefore have an influence on the fatigue life of bituminous mixtures. A methodology for a better calculation of the fatigue life of asphalt pavements is developed. It couples a structural approach and a material-based approach. This paper presents the material-based approach of the methodology. A database of laboratory fatigue tests with complex loadings has been built. An experimental plan whose variables are the independent shape parameters characterising a loading signal is defined and the synthetic complex loading signals constructed. The results of the fatigue tests enabled the calibration of a multi-linear fatigue model where the fatigue life is a function of the independent shape parameters characterizing the loading signal. Coupling the multi-linear fatigue model with a pavement model enables the calculation of the fatigue life of a pavement under different loading conditions.

1 Introduction

In France, main road pavements structures are designed to have a service life of 20 to 30 years. The fatigue performance of asphalt mixtures is a parameter for pavement design. At the material scale, the fatigue life of bituminous mixtures is determined in the laboratory. The standard fatigue test in Europe is the 2-point bending fatigue test. It consists of the application of a continuous sinusoidal signal on the top of a trapezoidal specimen clamped at its large base and the computation of the number of application of loading cycles before failure characterised by the loss of half of the initial stiffness. Different fatigue tests at different strain levels enable fitting the fatigue relationship of the mixture which is a function of the strain level, the only variable in the case of laboratory fatigue tests. The sinusoidal signal applied in the laboratory imitates the effect of a single wheel. Heavy trucks usually have multiple axle configurations, tandem (2 axles) and tridem (3 axles) in France and up to 8 axles in the international context [3]. In addition, it is environmentally beneficial to reduce the number of heavy trucks on pavements by increasing their weights for it reduces the emission of toxic gases in the atmosphere. Nevertheless, this results in longer trucks with multiple axle configurations. In the French design method, the effect of multiple axle loadings is simplified and taken into account via load equivalency factors using Miner's Law. Loading signals under multiple axle configurations measured at an experimental pavement differ in shape from the sinusoidal signal [6]. Additional parameters may therefore have an influence on the fatigue life of bituminous mixtures. The establishment of taxes on heavy weights makes it important to compute the effect of heavy weights on the fatigue life of asphalt pavements in a more accurate way.

The best way to assess the fatigue life of flexible pavements under multiple axle configurations is field measurements [11] or experimental pavement data. It is however difficult to find similar pavements cross sections subjected to significantly different axle configuration distributions and fatigue data using accelerated pavement testing with different axle configurations are not available. Several authors studied in the laboratory the effect of the loading shape on the fatigue life of bituminous mixtures by applying different loading signals (triangular, double-peak, rectangular, sinusoidal with different frequencies) on the specimen [1, 3, 5, 8 and 9]. In these tests, different shape factors varied at the same time (the loading area, the duration near the peak, the loading duration, number of peaks...). It was therefore impossible to compute the effect of each shape parameter on the fatigue life. In addition, all the loading signals applied in the laboratory were different in shape from real loading signals. This paper presents a method aiming at a better computation of the fatigue life of asphalt pavements under multiple axle loadings.

2 Methodology

The methodology proposed here for the computation of the fatigue life of asphalt pavements under real loading signals couples a structural approach and a material-based approach. It is illustrated in Figure 1, each step having a number. The structural approach consists of the characterisation of a database of measured loading signals with a set of shape factors (1) and the computation of the independent shape parameters (2) that characterise a real loading signal by means of a principal components analysis (PCA). The independent shape parameters are the variables of an experimental plan of laboratory fatigue tests (3). The values taken by each variable are inspired from their variation ranges in the database of measured signals. Mathematically built synthetic signals are then constructed and applied in the laboratory in 2-point bending fatigue tests (3) (material-based approach). A multi-linear regression made on the results of the fatigue tests enabled the computation of a multi-linear fatigue model (4) that gives the fatigue life as a function of the independent shape parameters characterising a loading signal. Coupling this fatigue model with a pavement model [2] (5) enables the computation of the fatigue life of pavements (6) under different loading conditions.

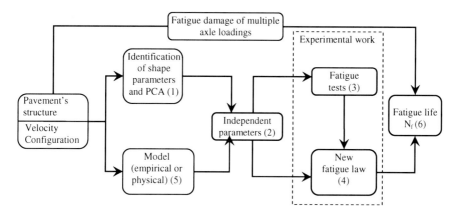

Fig. 1. Illustration of the methodology

3 The Experimental Program

The structural approach of the methodology consists of the characterisation of loading signals measured at an experimental pavement, at the IFSTTAR's accelerated pavement testing facility, under different loading conditions (pavement structure, temperature, axle configurations and vehicle speed) with a set of shape parameters. The horizontal strains (longitudinal and transverse) measured at the bottom of the asphalt base layer ($\approx$ 1750 loading signals) are considered [7] as critical signals for pavement design. The values taken by the shape parameters are evaluated for the database of measured loading signals. A principal component analysis enabled the identification of four independent shape parameters: the strain level (ε), the number of peaks (N_p), the area under the loading signal divided by the strain level and by the duration of the loading signal ($\hat{A}_n$) and the duration of the loading signal divided by the number of peaks ($\overline{D}$). These independent shape parameters are the variables of the experimental program of laboratory fatigue tests. The values taken by each variable in the experimental plan are inspired from their variation ranges in the database of measured signals for a temperature of the base course close to the temperature at which the laboratory fatigue tests will be done (20°C). Three values are considered for the variable ε : 347 µm/m, 240 µm/m and 166 µm/m (displacement 450 µm, 310 µm and 166 µm respectively). Three values are chosen for the variable number of peaks (N_p): 1 (single axle), 2 (tandem) and 3 (tridem). Two values are chosen for the variable $\hat{A}_n$: 0.21 (the median of the values taken by the longitudinal signals of the database) and 0.42 (the median of the values taken by the transverse signals of the database). Two values are chosen for the variable $\overline{D}$: 0.105 and 0.25 corresponding to velocities varying between 20 km/h and 50 km/h. The experimental program is made of 12 synthetic signals (Table 1) (four for each configuration) which were applied at the three strain levels. Figure 2 illustrates the synthetic loading signals imitating longitudinal measured signals under a tridem. The material used in this experimental program is a 0/14 mm graded asphalt concrete of class 3 with 4.12 % of a pure bitumen 35/50.

Table 1. Shape parameters of the 12 synthetic signals

Signal	N_P	$\hat{A}_n$	$\overline{D}$ (s)	T (s)	Signal	N_P	$\hat{A}_n$	$\overline{D}$ (s)	T (s)
1	1	0.21	0.105	0.28	7	2	0.42	0.105	0.51
2			0.25	0.67	8			0.25	1.22
3	1	0.42	0.105	0.28	9	3	0.21	0.105	0.76
4			0.25	0.67	10			0.25	1.80
5	2	0.21	0.105	0.51	11	3	0.42	0.105	0.76
6			0.25	1.22	12			0.25	1.80

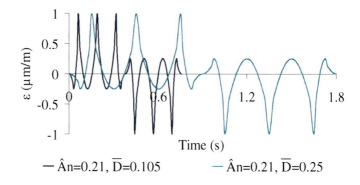

— $\hat{A}_n=0.21$, $\overline{D}=0.105$ — $\hat{A}_n=0.21$, $\overline{D}=0.25$

Fig. 2. Synthetic signals imitating longitudinal measured signals under a tridem

4 Results of the Fatigue Tests and Multi-linear Fatigue Model

Two-point bending fatigue tests were performed using the 12 synthetic signals at the three strain levels. Three replications were made for the displacement levels of 450 μm and 310 μm. One replication is made for the displacement level of 215 μm (tests having a long duration). This results in 84 tests. The failure criterion is the loss of half of the initial conventional stiffness. The fatigue lives obtained range between 2000 and 1 million loading cycles and the test durations range between 25 minutes and 15 days. The logarithm of the number of applications of loading cycles before failure (log Nf) for the 12 unit synthetic signals at the 3 displacement levels are plotted in Figure 3.

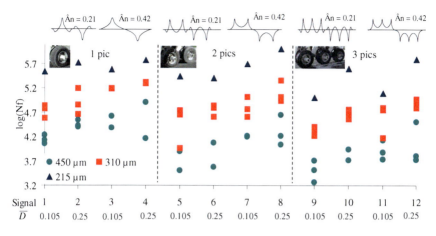

Fig. 3. Fatigue life results

4.1 Effect of the Independent Shape Parameters on the Fatigue Life

The results plotted in the Figure 3 show that the four independent shape parameters have an influence on the fatigue life. The fatigue life (logarithm of the number of loading cycles before failure) decreases with the increase of the displacement level. It decreases when the number of peaks increases. For example, the damage (D = 1/Nf) of a two peak signal with $\hat{A}_n = 0.21$ and $\overline{D} = 0.105$ is 1.49 times the damage of a one peak signal having the same values of $\hat{A}_n$ and $\overline{D}$. The multiple axle configurations are therefore more damaging than single axle configurations when the corresponding loading signals have the same values of the parameters ε, $\hat{A}_n$ and $\overline{D}$. The results also show that the fatigue life increases when $\hat{A}_n$ increases. A one peak signal with $\hat{A}_n = 0.21$ and $\overline{D} = 0.105$, for example, is 1.91 times more damaging than the one peak signal having the same $\overline{D}$ and $\hat{A}_n = 0.42$. In the experimental program of this paper, the lower value of $\hat{A}_n$ corresponds to synthetic signals imitating real signals measured in the longitudinal direction and the higher value of $\hat{A}_n$ corresponds to synthetic signals imitating real signals measured in the transverse direction. As a consequence, the longitudinal loading signals are more damaging than transverse loading signals having the same values of the parameters ε, N_P and $\overline{D}$. Figure 3 also shows that the fatigue life increases when $\overline{D}$ increases. For example, the one peak loading signal having $\hat{A}_n = 0.21$ and $\overline{D} = 0.105$ is 1.49 times more damaging than the one peak loading signal having $\hat{A}_n = 0.21$ and $\overline{D} = 0.25$. For the same pavement structure and the same axle configuration, the increase of $\overline{D}$ corresponds to a decrease of the vehicle speed. A fast vehicle (lower value of $\overline{D}$) is therefore more damaging than a slow vehicle (high value of $\overline{D}$) if the corresponding loading signals have the same values of ε, N_P and $\hat{A}_n$.

The observation of measured signals show that the different shape parameters do not change independently. An increase in the vehicle speed for example does not only decrease the value of the parameter $\overline{D}$ but also decreases the value of the strain level. These two effects may compensate. In addition, in the same loading conditions, the transverse signals usually have higher strain levels than those of the longitudinal signals while they are less damaging than longitudinal signals having the same strain level. The two effects may also compensate. A more complete fatigue model that takes into account the effect of the different shape parameters on the fatigue life is therefore necessary.

4.2 The Multi-linear Fatigue Model

The classical fatigue model, computed from laboratory fatigue tests with sinusoidal signal applied at different strain levels, is a linear function between the logarithm of the fatigue life and the logarithm of the strain level: $\log(Nf) = a \log(\varepsilon) + b$, where a is the slope usually found to be around 5. For more complex loading signals, additional parameters are found to have an impact on the fatigue life. An advanced model should therefore include other parameters. Using the experimental program, $N_P, \hat{A}_n$ and $\overline{D}$ have been incorporated in the fatigue model. A multi-linear regression analysis was made on the results of the experimental plan and allowed the computation of the factors of the multi-linear fatigue model. This model writes:

$$\log(Nf) = a \log(\varepsilon) + b \log(N_P) + c \hat{A}_n + d \overline{D} + e. \qquad (1)$$

The values of the coefficients of the multi-linear fatigue model as well as their standard errors, after the elimination of suspect experimental data using Peirce's criterion [10] are given in Table 2. The coefficient of correlation R has a value of 0.91.

Table 2. Coefficients of the multi-linear fatigue model

	a	b	c	d	e
Value	-4.58	-0.84	1.31	1.76	15.22
Standard error (%)	4	13	15	17	3

The signs of the coefficients confirm the effects of the shape parameters observed: the fatigue life decreases when the strain level ε and the number of peaks N_P increase. It increases when the parameters $\hat{A}_n$ and $\overline{D}$ increase. Figure 4 represents a plot of the fatigue life (log (Nf)) given by the model vs. the experimental results. It shows that the multi-linear fatigue model can predict the experimental results of the fatigue life with a confidence interval of approximately 10% on log(Nf).

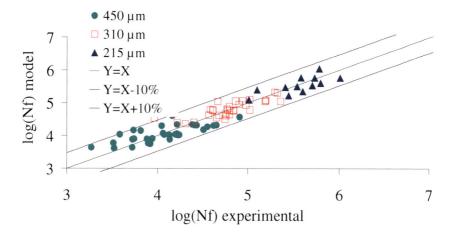

Fig. 4. Fatigue life predicted by the multi-linear fatigue model versus the experimental fatigue life

5   Computation of the Fatigue Life of the Pavement

At the pavement scale, the fatigue life under different loading conditions can be computed by coupling the multi-linear fatigue model with a pavement model. A viscoelastic pavement model is used for the simulations: ViscoRoute 2.0 © [2]. It uses the rheological model of Huet-Sayegh [4]. The simulation of a loading condition using a pavement model enables the computation of the corresponding loading signals. The shape parameters of the loading signals (the input parameters of the multi-linear fatigue model) are then calculated and used in the new fatigue life relationship. The effect of the axle configuration on the fatigue life of asphalt pavements can for example be evaluated. Three axle configurations (single wheel, tandem and tridem) carrying the same load (90 kN), moving with a speed of 70 km/h on a 16cm thick pavement at a temperature of 20°C are simulated (Figure 5). The corresponding loading signals in the longitudinal direction (direction of traffic) and transverse direction (perpendicular to traffic) are determined. The fatigue life under each configuration is then calculated using the multi-linear fatigue model. Examples of loading signals under the tridem configuration in the longitudinal and transverse directions are illustrated in *Figure 6*.

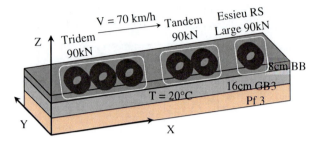

Fig. 5. The simulation cases

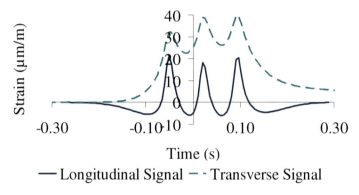

Fig. 6. Loading signals under the tridem configuration

Table 3 gives the number of loading cycles and the total load carried before failure obtained under the three configurations in the longitudinal and transverse directions. Figure 7 illustrates the ratio of the carried load before failure of an axle configuration in one direction and the carried load before failure under a single axle in the same direction. The results show that the signals in the transverse direction are more aggressive than the signals in the longitudinal direction and multiple axle configurations are less damaging than single axle. The tandem and tridem configurations allow carrying 4.9 and 15.7 times the load that allows carrying the single axle configuration.

Table 3. The fatigue life of the pavement under the three axle configurations

Configuration	Single axle		Tandem		Tridem	
Direction	Long	Trans	Long	Trans	Long	Trans
N_f (10^7)	1.2	1.3	28.2	6.6	135.1	20.9
Carried load before failure (10^6 t)	110	120	2500	590	12000	1900
Relative carried load before failure	1.0	1.0	22.8	4.9	109.3	15.7

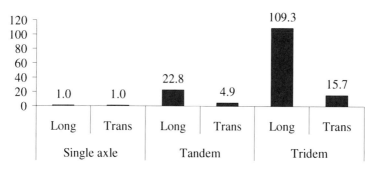

Fig. 7. Relative carried load before failure

6 Conclusion

The shape of the loading signal has an influence on the fatigue life of asphalt pavements. This paper presents a methodology to improve the fatigue life calculation of the fatigue life of asphalt pavements under multiple axle configurations. A principal components analysis of the shape parameters of a database of measured signals enabled the identification of four independent shape parameters. An experimental plan whose variables are these independent shape parameters is then defined and the complex synthetic signals constructed. Fatigue laboratory tests in which the synthetic signals were applied were made and the fatigue life under each synthetic signal computed. The results show that all the variables of the experimental plan have a non negligible influence on the fatigue life of asphalt pavements. The classical fatigue model expresses the fatigue life as a function of one variable: the strain level. A more complete fatigue model that takes into account the effects of all the independent shape parameters is necessary. A linear regression of the fatigue test results enabled the computation of the expression of a multi-linear fatigue model that writes the fatigue life as a function of the independent shape parameters characterizing a loading signal. Coupling this fatigue model with a pavement model enabled the computation of the fatigue life of a pavement under different loading conditions. An application on the effect of different axle configurations is presented. The results show that, for the same load on configuration, multiple axle configurations allow carrying more load than single axle configurations.

References

[1] Bodin, D., Merbouh, M., Balay, J.-M., Breysse, D., Moriceau, L.: In: 7th Int. RILEM Symp. on Advanced Testing and Characterization of Bituminous Materials, Rhodes (2009)

[2] Chabot, A., Chupin, O., Deloffre, L., Duhamel, D.: Road Materials and Pavement Design 11(2), 227–250 (2010)
[3] Chatti, K., El-Mohtar, C.: Transport Research Record 1891, 121–130 (2004)
[4] Duhamel, D., Chabot, A., Tamagny, P., Harfouche, L.: Bulletin de liaison des laboratoires des ponts et chaussées 258-259, 89–103 (2005)
[5] Francken, L.: La technique routière. Bruxelles 24, 1–26 (1979)
[6] Homsi, F., Bodin, D., Balay, J.-M., Yotte, S., Breysse, D.: In: 27èmes Rencontres Universitaires de Génie Civil, AUGC, Saint Malo (2009)
[7] Homsi, F., Bodin, D., Yotte, S., Breysse, D., Balay, J.-M.: European Journal of Environmental and Civil Engineering EJECE 758(5), 743–758 (2011)
[8] Kogo, K., Himeno, K.: Pavement cracking - Al Qadi. In: Scarpas, Loizos (eds.), pp. 509–517 (2008)
[9] Merbouh, M., Breysse, D., Moriceau, L., Laradi, N.: In: 25èmes rencontres universitaires de Génie Civil, Bordeaux (2007)
[10] Ross, S.: Journal of Engineering Technology (2003), http://mtp.jpl.nasa.gov/missions/start-08/science/piercescriterion.pdf
[11] Salama, H., Chatti, K., Lyles, R.: Journal of Transportation Engineering, 763–770 (2006)

Aggregate Base/Granular Subbase Quality Affecting Fatigue Cracking of Conventional Flexible Pavements in Minnesota

Yuanjie Xiao[1], Erol Tutumluer[1], and John Siekmeier[2]

[1] University of Illinois at Urbana-Champaign, Urbana, Illinois, USA
[2] Minnesota Department of Transportation, Maplewood, Minnesota, USA

Abstract. High quality aggregate materials are becoming increasingly scarce and expensive, and therefore optimizing the use of locally available materials is becoming an economic necessity. The research study highlighted in this paper aimed at optimizing the use of varying qualities of aggregate base/granular subbase materials found in Minnesota for achieving cost-effective conventional flexible pavement designs with satisfactory fatigue performances. The methodology consisted of establishing a comprehensive pavement structure sensitivity analysis matrix to include different pavement layer thicknesses and mechanistic material input properties for quality effects and then employing a validated nonlinear finite element program to compute asphalt tensile strains for the various sensitivity matrix variables. The contributions of the unbound aggregate base and granular subbase layers to pavement support and performance were evaluated from a mechanistic-empirical pavement design perspective by incorporating in the analyses cross-anisotropic stress-dependent layer modulus characterizations linked to two different aggregate quality levels (high and low). Aggregate base quality was found to significantly influence bottom-up fatigue cracking; whereas subbase material quality was somewhat important but not as influential as base material quality. Both initial base and subbase construction costs and rutting potential evaluation indicated that the use of marginal quality materials in either base/subbase courses could be cost-effective, depending on the actual pavement thickness and subgrade support conditions.

1 Introduction

Unbound aggregate materials are commonly used to construct flexible pavement foundation layers. High quality crushed aggregates are becoming increasingly scarce and expensive in many parts of Minnesota where local shortages have hindered road construction/maintenance applications particularly in urban areas. Further, various locally available aggregate materials, classified as low quality according to traditional testing techniques and specifications, are likely to offer significant opportunity to be utilized in road construction and perform satisfactorily for intended design traffic levels and operating environments. Although several such applications have been successfully demonstrated to date (*1,2*), challenges still remain regarding how to best utilize different qualities of locally available aggregate materials as road bases/subbases from a mechanistic-empirical (M-E) pavement design perspective. It is worth mentioning that herein "quality" refers exclusively to those characteristics that impact M-E pavement design function, i.e., resilient modulus (M_R), shear strength, and permanent deformation properties. As illustrated in Figure 1, several different sources of Minnesota aggregate base/granular subbase materials can be classified as either better or lower quality depending on the modulus and strength (peak deviator stress at failure obtained from repeated load triaxial tests) properties in relation to design values specified in MnPAVE program, the one used by Minnesota Department of Transportation (MnDOT) for flexible pavement analysis and design. Note that certain uncrushed gravel, often regarded as lower quality by empirical classification, is classified of better quality by mechanistic classification (see Figure 1).

Using local materials could be quite cost-effective for low-volume roads in Minnesota as well as other places, as indicated in recent research studies (3,4). Accordingly, the objective of this paper is to supplement recent findings on the bottom-up fatigue cracking performance governed by the critical response horizontal tensile strain at bottom of hot mix asphalt (HMA) and address some important issues which were not taken into account previously (3). Specifically, cross-anisotropic and stress-dependent aggregate material modulus characterizations are incorporated into nonlinear finite element analyses by considering compaction-induced residual stresses applied and secondly, initial construction costs of base and subbase courses dictated by material qualities are considered for optimizing material quality combinations. In addition, rutting potentials of base/subbase courses and subgrade soils are also evaluated from the shear strength perspective. This is justified by previous Mn/ROAD forensic case studies (5) where catastrophic shear failure and thus excessive rutting cases were observed due to low quality base/subbase layers and the modulus-strength relationships were found to be non-unique for most aggregate base and especially subbase materials based on the analyses of laboratory repeated load triaxial test data.

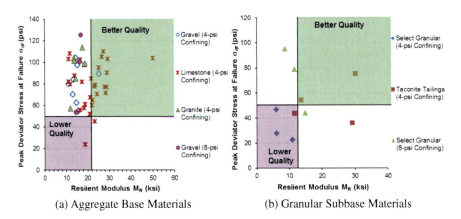

Fig. 1. Mechanical Quality Classification for Various Minnesota Aggregate Base (a) and Granular Subbase (b) Materials at Near Optimum Moisture Conditions

2 Finite Element (Fe) Simulations of Pavement Structures

2.1 Geometry of Pavement Structures and FE Model

Note that the comprehensive factorial matrix established for the previous aggregate quality sensitivity analysis (*3*) is used here with some minor modifications to reduce the total pavement structure cases involved. The conventional flexible pavement structures, as shown in Figure 2(a), comprised of AC surface, unbound aggregate base (UAB), unbound granular subbase (UGS), and natural subgrade layers with varying thicknesses for each layer. The axisymmetric nonlinear FE program GT-PAVE was employed to calculate resilient responses (stress, strain, and deformation) at any location in different pavement structures (*6*). The axisymmetric FE mesh consisting of isoparametric eight-node quadrilateral elements used by GT-PAVE is illustrated in Figure 2(b). Note that the mesh of a total of 800 elements was very fine near the load and progressively coarser as the distance from the load increased. Both the body force (overburden) and the wheel load were applied in 10 increments using a combined incremental and iterative procedure described in detail elsewhere (*6*). A static single wheel load of 40 kN (9 kips) was applied as a uniform pressure of 689 kPa (100 psi) over a circular area of radius 136 mm (5.35 in.). A fixed boundary was assumed at the bottom of the subgrade at a depth of 11 m (420 in.). A constant compressive horizontal residual stress of 21 kPa (3 psi) was assumed to exist initially throughout the base and subbase courses before the wheel load was applied in order to consider the benefits of adequate compaction. The inclusion of residual stresses in the analysis was reported to improve the predictive ability of the anisotropic model by realistically assigning the moduli in a zone of little load influence (*7*). One single season analysis, i.e., the standard Fall season, was performed for Beltrami County in Minnesota according to the seasonal property input guidelines in MnPAVE analyses.

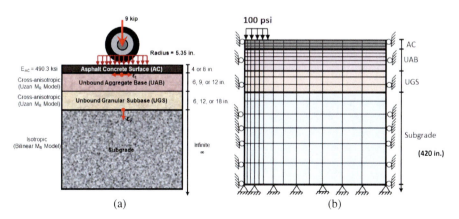

Fig. 2. Pavement Layer Thicknesses Considered (a) and the GT-PAVE FE Mesh (b) [1 in. = 25.4 mm; 1 psi = 6.89 kPa]

2.2 Material Properties

The hot mix asphalt (HMA) with PG58-34 asphalt binder assumed was simplified as isotropic and linear elastic with elastic/resilient modulus (M_R) of 3380.5 MPa (490.3 ksi) obtained from MnPAVE program and Poisson's ratio (v) of 0.35. An Uzan type stress-dependent modulus model (7) was used to characterize anisotropic resilient moduli of aggregate base/granular subbase materials with two representative quality levels, i.e., High (H) and Low (L) (3). Since no test results with lateral strain data were readily available to model the resilient horizontal and shear moduli by the same stress dependent functional form as the vertical one, the simplified procedure developed by Tutumluer (8) was adopted here to estimate the horizontal and shear modulus model parameters from the experimentally determined vertical modulus model parameters. Using this procedure, anisotropic modulus ratios, i.e., horizontal to vertical modulus ratio and shear to vertical modulus ratio, can be formulated by different constants using relationships derived from previous test data of different aggregates, as shown in Eqn (1). Table 1 lists the anisotropic modulus model parameters determined accordingly.

$$M_R^H = K_1 \left(\frac{\theta}{p_0}\right)^{K_2} \left(\frac{\sigma_d}{p_0}\right)^{K_3} ; \quad M_R^V = K_4 \left(\frac{\theta}{p_0}\right)^{K_5} \left(\frac{\sigma_d}{p_0}\right)^{K_6} ; \quad G_R^V = K_7 \left(\frac{\theta}{p_0}\right)^{K_8} \left(\frac{\sigma_d}{p_0}\right)^{K_9}$$

$$K_7 = -90.92 + 0.27 K_4 + 305.34 K_5 + 158.22 K_6$$

$$\frac{K_7}{K_4} = 0.187 + 1.079 \left(\frac{K_1}{K_4}\right) \tag{1}$$

$$K_8 - K_5 = -(K_9 - K_6) = 0.2$$

$$K_2 - K_5 = -(K_3 - K_6) = 2.5$$

where θ and σ_d are bulk stress ($\sigma_1 + 2\sigma_3$) and deviator stress ($\sigma_1 - \sigma_3$) in triaxial conditions, respectively; p_0 is the unit reference pressure (1 kPa or 1 psi); K_i are regression constants from repeated load triaxial test data, respectively.

Besides the anisotropic modulus model parameters for those two representative aggregate quality levels, also shown in Table 1 are other essential inputs for GT-PAVE FE runs. For nonlinear subgrade soil characterization, the bilinear or arithmetic model by Thompson and Elliot (9) was chosen to express the modulus-deviator stress relationship. The value of the resilient modulus at the breakpoint in the bilinear curve, E_{Ri}, was used to classify fine-grained soils as being soft, medium, or stiff. The structural support of the underlying subgrade was studied as an important factor by assuming two different levels, i.e., weak (W) and stiff (S). The subgrade bilinear model parameters for these two types of subgrade soils are listed in Table 1. Table 2 lists the Mohr-Coulomb shear strength properties currently specified in MnPAVE program and also used in this study as reference values for aggregate base and subbase materials in different seasons since such information was not readily available from the laboratory databases collected.

Table 1. M_R Model Parameters Used in GT-PAVE for Base, Subbase, and Subgrade

Layer Type		Aggregate Base		Granular Subbase	
Quality		High Quality (H)	Low-Quality (L)	High Quality (H)	Low-Quality (L)
Horizontal Modulus M_R^H	K_1 (psi)	780.9	238.5	1021	247.82
	K_2	3.723	3.522	3.278	3.282
	K_3	-3.346	-2.932	-3.035	-2.64
Vertical Modulus M_R^V	K_4 (psi)	8360.7	1259.3	12526.5	1707.1
	K_5	1.223	1.022	0.778	0.782
	K_6	-0.846	-0.432	-0.535	-0.140
Shear Modulus G_R^V	K_7 (psi)	2406	492.8	3444.1	586.6
	K_8	1.423	1.222	0.978	0.982
	K_9	-1.046	-0.632	-0.735	-0.34
Poisson's Ratios	v^h	0.3			
	v^v	0.1			
Layer Type		Subgrade			
Quality		Weak (W)		Strong (S)	
Bilinear Model Parameters	E_{Ri} (ksi)	1		5	
	σ_{di} (psi)	6			
	K_3	1110			
	K_4	178			
	σ_{dll} (psi)	2			
	σ_{dul} (psi)	21			
Poisson's Ratio v		0.45			

Table 2. MnPAVE Shear Strength Properties for Base and Subbase Materials

Parameter	Aggregate Base				Granular Subbase	
	Class 5		Class 6		Class 3/ Class 4	
	Summer/Fall	Spring	Summer/Fall	Spring	Summer/Fall	Spring
Cohesion c (psi)	6.55	6	7.34	6	8.38	6
Friction Angle (°)	43.7	24	48.2	24	39.2	24

Note: 1 psi = 6.89 kPa.

3 Results and Discussion

3.1 Effect of Aggregate Base/Granular Subbase Quality on Fatigue Cracking

To investigate the effects of unbound aggregate base and granular subbase material quality on bottom-up fatigue cracking, horizontal tensile strain responses at bottom of HMA for two different levels of material quality (i.e., high and low) are plotted in Figure 3. Note that each data point represents one individual pavement structure analyzed. The name convention used for the different simulations refers to the letter designating quality: for example, "H-L" denotes the combination of high-quality aggregate base (H) and low-quality granular subbase (L).

As shown in Figure 3(a), high quality aggregate base materials significantly reduced the HMA horizontal tensile strains predicted when compared to low quality base cases for the same pavement structure. This is evident from the data points scattered below the equality line. Even lower strain values are more profound for the thicker aggregate base courses (from 6 in. to 9 and 12 in.). On the other hand, granular subbase material quality is much less influential than base material quality, as illustrated in Figure 3(b). In other words, aggregate base material quality alone is quite an important factor governing bottom-up fatigue cracking; whereas subbase material quality alone is also somewhat important but not as influential as base material quality, evidenced by the data points scattered towards more around the equality line.

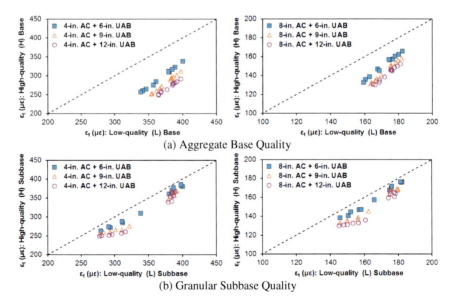

Fig. 3. Effects of Aggregate Base (a) and Granular Subbase (b) Material Quality on Predicted Horizontal Tensile Strain (ε_t) at Bottom of HMA

Investigation of optimal base and subbase material quality combinations in conventional flexible pavements is the subject of this paper to achieve long fatigue life, yet avoid shear failure (rutting) in those unbound granular layers and subgrade soils and reduce construction costs. To address the first issue, i.e., optimal material quality combinations for longer fatigue life, all the pavement structures analyzed were grouped based on base/subbase material quality combinations. The corresponding allowed repetitions for fatigue (N_f) were calculated from the horizontal tensile strain using the currently employed MnPAVE fatigue transfer function (3). The calculated results indicated that for the two asphalt thicknesses, material quality combinations of L-H and L-L more or less have similar allowed repetitions for fatigue, especially when the base thickness is greater than 154 mm (6 in.). However, H-L combination consistently exhibits larger N_f than L-H combination. Thus, the important role of base quality in controlling fatigue cracking is clearly apparent.

3.2 Material Quality Related Base and Subbase Construction Cost

At places where aggregate supply faces local shortage/depletion, high quality crushed aggregates often need to be imported and hauled from longer distances and thus can become expensive, thus making base and subbase construction cost associated with different combinations of material quality and layer thicknesses an important decision-making factor, especially for low volume roads. To this end, a simple initial construction cost analysis rather than more complicated life cycle cost analysis (LCCA) is performed in this study. Several default values were reasonably taken for analysis parameters as follows: (a) 70-mile and 30-mile distances for hauling high quality and more locally available low quality base/subbase materials, respectively; (b) a 16-ft pavement width (12-ft lane width and 4-ft shoulder used for most rural state trunk highways in Minnesota) and 1-mile unit length; (c) 135-pcf and 120-pcf density for base and subbase, respectively; (d) the unit costs of $2.5 and $1.5 per cubic yard (loose volume) for purchasing high quality and low quality base/subbase materials (which roughly reflects the fact that MnDOT currently pay the same for different quality base/subbase materials but get much different pavement life expectancies); (e) unit construction costs of $3.0 and $2.4 per cubic yard for base and subbase; and (f) material transportation cost of $0.3 per ton per mile.

Figure 4 presents the calculated initial base & subbase construction costs for all the pavement sections analyzed. Note that the four data points within each weak and strong subgrade E_{Ri} group (1 or 5 ksi), as shown in Figure 4, represent from left to right the material quality combinations denoted by H-H, H-L, L-H, and L-L, respectively. Also note that no asphalt or subgrade related costs were included in this simple cost analysis. As shown in Figure 4, H-L quality combination yields initial construction costs less than L-H when subbase thickness is greater than base thickness, making it more cost-effective than L-H quality combination (in terms of higher fatigue repetitions and lower initial costs).

3.3 Base/Subbase and Subgrade Rutting Potential Evaluation

Based on the previous analyses concerning the HMA horizontal tensile strain, allowed repetitions for fatigue, and initial base & subbase combined construction costs, it is found that low quality materials can be cost-effectively used in subbase layers without significantly compromising fatigue cracking performance, which is attributed to the relatively less significant role subbase material quality plays in fatigue cracking. However, from the perspectives of subgrade protection and avoiding subbase rutting, the feasibility of such designs should be checked. Since neither well-calibrated rutting prediction models nor relevant data sources are available, no attempt was made here to predict rut accumulation with load applications as an essential criterion for optimizing material utilization; instead, the rutting potential for base and subbase layers and shear strength requirement for subgrade soils were practically evaluated from Mohr-Coulomb failure criteria based stress ratios, i.e., τ_f/τ_{max} and $q/q_{failure}$ as shown in Figure 5 and the subgrade stress ratio (SSR), respectively.

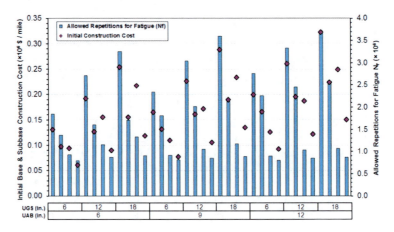

Fig. 4. Initial Base & Subbase Combined Construction Costs per Mile for Different Material Quality Combinations (4-in. AC).

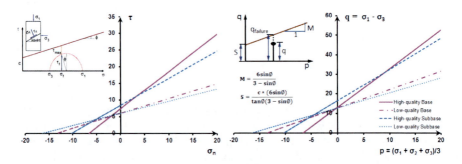

Fig. 5. Mohr-Coulomb Failure Criteria in Mohr Circles and p-q Diagram

Aggregate Base/Granular Subbase Quality Affecting Fatigue Cracking

According to Seyhan and Tutumluer (*10*), the shear stress ratio is not only an indicator of unbound aggregate performance under varying stress states, but also determines the maximum allowable working stress to control the permanent deformation or rutting behaviour of an unbound aggregate layer. The applied stresses within both base and subbase courses for each individual pavement structure analyzed were obtained from the GT-PAVE FE analyses; while the shear strength properties for both high and low quality base/subbase materials were taken from Table 2. Specifically, shear strength properties in MnPAVE summer/fall season are regarded as the minimum requirements for high quality base/subbase materials; whereas shear strength properties in spring thaw season in MnPAVE was conservatively assumed to represent low quality base/subbase materials. The subgrade stress ratio (SSR), defined as the ratio of deviator stress (σ_d) to unconfined compressive strength (Q_u), was used to indicate subgrade stability with SSR less than 0.7 being relatively stable.

Figure 6(a) shows the maximum stress ratios in both base and subbase layers for pavement structures with 4-in. asphalt surface and varying unbound aggregate material quality combinations. Note that the use of high quality materials in subbase course (L-H) considerably reduces subbase stress ratios without significantly increasing base stress ratios, applies much lower deviator stress to subgrade soils (see Figure 6b), and thus requires lower unconfined compressive strength for subgrade soils (see Figure 6c), as compared to the combination of high quality base and low quality subbase (H-L). The use of low quality materials in both layers are not recommended due to the high stress ratios.

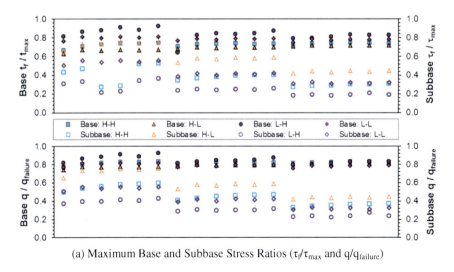

(a) Maximum Base and Subbase Stress Ratios (τ_f/τ_{max} and $q/q_{failure}$)

Fig. 6. Base & Subbase Stress Ratios (a) and Subgrade Stability Parameters (b & c) for Pavements with 4-in. AC and Varying Material Quality Combinations.

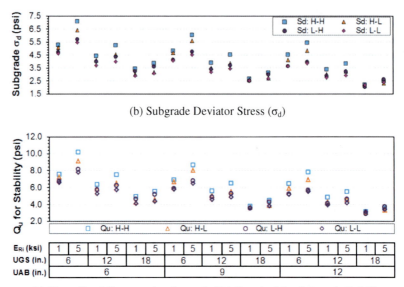

(b) Subgrade Deviator Stress (σ_d)

(c) Unconfined Compressive Strength (Q_u) Required for Subgrade Stability

Fig. 6. *(continued)*

4 Summary and Conclusions

This paper investigated the effects of unbound aggregate quality on conventional flexible pavement fatigue cracking performance in Minnesota using GT-PAVE nonlinear finite element (FE) analyses considering the most realistic stress-dependent and anisotropic granular base/subbase modulus characterizations. A factorial matrix of pavement layer thicknesses and high and low quality aggregate material properties was used for the FE analyses. In addition to hot mix asphalt horizontal tensile strain responses governing allowed repetitions for fatigue, initial construction costs for base and subbase layers were considered in a simplified cost analysis, as well as base, subbase and subgrade rutting potentials were evaluated from the Mohr-Coulomb failure criteria based stress ratios and subgrade stress ratio concepts, respectively.

Aggregate base material quality alone is quite an important factor governing bottom-up fatigue cracking; whereas subbase material quality alone is much less influential. For pavements with subbase thicknesses greater than base thickness, the use of "high-quality base and low-quality subbase (H-L)" can not only improve fatigue cracking performance, but also reduce the initial construction cost; however, it may not be adequate for protecting underlying weak subgrade and subbase from potential rut development. For pavements with base thicknesses greater than subbase thicknesses, the initial construction cost of H-L is larger than L-H, making the former possibly less cost-effective than the latter. Along with designing for maximum fatigue cracking performance, both the subgrade and granular base/subbase permanent deformation predictions must be used as the needed criteria for selecting optimal aggregate material quality combinations.

Acknowledgements. The authors acknowledge MnDOT Office of Materials & Road Research for the financial support under MnDOT H09PS07 study and providing the required databases. The contents of this paper do not necessarily reflect the official views or policies of MnDOT. This paper does not constitute a standard, specification, or regulation.

References

[1] Bullen, F.: Design and Construction of Low-Cost, Low-Volume Roads in Australia. Transportation Research Record, TRB, Washington, DC, pp. 173–179 (2003)
[2] Clyne, T.R., Johnson, E.N., Worel, B.J.: Use of Taconite Aggregates in Pavement Applications. Final Report MN/RC-2010-24, Minnesota Department of Transportation, MN, USA (2010)
[3] Xiao, Y., Tutumluer, E., Siekmeier, J.: Mechanistic-Empirical Evaluation of Aggregate Base/Granular Subbase Quality Affecting Flexible Pavement Performance in Minnesota. Transportation Research Record 2227, TRB, 97–106 (2011)
[4] Gautam, B., Yuan, D., Nazarian, S.: Optimum Use of Local Material for Roadway Base and Subbase. In: CD-ROM of the 89th TRB Annual Meeting, TRB (2010)
[5] Mulvaney, R., Worel, B.: MnROAD Cell 26 Forensic Investigation. Technical Report No. 2002-06, Minnesota Department of Transportation, MN, USA (2002)
[6] Tutumluer, E.: Predicting Behavior of Flexible Pavements with Granular Bases. Ph.D. Dissertation, Georgia Institute of Technology, Atlanta, GA, USA (1995)
[7] Uzan, J.: Characterization of Granular Materials. Transportation Research Record 1022, TRB, Washington, DC, pp. 52–59 (1985)
[8] Tutumluer, E.: State of the Art: Anisotropic Characterization of Unbound Aggregate Layers in Flexible Pavements. In: Proc. 1st Int. Conf. of the Eng. Mech. Inst. (2008)
[9] Thompson, M.R., Elliot, R.P.: ILLI-PAVE Based Response Algorithms for Design of Conventional Flexible Pavements. Transportation Research Record 10437, TRB, Washington, DC, pp. 50–57 (1985)
[10] Seyhan, U., Tutumluer, E.: Anisotropic Modular Ratios As Unbound Aggregate Performance Indicators. J. Mater. Civil Eng. 14(5), 899–1561 (2002)

Fatigue Performance of Asphalt Concretes with RAP Aggregates and Steel Slags

Marco Pasetto[1] and Nicola Baldo[2]

[1] University of Padua, Padua, Italy
[2] University of Udine, Udine, Italy

Abstract. The results are presented of an experimental investigation and a theoretical study on the fatigue behaviour of asphalt concretes, determined by the four-point bending test, according to the EN 12697-24 Annex D standard. The testing was performed on bituminous mixtures, with Reclaimed Asphalt Pavement (RAP) aggregates and Electric Arc Furnace (EAF) steel slags, used at different proportions (up to 70% of the weight of the aggregates), in partial substitution for natural limestone. Fatigue life was evaluated by means of the conventional approach, related to a 50% reduction in the initial stiffness modulus, as well as using more rational concepts, related to the macro-structural damage condition of the mixtures, in terms of dissipated energy and damage accumulation. With respect to the control mixture with limestone aggregates, the asphalt concretes with RAP aggregate and EAF slags presented improved fatigue properties and delayed macro-crack initiation.

1 Introduction

On the basis of rational mechanistic principles, the performance characterization of a bituminous mixture is fundamental for the design of flexible pavements. Nonetheless, it is known that mixtures' performances are not univocally defined, but are highly sensitive to different parameters, including temperature. At average operating temperatures, verification of the fatigue behaviour of the material is particularly important, i.e. with respect to repeated loading applications.

Laboratory fatigue tests can be conducted following many approaches. For example, the European EN 12697-24 Standard has five separate Annexes, each of which describes a different test protocol: a two-point bending test with trapezoid and prismatic samples (Annexes A and B); three- and four-point bending tests on prismatic beam specimens (C and D); repeated indirect tensile strength tests on cylindrical samples (E). Although modern laboratory equipment allows the various tests to be conducted with stress and strain control, the Annexes of the EN 12697-24 Standard involve just one of the two methods of application of the loading; more precisely Annexes B and E prescribe the stress control, while the others the strain control. Nevertheless, irrespective of the specifications of the Standard, the stress control tests are generally used for the fatigue study of thick pavements, while strain control tests are applied for flexible ones of the conventional type [Khalid, Carpenter].

In the stress control procedure, since stress is maintained constant, with a consequent progressive increase in the strain, the complete cracking of the sample is frequently reached at the end of the test. The failure condition is therefore clearly represented by the physical failure of the sample. However, there are other criteria of failure, for example associated to a 90% reduction of the initial stiffness modulus, or with increasing strain, up to a value double that of the initial one.

Vice versa, in the strain control tests, strain is maintained constant and a progressive reduction in the stress is registered. Consequently, at a high number of cycles, since the stress will be reduced to a very low value, it is unlikely that an evident crack will be found in the sample, which will therefore not be completely broken. For this reason, within the road scientific community, the criterion of failure for the strain control tests is generally established as a 50% reduction of the initial stiffness, or initial stress.

The cited criteria of failure, although defined by consistent variations of the mechanical parameter considered (stiffness modulus, rather than stress or strain) with respect to the initial conditions of the sample, are purely arbitrary and do not fully represent the state of internal damage in the material.

To overcome this problem, Pronk, for fatigue bending tests, introduced a rational criterion of failure, linked to the concept of dissipated energy [1], identifying the failure in correspondence to a number of loading cycles N_1 at which the micro-cracks coalesce, producing a macro-crack. N_1 therefore represents the triggering of that macro-crack, which then propagates in the material.

In his approach Pronk introduced an energy ratio Rn, defined as the ratio between the cumulative energy dissipated up to the n-th cycle and the energy dissipated at the n-th cycle.

In the strain control tests, the graphical representation of Rn with the varying of the number of cycles allows N_1 to be identified as the point at which Rn begins to show a non-linear trend. Vice versa, in the stress control tests, N_1 is identified as the peak point of Rn with the varying of the number of cycles. As already outlined by Artamendi and Khalid [2], the accurate identification of N_1 appears to be more subjective in the strain control test method than in that with constant stress control.

A radically different approach to the study of fatigue was proposed by Di Benedetto et al. [3], who focused attention on the evolution of the stiffness during bending tests, observing that it is possible to identify a phase, of predominant length, during which the modulus of the material shows a substantially linear reduction. Di Benedetto et al. therefore introduced a damage parameter, depending on the stiffness of the mixture, whose variation with the number of cycles allowed the development of damage from fatigue in the material to be described.

In the present research the fatigue behaviour of bituminous mixtures for base courses was investigated by means of the "strain control" four-point bending test (4PBT), interpreting the experimental data, as well as with the classic methodology based on a 50% reduction of the initial stiffness, also in the light of the approaches of Pronk and di Benedetto et al.

2 Materials Used

Three materials were used in the study: EAF slags, RAP and natural aggregates (crushed limestone and sand). Table 1 reports the grading composition of the bituminous mixtures and proportions of the components; five mixes were designed with an integrated slag-RAP-limestone lithic matrix (S0R2, S0R4, S3R0, S3R2, S3R4) and one, used as control, with only natural aggregate (S0R0).

Table 1. Aggregate type and particle size distribution of the mixtures (without slag S0, with 30% S3; without RAP R0, with 20% R2, with 40% R4)

Mix composition	Fraction [mm]	Quantity [%]					
		S0R0	S0R2	S0R4	S3R0	S3R2	S3R4
Crushed Limestone	5/10	25	15	12	29	21	14
	10/15	20	20	20	-	-	-
	15/25	12	12	11	-	-	-
Sand	0/2	40	30	13	38	26	12
RAP aggregate	0/10	-	20	40	-	20	40
EAF steel slag	10/15	-	-	-	10	18	18
	15/20	-	-	-	20	12	12
Filler (additive)	-	3	3	4	3	3	4

Table 2. Physical and mechanical characteristics of the bituminous mixtures (without slag S0, with 30% S3; without RAP R0, with 20% R2, with 40% R4)

Properties	Mixture					
	S0R0	S0R2	S0R4	S3R0	S3R2	S3R4
Bulk density (Kg/m^3)	2348	2452	2413	2589	2765	2651
ITS @ 25°C (MPa)	1.62	1.88	2.35	1.74	1.90	2.18
Sm @ 20°C, 10Hz (MPa)	5922	10483	12319	8939	9076	11761

A normal bitumen (50/70 dmm pen) was used for all the mixtures in the experiments. Table 2 reports bulk density, Indirect Tensile Strength (ITS) and Stiffness Modulus (Sm) by 4PBT, of the bituminous mixtures.

3 Fatigue Characterization

The four-point bending fatigue tests were conducted using the protocol described in Annex D of the European EN 12697-24 Standard as reference, in a regime of strain control, with a wave of sinusoidal loading without rest periods. The tests were all conducted at a temperature of 20 °C and frequency of 10 Hz, in a range of strain of between 200 and 500 μm/m. In addition to the data of stress and strain, the phase angle and dissipated energy (both cumulative, and relative to each loading cycle) were determined for each fatigue test.

The beam specimens necessary for conducting the fatigue tests, with dimensions of 400 mm x 50 mm x 60 mm, were cut from 300 mm x 400 mm x 50 mm slabs produced by a laboratory compacting roller, in accordance with the EN 12697-33 Standard.

The results of the fatigue tests were interpreted with three different approaches. In addition to the classical methodology based on a 50% reduction of the initial stiffness, the energy approach of Pronk was also applied, as well as that of Di Benedetto in terms of damage.

3.1 Fatigue Characterization Based on the Stiffness Reduction Approach

The classical fatigue curves, elaborated according to the initial value of strain ε_0 and number of cycles N_f, at which a 50% reduction of the initial stiffness is registered, are presented in Figure 1. The initial strain was evaluated at the 100th cycle (EN 12697-24, Annex D), since as generally recognized in the literature, this is the stage of the test when the material shows a stress-strain response that reliably represents the initial conditions, without yet being significantly affected by damage phenomena.

The regression analysis of the fatigue data was performed using a power law model of the type:

$$\varepsilon_0 = aN_f^b \tag{1}$$

where a and b are regression coefficients depending on the type of material. Table 3 reports the regression coefficients and coefficient of determination R^2.

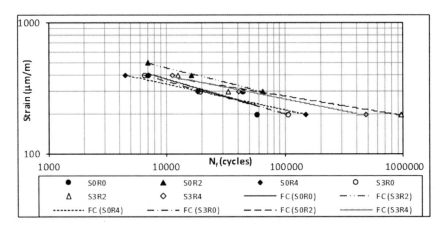

Fig. 1. Fatigue life N_f versus initial strain

Table 3. Fatigue curves - regression coefficients (N_f approach)

Mixture	a (μm/m)	b (-)	ε (10^6) (μm/m)	R^2 (-)
S0R0	4354.6	-0.2671	111	0.7680
S0R2	3713.2	-0.2280	159	0.9960
S0R4	2104.8	-0.1979	137	0.9998
S3R0	3521.4	-0.2485	113	0.9990
S3R2	1543.2	-0.1498	194	0.9559
S3R4	2134.5	-0.1819	173	0.9928

With reference to a fatigue resistance of 1,000,000 loading cycles (as indicated in Standard EN 12697–24, Annex D), and using Eqn. (1), it was possible to calculate the corresponding tensile strain ε (10^6), which was higher for the asphalts with EAF slags and RAP aggregate; in particular the highest value, 194 μm/m, was obtained for S3R2 (Table 3). However, it should be stressed that all the asphalts showed a reasonable fatigue resistance.

3.2 Fatigue Characterization Based on the Energy Ratio Approach

Pronk's energy approach is based on the calculation of the energy ratio Rn, defined as the ratio between the cumulative energy dissipated up to the n-th cycle and that dissipated at the n-th cycle, according to Eqn. (2):

$$R_n = \frac{\pi \sum_{i=0}^{n} \sigma_i \varepsilon_i sen\phi_i}{\pi \sigma_n \varepsilon_n sen\phi_n} \qquad (2)$$

where σ is the stress, ε the strain, φ the phase angle, i the generic i-th cycle, n the n-th cycle. The study of the evolution of the energy ratio during the test allows the number of cycles N_1 to be determined in correspondence to which macro-cracks form. Figure 2 presents an example, relative to the S3R4 mix, of the determination of N_1. It is possible that, with the varying of the number of cycles, the exact identification of the point in which Rn shows a non-linear trend depends in practice on the subjectivity of the researcher.

Figure 3 presents the fatigue curves in terms of N_1 and the initial strain value, for the various mixtures. Similarly to what was previously done for the classical fatigue curves, represented as a function of N_f, a power function, analogous to Eqn. (1), was also used in this case, substituting N_1 for N_f. Table 4 reports the coefficients of regression and determination, as well as the value of ε (10^6).

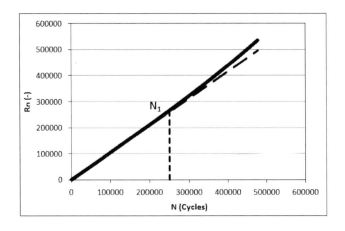

Fig. 2. Determination of failure N_1 for mix S3R4 at 200 μm/m

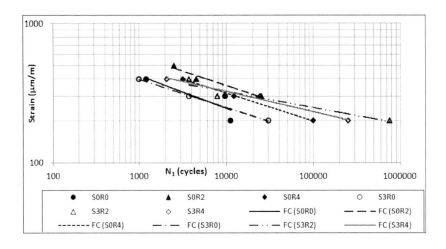

Fig. 3. Fatigue life N_1 versus initial strain

Table 4. Fatigue curves - regression coefficients (N_1 approach)

Mixture	a (μm/m)	b (-)	ε (10^6) (μm/m)	R^2 (-)
S0R0	2168.0	-0.2366	83	0.7151
S0R2	2488.0	-0.2105	136	0.9599
S0R4	2007.3	-0.2008	125	0.9990
S3R0	1610.5	-0.2029	98	0.9988
S3R2	941.93	-0.1156	191	0.9149
S3R4	1238.2	-0.1449	167	0.9879

Although the comparative analysis of the values of ε (10^6) relating to the different mixtures, using the conventional approach and that of Pronk, leads to a similar ranking of the various materials, the interpretation in terms of N_1 allows the comparison between asphalts in the same damage conditions, corresponding to the formation of macro-cracks, and is therefore more reliable and significant from a physical point of view. It can also be observed that the analysis with the energy approach leads to a more precautionary estimate of the fatigue life of the mixtures; in any case the values of ε (10^6) determined starting from N_1 are lower than those calculated with reference to N_f.

3.3 Fatigue Characterization Based on the Linear Damage Evolution Approach

Figure 4 presents an example of evolution of the stiffness, registered for the S3R4 mix in a 4PBT test at a strain of 200 μm/m; similar trends were obtained for the other mixtures investigated. It is possible to identify a phase, with a uniform slope and of predominant length, characterized by a linear reduction of the modulus.

The approach of Di Benedetto is based on this observation, which introduces a parameter of damage to the fatigue study, determined according to Eqn. (3):

$$D(N) = \frac{E_{00} - E(N)}{E_{00}} \tag{3}$$

where E(N) represents the stiffness value at the N-th cycle. Di Benedetto also considers the rate of damage evolution, determined with Eqn. (4):

$$\frac{dD}{dN} = -\frac{1}{E_{00}} \frac{dE}{dN} = -a_t \tag{4}$$

where E_{00} and dE/dN represent the stiffness value determined by the intercept, with the y axis, the curve of linear interpolation of the experimental data and the slope of the curve, respectively (Figure 4).

The ratios between the rate of damage evolution and initial strain are shown in Figure 5. A power function analogous to Eqn. (1) was used for interpolation of the data, substituting dD/dN for N_f. The coefficients of regression and determination are presented in Table 5, together with the value of dD/dN, evaluated at a strain of 100 μm/m.

The comparative analysis between the mixtures in terms of rate of damage evolution confirmed what had been revealed by the previous approaches, i.e. an improvement in the fatigue performance of the mixtures with RAP, more accentuated at 20%. The presence of EAF slags resulted as being advantageous independently of the RAP content.

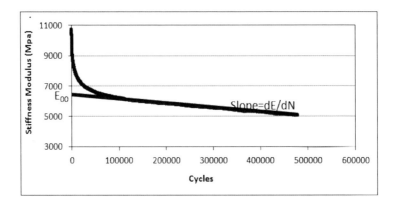

Fig. 4. Stiffness Modulus versus cycles for mix S3R4 at 200 μm/m

Table 5. Damage evolution analysis - regression coefficients

Mixture	a (μm/m)	b (-)	dD/dN (-)	R^2 (-)
S0R0	4547.3	0.2416	1.37543E-07	0.9038
S0R2	3281.3	0.1916	1.22305E-08	0.9976
S0R4	2123.0	0.1806	4.49311E-08	0.9999
S3R0	4039.9	0.2361	1.57119E-07	0.9982
S3R2	1420.8	0.1267	8.00663E-10	0.9498
S3R4	2155.0	0.1638	7.23263E-09	0.9932

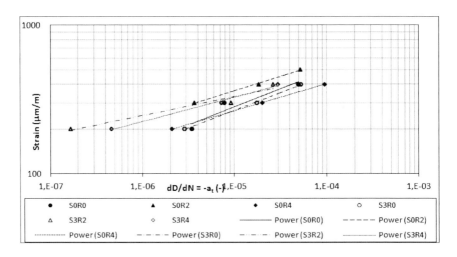

Fig. 5. Strain versus rate of damage evolution

4 Conclusions

The performance of bituminous mixtures for base courses, produced with different combinations of EAF slags and RAP aggregates, was investigated in terms of fatigue life by means of 4PBT tests, in a regime of strain control; the results were interpreted using three different approaches.

With respect to the control mixture, produced with just natural aggregate, all five asphalts with marginal materials demonstrated a clearly higher fatigue resistance; in particular, the use of 30% of EAF slags and 20% of RAP aggregate increased the fatigue life, expressed in terms of ε (10^6), from 74% to more than 100%, in relation to the criterion of failure considered (N_f and N_1 respectively).

The comparison between the mixtures conducted with the criteria N_f and N_1 led to a similar quality evaluation, but the energy criterion showed increases in fatigue life quantitatively higher than those of the classical approach, based on a 50% reduction of the initial stiffness.

The study was completed with an analysis based on the linear damage evolution model, which further supported what had emerged with the criteria N_f and N_1 regarding a lower rate of damage evolution of the bituminous mixtures containing a lithic matrix integrated with EAF slags and RAP aggergate.

References

[1] Pronk, A.C.: In: Proceedings of the 8th International Conference on Asphalt Pavement, Seattle, pp. 987–994 (1997)
[2] Artamendi, I., Khalid, H.: Fatigue Fract. Engng. Mater. Struct. (28), 1113–1118 (2005)
[3] Di Benedetto, H., Ashayer Soltani, M.A., Chaverot, P.: J. Assoc. Asphalt Paving Technologists. (65), 142–158 (1996)

Fatigue Characterization of Asphalt Rubber Mixtures with Steel Slags

Marco Pasetto[1] and Nicola Baldo[2]

[1] University of Padua, Padua, Italy
[2] University of Udine, Udine, Italy

Abstract. The paper discusses the results of a laboratory testing concerning the fatigue properties of asphalt rubber mixtures (by wet process), made with Electric Arc Furnace (EAF) steel slags (up to 93% of the weight of the aggregates). The experimental trial analyzed different bituminous mixtures, i.e. Stone Mastic Asphalt (SMA), base course and wearing course concretes, by means of the four-point bending test, according to EN 12697-24 Standard Annex D. The purpose was to evaluate the fatigue resistance of the mixtures, considering the dissipated energy approaches, which allow consistent material parameters to be identified, indicative of the damage accumulated in the asphalt mixes. Both aged and unaged samples were tested, in order to investigate the ageing effects. The asphalt rubber mixes presented better fatigue behaviour than the reference mixtures made with conventional or polymer modified bitumen.

1 Introduction

This paper describes the results of a trial studying the fatigue resistance performances of bituminous concretes for base courses and wearing courses (traditional and SMA type), made with steel slags and bitumen modified with crumb rubber or SBS polymers. The fatigue behaviour was also investigated in conditions of long-term ageing, in order to evaluate the effectiveness of the modification with crumb rubber using the wet process in reducing the fatigue damage in heavily oxidized mixtures. The experimental data, obtained from four point bending tests (4PBT), were interpreted both with the classical methodology based on a 50% reduction of the initial stiffness and with Pronk's energy approach [1].

2 Materials and Mix Design

Two granular materials were used in the study: EAF slags and limestone filler. The slags utilized are the main by-product of steel production based on the electric arc furnace (EAF) technology [2-4]; they were made available in 3 particle sizes: 0/4, 4/8, 8/14 mm. Table 1 reports the physico-mechanical properties of the steel slags, plus the test protocols adopted.

Table 1. Physical and mechanical characteristics of EAF slags

Physical ÷ Mechanical properties	Standard	EAF slags 0/4 mm	EAF slags 4/8 mm	EAF slags 8/14 mm
Los Angeles coefficient [%]	EN 1097-2	-	16	14
Equivalent in sand [%]	EN 933-8	86	-	-
Shape Index [%]	EN 933-4	-	1.9	2.9
Flakiness Index [%]	EN 933-3	-	4.2	6.4
Freeze/thawing [%]	EN 1367-1	-	0.1	0.1
Fine content [%]	EN 933-1	2.7	0.0	0.0
Grain density [Mg/m^3]	EN 1097-6	4.017	3.979	3.919
Water absorption [%]	EN 1097-6	0.510	0.307	0.112

Three mixes were designed: a Stone Mastic Asphalt mix (SMA), a Wearing Course Asphalt concrete (WCA) and a Base course Asphalt Concrete (BAC). The study of the grading curves was conducted with reference to the design grading envelopes included in SITEB – Italian Society of Bitumen Technologists [5]. Table 2 reports the grading composition of the bituminous mixtures and proportions of the components, while the grading curves of the mixes are presented in Table 3; the total amount of steel slag was 89%, 92%, 93% for BAC, WCA and SMA respectively.

Table 2. Aggregate type and particle size distribution of the mixtures

Mix composition	Fraction [mm]	Quantity [%]		
		SMA	WCA	BAC
EAF steel slag	0/4	45	70	50
	4/8	22	12	13
	8/14	22	10	30
Filler (additive)	-	11	8	7

Table 3. Design grading curve of the mixtures

Sieves size [mm]	Design grading curve [%]		
	SMA	WCA	BAC
20	100.0	100.0	100.0
15	99.9	99.9	99.9
10	85.2	93.3	79.8
5	43.8	58.2	43.0
2	24.6	28.9	22.0
0.4	15.3	14.5	11.7
0.18	12.7	10.8	9.0
0.075	8.5	6.9	5.9

Three different bitumens were used in the experiments for each of the mixes: crumb rubber modified bitumen (45 dmm pen), as well as hard and soft SBS polymer modified bitumen (52 dmm pen and 59 dmm pen, respectively). The polymer modified mixtures were investigated in order to allow a direct comparison with the corresponding asphalt rubber mixtures, with the same skeleton matrix, but made following the wet process technology [6]. The Marshall procedure was used for determining the optimal binder content, along with the indirect tensile strength test. The mixes with maximum Marshall Stability and maximum indirect tensile strength at 25 °C were considered optimal. Table 4 reports Optimum Bitumen Content (OBC), bulk density, Indirect Tensile Strength (ITS) at 25 °C and Stiffness Modulus (Sm) by 4PBT at 20 °C and 10 Hz, of the bituminous mixtures. The stiffness has been evaluated for both aged and unaged specimens, for each mixture. The asphalt rubber mixtures and those with hard and soft modified polymers are indicated by the letters "ar", "hm", "sm" next to the mixture acronym.

Table 4. Physical and mechanical characteristics of the bituminous mixtures

Mixture	OBC [%]	Bulk density [kg/m³]	ITS [MPa]	Sm [MPa]	Sm [MPa] aged samples
SMA/ar	6.0	3150	1.46	5438	5978
SMA/hm	6.0	3130	1.29	5386	6415
SMA/sm	6.0	3080	1.13	4405	5366
WCA/ar	5.0	3110	1.29	6102	8145
WCA/hm	5.0	3100	1.16	6573	8738
WCA/sm	5.0	3060	1.00	6349	7384
BAC/ar	4.0	3030	0.94	6779	7779
BAC/hm	4.0	2930	0.86	6798	7853
BAC/sm	4.0	2910	0.78	4992	6582

3 Fatigue Characterization

The fatigue characterization was conducted with reference to the protocol described in Annex D of the European Standard EN 12697-24 [7], relative to the four point bending test. These are tests of controlled strain (strain values between 200 μm/m and 600 μm/m), with a continual wave of sinusoidal loading without rest periods, conducted at a temperature of 20 °C and frequency of 10 Hz. For each test, in addition to the stress and strain values, the angle phase and energy dissipated at each loading cycle were also monitored, in order to be able to analyze the experimental data with an energy approach.

The beam specimens submitted to the bending tests, with dimensions of 400 mm x 50 mm x 60 mm, were cut from 300 mm x 400 mm x 50 mm slabs, densified using a laboratory compacting roller in accordance with the EN 12697-33 Standard. Some of the beam specimens were then exposed to accelerated long-term ageing, by means of conditioning in an oven at 85 °C for 5 days, in order to evaluate the effect of ageing on the fatigue performances of the mixtures and any benefits produced by the bitumen modified with crumb rubber, compared to binders modified with polymers.

3.1 Fatigue Characterization Based on the Stiffness Reduction Approach

Figures 1 and 2 present the fatigue curves in the standard format, which links the initial strain ε_0 (evaluated at the 100-th cycle) to the number of cycles (failure) N_f that correspond to a 50% reduction in the initial stiffness modulus, for the unaged and aged mixtures, respectively.

The regression analysis of the fatigue data was performed using a power law model of the type:

$$\varepsilon_0 = aN_f^b \qquad (1)$$

where a and b are regression coefficients depending on the type of material. Tables 5 and 6 report the regression coefficients and the coefficient of determination R^2.

With reference to a fatigue resistance of 1,000,000 loading cycles (as indicated in Standard EN 12697-24, Annex D), and using Eqn. (1), it was possible to calculate the corresponding tensile strain ε (10^6), which was always higher for the asphalt rubber mixtures, with respect to the corresponding polymer modified mixtures; in particular the highest value, 372 μm/m, was obtained for SMA/ar (Table 5).

In each type of mixture, the hard modified bitumen led to an improved fatigue life compared to the soft; the increase, although minimal in the case of the BAC concrete, is 24% and 15% for SMA and WCA, respectively.

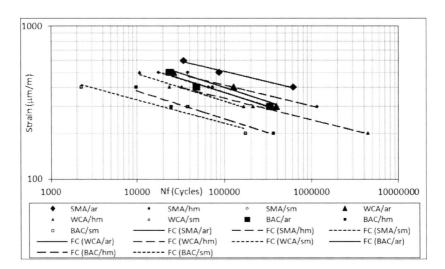

Fig. 1. Fatigue life N_f versus initial strain (SMA Stone Mastic Asphalt, WCA wearing course bituminous concrete, BAC base course bituminous concrete; ar asphalt rubber, hm hard modified bitumen, sm soft modified bitumen; FC fatigue curve).

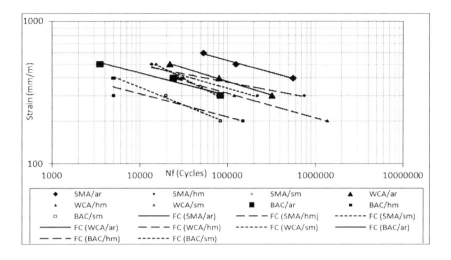

Fig. 2. Fatigue life N_f versus initial strain; aged mixtures (see Fig. 1 legend)

Table 5. Fatigue curves - regression coefficients (N_f approach)

Mixture	a [μm/m]	b [-]	ε (10⁶) [μm/m]	R^2 [-]
SMA/ar	2435.4	-0.1364	372	0.9792
SMA/hm	1979.6	-0.1362	302	0.9263
SMA/sm	2854.5	-0.1778	244	0.9995
WCA/ar	3381.7	-0.1859	259	0.9719
WCA/hm	1702.3	-0.1403	246	0.9985
WCA/sm	2573.0	-0.1801	214	0.9732
BAC/ar	3048.4	-0.1835	243	0.9652
BAC/hm	2032.5	-0.1825	164	0.9687
BAC/sm	1349.4	-0.1527	163	0.9294

Table 6. Fatigue curves, aged mixtures - regression coefficients (N_f approach)

Mixture	a (μm/m)	b (-)	ε (10⁶) (μm/m)	R^2
SMA/ar	3743.5	-0.17	358	09911
SMA/hm	1432.4	-0.117	284	0.9378
SMA/sm	2476.2	-0.173	227	0.9562
WCA/ar	3334.9	-0.189	245	0.9971
WCA/hm	2275.8	-0.172	211	0.9995
WCA/sm	9757.0	-0.309	137	0.9983
BAC/ar	1843.6	-0.157	211	0.9610
BAC/hm	1363.3	-0.161	147	0.8347
BAC/sm	3345.1	-0.247	110	0.9933

The increase in fatigue life due to the adoption of crumb rubber modified bitumen is much more substantial, in particular compared with the soft modified binder, with increases varying up to a maximum of 52%, in relation to the type of concrete.

A comparison of the data in Tables 4, 5 and 6 allows the increase in stiffness to be evaluated linked to the ageing of the concretes, which is followed by a reduction in the fatigue life. The concretes with soft modified bitumen were more affected by ageing (reductions of ε (10^6) of up to 36%), while variations of ε (10^6) of less than 13% were recorded for the asphalt rubber mixtures. The hard modified bitumen, compared to that modified with crumb rubber, showed similar effects, if slightly worse, in relation to the type of mixture.

3.2 Fatigue Characterization Based on the Energy Ratio Approach

The data gathered in the fatigue tests were also analyzed with Pronk's energy approach [1], recently also utilized by Artamendi and Khalid [8], in which a fundamental role is assumed by the energy ratio Rn, calculated as the ratio between the cumulative energy dissipated up to the n-th cycle and that dissipated in the n-th cycle, according to Eqn. (2):

$$R_n = \frac{\pi \sum_{i=0}^{n} \sigma_i \varepsilon_i sen\phi_i}{\pi \sigma_n \varepsilon_n sen\phi_n} \quad (2)$$

where σ, ε, ϕ, i and n represent the stress, strain, phase angle, generic i-th cycle and n-th cycle, respectively. The criterion of failure in this approach is the formation of macro-cracks in correspondence to the number of cycles N_1, so Rn assumes a non-linear trend. Figure 3 reports an example of the determination of N_1 for the asphalt rubber mix type BAC.

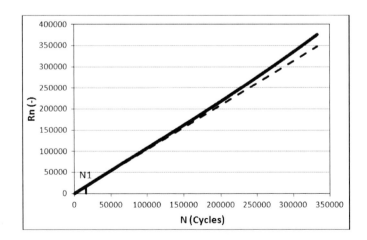

Fig. 3. Determination of failure N_1 for mix BAC/ar at 300 μm/m

In the analysis of the evolution of Rn with the varying of the number of cycles, the exact identification of N_1, although precisely defined in theory, results as being significantly influenced by the subjectivity of the researcher.

Figures 4 and 5 show the fatigue curves elaborated as a function of N_1 and the value of initial strain, for the unaged and aged mixtures, respectively.

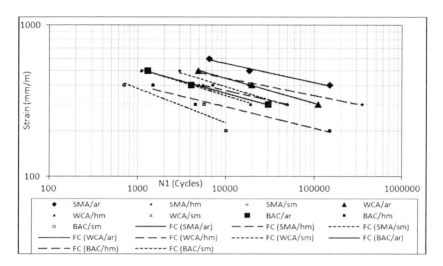

Fig. 4. Fatigue life N_1 versus initial strain (see Fig. 1 legend)

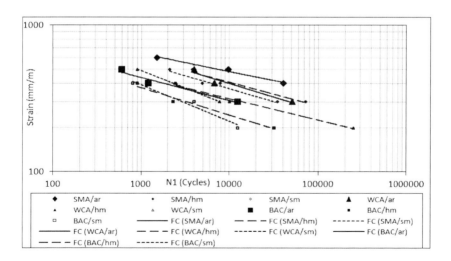

Fig. 5. Fatigue life N_1 versus initial strain; aged mixtures (see Fig. 1 legend)

For the analytical interpolation, a power function was again used, analogous to Eqn. (1), substituting N_1 for N_f. Tables 7 and 8 report the coefficients of regression and determination, as well as the value of ε (10^6), for the unaged and aged mixtures, respectively.

Table 7. Fatigue curves - regression coefficients (N_1 approach)

Mixture	a [μm/m]	b [-]	ε (10^6) [μm/m]	R^2 [-]
SMA/ar	1776.8	-0.126	312	0.9835
SMA/hm	1302.6	-0.116	262	0.9800
SMA/sm	2014.0	-0.178	172	0.9794
WCA/ar	2019.2	-0.164	209	0.9999
WCA/hm	1225.3	-0.130	203	0.9999
WCA/sm	1765.0	-0.178	151	0.9886
BAC/ar	1555.0	-0.160	171	0.9934
BAC/hm	1067.1	-0.142	150	0.9623
BAC/sm	1824.7	-0.227	79	0.8364

Table 8. Fatigue curves, aged mixtures - regression coefficients (N_1 approach)

Mixture	a [μm/m]	b [-]	ε (10^6) [μm/m]	R^2 [-]
SMA/ar	1483.6	-0.122	275	0.9817
SMA/hm	1834.8	-0.163	193	0.9565
SMA/sm	1863.7	-0.176	164	0.9771
WCA/ar	2182.6	-0.184	172	0.9377
WCA/hm	1213.2	-0.146	161	0.9842
WCA/sm	2537.6	-0.238	95	0.9995
BAC/ar	1299.8	-0.157	149	0.9478
BAC/hm	1348.9	-0.186	103	0.9717
BAC/sm	2158	-0.247	71	0.9620

The analysis of the values of ε (10^6), determined with Pronk's approach, allows the different types of mixtures to be discriminated in an analogous way to the standard fatigue characterization, also with regard to the aged concretes.

The better performance of the modification with crumb rubber is therefore fully confirmed for all three types of concrete. Nonetheless, the energy analysis can be considered more reliable, in that the criterion of failure used allows the fatigue performances of the bituminous concretes to be compared in the same damage conditions.

The energy interpretation leads to an estimate of values of ε (10^6), which is from around 10% to more than 100% lower than that determined with the criterion of reduction of the stiffness, depending on the type of concrete.

4 Conclusions

The fatigue tests, conducted on bituminous concretes of the SMA, WCA, BAC type (all made with EAF steel slags), have demonstrated the clearly better performance of the asphalt rubber mixtures compared with those of similar type but made with SBS polymer modified bitumen.

The effectiveness of the crumb rubber modified bitumen in the increase of fatigue life was also clear for the mixtures in conditions of post-ageing.

The interpretation of the data from the 4PBT tests on strain control, according to the criteria N_f and N_1, led to a similar comparative evaluation of the mixtures from the qualitative point of view; nonetheless, with the energy approach, a more precautionary estimate was obtained of the fatigue life of the concretes.

Among the different types of mixtures, the one for base courses (BAC) showed the greatest benefits from the use of the crumb rubber modified bitumen; with reference to the energy analysis, the increase in fatigue life in terms of ε (10^6), with respect to the same mixture with the soft SBS modified bitumen, was more than 100%, in both aged and unaged conditions.

References

[1] Pronk, A.C.: In: Proceedings of the 8th International Conference on Asphalt Pavement, Seattle, pp. 987–994 (1997)
[2] Pasetto, M., Baldo, N.: J. Hazard. Mater. (181), 938–948 (2010)
[3] Pasetto, M., Baldo, N.: Constr. Build. Mater. 25(8), 3458–3468 (2011)
[4] Pasetto, M., Baldo, N.: Mater. Struct (2011) (in press), doi: 10.1617/s11527-011-9773-2
[5] SITEB, Capitolato d'appalto per pavimentazioni stradali con bitume modificato, Roma (2000) (in Italian)
[6] Pasetto, M., Baldo, N.: In: Proceedings of the 5th International Conference on Road & Airfield Pavement Technology, cd-rom, Seul (2005)
[7] EN 12697-24, Bituminous mixtures - Test methods for hot mix asphalt - Part 24: Resistance to fatigue; Annex D: Four-point bending test on prismatic shaped specimens
[8] Artamendi, I., Khalid, H.: Fatigue Fract. Engng. Mater. Struct. (28), 1113–1118 (2005)

Fatigue Cracking of Gravel Asphalt Concrete: Cumulative Damage Determination

F.P. Pramesti, A.A.A. Molenaar, and M.F.C. van de Ven

Road and Railway Engineering, Faculty of Civil Engineering and Geosciences,
Delft University of Technology

Abstract. The aim of this paper is to analyse the fatigue performance of two accelerated pavement test sections. Gravel Asphalt Concrete (GAC) has been used as asphalt mixture for these sections. In order to be able to do so four point bending fatigue tests were carried out to obtain the fatigue characteristics. These tests were done in the same frequency range as loading frequency (loading time) applied in the accelerated pavement test. The fatigue of both GAC sections is represented by means of Miner's cumulated damage ratio and the observed amount of cracking. It will be shown that there is a poor match between Miner's damage ratio and the observed amount of cracking. A calibration factor was developed to "match" theoretical life predictions with observed pavement performance. These calibration factors are considered to be useful for the prediction of pavement fatigue life in practice.

1 Introduction

In order to meet the requirements of resistance against deformation and cracking as well as durability, asphalt mixes ought to acquire certain properties. Fatigue is one of the material properties which is essential to be examined. Edwards [1] already ascertained that in cases when service loadings are variable in nature it is necessary to employ a cumulative damage rules. To predict fatigue cracking in flexible pavement, damage is commonly cumulated based on the Miner's law, one of the most popular rules of this kind.

Cumulated fatigue damage due to traffic loading of flexible pavements is usually defined as the ratio of applied nr of load repetitions n over the allowable nr of load repetitions N. This n/N ratio however needs to be related to the percentage of the pavement surface that shows cracking. There is however very limited information available on how this cumulated damage n/N is correlated to the amount of cracking. As Sun et al [2] described that no matter which kind of predictive model of fatigue cracking is adopted, in the literature no experiment-based empirical evidence or procedure is currently available to convert damage in terms of the number of load repetitions obtained from these predictive models to percentage fatigue cracking, which is very important in providing a meaningful interpretation of damage in practice. Having pavement damage information that had been collected by The Road and Railroad Research Laboratory (RRRL) of Delft University of Technology (DUT) from accelerated pavement testing for test sections made of gravel asphalt concrete, give us opportunity to provide this kind of information, specifically to relate n/N ratio and percentage of cracking on the pavement surface.

The objective of this paper is to develop a relationship between the cumulated damage ratio n/N and the amount of cracking observed test sections made of GAC (until recently GAC was used widely for bituminous base course layers). For this purpose, samples of GAC were made to be tested in the Four Point Bending Test (FPBT) in which cyclic loading was performed to obtain a relationship between the number of cycles to fatigue versus the applied tensile strain level. Then fatigue lines were established with 5°C increments for temperatures ranging from 0°C to 30°C. The tensile strain level, ε, occurring in the test sections under the applied test load (load, tire pressure, temperature, loading speed) was calculated using BISAR. Then we calculated the ratio of actual nr of load repetitions applied to the pavement during certain period, n_i, to the allowable nr of load repetitions as determined from the tensile strain and the fatigue characteristics, N_i. Cumulative pavement damage was defined as the ratio $\Sigma\, n_i / N_i$.

2   Accelerated Pavement Testing Facility Lintrack

Lintrack is an accelerated pavement testing facility, owned by RRRL of DUT and the Road and Hydraulic Engineering Division (RHED) of the Dutch Ministry of Transport Public Works and Water Management, which simulates the effects of heavy vehicles. Groenendijk [3] has described in detail tests that have been conducted by means Lintrack and only a very short summary will be given hereafter. In 1991 4 identical lanes (test lane I to IV) were built on the test field of the RRRL. These pavements were full depth asphalt structures consisting of 0,15 m of hot mix Gravel Asphalt Concrete (GAC). Each section was 16 m long and 4 m wide. After testing of lane I, it was decided to perform another test with the same load condition but on a thinner construction. Therefore test lane II was reduced in May 1995 from 0.15 to 0.075 m thickness. This lane is called "lane Va". [3]

3   Experimental Program

To obtain the fatigue characteristics of (GAC) the following experiment was carried out. According to the prevailing Dutch standards [4], the GAC for the Lintrack sections was a GAC 57 type 0/32 for traffic class IV. Obtaining specimens with similar volumetric and mechanical properties as the GAC used in

Fatigue Cracking of Gravel Asphalt Concrete: Cumulative Damage Determination   741

the Lintrack experiments is a must in order to be able to make a good comparison between what the predicted and observed damage. Since the old pavement test sections had been removed, new samples based on the original GAC mix design were produced. Because the average air void content of Lintrack test sections 1 and Va is 4,475%, this meant that specimens had to be produced with an air void content between 4% to 5 %.

FPBT were performed to obtain the stiffness and fatigue characteristics of the specimens. FPBT stiffness tests have been done at 5°C, 10°C, 15°C, 20°C, 25°C and 30°C at frequencies of 0.5Hz, 1Hz, 2Hz, 4Hz and 8Hz. The FPB fatigue tests have been performed at 5°C, 20°C and 30°C. The loading frequencies at 5°C and 20°C were 3Hz and 8Hz, while for 30°C the tests were only done at 8 Hz. The fatigue tests were carried out in the constant displacement mode.

Analyzing the measured transversal and longitudinal strain signals for Lintrack test section 1, Bouman [5] and Groenendijk [6] concluded that an equivalent frequency of loading of 3 Hz could be assumed for the transversal strains and 7 Hz for the longitudinal strains. Bhairo [7] reported equivalent frequencies for Lintrack test section 5a as 4 Hz and 8 Hz for the transversal and the longitudinal direction respectively. The difference is trivial and in actual fact can be ignored. Therefore the analysis in this paper is conducted for 3 Hz and 8 Hz.

Monotonic Uniaxial Tension and Compression Tests were carried out to obtain information of the tensile and compression strength at various strain rates and temperatures. The test has been done at 5°C, 20°C and 30°C.

4 Calculation of GAC Fatigue Line

The FPBT provided information about the fatigue behaviour of the asphalt concrete at different temperatures and loading frequencies. The failure point N_f was defined as the number of cycles where the stiffness is half the initial value. The obtained fatigue lines are shown in figure 1 and can be expressed by equation 1.

$$N_f = c\varepsilon^k \quad \text{or} \quad \text{Log} N_f = \log c + k \log \varepsilon \tag{1}$$

Where N_f is number of constant strain applications until the specimen reaches half of its initial stiffness, ε is applied strain (μm/m), while c and k are regression constants. Figure 1 also shows the regression equations for the different fatigue lines.

The pavement temperature during Lintrack performance test varied from -7°C to 30°C. Meanwhile FPBT was conducted at temperatures of 5°C, 20°C and 30°C and loading frequencies of 3Hz and 8 Hz. In order to be able to estimate the fatigue relation for conditions which differed from the test conditions, a relationship was developed between the constants of the fatigue relations on one hand and the stiffness of the mixture on the other. In order to get a more reliable relationship, the data set which is mentioned above was extended by adding GAC

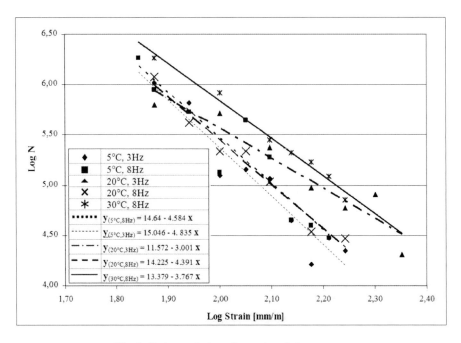

Fig. 1. Fatigue relation of gravel asphalt concrete

Table 1. Constant c & k for other temperatures

Temp	3 Hz			8Hz		
°C	Smix (Mpa)	k	c	Smix (Mpa)	k	c
0	21965	-4,979	15,552	23586	-5,119	15,859
5	17862,96	-4,835*	15,046*	20031,9	-4,584*	14,640*
10	13761,22	-4,272	13,993	16478,12	-4,506	14,509
15	10176,1	-3,963	13,312	12426,14	-4,157	13,740
20	7961,33	-3,001*	11,572*	10313,18	-4,391	14,225*
25	5091,93	-3,525	12,346	6930,39	-3,683*	12,696
30	3628,07	-3,398	12,068	5052,72	-3,767*	13,379*

* Constants from GAC fatigue line calculation as shown in figure 1.

FPBT data which were collected at the time of the construction of the Lintrack sections. For these later data the reader is referred to van de Ven [8] and Wattimena [9] and Groenendijk [3]. The regression analysis gave the following equations to predict the constants "c" and "k" (equations 2 and 3). Table 1 show the c and k values as estimated for different temperatures and loading frequencies.

$$k = \left(0.0862 \times \frac{S_{mix}}{1000}\right) - 3.0857, \quad R^2 = 0.6694 \qquad (2)$$

$$c = (-2.2036 \times k) + 4.5796, \quad R^2 = 0.9642 \qquad (3)$$

5 Results and Discussion

As mentioned before, the cumulative amount of damage was calculated using equation 4.

$$D = \sum_{i=1}^{m} \frac{n_i}{N_i} \qquad (4)$$

Where n_i is the number of load repetitions applied during period i and N_i the allowable number of load repetitions for period; m indicates the number of periods and is equal to the nr of temperature classes shown in table 1. During the Lintrack test there are rest periods in between the successive loadings. It is well known that due to the visco-elastic behaviour of AC, this condition -especially at high temperature- allows micro cracks to heal. Therefore the number of allowable load repetitions that calculated from laboratory test results need to be multiplied with a "healing factor" (H). Besides this correction, another correction factor called "lateral wander" (LW) must be taken into account. This correction considers the fact that the real traffic loads are applied with a certain lateral wander pattern over cross section of the lane. Therefore the N_i should be calculated by means of equation 5 [7].

$$N_i = N_{fat,lab} * H * LW \qquad (5)$$

LW is calculated using the actual lateral wander of the traffic and the lateral wander reduction chart of RHED [7]. This chart shows that the magnitude of LW depends on the thickness and stiffness of the top layer as well as the stiffness of the subgrade.

H is the correction factor for healing for which values are reported in literature ranging from 1 to 20. The value depends on the amount and type of bitumen used in the mixture[10] as well as the loading time to rest period ratio. It is generally accepted in the Netherlands to take H = 4 for this type of GAC mixtures.

Table 2 shows the results of the damage calculations for Lintrack section 1. The cumulative damage of Lintrack section 1 and 5a are given in table 3.

Figures 2 and 3 show the development of the damage ratio as well as the development of the amount of cracking that were visible at the pavement surface.

During the tests also strain measurements were made. On five locations of each section, longitudinal and transversal strain gauges were installed at the bottom of the asphalt layer. It is obvious that the increase in the measured tensile strain is also telling something about damage development. Information on this is given in figure 4.

Table 2. Cumulative damage for Lintrack I at frequency 3 Hz

Temp	Smix	ε			$N_{fat,lab}$	H	LW	N_i	applied nr of load repetitions	Damage
°C	MPa	μm/m	k	c	cycles			cycles	cycles	n_i/N_i
0	21965	120	-4.979	15.552	159799	4	1.96	1253326	436,000	0.348
5	17862	136	-4.835	15.046	53864	4	2.08	448867	1,190,500	2.652
10	13761	160	-4.272	13.993	38275	4	2.22	340223	1,090,000	3.204
15	10176	192	-3.963	13.312	18370	4	2.38	174953	681,000	3.892
20	7961	223	-3.001	11.572	33411	4	2.56	342677	496,500	1.449
25	5091	294	-3.525	12.347	4457	4	2.86	50938	106,000	2.081
30	3628	361	-3.398	12.068	2374	4	3.13	29675	0	0.000
									Sum	**13.63**

Table 3. Total cumulative damage for Lintrack I and Va

Lintrack Section 1		Lintrack Section 5a	
Frequency (Hz)	n/N	**Frequency** (Hz)	n/N
3	13.63	3	18.49
8	10.76	8	15.79

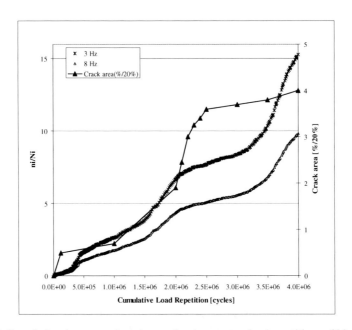

Fig. 2. Cumulative damage and crack area development vs load repetitions of Lintrack I

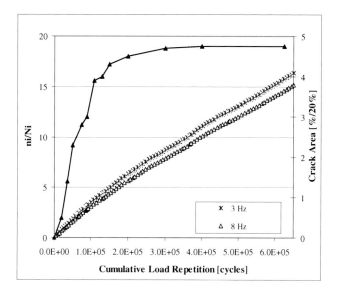

Fig. 3. Cumulative damage and crack area development during loading repetition of Lintrack Va

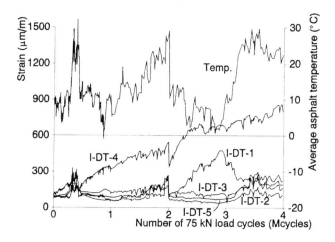

Fig. 4. Increase in strain levels vs nr of load repetitions in section 1

From figure 4 it is clear that in section 1 only transversal strain gauge IDT 4 was measuring an increase in tensile strain from the very beginning of the test. IDT 1 only showed an increase after about $2 \ast 10^6$ repetitions while IDT 2, 3 and 5 showed an increase after approximately $3.2 \ast 10^6$ repetitions. If we assume that the moment at which the tensile strain increases is the moment of initiation of cracking, then it is quite clear that not all visible cracking is classical fatigue cracking which is growing from the bottom of the asphalt layer to the top! This has been further substantiated by plotting the occurrence of cracking in relation to the location of the transversal strain gauges (see figure 5).

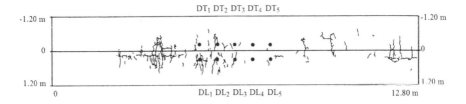

Fig. 5 Shows the total crack pattern Lintrack section 1 after 2,5 million load repetition in relation to the location of the strain gauges

When combining figure 5 with figure 4 it becomes clear that at position IDT4 and IDT5 no cracking is visible after 2.5 million load repetitions while at the other locations cracking is visible.

Because the high values for the n/N ratio did not compare well with the observed increase in tensile strain, an additional analysis was made in order to understand the reasons for this mismatch. One aspect that might have influenced considerably the behaviour of the test sections is the fact that the loads were not continuously applied on both test sections. Because of maintenance of the equipment, Lintrack was sometimes out of order for a considerable period of time. Figures 6 shows how the load applications were applied on section 1. As one will observe some very long rest periods occurred on section 1.

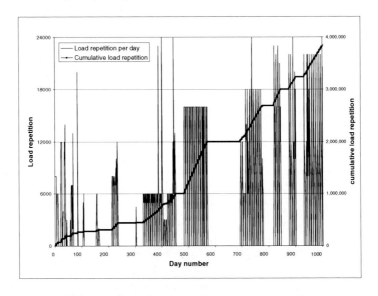

Fig. 6. Number of load repetition per day (Lintrack I)

Work done by Qiu [11] has shown that complete healing can occur when the rest periods are very long. Therefore figures 2 and 3 are plotted again assuming that damage in the mixture had completely healed after a long rest period implying that the ratio n/N returned to zero after such a period. The results are shown in figures 7 and 8.

Figure 7 shows that approximately 10% cracking was observed on section 1 after the first 110 days of testing. At that moment the n/N ratio obtained a value of approximately 0.5. Cracking rapidly increased when testing was resumed after 480 days. It appears from figure 7 that during the next 90 days (from 480 – 570 days) of testing, the amount of cracking increased from approximately 13% to 39% while n/N increased from 0 to approximately 3.3 (average of 3 Hz and 8 Hz line).

Figure 8 provides the same information for section 5a. The calculated damage as well as the amount of cracking increased rapidly from the beginning of the test.

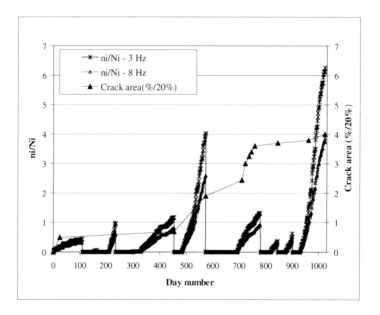

Fig. 7. Development of damage represented as n/N and crack area assuming full healing after a long rest period (Lintrack I)

In case of section 5a, the n/N ratio was approximately 1 when approximately 10% cracking was obtained. From these observations it can be concluded that at a damage ratio of n/N = 1, the pavement can certainly not be considered as completely failed. Furthermore it seems that calibration factors matching predicted with observed damage are dependent on the thickness of the asphalt layer. It also seems that thicker pavements may show a significant amount of visible damage at the pavement surface while the amount of damage according to Miner's law is still limited.

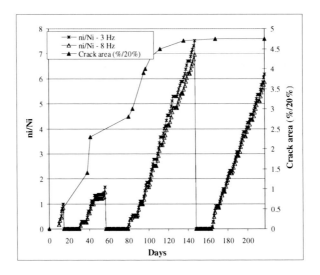

Fig. 8. Development of damage represented as n/N and crack area assuming full healing after a long rest period (Lintrack Va)

6 Conclusions

In this paper, fatigue performances of two accelerated pavement test sections are discussed. From this study it can be concluded that damage ratios calculated using a classical fatigue relation and Miner's law do not match with the amount of damage (cracking) observed on the pavement surface. Calibration factors needed to match predicted and observed damage seem to be dependent on the thickness of the asphalt layer. Thicker asphalt pavements seem to show more surface damage at equal Miner ratio values than thinner pavements. Furthermore advanced types of analysis are needed to explain better the damage development as observed.

References

[1] Edwards, P.R.: Cumulative Damage in Fatigue with Particular Reference to the Effects of Residual Stresses. Her Majesty's Stationery Office, London (1971)
[2] Sun, L., Hudson, R.P.E., Zhang, Z.: Empirical-Mechanistic Method Based Stochastic Modeling of Fatigue Damage to Predict Flexible Pavement Cracking for Transportation Infrastructure Management. Journal of Transportation Engineering, ASCE, 109–117 (March/April 2003)
[3] Groenendijk, J.: Accelerated Testing and surface cracking of asphaltic concrete pavements, PhD, Delft University of Technology, Delft The Netherlands (1998)
[4] C.R.O.W., Standard RAW bepalingen 1990 (RAW standard conditions of contract of work of civil engineering construction 1990), Dutch centre for research & Contract standardization in Civil & Traffic Engineering (1990)

[5] Bouman, S.R., et al.: Lintrack Response Measurements; Comparison of Measured and Predicted Asphalt Strain (part I and II), Report nr 7-91-209-17 and 7-91-209-18, Delft University of Technology, Delft, The Netherlands(1991) (in Dutch)
[6] Groenendijk, J.: Equivalence between a Practical-Loading Pulse and Loading Frequency for Four Point Bending Test in Retrospect to the Comparison of Measured and Calculated Asphalt Stiffness, Internal report, Delft University of Technology, Delft, The Netherlands (1992) (in Dutch)
[7] Bhairo, P.D.: Comparison of the predicted and Observed Pavement Life of LINTRACK Test Lane Va, Delft University of Technology, Delft (1997)
[8] Ven, M.F.C.V.D.: Fatigue Testing GAC test section II (Lintrack DUT), Report 91489, NPC, Hoevelaken (1991) (in Dutch)
[9] Wattimena, J.S.: Fatigue Testing GAC (Lintrack), Report 91482, NPC, Hoevelaken (May 1991) (in Dutch)
[10] Molenaar, A.A.A.: Design of Flexible Pavement, Lecture Note CT 4860 Structural Pavement Design, Delft University of Technology, Delft (2007)
[11] Qiu, J., et al.: Investigating the Self Healing Capability of Bituminous Binders. Int. Journal Road Materials and Pavement Design, ICAM 10 (2009)

Fatigue Resistance and Crack Propagation Evaluation of a Rubber-Modified Gap Graded Mixture in Sweden

Waleed Zeiada[1], Mena Souliman[1], Jeffrey Stempihar[1], Krishna P. Biligiri[1,2], Kamil Kaloush[1], Safwat Said[2], and Hassan Hakim[2]

[1] School of Sustainable Engineering and the Built Environment, Arizona State University, Tempe, AZ 85287-5306, USA
[2] Swedish National Road and Transportation Research Institute, SE-58 195, Linköping, Sweden

Abstract. The main purpose of this study was to document the laboratory experimental program results conducted at Arizona State University (ASU) and the Swedish National Road and Transportation Research Institute (VTI) to obtain material properties and performance characteristics for a "reference-gap", "polymer-modified gap", and "rubber-modified gap" graded mixtures placed on the Swedish Malmo E6 Highway. The advanced material characterization tests of interest to this paper included: bending beam for fatigue cracking evaluation and C* line integral test along with Wheel Tracking Tests (WTT) to evaluate crack propagation. The test results were used to compare the performance of the rubber-modified gap graded mixture to a polymer-modified as well as a reference-gap mixture. The results showed that the expected fatigue life for the rubber-modified gap graded mixture was the highest followed by the polymer-modified and then the reference-gap mixture. Furthermore, the crack propagation test results showed that the rubber-modified gap graded mixture had higher resistance to crack propagation; also, it was observed that rubber-modified mix satisfied Swedish requirements of the corresponding wear layer coatings. To make an overall assessment and verify the laboratory results, it was recommended to conduct a multi-year continuous field monitoring and laboratory evaluation of the test sections.

1 Introduction

Load-associated fatigue cracking is considered to be one of the most significant distress modes in flexible pavements besides thermal cracking and rutting. Different tests and analysis methodologies have been developed over the past few decades for measuring the fatigue behavior of asphalt concrete mixtures. The prediction quality of the fatigue life using any test method will depend on how exact the method is to simulate the condition of loading, support, stress state and environment [1].

Asphalt rubber pavements have already gained interest in Europe. The Swedish Transport Administration (STA) became interested in placing rubber-modified pavement test sections on a few highways since 2007 [2]. At the end of 2009, about 15 test sections had been constructed, using approximately 57,000 tonnes covering about 100 lane-km. The majority of the rubber-modified pavement sections have been tested and evaluated mainly for noise and rolling resistance [3]. So far, there is not adequate information about the fatigue behavior of the Swedish rubber-modified mixtures pertinent to its regional climatic conditions.

Arizona State University (ASU) and STA undertook two joint collaborative projects during 2008-09 to understand the fundamental materials properties of the different gap graded, asphalt mixtures [2, 4]. As part of these two projects, advanced mixture material characterization tests were performed that included rutting evaluation, fatigue and thermal cracking evaluation, and crack propagation phenomenon assessment. Also, binder consistency tests were performed to complement mixture tests.

This paper presents results of the fatigue evaluation and crack resistance tests performed on the three variants of gap-graded, asphalt mixtures: reference, polymer-modified and rubber-modified. The three gap-graded mixes were placed on E18 highway between the interchanges Järva Krog and Bergshamra in the Stockholm area of Sweden. In addition, this paper documents the comparative results of the wheel tracking tests performed on the reference and rubber-modified gap graded mixtures at VTI, Swedish National Road and Transport Research Institute in Sweden.

2 Objective

The main objective of this study was to compare the load-related crack behavior of three types of gap graded mixtures: reference, polymer-modified and rubber-modified mixes placed in the Stockholm Area of Sweden. The results were compared / ranked amongst each other to evaluate the anticipated performance of these mixes in the field.

3 Description of the Project and Mixtures

The designated road section within the construction project had three different asphalt gap graded mixtures: a reference mix (designation: ABS 16 70/100) used as a control, a polymer-modified mix (designation: ABS 16 Nypol 50/100-75) that normally contained 3 to 6% polymer, and a rubber-modified mix (designation: GAP 16) that contained approximately 20 percent ground tire rubber.

4 Mixtures

The three variants of asphalt gap graded mixtures and the associated binders were sampled from the project sites during construction. At the ASU laboratories, rectangular beam specimens were prepared for four-point bending fatigue testing and cylindrical gyratory samples were manufactured for crack propagation tests. In addition, asphalt slabs were manufactured at VTI to perform wheel tracking tests.

The field compaction / air voids for the three mixtures were about 3%. The original mix designs were done using the Marshall Mix design method. The in-situ mixture properties of the Stockholm pavement test sections are reported in Table 1. Table 2 shows the reported average aggregate gradations for the each mixture. The base bitumen used was Pen 70/100. The polymer bitumen was designated Nypol 50/100-75 and rubber-modified was called GAP 16.

Table 1. Mixture Characteristics, Stockholm Highway

Mix	Binder Content (%)	Air Voids (%)	Max. Theoretical Density (G_{mm})
Reference ABS 16 70/100	5.9	2.6	2.4642
Polymer ABS 16 Nypol 50/100-75	5.9	2.6	2.4558
Rubber GAP 16	8.7	2.4	2.3588

Table 2. Average Aggregate Gradations, Stockholm Highway

	Sieve Size (mm)	Reference-Gap	Polymer-Modified	Rubber-Modified
	22.4	100	100	100
Gradation	16	98	98	98
(% Passing by mass	11.2	65	65	68
of each sieve)	8	38	38	44
	4	23	23	24
	2	21	21	22
	0.063	10.5	10.5	7.5

5 Mixture Characterization

5.1 Beam Fatigue Test

The flexural fatigue test has been used by various researchers to evaluate the fatigue performance of pavements [5-7]. Flexural fatigue tests were conducted according to the AASHTO T321 [8]. In this study beams were prepared using vibratory loading applied by a servo-hydraulic loading machine. A beam mold was manufactured at ASU with structural steel that is not hardened. The inside dimensions of the mold are 12 mm larger than the required dimensions of the beam after sawing in each direction to allow for a 6 mm sawing from each face. The beams are saw-cut from the compacted specimens to the required dimensions of 63.5 mm wide, 50.8 mm high and 381 mm long.

The following conditions were used:

- Air voids as compacted in the laboratory: 3.0 ±0.5% for the three mixtures.
- Load condition: Constant strain level, at least 5 levels of the range 325-1300 με.
- Test temperature: 21.1 °C

Initial flexural stiffness was measured at the 50th load cycle. Fatigue life or failure under control strain was defined as the number of cycles corresponding to a 50% reduction in the initial stiffness. Figure 1 shows a comparison of the number of cycles to failure, N_f for the three mixtures at 21.1 °C. It is observed that the rubber-modified mix has the greatest fatigue life, followed by the polymer-modified mix, and the reference-gap mix has the least fatigue life amongst the three mixtures. Initial stiffness values were not similar across all mix specimens and thus the relationships should be used to compare fatigue data as general trend lines.

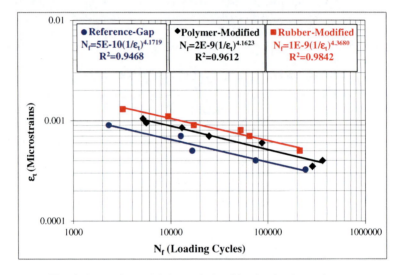

Fig. 1. Comparison of fatigue relationships for the three mixtures

5.2 Crack Propagation Test – C* Line Integral

The C* parameter was first applied to fracture mechanics in 1976 by Landes and Begley [9]. C* can be described as an energy rate line integral describing the stress and strain rate field surrounding a crack tip. Abdulshafi (1983) originally applied the C* line integral (energy approach) to asphalt concrete in attempt to predict fatigue life and used a notched disk specimen to evaluate C* in the laboratory [10]. It was also used in subsequent studies to evaluate different modifiers for asphalt concrete [11]. The relationship between the J-integral and C* allows for experimental measurement of C* from laboratory testing. The J parameter is defined as the energy difference between two specimens that have incrementally differing crack lengths for the same applied load and can be considered a path-independent energy rate line integral (Eqn. 1).

$$J = -\frac{dU}{da} \quad (1)$$

Where dU/da represents the change in potential energy (U) with respect to change in crack length (a). Similarly, C* can be defined as an energy rate or power difference between bodies with different crack lengths under identical loading conditions. C* can be described mathematically according to Eqn. 2 [9].

$$C^* = -\frac{\partial U^*}{\partial l} \quad (2)$$

Where l is crack length and U* is power or energy rate for a given load (P) and displacement rate(û), given by Eqn. 3.

$$U^* = \int_0^{\hat{u}} P d\hat{U} \quad (3)$$

In this study, test specimens similar to those used in the Superpave IDT test were prepared from Superpave gyratory samples in the laboratory. Specimens measured 150 mm in diameter with 40 mm thickness. A right-angle notch was cut into each disk to seat the loading apparatus and a small vertical cut was introduced to simulate crack initiation as depicted in Figure 2. Loading was applied at constant displacement rates of 0.3, 0.45, 0.6, 0.75 and 0.9 mm/min and tests were carried out at 21 °C.

Crack propagation as a function of time was monitored visually using a series of equally spaced tick lines marked on the face of each specimen. Test data was used to determine load as a function of displacement rate for differing crack lengths. A power or energy rate (U*) was computed as the area under these load displacement rate curves. Next, U* was plotted versus crack length for all displacement rates used during testing. The slope of these curves represents the C*-integral. Finally, C* is plotted as a function of crack growth rate (a*).

The relationship between C* and crack growth rate for the reference-gap, polymer- modified and rubber-modified gap mixtures are presented in Figure 3. A steeper slope value indicates higher resistance to crack propagation. The slope value of the rubber-modified mixture (0.088) is double that of the conventional (0.041) and nearly three times greater than the polymer modified mixtures (0.03). Thus, the energy difference required to increase the crack propagation rate from low to high in the rubber-modified mix is much higher than the polymer-modified and reference-gap mixtures.

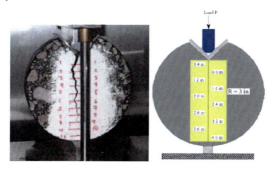

Fig. 2. Typical C* test setup

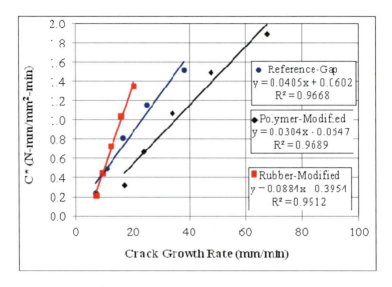

Fig. 3. Comparison of C* versus crack growth rate

5.3 Wheel Tracking Tests

Crack propagation phenomenon was also investigated using the VTI's Wheel Tracking Test (WTT) apparatus. The intent of including the WTT results in this paper is to verify the findings obtained from ASU's C* crack propagation test results for the same gap-graded mixtures. The reference and rubber- modified mixtures used to prepare the wheel tracking test slabs were sampled from another project that had similar aggregate gradations and volumetric properties using the same aggregate types and binders [12]. The equipment used is described as an "extra-large size device" in the testing standard EN 12697-22. The test plates were placed in the WTT on a tack coat surface to generate cracks through the coating. Two strain gauges were mounted in the lower and upper edges of the surface so that deformations could be registered. Testing was carried out separately on both the mix types with three sample replicates. Each slab had a dimension of 50 cm x 70 cm x 4 cm (with a 30 kg mass). The test temperature was 5 °C and the load was adjusted for a desired deformation until cracks propagated to the surface for a finite number of passes. Figure 4 shows the tested slab and the WTT apparatus.

The crack elongation was recorded by double strain gauges in the bottom and top of each sample plate. For each test, the initial elongation (~strain after 100 passages) and the difference in the number of passages from when the crack starts on the lower side until it reaches the upper edge of sample plate (this difference is a measure of how quickly the cracks developed) were recorded. Table 3 includes the WTT results for both rubber-modified and reference-gap graded mixtures.

Figure 5 illustrates the relationship between the initial strain and the difference in number of wheel passages. The results show that rubber-modified has less susceptibility to crack propagation than the reference mixture.

Table 3. Crack Propagation Test Results using the WTT Apparatus

Mix Type	Load (kN)	Initial Strain ($\mu\varepsilon$)	Number of Passes		Difference
			Lower side	Upper side	
Rubber-Modified	15	494	5550	21500	15950
	11.3	455	19000	38000	19000
	4	244	66000	120000	54000
Reference-Gap	15	404	5450	8375	2925
	9	280	16250	33500	17250
	7.5	211	67000	99000	32000

Fig. 4. (a) Strain gauges instrumentation; (b) WTT equipment

5.4 Ranking of Mixtures

Based on the results from the three different tests, the three mixtures were ranked according to their performance. In this analysis, the results of the three mixtures were normalized according to the reference-gap graded mixture. For both beam fatigue and wheel tracking tests, the mixtures were compared by calculating the number of cycles until failure and the difference of passages number at strain value of 400 micro-strains. For the crack propagation-C* line integral test, the three mixtures were compared according to the slope of the C*-crack growth rate relationship. The normalized ranking results are shown in Table 4. The rubber-modified mixtures provided the best performance in resisting fatigue cracks evolution compared to the other two mixtures.

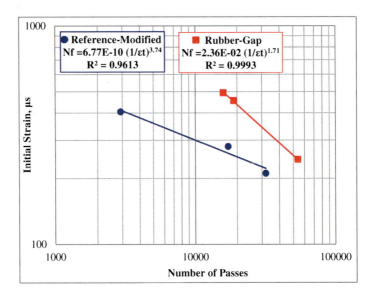

Fig. 5. Initial deformation versus number of passes relationship of reference-gap rubber-modified mixtures

Table 4. Ranking of the three mixtures for the different tests

Test	Reference-Gap	Polymer-Modified	Rubber-Modified
Beam Fatigue Test (N_f at 400 µs)	1.00	3.75	8.75
Wheel Tracking Test (N_f at 400 µs)	1.00	--	7.70
C* Line Integral Test (C*-a relationship Slope)	1.00	0.75	2.20

-- No WTT test was conducted for the polymer-modified mixture.

6 Discussion

Three different tests were conducted to compare the load-related crack behavior of three types of gap graded mixtures. The tests included: bending beam for fatigue cracking evaluation, and C* line integral test along with Wheel Tracking Tests (WTT) to evaluate crack propagation. The test results were used to compare the performance of the rubber-modified gap graded mixture to a polymer-modified as well as a reference gap mixture.

The fatigue results clearly showed that the rubber-modified mix has the greatest fatigue life, followed by the polymer-modified mix, and the reference mix has the least fatigue life. The crack propagation results showed that the crack propagation resistance for the rubber-modified gap mixture is two times higher compared to the reference-gap mixture, and three times higher compared to the polymer-modified gap mixture. The WTT crack propagation results also confirmed that the rubber-modified mix required much higher number of passes for cracks to propagate from the bottom of the pavement slab to the top surface when compared to the reference mixture.

Furthermore, the binder consistency test results (penetration, softening point, and rotational viscosity) of the three binders supported the outcomes of mixture tests. The conventional consistency tests were conducted on the virgin, polymer-modified, and rubber-modified binders at tank conditions. Figure 6 shows a comparison of the viscosity-temperature relationship for the three binders where the intercepts (Ai) and the slope (VTSi) of these relationships are also noted on the plot. It is clearly observed that the polymer-modified binder has a higher viscosity values across a wide range of temperatures compared to the virgin binder. At the same time, the rubber-modified binder has much lower slope with increasing temperature than the virgin and polymer-modified binders. This behavior is highly desirable for better performance at higher and lower temperatures.

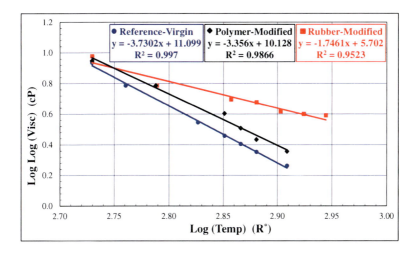

Fig. 6. Viscosity – temperature relationship of Stockholm highway binders

Acknowledgments. The authors would like to thank Mr. Thorsten Nordgren, TRV–Trafikverket: Mr. Leif Viman, VTI – Swedish National Road and Transport Research Institute, and Dr. Gunilla Franzen, VTI – Swedish National Road and Transport Research Institute for thier valuble assistance and support in this study.

References

[1] Tangella, S.R., Craus, J., Deacon, J.A., Monismith, C.L.: Summary Report of Fatigue Response of Asphalt Mixtures, Technical Memorandum No. TM-UCB-A-003A-89-3M, prepared for SHRP Project A-003A, Institute of Transportation Studies, University of California, Berkeley (1990)

[2] Kaloush, K.E., Biligiri, K.P., Zeiada, W.A., Rodezno, M.C., Souliman, M.I.: Laboratory Pavement Performance Evaluation of Swedish Gap Graded Asphalt Concrete Mixtures – Malmo E-06 Highway, Final Report Submitted to Swedish Road Administration, Vägverket, 405 33 Göteborg, Kruthusgatan 17, Sweden (2008)
[3] Nordgren, T., Preinfalk, L.: Asphalt Rubber - a new concept for asphalt pavements in Sweden? Progress report, Swedish Transport Administration, 405 33 Gothenburg, Sweden (February 2009)
[4] Kaloush, K.E., Biligiri, K.P., et al.: Laboratory Evaluation of Rubber & Polymer Modified Bituminous Mixtures Constructed in Stockholm (E18 Highway between the Järva Krog & Bergshamra Interchanges), Final Report Submitted to Swedish Transport Administration, Vägverket, 405 33 Göteborg, Kruthusgatan 17, Sweden (2010) (December 2009/January 2010)
[5] Harvey, J., Monismith, C.L.: Effect of Laboratory Asphalt Concrete specimen Preparation Variables on Fatigue and Permanent Deformation Test Results Using Strategic Highway Research Program A-003A Proposed Testing Equipment, Record 1417, Transportation Research Board, Washington, DC (1993)
[6] Tayebali, A.A., Deacon, J.A., Monismith, C.L.: Development and Evaluation of Surrogate Fatigue Models for SHRP, A-003A Abridged Mix Design Procedure. Journal of the Association of Asphalt Paving Technologists 64, 340–366 (1995)
[7] Witczak, M.W., Mamlouk, M., Abojaradeh, M.: Flexural Fatigue Tests, NCHRP 9-19, Subtask F6 Evaluation Tests, Task F Advanced Mixture Characterization. Interim Report, Arizona State University, Tempe, Arizona (2001)
[8] American Association of State Highway and Transportation Officials, Determining the Fatigue Life of Compacted Hot-Mix Asphalt (HMA) Subjected to Repeated Flexural Bending. Test Method T321-03, AASHTO Provisional Standards, Washington, DC (2003)
[9] Landes, J.D., Begley, J.: A fracture mechanics approach to creep crack growth. Mechanics of Crack Growth. In: Proceedings of the Eighth National Symposium on Fracture Mechanics, vol. 590, pp. 128–148 (1976)
[10] Abdulshafi, O.: Rational Material Characterization of Asphaltic Concrete Pavements, Ph.D. Dissertation, Ohio State University, Columbus, OH (1983)
[11] Abdulshafi, A., Kaloush, K.E.: Modifiers for Asphalt Concrete. ESL-TR-88-29, Air Force Engineering and Services Center, Tyndall Air Force Base, Florida (1988)
[12] Viman, L.: Rubber asphalt – laboratory experiments. VTI (Swedish National Road and Transport Research Institute). SE-581 95 Linköping Sweden (2009), http://www.vti.se/sv/publikationer/gummiasfaltlaboratorieforsok/

On the Fatigue Criterion for Calculating the Thickness of Asphalt Layers

Moshe Livneh

Transportation Research Institute, Technion-Israel Institute of Technology

Abstract. The Israeli Flex-Design program for calculating the thicknesses of flexible pavement layers makes use of the asphalt fatigue equation to determine the thickness of the upper asphaltic layers. In this procedure, the values of the calculated asphalt layers decrease considerably with the increase in the granular basecourse modulus values for the same traffic volume. It has been shown, however, that the determination of the modulus of the granular basecourse by the Flex-Design program is not compatible with recent findings, thus leading to possible erroneous thicknesses for the upper asphaltic layers. In order to minimize these errors, the Flex-Design program has been accompanied by a limiting criterion for determining the minimum thickness of the upper asphaltic layers. This limiting criterion, which is a function of the number of design ESALs, actually leads to another accompanying limiting criterion for determining the maximum modulus of the granular basecourse, which is again a function of the number of design ESALs. These findings lead to the conclusion that the use of a pre-defined pavement-design catalogue for determining the upper asphaltic layer thickness is the most preferable way of proceeding. This catalogue exists in nearly all European countries. The same conclusion applies to the design of perpetual pavements (zero-maintenance pavements), for which a limiting maximum tensile strain of 70µS under an equivalent axle load of 80 kN exists. To recall, this strain is developed at the bottom of the asphalt layers.

1 Introduction

In Israel, the Flex-Design program developed by Uzan [1, 2] is used for determining the thicknesses of flexible pavement layers. In this program, the thickness of the upper-bound (asphaltic) layers is calculated by means of the modified Finn et al. fatigue mechanism for asphaltic mixtures, the number of applications to failure being a function of (a) the tensile strain developed at the bottom of the asphalt layers, (b) the asphalt modulus of elasticity at the asphalt design temperature, and (c) the thickness of the asphalt layers (see also Eqn. (1) in the next section). To recall, the original Fin et al. asphalt-fatigue equation [3] was modified for the Flex-Design program in order to take into account the crack-propagation phase beyond the original crack-initiation phase [4] (i.e., the entire bottom-up crack mechanism).

In addition, the design of perpetual pavement structures incorporates a different criterion for calculating the total asphalt-layer thickness. As commonly agreed, this specific criterion calls for a horizontal (tensile) strain that develops at the bottom of the asphalt layers under the equivalent axle load of 80 kN, not exceeding 70 microstrain [5].

It is obvious that only reliable input values required for these two procedures, including the underlying granular layer modulus, will lead to a reliable asphalt thickness. Thus, it is worthwhile mentioning that although subgrade and paving materials moduli can be obtained from laboratory tests, unbound granular materials have been found to exhibit moduli that are nonlinear or stress dependent. As a result, various agencies and researchers have developed techniques to incorporate some aspects of this nonlinearity directly into elastic layered solutions. In general, these procedures can be grouped into empirical relationships, iterative layered approaches, and finite element solutions. These various procedures yield a wide range of results, but they affect mainly the fatigue behavior of the asphalt layers or their horizontal strain development.

Among the procedures for calculating the granular layer moduli, the FAA's recent FAARFIELD software [6] incorporates the 1977 USCOE method [7] for computing the modulus of non-stabilized layers. Furthermore, the current Israeli Flex-Design software incorporates the earlier, 1975 USCOE method [8] for computing the same modulus. In addition to these methods, the newly developed software of the Mechanistic-Empirical Pavement Design Guidelines (MEPDG) [9] incorporates a non-linear constitutive model for granular materials along somewhat the same lines as the Asphalt Institute's 1983 DAMA software [10] or the more recent KENLAYER program [11]. The granular modulus calculated from this program was found, in contrast to the other methods, (a) to decrease slightly with the increase in granular layer thickness and (b) to increase with the increase in subgrade modulus at a lesser rate than the linear mode. The same results have recently been found in [12, 13].

The outcome of the previous argumentations leads to the conclusion that it is only logical to evaluate the usefulness of calculating the thickness of the upper-bound (asphaltic) layers with the aid of the granular layer modulus. Thus, with this background, the objectives of the present paper were formulated as follows: (a) to evaluate the calculated total asphalt layers as a function of the underlying granular layer modulus according both to the modified Finn et al. fatigue equation of the Flex-Design program and to a Tanzanian study described in [14]; (b) to evaluate the calculated total asphalt layers as a function of the underlying granular layer modulus according to the criterion of the horizontal strain governing perpetual pavements; (c) to discuss and compare the existing approaches for calculating the granular layer modulus according to the Flex-Design program and others; and (d) to present relevant recommendations for calculating the total asphalt layers as a function of the traffic loadings. The sections to follow will detail the process of attaining these four objectives and their associated conclusions.

2 Asphalt-Thickness Sensitivity for Conventional Pavements

The previous section mentioned that the maximum tensile (horizontal) strain criterion in the Flex-Design program forms the basis for computing the total asphaltic layer when following the Fin et al. asphalt-fatigue equation [3], modified to take into account the crack-propagation phase beyond the original crack-initiation phase [4]:

$$log \ \Sigma ESAL = -3.13 + \frac{H_A}{380} - 3.291 \times log \ \varepsilon_t - 0.854 \times log \ E_A \quad (1)$$

where $\Sigma ESAL$ denotes the anticipated sum of 80kN AASHTO-equivalent single-axle loads as calculated from the AASHTO method for anticipated traffic during the design period; H_A denotes the asphalt-layer thickness, in mm; ε_t denotes the maximum tensile (horizontal) strain at the bottom of the asphalt layer; E_A denotes the modulus of elasticity of the asphalt layer at the design temperature, in MPa.

According to Eqn. (1), the calculation of the maximum tensile (horizontal) strain at the bottom of the asphalt layer makes it possible to predict the H_A-$\Sigma ESAL$ relationship. This is done in the Flex-Design program, leading to the required asphalt-layer thickness for any given pavement structure. In a previous paper [15], it was shown that the required asphalt-layer thickness derived from the Flex-Design program can also be direct-calculated from the following equation based on the Ullidtz approximate equation [16] for calculating ε_t:

$$H_A = [18.493 \times (log \ E_B)^2 - 48.482 \times log \ E_B + 106.63] \times log \ \Sigma ESAL - 280.02 \times (log \ E_B)^2 + 875.80 \times log \ E_B - 1012.3 \quad (2)$$

where H_A denotes the asphalt-layer thickness, in mm; $\Sigma ESAL$ denotes the anticipated sum of 80 kN AASHTO-equivalent single-axle loads acting along the design period, calculated for a given traffic based on the AASHTO method; E_B denotes the granular basecourse modulus, in MPa; E_A denotes the modulus of elasticity of the asphalt layer at the designated standard temperature for the asphalt-fatigue criterion (25^0C), which is equal to 3,000 MPa.

The impact of the granular basecourse modulus (E_B) on the calculated asphaltic layer (H_A in Eqn. (2)) for two $\Sigma ESAL$ levels (10^7 and 10^8) is shown in Figure 1. This figure indicates the importance of presenting a correct evaluation of the granular basecourse in the determination of the total asphalt-layer thickness. For example, a reduction in the granular basecourse modulus from 400 MPa to 200 MPa increases the total asphalt-layers thickness from 110 mm to 170 mm for an $\Sigma ESAL$ level of 10^7, or from 220 mm to 270 mm for an $\Sigma ESAL$ level of 10^8.

Another example of the impact of the granular basecourse modulus on the asphalt fatigue failure is given in Figure 2, taken from [14]. In contrast to Figure 1, Figure 2 depicts the impact on performance (i.e., the allowable number of EASLs)

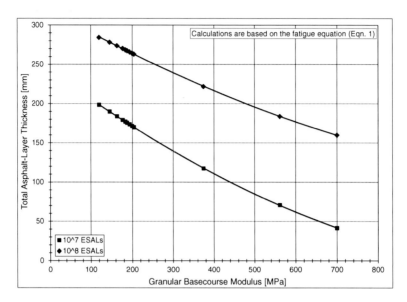

Fig. 1. Example of asphalt layer calculated according to Eqn. (2) based on Flex-Design versus granular basecourse moduli for two ΣESAL levels.

for each of *all* the design input parameters, including that of the granular basecourse modulus. The flexible pavement structure of Figure 2 that was examined comprised a mean thickness of 65 mm of asphalt concrete (AC) over a 150-mm crushed stone base and a 225-mm granular subbase on top of subgrade soil. Further, the mean stiffness of these layers was as follows: 5,500 MPa for the asphalt concrete, 350 MPa for the crushed stone base, 110 mm for the granular subbase, and 70 MPa for the subgrade.

In Figure 2, the effect of the design parameters on pavement performance was expressed in the form of a ratio of the percentage change in specific output results (measured by their coefficient of variances) to the percentage change in input design parameters (measured, again, by their coefficient of variances), holding all other design parameters constant at mean values. Here, it should be added that the fatigue equation used in Figure 2 was adopted from the South African Mechanistic Design Method [17], which is different from Eqn. (1).

Finally, Figure 2 indicates that the design parameters with the greatest influence on fatigue pavement performance were the tire pressure, crushed base modulus, AC thickness, and axle load. Conversely, the subbase thickness subgrade modulus and the subbase modulus were least sensitive as far as fatigue performance was concerned. This finding contrasts with the behavior of the Flex-Design outputs and supports the conclusions given in [15, 18].

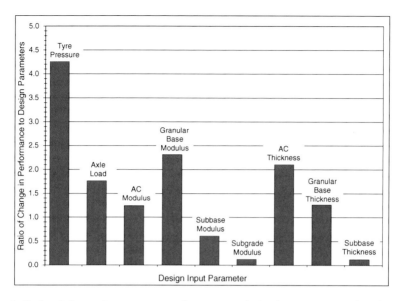

Fig. 2. Ratio of change in pavement performance to design input parameter based on the fatigue-failure criterion [14]

3 Asphalt-Thickness Sensitivity for Perpetual Pavements

The determination of the total asphaltic layer thickness in a perpetual pavement structure follows the bottom maximum tensile strain criterion. As mentioned before, this criterion calls for the horizontal (tensile) strain that develops at the bottom of the asphalt layers under the equivalent axle load of 80 kN not to exceed 70 microstrain [5].

Using, again, the Ullidtz approximate equation [16] for calculating the bottom horizontal stain (ε_t), the variation in the required asphalt thickness with the underlying granular layer modulus that leads to the maximum permissible 70 microstrain for the bottom horizontal strain can be calculated. This variation is shown in Figure 3.

Figure 3, which is based on the same underlying granular layer modulus values of Figure 1, indicates once more the importance of presenting a correct evaluation of the underlying granular layer when determining the total asphalt-layers thickness. For example, a variation in the underlying granular layer modulus from 400 MPa to 200 MPa increases the total asphalt-layer thickness from 240 mm to 300 mm.

Figure 3 also includes the calculated number of ESALs to failure, for which the given asphalt thickness and its accompanying given granular layer modulus are compatible with the asphalt-fatigue equation (i.e., Eqn. (1)). The values obtained are higher than 10^8 ESALs because the calculated asphalt thicknesses of Figure 3 are also higher than those of Figure 1, which are associated with the 10^8 ESAL level.

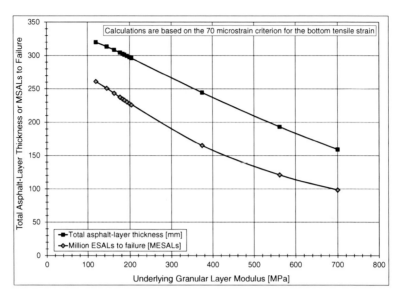

Fig. 3. Example of calculated asphalt layer according to the bottom tensile strain criterion (70µS) versus granular layer modulus and their accompanying ΣESAL to failure

Finally, it should be added at this juncture that if the term of $H_A/380$ is omitted from Eqn.(1), the new calculated number of ESALs to failure of Figure 3 decreases to 0.38×10^8 ESALs for all calculated points in the figure. This reduced ESAL level is believed to be a more reliable value for the perpetual pavement structures in which no crack-propagation phase beyond the original crack-initiation phase is to be considered. In other words, in this kind of pavement, the entire bottom-up crack mechanism is expected not to develop.

4 Minimum Asphalt Design Thickness

In order to bypass the difficulties associated with predefined equations for calculating the moduli of the granular subbase and basecourse layers, the Flex-Design methodology defines the following minimum design thickness for the total asphaltic layers as a function of the design ΣESAL only:

$$H_{Amin} = -550 + \frac{1,279}{log(\Sigma ESAL)} + 73 \times log(\Sigma ESAL) \quad (3)$$

where ΣESAL denotes the anticipated sum of 80kN AASHTO-equivalent single-axle loads as calculated from the AASHTO method for anticipated traffic during the design period; H_{Amin} denotes the minimum thickness of the total thickness of the asphaltic layers, in mm.

Eqn. (3) actually dictates the existence of maximum values for the modulus of the granular basecourse layer as being a function of the design ΣESAL. These maximum values are in given in Figure 4.

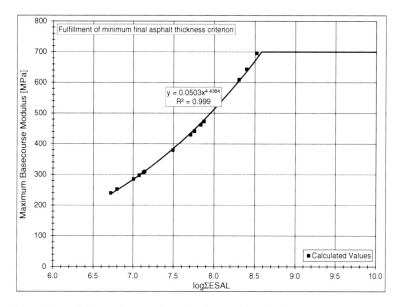

Fig. 4. Variation of the maximum values for the modulus of the granular basecourse layer with the design ΣESAL as derived from the Flex-Design program

Figure 4 indicates that for low and medium levels of the design ΣESAL, the maximum values for the modulus of the granular basecourse layer is lower than the 700 MPa dictated by the Flex-Design program. The search continues, however, for a more reliable value of the modulus of the granular basecourse layer that may be still lower than those given in Figure 4. Obviously, when a more reliable value is found, the final calculated total thickness of the asphaltic layers becomes higher than the minimum value dictated by the Flex-Design program. This issue is discussed in the next section.

4.1 General Comments on the Granular Modulus

Figure 5 depicts an example of calculated moduli obtained for the granular layers according to (a) the Flex-Design Program and (b) the Asphalt Institute's DAMA Program for the inputs shown in the figure. The Flex-Design outputs are seen to exhibit higher moduli values for a majority of cases than those shown in Figure 4. For these cases, the limiting moduli values of Figure 4 are important for a reliable determination of the total asphaltic layer thickness.

In addition to the above findings, Figure 5 also indicates that the moduli values obtained by the Asphalt Institute's DAMA program for the given example are of a constant nature of 200 MPa, which is lower than the values given in Figure 4. Thus, the use of this lower moduli value will lead to higher values of the total thickness of the asphaltic layers than those specified by the Flex-Design program (Eqn. (3)).

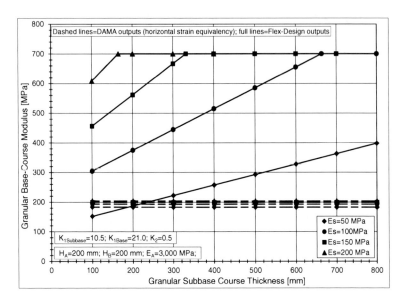

Fig. 5. Variation in the calculated granular basecourse modulus with the granular subbase according to (a) the Flex-Design Program and (b) the Asphalt Institute's DAMA Program for the inputs shown in the figure

At this juncture, it is important to state that the moduli outcomes obtained from the Asphalt Institute's DAMA program are more reliable than those obtained by the Flex-Design program. More details concerning this issue are given by the author elsewhere [15, 18].

Furthermore, it is necessary to mention that the new Mechanistic-Empirical (M-E) design of flexible pavements, termed MEPDG, also calculates the granular layers moduli in a similar way to that of DAMA, but this time with a more advanced constitutive model for the granular materials. In addition, level 3 of the MEPDG method (i.e., the typical values method) suggests granular modulus values as follows: (a) a maximum modulus value of 290 MPa for a GW material and a typical modulus value of 283 MPa; (b) a maximum modulus value of 276 MPa for a GP material and a typical modulus value of 262 MPa. The latter modulus value is at the same order of magnitude as the 200 MPa value specified by [19] for the perpetual pavement structures. To conclude, the MEPDG procedure, too, leads to granular modulus values smaller than those associated with the Flex-Design program.

5 Conclusions

The paper demonstrates that variations in the granular basecource modulus affect considerably the outputs of the total asphaltic layer thickness for both conventional and perpetual pavements. Thus, it is of utmost importance to input the most reliable values for this modulus. The values obtained by the Flex-Design method were shown, for the majority of cases, to be higher than those considered more realistic, thus leading to reduced values of asphalt thicknesses.

Moreover, when the non-linear and stress-dependent relationship of the resilient modulus of a granular material is applied (as is done in the old DAMA program or the new MEPDG program), more realistic results may be obtained for the granular basecourse modulus and, thus, for the required asphalt thickness.

In sum, at this stage of knowledge, it seems that the Level 3 method (i.e., the typical values method) of the new MEPDG procedure is more suitable for the Flex-Design and other software that calculate the required asphalt layer according to the maximum tensile (horizontal) strain occurring at the bottom of the asphalt layer.

Finally, in cases in which no advanced constitutive model for the granular materials is applied, these findings lead to the conclusion that the use of a predefined pavement-design catalogue for determining the upper asphalt-layer thickness is the most preferable way of proceeding. To recall, such a catalogue exists in nearly all European countries [20].

References

[1] Uzan, J.: Journal of Transportation Engineering 111(5), 561–569 (1985)
[2] Uzan, J.: Transportation Research Record (1539), 110–115 (1996)
[3] Finn, F., Saraf, C., Kulkarni, R., Nair, K., Smith, W., Abdullah, A.: The Use of Distress Prediction Subsystems for the Design of Pavement Structures. In: Proceedings of 4th International Conference on the Structural Design of Asphalt Pavements, vol. 1, pp. 3–38. Michigan, Ann Arbor (1977)
[4] Uzan, J., Zollinger, D.G., Lytton, R.L.: The Texas Flexible Pavement System (TFPS), Report FHWA/TX-91/455-1, Texas Transportation Institute, Texas A&M University, College Station, Texas (1991)
[5] Athanasopoulou, A., Kollaros, G.: Pavements Can Last Longer. In: Proc. 5th International Conference on Bituminous Mixtures and Pavements, Thessaloniki, Greece, pp. 1334–1343 (2011)
[6] US Department of Transportation, Federal Aviation Administration, Airport Pavement Design and Evaluation, Advisory Circular, AC 150/5320-6E, Washington, DC (2009)
[7] Barker, W.R., Brabston, W.N., Chou, Y.T.: A General System for the Structural Design of Flexible Pavements. In: Proc. 4th International Conference on the Structural Design of Asphalt Pavements. Michigan, Ann Arbor (1977)
[8] Barker, W.N., Barker, W.R., Harvey, G.G.: Development of a Structural Design Procedure for All-Bituminous Concrete Pavements for Military Roads, Technical Report S-75-10, U.S. Army Corps of Engineers Waterways Experiment Station, Vicksburg, Mississippi (1975)
[9] National Cooperative Highway Research Program, Guide for Mechanistic-Empirical Design of New and Rehabilitated Pavements Structures, NCHRP Project 1-37A, Transportation Research Board, Washington, DC (2004)
[10] Hwang, D., Witczak, M.W.: Program DAMA (Chevron) User's Manual. Department of Civil Engineering, University of Maryland, College Park, Maryland (1979)
[11] Haung, Y.H.: Pavement Analysis and Design. Prentice Hall, Englewood (2004)

[12] Sahoo, P.K.: Moduli of Granular Layers and Subgrade Soils for Flexible Pavements, Unpublished Ph.D. Thesis, Civil Engineering Department, Indian Institute of Technology, Kharagpur, India (2008)
[13] Sahoo, U.C.: Performance Evaluation of Low Volume Roads, Unpublished Ph.D. Thesis, Civil Engineering Department, Indian Institute of Technology, Kharagpur, India (2009)
[14] Mfinanga, D.A., Salehe, J.: International Journal of Pavements (IJP) 7(1-2-3), 38–50 (2008)
[15] Livneh, M.: On the Variation of Required Asphalt Thickness with Subgrade Modulus: A Reality or an Illusion? In: Proc. 5th International Conference on Bituminous Mixtures and Pavements, Thessaloniki, Greece, pp. 329–338 (2011)
[16] Ullidtz, P.: Pavement Analysis. Elsevier, Amsterdam (1987)
[17] Freeme, C.J., Maree, T.Y., Viljeon, A.W.: Mechanistic Design of Asphalt Pavements and Verification Using the Heavy Vehicle Simulator. In: Proc. 5th International Conference on the Structural Design of Asphalt Pavements, Delft, Holland, pp. 156–173 (1982)
[18] Livneh, M.: Some Findings Concerning the Determination of Granular-Base Moduli for Flexible-Pavement Thickness Design. In: Proc.1st International Conference on Road and Rail Infrastructure, Opatija, Croatia, pp. 271–282 (2010)
[19] Sides, A., Uzan, J.: Examination of the Implementation of Designing Perpetual Pavements in Israel, Research Report, Transportation Research Institute, Technion-Israel Institute of Technology, Haifa (2010) (in Hebrew)
[20] Darter, M.I., Von Quintus, H., Owusu-Antwi, E.B., Jiang, J.: Systems for Design of Highway Pavements, NCHRP Project 1-32, Eres Consultants, Inc., Champaign, Illinois (1997)

Acoustic Techniques for Fatigue Cracking Mechanisms Characterization in Hot Mix Asphalt (HMA)

Malick Diakhaté[1], Nicolas Larcher[2], Mokhfi Takarli[2], Nicolas Angellier[2], and Christophe Petit[2]

[1] Université de Bretagne Occidentale, LBMS, France
[2] Université de Limoges, Groupe d'Etude des Matériaux Hétérogènes, Equipe Génie Civil et Durabilité, France

Abstract. This article deals with the investigation of the damage process in Asphalt Concrete (AC) using Acoustic Emission technique (AE). The AE response, particularly how it can change as a function of applied stresses, is known to be promising for micro-cracking detection. AE events are correlated with mechanical damage analysis, and allow determining fatigue cracking mechanisms. Nevertheless, the pertinence of AE approaches is closely related to the comprehension of wave propagation phenomenon within the tested material. In fact, wave propagation is affected by attenuation and velocity variation due to the viscoelastic behavior of the AC. For understanding this phenomenon, ultrasonic techniques can be used for determining both the attenuation curves and variation of the propagation velocity.

1 Introduction

Understanding failure mechanisms of construction material as well as their damage evolution are two key factors to improve design tools of structures. Depending on failure modes to be highlighted and studied, several tests methods, and analysis tools have been developed, in particular AE. This latter is an experimental tool well suited for characterizing material behaviour by monitoring fracture process. Despite the wide use of AE technique to characterize and monitor damage evolution of composite materials (often with a brittle behavior at low temperatures) [1-5] very few studies focused on using AE technique to characterize AC behaviour.

Li et al. [1] interested in the location of acoustic events within AC specimens. For purposes of their experimental campaign, pre-notched specimens were tested in tree-point bending using a Semi-Circular Bend (SCB) device. Monotonic tests were performed inside a climatic chamber at –20°C. The authors determined that the acoustic events are located on the fracture path. Moreover, a comparison between the load vs. displacement curve and the cumulative AE events vs. testing time curve showed an increase of the acoustic activity when the force values (post peak stage) range from 70% to 90% of the strength. AE events located before the peak load define the Fracture Process Zone (FPZ) which is a key factor in the determination of material fracture process.In fact, formation of the FPZ is the consequence of the formation of micro cracks, a few of which later link up to form a macro crack. The formation and growth of cracks are associated with the release of elastic strain energy in the form of acoustic emission waves [2]. Li et al. [3] also investigated particularly the size variation of the FPZ based on various parameters. The results showed that the size of the FPZ is mainly affected by differents parameters such as temperature, type of aggregates and air voids. It also showed that the binder percentage doesn't affect the FPZ size. In the energetic approach, more than 50% of events are little energetic and account for about 5% of total energy, whereas only 3% of events are great energetic and account for 40% of total energy. Thus, in the pre-peak stage (load lower than the strength), the energy level of AE events is globally low, and is related to the FPZ as well as the micro cracking.

Fatigue tests have been conducted by Seo et al. [4] and showed the influence of rest period between loading cycles on mechanical behavior of AC. In fact, both the cumulative AE energy and cumulative AE counts increase when the fatigue test includes rest periods between loading cycles. It indicates that rest period increases the fatigue lifetime which is related to the healing phenomenon. In addition, the Kaiser effect is commonly related to damage state in the material. However, the fatigue tests whitout rest periods didn't allow to systematically observe this effect.

Apeagyei et al. [5] studied the thermal damage in pavement binder using a rapid cooling test, for temperature ranging from 15°C to –55°C. The curve signal energies vs. events showed an energetic peak which is the necessary energy for crack propagation. The comparison between the cracking temperature obtained by traditional method (Bending Beam Rheometer) and the cracking temperature obtained by acoustic emission gives good agreement, this cracking temperature ranges between -20°C to -30°C according to the type of AC. Similar results on Recycled Asphalt Pavement (RAP) are reported by Behzad et al. [6].

The work presented herein is a first experimental attempt where acoustic emission technique is used to characterize AC behaviour under a double shear cyclic loading. The AE activity within the asphalt concrete material was monitored using an AE system. Both the mechanical properties and the acoustic properties of the material are evaluated and compared. Let's notice that AC material exhibits a structural heterogeneity and thermo viscoelastic behaviour, and due to this particularity, wave propagation within this media is accompanied by attenuation and velocity variation.

2 Material and Experimental Techniques

2.1 Material and Sample Preparation

The HMA used in this experimental campaign was manufactured in the laboratory. The mixture is a dense-graded AC with a maximum aggregate size of 6 mm and a penetration grade 35/50 binder (6.85% bitumen content). A mould (measuring 600 mm long, 40 mm wide and 150 mm high) was used to prepare the slab. Once the hot mix asphalt (160°C) had been poured into the mould, the French rolling wheel compactor compacted the mixture to a thickness of 150 mm, in accordance with European standard NF EN 12697-33. The slab was then allowed to cool down to room temperature for several days. A first sawing process allows extracting 18 specimens with the specified geometry (i.e. L x H x W: 124 mm x 70 mm x 50 mm). A specific device was subsequently used to glue steel plates onto the corresponding specimen faces. This operation served to ensure the best alignment of the steel plates in relation to the double shear loading configuration. A minimum rest time of 1 day was observed to allow for the glue to set. A second sawing process aimed at creating four notches. Each notch of 10 mm in depth symbolizes a crack initiation within the AC.

2.2 Loading Device: Double Shear Test

The testing device used in this study was developed in a previous experimental work devoted to the shear fatigue behavior of tack coats in the pavement structures [7-8]. Its working is based on the double shear test principle, which has been successfully used in a previous study on the investigation of the mode II fatigue crack propagation under shear loading [9]. The double shear test is performed on a specimen in such a way that its lateral parts are fixed during the test, and its central part is subjected to a sinusoidal and symmetric loading function. Through a numerical study, this loading configuration allowed investigating the crack propagation within the bituminous material under a shear loading. The use of mark tracking technique (optical method) in an experimental study confirmed this numerical finding [8]. The double shear testing device is mounted on a servo-hydraulic frame (Figure 1). The upper part of the device is connected to the jack, and the lower part is connected to the load cell.

The fatigue test was carried out under force control at 10 Hz. The force values were measured by the load cell (±100 kN), and the relative displacement values, between the side parts and the central one of the specimen, were measured by an extensometer (±1 mm).

A data acquisition system recorded the force and displacement values during the fatigue test. Additionally to the previous experimental conditions, a climatic chamber was placed around the double shear testing device (Figure 1) in order to carry out the test at a controlled temperature (10°C in this study).

Prior to the beginning of the test, a homogenization of the temperature within the specimen was performed for at least 6 hours.

2.3 Acoustic Emission Principles and Instrumentation

Acoustic emission may be defined as transient elastic waves generated by the rapid release of strain energy in a material. A number of micro and macro

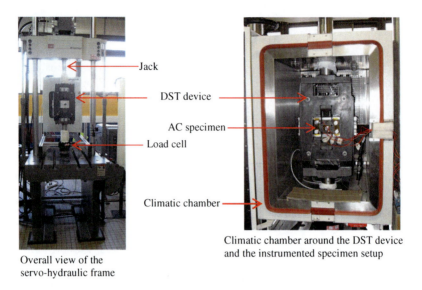

Fig. 1. Photography of the instrumented specimen setup for the fatigue test

processes contribute to both the deformation and the deterioration of a material under strain, resulting to a series of acoustic events. Thus, the events released by the material contain information regarding the general deformation process. In the work reported herein, analysis of AE signals was done according to an estimation of events count and cumulative energy.

During the double shear test, AE event signals were monitored and recorded using an Euro Physical Acoustics (EPA) system:

- Five piezoelectric transducers (R15 model from Physical Acoustics corporation), with a band characteristic from 50 to 200 kHz, and 150 kHz resonant frequency, were mounted on the specimen, and coupled to the asphalt concrete with a special adhesive tape, according to the configuration presented in Figure 2. This configuration is chosen to try locating the events within the XY plane of the specimen, and to follow the crack propagation in the vicinity of the notches. The R15 model operates at a temperature between -65°C and 175°C, and weighs 34 grams.

- Pre-amplification of the AE signals was provided by five preamplifiers (IL40S model) with a gain set for 40 dB.

- AE signals were sampled at 20 MHz and filtered with amplitude threshold about 40 dB. It is clear that the detected events depend on the value of this threshold. Before the loading test, the propagation velocity of the longitudinal waves was determined by generating an elastic wave using the conventional pencil lead breaking. The attenuation curve is performed by using the AST procedure (Auto Sensor Tests).

- A signal conditioner and software that allow recording the AE features in a computer for further analysis.

Fig. 2. Photography of a specimen instrumented with AE sensors

3   Results and Discussion

3.1   Mechanical Analysis

In this paragraph, the analyzed mechanical property is the shear stiffness modulus of the AC. One interests in its evolution during the fatigue test. At each cycle of the test, the shear stiffness $K_{S,k}^*$ correlates the applied shear stress with the resulting relative tangential displacement at both sides of the central part of the specimen [7] and can be expressed as a complex number as described in Eq. (1):

$$\begin{cases} K_{S,k}^*(iw) = \dfrac{\text{Re}\left(F_k^*(t).e^{iwt}\right)}{S \times \text{Re}\left(u_{k1}^*(t).e^{(iwt+\varphi_k)} + u_{k2}^*(t)\right)} = \dfrac{\Delta \tau_k}{\Delta u_k} \cdot e^{i\varphi_k} \\ \Delta K_{S,k} = \left|K_{S,k}^*(iw)\right| = \dfrac{\Delta \tau_k}{\Delta u_k} \end{cases} \quad (1)$$

With (at the cycle # k):

- $\Delta K_{S,k}$: shear stiffness modulus of the specimen (MPa.mm^{-1});
- ΔF_k: amplitude of the applied shear force (N);
- Δu_k: amplitude of the measured relative displacement (mm);
- φ_k: phase angle between shear force and the relative displacement signals (°);
- S: sheared cross-sections at both sides of the central part of the specimen (mm²).

The evolution in the normalized values of shear stiffness modulus ΔK_S (compared with the initial value of the shear stiffness modulus) is plotted against the number of loading cycles (Figure 3). As expected, this modulus decreases during the test, and its trend can be divided into three main stages. At first, ΔK_S decreases quite quickly and lost around 15% of its initial value. During the second stage, the damage grows, as the micro-cracks occur; ΔK_S decreases slowly. In the third stage, shorter than the second, ΔK_S decreases quickly, which implies both the coalescence and rapid propagation of macroscopic cracks within the asphalt concrete.

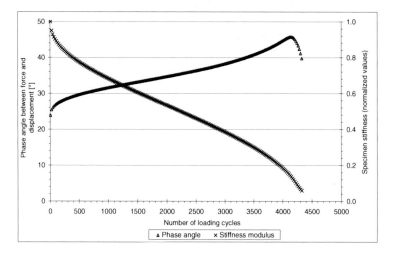

Fig. 3. Evolution in specimen stiffness modulus and phase angle during the test

To further the analysis of fatigue test results, evolution of dissipated energy per each loading cycle is investigated. In this study, the formula (Eq. (2)) is proposed to evaluate the dissipated energy at the cycle #k ($W_{D,k}$):

$$W_{D,k} = \pi \cdot \Delta F_k \cdot \Delta u_k \cdot \sin(\varphi_k) \qquad (2)$$

Moreover, the approach of dissipated energy can be used to evaluate the number of loading cycles to failure. For instance, Rowe [10] proposed the assessment of the material failure based on the criterion « Dissipated Energy Ratio » DER, as described in Eq. (3).

$$DER = \frac{k \cdot W_{D,1}}{W_{D,k}} \qquad (3)$$

Both the dissipated energy per cycle and the DER are plotted against the number of loading cycles (Figure 6).

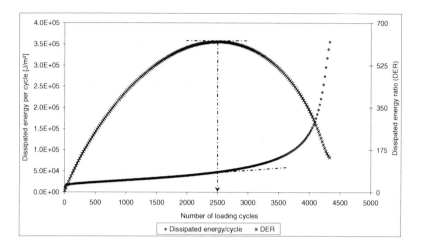

Fig. 4. Dissipated energy per cycle and DER vs. number of loading cycles

From the two previous figures (Figs. 3 and 4), the number of loading cycles to failure of the AC can be assessed by means of the following criteria:

- Conventionally, the number of loading cycles to failure is defined as the required number to decrease by one-half the initial value of the specimen stiffness modulus (Figure 3). Under this conventional criterion, the number of loading cycles to failure (N_{50}) is approximately 2300 cycles;

- Based on the dissipated energy ratio proposed by Rowe (see Eq. (3) and Figure 4), the number of loading cycles to failure (N_{DER}) is about 2500 cycles.

3.2 Acoustic Emission Analysis

Figure 5 displays the evolutions in both cumulative number of events and acoustic energy plotted against the testing time. From this figure, it is observed that events with high energy levels occur at the end of the fatigue test, in other words, during the macro-cracking stage of the asphalt concrete (third stage, Figure 3). Moreover, analysis of both parameters cumulative number of events and acoustic energy shows that failure process under loading can be divided into several stages.

The change from one stage to another is accompanied by a high increase of number of located events with an intensification of their corresponding acoustic energy levels. The next paragraph presents a deep analysis of the emission acoustic results, and a further comparison of both acoustic parameters and mechanical parameters.

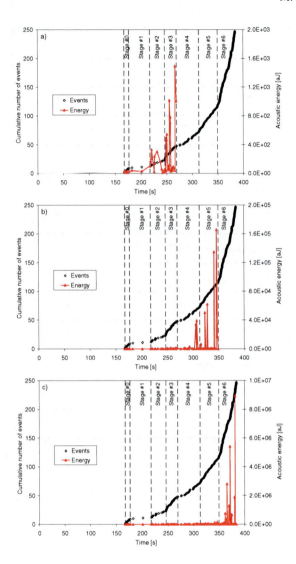

Fig. 5. Number of located events and acoustic energy vs. time; a) stages 0 to 3; b) stages 0 to 5; c) stages 0 to 6

3.3 Correlation between Mechanical and AE Results

Analysis of both mechanical and acoustic test results (Figure 6) leads us to state the following findings:

- By setting the signal threshold for 40 dB, the located events start appearing when a 42% decline in initial value of the specimen stiffness modulus is reached. These early events are the warning signs of an initiation failure within the material (Stage #0, Figure 6);

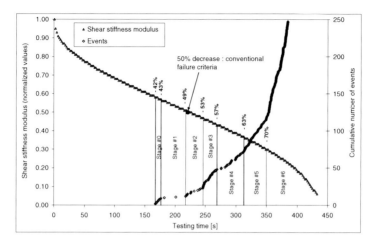

Fig. 6. Evolution in stiffness modulus and number of events during test

- When the initial value of the stiffness modulus decreases by 42% to 49%, the number of located events increases slightly, and their corresponding acoustic energies are relatively low (stages #0, and #1, Figs. 5-a, and 6);

- A critical threshold can be identified as a 49% drop in initial value of the specimen stiffness modulus, since it corresponds to a highly and significant increase in acoustic energy levels of the located events (stage #2, Figs. 5-a, and 6). Moreover, this critical threshold also coincides with the conventional failure criterion of asphalt concrete mixes (drop by one-half in the initial value of the specimen stiffness modulus);

- Beyond the stage #2, the located events rate considerably increases with time. At each stage (#2 to #6, Figs. 5-c, and 6), test results analysis clearly shows an increase in the acoustic energy level of located events. This finding may be a result of a gradually crack propagation within the asphalt concrete; indeed, the micro-crack starts propagating through the asphalt mastic before being stopped by an aggregate. Then, to get over this obstacle, the crack gets around the aggregate (cohesive or adhesive failure). The change in slope value of the curve, which correlates the cumulative number of events with the testing time (or number of loading cycles), can give an idea on the failure type (cohesive or adhesive). In fact, a low slope value can correspond to crack propagation within the asphalt mastic (cohesive failure). Moreover, it's clear that cohesive failure leads to appearance of located events with low acoustic energy level. However, additional tests should be carried out to evaluate the repeatability of the test, strengthen findings from this preliminary investigation, and to get enough results for identifying acoustic parameters specific to different failure mechanisms;

- The number of located events roughly increases beyond a 70% decrease in initial value of the specimen shear stiffness modulus. During this last stage, acoustic events exhibit acoustic energy levels greater than those recorded in the previous stages. This finding may correspond to a change in scale (from micro to macro) of the failure.

4 Conclusion and Perspectives

In this study, we observed that the AE technique is an efficient tool for characterizing the fatigue cracking mechanisms in AC. The events start appearing when about 40% decline in initial value of the specimen stiffness modulus is reached. A critical threshold can be identified as about 50% drop in initial value of the specimen stiffness modulus, since it corresponds to a highly and significant increase in acoustic energy level. Studying cumulative events, several stages were defined in fatigue test which corresponds to different mechanisms such as micro or macro cracking. Nevertheless, these findings should be moderated because the threshold value determines the rate of the acoustic emission activity.

In future works, it's necessary to confirm the correlation between AE events and mechanical criteria. Moreover, wave propagation phenomenon should be investigated by analyzing evolution in wave attenuation as well as wave velocity during the fatigue test. In addition, initial heterogeneity and viscoelastic behavior (heating) of AC, the loading mode, the induce damage lead to a complex variation of acoustic properties. Comprehension of acoustic properties variation during the fatigue test allows a good evaluation of acoustic amplitude, energy and location to study FPZ phenomena.

References

[1] Li, X., Marasteanu, M., Iverson, N., Labuz, J.: Observation of crack propagation in asphalt mixtures with acoustic emission. Trans. Res. Bo. (1970), 171–177 (2006)
[2] Muralidhara, S., Raghu Prasad, B.K., Eskandari, H., Karihaloo, B.L.: Fracture process zone sizeand true fracture energy of concrete using acoustic emission. Const. Buil. Mat. (24), 479–486 (2010)
[3] Li, X., Marasteanu, M.: The fracture process zone in asphalt mixture at low temperature. Eng. Frac. Mech. (77), 1175–1190 (2010)
[4] Apeagyei, A.K., Buttlar, W.G., Reis, H.: Jour. Brit. Inst. for NDT 51(3), 129–136 (2009)
[5] Behzad, B., Eshan, V.D., Sarfraz, A., Buttlar, W.G., Reis, H.: Trans. Res. Bo., 14 (2011)
[6] Seo, Y., Kim, Y.R.: Using acoustic emission to monitor fatigue damage and healing in asphalt concrete. J. of Civil Eng. 12(4), 237–243 (2008)

[7] Diakhaté, M., Millien, A., Petit, C., Phelipot-Mardelé, A., Pouteau, B.: Experimental investigation of tack coat fatigue performance: Towards an improved lifetime assessment of pavement structure interfaces. Const. Buil. Mat. 25, 1123–1133 (2011)
[8] Diakhaté, M.: Fatigue et comportement des couches d'accrochage dans les structures de chaussée, Thèse de doctorat, Université de Limoges, France (2007)
[9] Petit, C., Laveissière, D., Millien, A.: Modelling of reflective cracking in pavements: fatigue under shear stresses. In: Proceedings of the 3rd International Symposium on 3D FE for Pavement Analysis, Design and Research, pp. 111–123 (2002)
[10] Rowe, G.M.: Application of dissipated energy concept to fatigue cracking in asphalt pavements, Ph.D. thesis, University of Nottingham, UK (1996)

Fatigue Characteristics of Sulphur Modified Asphalt Mixtures

Andrea Cocurullo, James Grenfell, Nur Izzi Md. Yusoff, and Gordon Airey

NTEC, Civil Engineering, University of Nottingham, UK

Abstract. The use of solid sulphur pellets has been shown to successfully extend and modify bitumen in an asphalt mixture with the subsequently modified asphalt mixture showing improved performance properties in terms of enhanced stiffness and increased resistance to permanent deformation. However, with the increase in stiffness comes the possible reduction in the fatigue and fracture properties of the sulphur modified asphalt mixture. Due to these concerns, the fatigue properties of sulphur modified asphalt mixtures have been investigated using both diametral, indirect tensile fatigue tests (ITFT) and two-point bending fatigue tests (2PB) under both controlled stress (load) and controlled strain (deformation) conditions. The fatigue tests have been undertaken at a temperature of 10°C and have used a range of fatigue failure definitions to quantify the laboratory fatigue cycles to failure. The fatigue results of the sulphur modified asphalt mixtures have been compared with a standard dense bitumen macadam (DBM) control mixture made using the same gradation and the same 40/60 penetration grade base binder. The fatigue comparisons were performed using the traditional strain criterion as well as a stress criterion. It can be seen from the results, that despite the significant stiffness gain, the fatigue properties of the sulphur modified asphalt mixtures compare well with the fatigue properties of the control mixtures. However, results were found to depend also on the testing configuration.

1 Introduction

Sulphur-extended asphalt was developed in the 1970s using hot liquid sulphur but was considered, due to the high price of sulphur, to be too expensive for use in road paving mixtures and also suffered from health and safety concerns. However, the increase in sour oilfield operations and the development of a more user-friendly sulphur based asphalt modifier in a pelletised form has renewed interest in the use of sulphur modified mixtures as a paving material.

Sulphur pellets are used as both a binder extender and an asphalt mixture modifier when added to an asphalt mixture [1]. The sulphur pellets are added to the hot asphalt mixture (aggregate and bitumen) at ambient temperatures during the mixing process. The addition of the modified sulphur pellets into the bitumen modifies the bitumen properties when the bitumen and sulphur combine at temperatures above the melting point of sulphur (120°C). For percentages marginally less than 20% by weight, all the sulphur is chemically combined or dissolved within the bitumen and acts as an extender, modifying the bitumen properties with a reduction in viscosity and an increase in ductility (extended bitumen softer and more ductile). At higher percentages of added sulphur, part of the sulphur not dissolved in the bitumen remains predominantly as free sulphur and when the blend (bitumen-sulphur or sulphur modified asphalt mixture) cools, it crystallises. Depending on the amount of sulphur, the crystallisation gives different levels of strengthening (stiffening). As the bitumen is extended by sulphur, the bitumen content is usually reduced. The most commonly used ratios of bitumen to sulphur are 60% to 40% and 70% to 30% by mass. As sulphur has approximately double the density of bitumen, the total binder (bitumen plus sulphur) content by mass in the asphalt mixture needs to be higher than for the pure bitumen asphalt mixture to achieve the same binder volume (mixture volumetrics). It takes several days for the paving mixture made with modified sulphur pellets to develop its final strength (stiffness) due to the progressive crystallisation of the modified sulphur in the mixture.

In general, the stiffness (complex modulus) of sulphur modified asphalt mixtures over all loading frequencies and temperatures is higher than that for conventional asphalt mixtures. In addition, the temperature and/or frequency dependency of the modified mixture is also reduced. The high temperature permanent deformation properties of the modified mixture are therefore improved. In general, the fatigue performance of the modified mixtures is reduced, although as the stiffness is increased this may not be a problem due to the combined effects of material properties, pavement structure and traffic loads. There is also evidence of a slight reduction in low temperature cracking resistance and potentially a slightly higher sensitivity to water damage.

As world-wide interest in sulphur modified mixtures grows, there is a need to expand the detailed constitutive (stress-strain dynamic mechanical analysis) and failure property analysis of the material. Initial work in this area is detailed in this paper.

2 Experimental Procedure

2.1 Materials and Mixture Design

Bitumen. A Venezuelan standard 40/60 pen bitumen was used in the study. The conventional bitumen properties of penetration [2] and softening point [3] are 47 dmm and 53.0°C respectively.

Additives. Two additives were used in the study to produce the sulphur modified mixture:

o workability additive blended with the bitumen in the proportion of 1.5% by mass of bitumen;
o modified sulphur pellets added during the asphalt mixing process in the proportion of 30% by mass of total binder (i.e. bitumen and additives).

The conventional bitumen properties of penetration [2] and softening point [3] of the 40/60 pen bitumen blended with the workability additive are 33 dmm and 72.6°C respectively.

Aggregates. Porphyritic andesite aggregates were used during the study. The mixture design for the control and sulphur modified asphalt mixtures was based on a 0/20 mm size dense binder course according to BS 4987-1: 2005 [4].

Asphalt Mixtures. Two asphalt mixtures were used in this study:

o control mix (4.7% of conventional 40/60 pen bitumen by mass of total mixture);
o sulphur modified mix (30% modified sulphur pellets to 70% bitumen, blended with workability additive, by mass). The volume of total binder (i.e. bitumen and additives) was maintained equal to the volume of bitumen in the control mix [5,6]. As sulphur has approximately double the density of bitumen, this led to a total mass of binder of 5.5% by mass of total asphalt mixture.

The maximum densities of both mixture designs were measured according to BS EN 12697-5: 2009 [7].

2.2 Specimen Preparation

The asphalt mixtures were mixed and compacted in the laboratory conforming to BS EN 12697-35: 2004 [8] and BS EN 12697-33: 2003 [9] respectively, to achieve target air void contents of 6 ± 1% within the specimens. The temperature of mixing and compaction were 160 ± 5 and 140 ± 5°C for the control and the sulphur modified mixtures respectively. The temperature range for the modified mixture guarantees workability and allows management of the sulphur-based emissions [1]. Due to the progressive crystallisation of the modified sulphur in the mixture, it can take several days for the modified asphalt mixture to develop its final mechanical properties [1]. According to previous research work on sulphur modified mixtures, laboratory prepared specimens were allowed to mature for 14 days before measuring their mechanical properties. To speed up the testing programme, an accelerated curing regime consisting of 24 hours at 60°C on the whole slab was used. Slabs (305 mm by 305 mm by 60 mm) were cored and cylindrical specimens (100 mm in diameter and 40 mm high) were obtained by trimming the top and the bottom of each core. The trapezoidal specimens (B=70, b=25, e=25 and h=250 mm according to BS EN 12697-24: 2004 [10]) were manufactured from slabs measuring 305 mm by 305 mm by 80 mm by using a masonry saw and a purpose-built clamping device. After the volumetric characteristics of the specimens were determined according to BS EN 12697-6: 2003 [11] and BS EN 12697-8: 2003 [12], they were then stored at room temperature on one of their flat faces for up to two weeks from compaction and then stored in a dry atmosphere at a temperature of 5°C to prevent distortion. Specimens were only removed from this controlled environment to be conditioned and tested at the required temperature.

2.3 Testing Procedure

Test applying Indirect Tension to Cylindrical Specimens (IT-CY). The IT-CY test, carried out conforming to BS EN 12697-26: 2004 [13], was used to determine the stiffness modulus of the cylindrical specimens in order to quickly assess the effect of curing time on stiffness.

Two-Point Bending Test (2PB). The two-point bending cantilever configuration was used to perform the fatigue conforming to BS EN 12697-24: 2004 [10]. These tests were used to widen the understanding of the fatigue properties of control and sulphur modified mixtures (see Figure 1).

Indirect Tensile Stiffness Modulus (ITSM). The ITSM was carried out to measure the tensile stiffness of asphaltic cores at a given stress level (as opposed to the specific strain level as prescribed in the IT-CY test) that was used to apply loading to a specimen in the ITFT detailed below. This test is detailed in DD 213: Determination of the indirect tensile stiffness modulus of bituminous mixtures [14].

Indirect Tensile Fatigue Test (ITFT). The ITFT characterises the fatigue behaviour of bituminous mixtures under controlled load test conditions. The ITFT was carried out at 10°C over a range of stress levels to produce fatigue lines for the different mixtures (see Figure 1). This test was carried out according to the older British method and is detailed in DD ABF Method for the determination of the fatigue characteristics of bituminous mixtures using indirect tensile fatigue [15].

3 Results and Discussion

3.1 Crystallisation Study

One slab for each mixture was produced using the same compaction effort. Five cores were then taken from each slab (mixture type) and used for the crystallisation study. The addition of the sulphur pellets had a positive influence on the compactability of the asphalt mixture slab. The stiffness values for these specimens are greater compared to the control mixture. This is probably due to a lower air void content but also the early stages of sulphur crystallisation after 9 days.

IT-CY stiffness values for five specimens each of the control mix and the sulphur modified mix at 20°C were determined every 3 to 4 days over a period of 75 days to investigate the crystallisation (increasing stiffness) effect of the sulphur modified mixture when cured at room temperature.

The results of the individual specimens are shown in Figure 2 together with a trend line in terms of stiffness change with time. The control mixture shows no significant change in stiffness with time as expected. The sulphur modified mixture shows a continuous trend of increasing stiffness which begins to plateau

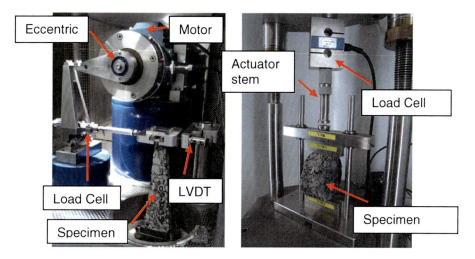

Fig. 1. The equipment and specimen set-up for the two-point bending fatigue test (2PB) and the Indirect Tensile Fatigue Test (ITFT)

after approximately 30 days. The increase is approximately 50% compared to the initial stiffness readings taken after 9 days. The grey band in the figure represents the stiffness range (repeated for 5 specimens) that can be developed in the sulphur modified mixture by using the accelerated curing regime of 60°C for 24 hours once. This band is in good agreement with the 30 day plateau showing the applicability of the accelerated curing regime.

3.2 Two-Point Bending Fatigue

The fatigue performance of the sulphur modified mixture compared to the control mixture based on the 2PB controlled stress fatigue tests is shown in Figures 3 and 4. The fatigue functions are based on a relationship between fatigue life to failure (defined in the plot as N_f 50% [16]) and applied (initial and constant) stress level. This stress fatigue criterion shows the stiffer sulphur modified mixture having a greater life compared to the control mixture.

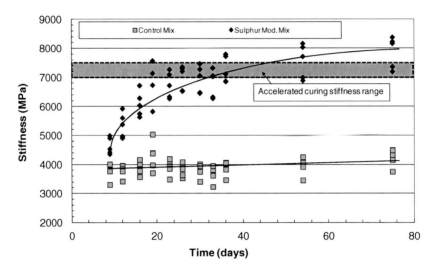

Fig. 2. Stiffness evolution for sulphur crystallisation study after 75 days

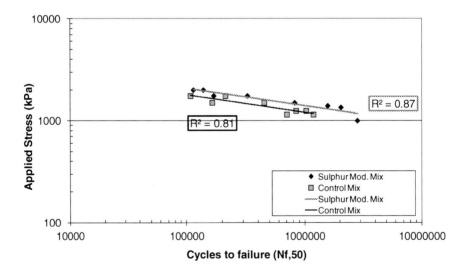

Fig. 3. Stress fatigue criterion for both mixtures based on 2PB fatigue tests carried out in stress control at 10°C and 25 Hz using N_f 50%

The fatigue results have also been plotted against initial strain level in terms of the standard strain criterion and show a very comparable fatigue performance for both the control and sulphur modified asphalt mixtures.

The fatigue performance of the mixtures was also compared using the 2PB fatigue test in strain control. This is also shown in Figure 4 based on a relationship between fatigue life to failure (defined in the plot as N_f 50%) and applied strain. This strain fatigue criterion again shows a very comparable fatigue performance for both the control and sulphur modified asphalt mixtures.

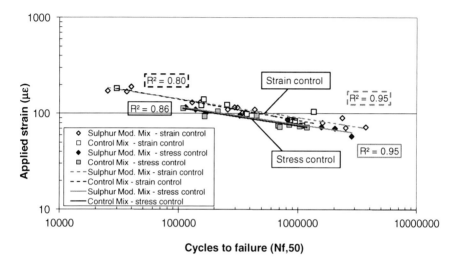

Fig. 4. Strain fatigue criterion for both mixtures based on 2PB fatigue tests carried out in both stress and strain control at 10°C and 25 Hz using N_f 50%

3.3 Indirect Tensile Fatigue Testing

Further fatigue testing was carried out using ITFT. Again if we consider applied stress, the sulphur modified mix performs better than the control mix (see Figure 5). However, if the results are plotted as a function of initial strain, (see Figure 6), the sulphur modified mix performs much worse than the control mix. This is due to the large difference in the stiffnesses of the two mixtures in the indirect tensile mode of loading. If we look at Table 1, it can be seen that the stiffness for the mixtures as generated from the 2PB testing are similar. This stiffness value is based on a combination of the compressive and tensile moduli. However, the stiffnesses from the IT-CY test differ significantly between mixtures. This comes from the stiffening effect of the sulphur modifier, which stiffens the binder. The stiffening of the binder has a much larger effect on the tensile stiffness, which relies on the binder for strength. The stiffness in 2PB is more a combination of the compressive and tensile stiffness, where the compressive stiffness relies much more on aggregate interlock for strength and so the two mixtures behave similarly.

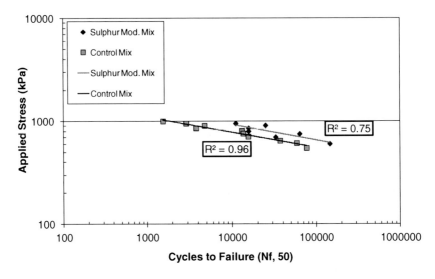

Fig. 5. Stress fatigue criterion for both mixtures based on ITFT fatigue tests carried out in stress control at 10°C using N_f 50%

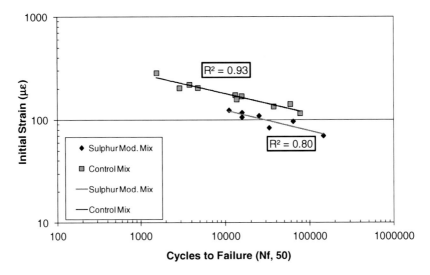

Fig. 6. Strain fatigue criterion for both mixtures based on ITFT fatigue tests carried out in stress control at 10°C using N_f 50%

Table 1. Comparison of the average stiffness values of both mixtures tested by IT-CY (10°C, 124 ms rise time) and 2PB (10°C, 25 Hz)

Average Stiffness from IT-CY (MPa)			Average Stiffness from 2PB (MPa)		
Control Mix	Sulphur Mod. Mix	% improvement	Control Mix	Sulphur Mod. Mix	% improvement
11375	15694	38	15692	17528	12

4 Conclusions

The crystallisation study shows that the sulphur modified mixture initially continues to increase in stiffness and reaches a plateau after approximately 30 days. An increase in stiffness of about 50% is reached from the initial measurements that were taken after 9 days. As expected the control mixture did not exhibit any change in stiffness during this study. The accelerated curing protocol of 60°C for 24 hours was shown to raise the stiffness of the sulphur modified mixture to an equivalent level of the 30 day plateau.

In terms of the 2PB fatigue response of the control and sulphur modified mixtures, the results show that if stress is considered as the fatigue criterion, the sulphur modified mixture performs better than the control mixture. However, in terms of a strain fatigue criterion, both mixtures have an almost identical fatigue response. In addition, the strain criterion also allows the fatigue response to be predicted with great confidence over a wide range of strain levels and fatigue lives. When ITFT is considered, the sulphur modified mix is shown to exhibit a better fatigue response when plotted as a function of stress level, but due to its much larger stiffness in tension, performs less well compared to the control mix when plotted as a function of initial strain.

References

[1] Strickland, D., Colange, J., Martin, M., Deme, I.: Performance properties of paving mixtures made with modified sulphur pellets. In: Proceedings of the International ISAP Symposium on Asphalt Pavements and Environment, ISAP, Zurich, pp. 64–75 (2008)
[2] British Standards Institution, Methods of test for petroleum and its products, Bitumen and bituminous binders - Determination of needle penetration, BS EN 1426, London (2000)
[3] British Standards Institution, Methods of test for petroleum and its products, Bitumen and bituminous binders - Determination of softening point, Ring and ball method, BS EN 1427, London (2000)
[4] British Standards Institution, Coated macadam (asphalt concrete) for roads and other paved areas – Part 1: Specification for constituent materials and for mixtures, BS 4987-1, London (2005)

[5] Bailey, H.K., Allen, R., Strickland, D., Colange, J., Gilbert, K.: Innovative Sulphur Technology Applied to European Asphalt Mixtures. In: Proceedings of 6th International Conference on Maintenance & Rehabilitation of Pavements & Technological Control (MAIREPAV6), Turin, vol. 1, pp. 499–508 (2009)

[6] McBee, W.C., Sullivan, T.A., Izatt, J.O.: In: FHWA-IP-80-14: State-of-the-Art Guideline Manual for Design, Quality Control, and Construction of Sulfur-Extended-Asphalt (SEA) Pavements, Implementation Package, Federal Highway Administration, Washington D.C (1980)

[7] British Standards Institution, Bituminous mixtures – Test methods for hot mix asphalt – Part 5: Determination of the maximum density, BS EN 12697-5, London (2009)

[8] British Standards Institution, Bituminous mixtures – Test methods for hot mix asphalt – Part 35: Laboratory mixing, BS EN 12697-35, London (2004)

[9] British Standards Institution, Bituminous mixtures – Test methods for hot mix asphalt – Part 33: Specimen prepared by roller compactor, BS EN 12697-33, London (2003)

[10] British Standards Institution, Bituminous mixtures – Test methods for hot mix asphalt – Part 24: Resistance to fatigue, BS EN 12697-24, London (2004)

[11] British Standards Institution, Bituminous mixtures – Test methods for hot mix asphalt – Part 6: Determination of bulk density of bituminous specimens, BS EN 12697-6m, London (2003)

[12] British Standards Institution, Bituminous mixtures – Test methods for hot mix asphalt – Part 8: Determination of void characteristics of bituminous specimens, BS EN 12697-8, London (2003)

[13] British Standards Institution, Bituminous mixtures – Test methods for hot mix asphalt – Part 26: Stiffness, BS EN 12697-26, London (2004)

[14] British Standards Institution, Method for Determination of the indirect tensile stiffness modulus of bituminous mixtures, Draft for development DD213 (1993)

[15] British Standards Institution, Method for the determination of the fatigue characteristics of bituminous mixtures using indirect tensile fatigue, Draft for development DD ABF (2003)

[16] Cocurullo, A., Airey, G.D., Collop, A.C., Sangiorgi, C.: Indirect Tensile versus Two Point Bending Fatigue Testing. ICE Transport 161(TR4), 207–220 (2008)

Effect of Moisture Conditioning on Fatigue Properties of Sulphur Modified Asphalt Mixtures

Andrea Cocurullo, James Grenfell, Nur Izzi Md. Yusoff, and Gordon Airey

NTEC, Civil Engineering, University of Nottingham, UK

Abstract. Sulphur modified asphalt mixtures using pelletised sulphur as the bitumen extender/modifier have shown enhanced mechanical performance in terms of stiffness, resistance to permanent deformation and potentially even fatigue. However, concerns still exist in terms of the long-term durability of the modified asphalt mixtures in relation to their moisture damage susceptibility. A moisture damage protocol has therefore been designed specifically for sulphur modified asphalt mixtures. The dynamic mechanical and fatigue properties of the moisture damaged sulphur modified asphalt mixtures were then investigated using stiffness tests applying indirect tension to cylindrical specimens, two-point bending DMA and fatigue tests. The moisture damage protocol involves placing the specimens in a water bath at 85°C for 9 days for both the sulphur modified asphalt mixtures and a control dense bitumen macadam (DBM) asphalt mixture. Initial results show significant stiffness loss for the moisture conditioned sulphur modified asphalt mixtures and a reduction in fatigue performance relative to the unconditioned asphalt mixtures. Nevertheless non-permanent effects of moisture conditioning have been observed which led to partial recovery of the material properties.

1 Introduction

Sulphur-extended asphalt was developed in the 1970s using hot liquid sulphur but was considered, due to the high price of sulphur, to be too expensive for use in road paving mixtures and also suffered from health and safety concerns. However, the increase in sour oilfield operations and the development of a more user-friendly sulphur asphalt modifier in a pelletised form has renewed interest in the use of sulphur modified asphalt mixtures as a viable paving material. The pellets are used as both a binder extender and an asphalt mixture modifier when added to an asphalt mixture [1]. The pellets are added to the hot asphalt mixture (aggregate and bitumen) during the mixing process. The addition of the modified sulphur pellets into the bitumen modifies the bitumen properties when the bitumen and sulphur combine at temperatures above the melting point of sulphur (120°C). For percentages marginally less than 20% by weight, all the sulphur is chemically combined or dissolved within the bitumen and acts as an extender, modifying the bitumen properties with a reduction in viscosity and an increase in ductility (extended bitumen softer and more ductile). At higher percentages of added sulphur, part of the sulphur not dissolved in the bitumen remains predominantly as free sulphur and when the blend cools, it crystallises. Depending on the amount of pellets, the crystallisation gives different levels of strengthening (stiffening). As the bitumen is extended by sulphur, bitumen content is usually reduced. The most commonly used ratios of bitumen to modified sulphur are 60% to 40% and 70% to 30% by mass. As sulphur has approximately double the density of bitumen, the total binder (bitumen plus sulphur) content by mass in the asphalt mixture needs to be higher than for the pure bitumen asphalt mixture to achieve the same binder volume (mixture volumetrics). It takes several days for the paving mixture made with modified sulphur pellets to develop its final strength (stiffness) due to the progressive crystallisation of the modified sulphur in the mixture.

Although the use of solid sulphur pellets has been shown to produce mixtures with increased strength, stiffness and high temperature permanent deformation performance compared to conventional asphalt mixtures, questions still persist over the fatigue and fracture properties of these 'stiffer' modified asphalt mixtures as well as their durability related to moisture damage susceptibility.

The overall aim of this research study was to use stiffness tests applying indirect tension to cylindrical specimens, dynamic mechanical analysis measurements and fatigue material characterisation tests in two-point bending configuration to characterise the performance of both a control and a sulphur modified mixture when subjected to moisture conditioning. The first results show how complex the characterisation of moisture sensitivity in the laboratory is. Concerns about short and long-term effects (after accelerated moisture conditioning) on stiffness and fatigue characteristics of the materials lead to the need for further investigation.

2 Experimental Procedure

2.1 Materials and Mixture Design

Bitumen. A Venezuelan standard 40/60 pen bitumen was used in the study. The conventional bitumen properties of penetration [2] and softening point [3] are 47 dmm and 53.0°C respectively.

Additives. Two additives were used in the study to produce the sulphur modified mixture:

o   workability additive blended with the bitumen in the proportion of 1.5% by mass of bitumen;
o   modified sulphur pellets added during the asphalt mixing process in the proportion of 30% by mass of total binder (i.e. bitumen and additives).

The conventional bitumen properties of penetration [2] and softening point [3] of the 40/60 pen bitumen blended with the workability additive are 33 dmm and 72.6°C respectively.

Aggregates. Porphyritic andesite aggregates were used during the study. The mixture design for the control and sulphur modified asphalt mixtures was based on a 0/20 mm size dense binder course according to BS 4987-1: 2005 [4].

Asphalt Mixtures. Two asphalt mixtures were used in this study:

o control mix (4.7% of conventional 40/60 pen bitumen by mass of total mixture);
o sulphur modified mix (30% modified sulphur pellets to 70% bitumen, blended with workability additive, by mass). The volume of total binder (i.e. bitumen and additives) was maintained equal to the volume of bitumen in the control mix ([5] and [6]). As sulphur has approximately double the density of bitumen, this led to a total mass of binder of 5.5% by mass of total asphalt mixture.

The maximum densities of both mixture designs were measured according to BS EN 12697-5: 2009 [7].

2.2 Specimen Preparation

The asphalt mixtures were mixed and compacted in the laboratory conforming to BS EN 12697-35: 2004 [8] and BS EN 12697-33: 2003 [9] respectively, to achieve target air void contents of 6 ± 1% within the specimens. The temperature of mixing and compaction were 160 ± 5 and 140 ± 5°C for the control and the sulphur modified mixtures respectively. The temperature range for the modified mixture guarantees workability and allows management of the sulphur-based emissions [1]. Due to the progressive crystallisation of the modified sulphur in the mixture, it can take several days for the modified asphalt mixture to develop its final mechanical properties [1]. According to previous research work on sulphur modified mixtures, laboratory prepared specimens were allowed to mature for 14 days before measuring their mechanical properties. To speed up the testing programme, an accelerated curing regime consisting of 24 hours at 60°C on the whole slab was used. Slabs (305 mm by 305 mm by 60 mm) were cored and cylindrical specimens (100 mm in diameter and 40 mm high) were obtained by trimming the top and the bottom of each core. The trapezoidal specimens (B=70, b=25, e=25 and h=250 mm according to BS EN 12697-24: 2004 [10]) were cut from slabs measuring 305 mm by 305 mm by 80 mm by using a masonry saw and a purpose-built clamping device. After the volumetric characteristics of the specimens were determined according to BS EN 12697-6: 2003 [11] and BS EN 12697-8: 2003 [12], they were then stored at room temperature on one of their flat faces for up to two weeks from compaction and then stored in a dry atmosphere at a temperature of 5°C to prevent distortion. Specimens were only removed from this controlled environment to be conditioned and tested at the required temperature.

2.3 Testing Procedure

Test applying Indirect Tension to Cylindrical Specimens (IT-CY). The IT-CY test, carried out conforming to BS EN 12697-26: 2004 [13], was used to determine the stiffness modulus of the cylindrical specimens in order to quickly assess the effect of moisture on stiffness and thus define a suitable moisture conditioning protocol for more fundamental testing.

Two-Point Bending Test (2PB). The two-point bending cantilever configuration was used to perform both Dynamic Mechanical Analysis (DMA) and fatigue conforming to BS EN 12697-26: 2004 [13] and BS EN 12697-24: 2004 [10] respectively. These tests were used to widen the understanding of the effects of moisture on the mechanical and fatigue properties of control and sulphur modified mixtures.

3 Moisture Conditioning Protocol: First Results and Discussion

Moisture damage significantly influences the durability of asphalt mixtures. A reduction of cohesion results in a reduction of the strength and stiffness of the mixture and thus a reduction of the pavement's ability to support traffic-induced stresses and strains. Failure of the bond between the bitumen and aggregate (stripping) also results in a reduction in pavement support.

Previous research showed that the addition of sulphur-modified pellets can cause up to a 10% reduction in both the retained Marshall stability and the tensile strength ratio of the modified mixtures in comparison to control mixtures [1]. Further research confirmed an increase in the moisture susceptibility of sulphur modified mixtures compared to the control mixture when determining tensile strength ratio and when performing Hamburg wheel-tracking testing [14]. Other research witnessed comparable moisture resistance of a sulphur modified mixture compared to conventional mixes when an anti-stripping agent was used [15].

This current research was conducted in order to assess the mechanical performance, in terms of stiffness and behaviour under fatigue, of conditioned and unconditioned modified mixtures compared to conventional mixture performance.

Based on previous comparisons of moisture sensitivity procedures [16] and on AASHTO T 283-07 [17], a cyclic moisture sensitivity conditioning procedure was devised and a preliminary investigation was conducted on the stiffness (IT-CY) evolution at 60°C and 85°C for the sulphur modified mix. Based on these results it was possible to define the best temperature and conditioning time needed to see significant effects of moisture on the mechanical properties of the sulphur modified material and the study carried on with 2PB rheological DMA and fatigue testing for both the control and the modified mixtures.

3.1 Preliminary IT-CY Testing

The first part of the investigation was carried out on the sulphur modified mixture. A set of two specimens was subjected to repeated moisture conditioning cycles. The unconditioned stiffness was determined on the samples in accordance to BS EN 12697-26 [13] at 20°C (IT-CY_u). The test temperature was chosen based on previous tests that showed that the moisture conditioning effect on sulphur modified mixtures is more evident at higher temperatures. The samples were placed in a vacuum vessel and covered with water at room temperature. The apparatus was then sealed and a partial vacuum of 68.0 ± 3.3 kPa was applied for 30 ± 1 minutes. The specimens were then removed from the vacuum vessel and were placed in a hot water bath at 85 ± 1°C for a period of time of at least 24 ± 1 hours per cycle. The water bath temperature was chosen after conducting some trials at lower temperature (60°C) and witnessing no significant stiffness decrease even after eleven days.

After every cycle the samples were removed from the hot water bath and placed in a water bath set at the test temperature (20 ± 0.5°C) for at least 3 hours (former tests done on dummy specimens fitted with both a surface and an internal thermocouple showed that a stable temperature of 20°C was reached in less than 2 hours). The specimens were then removed from the water bath, surface dried and the conditioned IT-CY was determined for each conditioning cycle (IT-CY_{ci}).

The results in terms of stiffness ratio (IT-$CY_{Ratio,ci}$=IT-CY_{ci}/IT-CY_u) were plotted against time (hours of conditioning) as depicted in Figure 1.

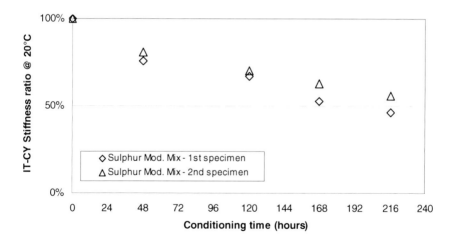

Fig. 1. Moisture conditioning cycles at 85°C: IT-CY ratio at 20°C for the sulphur modified mixture

The results showed that after nine days (216 hours) of conditioning the decrease in the conditioned IT-CY was approximately 50% of the initial (unconditioned) measured value. The chosen temperature of 85°C was considered appropriate for this study since it gave a good indication of the moisture conditioning effects on the material. Nine days were also considered suitable for conditioning the samples as this allowed sufficient stiffness decrease with respect to the unconditioned measured values. This clear modification of the mixture properties in terms of IT-CY seemed promising for carrying on with the next testing stage which was also extended to the control material.

3.2 Two-Point Bending Tests

Two sets of specimens for the control and the sulphur modified mixtures were selected. One set for each mixture was tested unconditioned whereas the remaining two sets were subjected to moisture conditioning at 85 ± 1°C for nine days (± 2 hours). The conditioned samples were brought to 20°C by placing a cooling unit into the water bath first before the samples were removed from the bath (in order to prevent distortion) and they were left to dry out at a constant temperature of 20°C. After 24 hours the geometry of all the samples was checked in order to assess whether or not any significant distortion had occurred. It was found that for the moisture conditioned samples the maximum increase in the measured dimensions was 1.2 % of the thickness whilst the average expansion was below 0.6 %. These findings were considered as acceptable for this study.

The specimens were then glued to the metal plates and conditioned at the test temperature for at least 4 hours prior to testing. DMA and fatigue analysis were undertaken and the results obtained on the conditioned and unconditioned specimens were compared.

DMA Analysis. The rheological DMA experimental data were obtained on subsets of two samples tested at three temperatures (-20, 10, 30°C) and eight frequencies (1, 2, 5, 10, 15, 20, 25, 30 Hz). The analysis was conducted on unconditioned and moisture conditioned (within 4-5 days from conditioning) control and sulphur modified mixtures and the results were used to generate complex modulus master curves using the 2S2P1D model [18]. A reference temperature (T_{ref}) of 10°C was used in this study (Figure 2).

The repeats for both mixtures provided consistent parameters for the models with the unconditioned sulphur modified mixture showing a higher stiffness (complex modulus) compared to the unconditioned control. However the moisture conditioning only had a significant effect on the modified mix above all at higher temperatures (lower frequencies). The control mix did not seem to be affected by the moisture conditioning procedure used in the study.

There was a concern that sulphur may dissolve and crystallise in and out of the binder during temperature changes. Thus there could be a possibility that some of the moisture damage witnessed in terms of stiffness reduction might be attributed to changes in sulphur crystallisation rather than actual moisture damage. This led to the need for further investigation. Therefore the DMA results obtained directly after 9 days of moisture conditioning (plus 4 to allow the specimens to dry out) were compared to the DMA results obtained on the same pair of specimens tested

Effect of Moisture Conditioning on Fatigue Properties

after a further 2 weeks (19 days of total recovery period). It can be seen from the graph that after two weeks the sulphur modified material recovers in terms of stiffness.

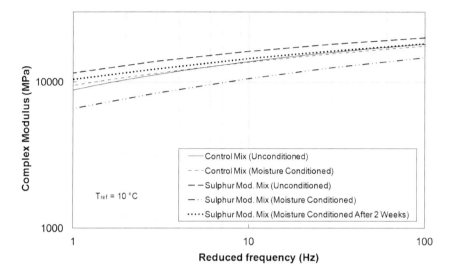

Fig. 2. Complex modulus master curves

Fatigue. Two-point bending controlled strain fatigue tests were undertaken under standard conditions (at 10°C and at 25 Hz) for the unconditioned and moisture conditioned mixtures. The tests were conducted after 25 to 27 days (long-term rest period) from conditioning for the control mix and after 10 to 12 days (mid-term rest period) for the modified mix. The initial stiffness, calculated at the 100th cycle, was recorded for each test and it was plotted against the applied strain level (Figure 3). It can be seen that no moisture effect was recorded after the long-term rest period for the control mixture whereas for the modified mixture an average decrease in stiffness of 25% was shown after a mid-term rest time. Agreement with these results was obtained from the DMA under the same temperature and frequency conditions.

Figure 4 shows the results (in terms of 2PB stiffness) from both the fatigue and the DMA testing. Each DMA data point averages two replicates, whereas every FATIGUE data point averages eight to ten replicates (initial stiffness measured at the beginning of the fatigue test). It can be seen that the moisture conditioning produced a decrease in stiffness of 31 % in the modified material in the short-term and then a progressive recovery in the mid and long-term (25% and 15% stiffness decrease respectively). Testing carried out on the control mix produced a short-term reduction in stiffness of 7% and a total recovery in the long-term.

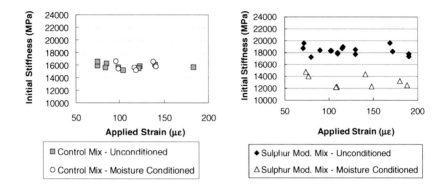

Fig. 3. Initial 2PB Stiffness for the control (left) and the sulphur modified (right) mixtures, both unconditioned and moisture conditioned

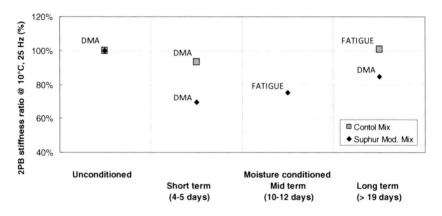

Fig. 4. 2PB (DMA & FATIGUE) stiffness ratio at 10°C and 25 Hz for unconditioned and moisture conditioned mixtures

The fatigue results for the control mixture are depicted in Figure 5. Although only 6 moisture conditioned test specimens (long-term recovery) were tested the results indicate that there are no clear effects of moisture on fatigue. Therefore a single regression line was plotted by fitting the overall results. Further tests would be necessary to achieve a higher coefficient of determination (R^2).

The fatigue results for the modified mix are shown in Figure 6. The materials properties in terms of resistance to fatigue were affected by moisture conditioning, although the tests were carried out after a mid-term recovery period.

Some extra tests will be needed to see whether the recovery in terms of stiffness in the mid and long-term is a peculiarity of sulphur modified materials, which could be explained by the crystallisation changes. Questions still persist over short-term effects of the moisture conditioning procedure on asphalt mixtures and potential affects on the fatigue performance.

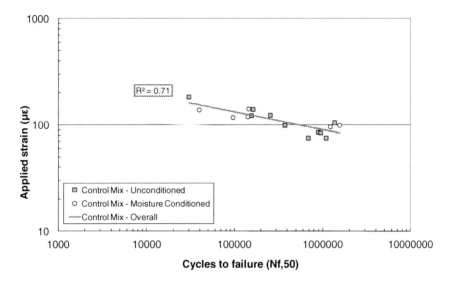

Fig. 5. 2PB controlled strain fatigue tests at 10°C and 25 Hz for unconditioned and moisture conditioned control mixture

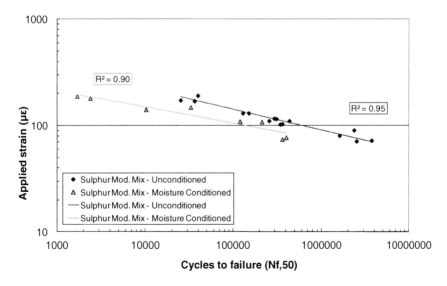

Fig. 6. 2PB controlled strain fatigue tests at 10°C and 25 Hz for unconditioned and moisture conditioned sulphur modified mixture

4 Conclusions

From the DMA results, a conditioning temperature of 85°C for 9 days does seem to modify the properties of the sulphur modified mixture (mainly at higher temperatures and lower frequencies) in terms of stiffness in the short-term. The short-term effect is also confirmed by the IT-CY tests carried out during the preliminary study (no rest periods). However, it has been shown that mid-term and long-term rest periods applied after conditioning seem to progressively eliminate the effect of moisture on stiffness modulus. The fatigue tests carried out in the mid-term still show a significant effect of the moisture conditioning of the material performance although long-term fatigue testing would be needed to see whether a total recovery in stiffness would result in a total fatigue performance recovery.

The DMA as well as the fatigue test results conducted on the control mixture showed no significant effects of moisture.

References

[1] Strickland, D., Colange, J., Martin, M., Deme, I.: Performance properties of paving mixtures made with modified sulphur pellets. In: Proceedings of the International ISAP Symposium on Asphalt Pavements and Environment, ISAP, Zurich, pp. 64–75 (2008)
[2] British Standards Institution, Methods of test for petroleum and its products, Bitumen and bituminous binders - Determination of needle penetration, BS EN 1426, London (2000)
[3] British Standards Institution, Methods of test for petroleum and its products, Bitumen and bituminous binders - Determination of softening point, Ring and ball method, BS EN 1427, London (2000)
[4] British Standards Institution, Coated macadam (asphalt concrete) for roads and other paved areas – Part 1: Specification for constituent materials and for mixtures, BS 4987-1, London (2005)
[5] Bailey, H.K., Allen, R., Strickland, D., Colange, J., Gilbert, K.: Innovative Sulphur Technology Applied to European Asphalt Mixtures. In: Proceedings of 6th International Conference on Maintenance & Rehabilitation of Pavements & Technological Control (MAIREPAV6), Turin, vol. 1, pp. 499–508 (2009)
[6] McBee, W.C., Sullivan, T.A., Izatt, J.O.: In: FHWA-IP-80-14: State-of-the-Art Guideline Manual for Design, Quality Control, and Construction of Sulfur-Extended-Asphalt (SEA) Pavements, Implementation Package, Federal Highway Administration, Washington, DC (1980)
[7] British Standards Institution, Bituminous mixtures – Test methods for hot mix asphalt – Part 5: Determination of the maximum density, BS EN 12697-5, London (2009)
[8] British Standards Institution, Bituminous mixtures – Test methods for hot mix asphalt – Part 35: Laboratory mixing, BS EN 12697-35, London (2004)
[9] British Standards Institution, Bituminous mixtures – Test methods for hot mix asphalt – Part 33: Specimen prepared by roller compactor, BS EN 12697-33, London (2003)
[10] British Standards Institution, Bituminous mixtures – Test methods for hot mix asphalt – Part 24: Resistance to fatigue, BS EN 12697-24, London (2004)

[11] British Standards Institution, Bituminous mixtures – Test methods for hot mix asphalt – Part 6: Determination of bulk density of bituminous specimens, BS EN 12697-6, London (2003)
[12] British Standards Institution, Bituminous mixtures – Test methods for hot mix asphalt – Part 8: Determination of void characteristics of bituminous specimens, BS EN 12697-8, London (2003)
[13] British Standards Institution, Bituminous mixtures – Test methods for hot mix asphalt – Part 26: Stiffness, BS EN 12697-26, London (2004)
[14] Taylor, A.J., Tran, N.H., May, R., Timm, D.H., Robbins, M.M., Powell, B.: Laboratory Evaluation of Sulfur-Modified Warm Mix. Journal of the Association of Asphalt Paving Technologists 79, 403–442 (2010)
[15] Cooper, S.B., Mohammad, L.N., Elseifi, M.: Laboratory Performance Characteristics of Sulfur-Modified Warm-Mix Asphalt. Journal of Materials in Civil Engineering 23(9), 1338–1345 (2011)
[16] Airey, G.D., Choi, Y.-K.: State of the Art Report on Moisture Sensitivity Test Methods for Bituminous Pavement Materials. International Journal of Road Materials and Pavement Design 3(4), 355–372 (2002)
[17] American Association of State Highways and Transportation Officials, Standard Method of Test for Resistance of Compacted Asphalt Mixtures to Moisture-Induced Damage, AASHTO Designation: T 283-07 (2007)
[18] Olard, F., Di Benedetto, H.: The 2S2P1D Model and Relation between the Linear Viscoelastic Behaviours of Bituminous Binders and Mixes. Road Materials and Pavement Design 4(2), 185–224 (2003)

Fatigue Investigation of Mastics and Bitumens Using Annular Shear Rheometer Prototype Equipped with Wave Propagation System

M. Buannic[1,*], H. Di Benedetto[2], C. Ruot[3], T. Gallet[3], and C. Sauzéat[2]

[1] TOTAL France & Département Génie Civil et Bâtiment (Université de Lyon, ENTPE, CNRS), rue Maurice Audin, Vaulx-en-Velin Cedex, 69518, France
mael.buannic@entpe.fr
[2] Département Génie Civil et Bâtiment (Université de Lyon ENTPE), rue Maurice Audin, Vaulx-en-Velin Cedex, 69518, France
{herve.dibenedetto,cedric.sauzeat}@entpe.fr
[3] TOTAL France (Centre de recherche de Solaize), chem. canal, Solaize, 69360, France
{carole.ruot,thibaud.gallet}@total.com

Abstract. A research project on fatigue behavior of bitumens and mastics is developed at University of Lyon, ENTPE/DGCB in collaboration with TOTAL Company. An innovative device, the Annular Shear Rheometer (ASR), is used to perform advanced experimental investigation. It allows practicing fatigue tests, which could be considered as homogenous, on larger scale specimen than traditional other devices. This apparatus allows measuring the linear viscoelastic (LVE) shear complex modulus (G^*) of bituminous materials for small strain amplitudes, and the non linear modulus G_e^* for higher strain levels applied during fatigue tests.

Different types of materials, including pure bitumens and mastics (phase composed of bitumen and filler) are tested. A new measurement dynamic system consisting of wave propagation, equips the ASR. This system allows sending ultrasound at high frequencies (between 100 and 600 kHz). In this paper, the ASR and the wave propagation system, as well as the fatigue testing procedure and the tested materials are presented. During fatigue tests, the evolution of complex modulus from cyclic and dynamic and non linear modulus is observed.

1 Introduction

Bituminous materials are used in multiple areas, especially in road pavements. Repeated loadings in pavement layers cause damage. Durability of road pavement materials is a recurring problem which is an important research topic. Different studies on bituminous mixtures and binders fatigue ([14]; [3]; [16]; [17]) were performed in order to characterize the behavior of these materials with respect to fatigue. This paper presents the experimental device, including the Annular Shear

---

[*] Corresponding author.

Rheometer (ASR) and dynamic waves propagation system. The testing procedure and specific results are developed. The two presented tests have the same characteristics (bitumen, filler, amplitude …). Repeatability is checked. Moreover a comparison between the 2S2P1D model and experimental values is achieved. Then observed fatigue characteristics are discussed.

2 Experimental Campaign

2.1 Presentation of the Annular Shear Rheometer

The principle of the annular shear rheometer (ASR) consists in applying sinusoidal shear stress or sinusoidal shear strain (distortion) on a hollow cylinder of bitumen or mastic, at different temperatures and frequencies. The sample has a rather large size: 5 mm thickness, 95 mm inner diameter and 40 mm height. With such dimensions, the test is homogenous as a first approximation even with aggregate sizes up to 1 millimetre. A schematic view of the apparatus is presented in Figure 1.

The outer duralumin hollow cylinder is screwed on the piston of a 50 kN capacity hydraulic press. The core is linked to a fixed load cell. A sinusoidal cyclic loading is applied in stress or strain mode by means of the controlled hydraulic press movements. Three displacement transducers placed at 120° around the sample measure strain. The mean value of the three measured displacements is used to control the strain. The transducers measure the relative displacements between outer and inner lateral surfaces of the bituminous sample.

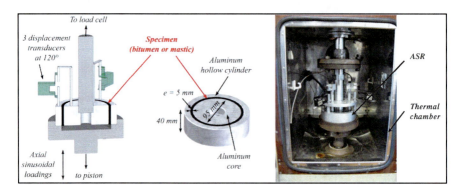

Fig. 1. Schematic view of the annular shear rheometer or ASR (left); picture of the apparatus placed in a thermal chamber (right)

The ASR allows measuring the complex shear modulus (G*). For "small" strain amplitudes, the behavior is linear viscoelastic (LVE) and the modulus is called G^*_{LVE}. For higher strain levels applied during fatigue tests, the measured "equivalent complex modulus" is called G_e^* [16; 17].

Expression of the complex shear modulus is given by equation(1):

$$G^* = |G^*|e^{j\phi} = G_1 + jG_2 \qquad (1)$$

$|G^*|$ is the norm of the complex shear modulus, ϕ is its phase angle, j is the complex number defined by $j^2=-1$. G_1 is the storage modulus and G_2 is the loss modulus. As the ASR is placed in a thermal chamber, G^* can be measured at different temperatures T (from -20°C to 80°C). It can also be measured on a large range of frequencies f (from 0.03 Hz to 10 Hz). $|G^*|$ is the ratio between the amplitudes of distortion γ_A and shear stress τ_A, where $\gamma(t)=\gamma_A \sin(\omega t-\phi)$ is the expression of the distortion signal, $\tau(t)=\tau_A \sin(\omega t)$ is the expression of the shear stress signal and $\omega=2\pi f$ is the pulsation.

A temperature sensor is used for the temperature measurements inside the sample.

2.2 Waves Propagation System

The waves propagation system is used to measure propagation rate "C_S" of shear waves in the bitumen or mastic sample. The propagation speed C_S of a shear wave at a given frequency $f=\omega/2\pi$ in a Linear Viscoelastic (LVE) medium at a temperature T is a function of the value of the LVE complex modulus $G^*(\omega,T)$ at this frequency and temperature and ρ, the density of the material (Eqn 2). If the phase lag at the wave frequency ($f=\omega/2\pi$) is known, measuring the speed of a wave allows to get the value of $|G^*(\omega,T)|$ from equation 2. An interest of such a device is to allow obtaining complex modulus at high frequencies, which is unreachable with classic rheometers. The system has already provided interesting results on bituminous mixture samples [26; 23] and bituminous mastics [11; 16; 17].

$$C_S = \frac{1}{\cos\left(\dfrac{\phi(\omega,T)}{2}\right)} \sqrt{\dfrac{|G^*(\omega,T)|}{\rho}} \qquad (2)$$

Figure 2 shows pictures of the developed waves propagation system. Two piezoelectric sensors are used. The first one is an emitter, fixed on the inner mold. An electric sinusoidal signal, produced by a wave generator, is applied to this sensor, which makes it vibrates. Any excitation frequency can be taken between 200 kHz and 500 kHz, the frequency range of the sensor. In this study we chose 500 kHz. A mechanical shear wave is emitted at the chosen frequency and propagates through the sample. When this wave touches the second sensor, that acts as receiver, it makes it vibrates (Figure 3). The electric signal is then recorded, which enables to obtain the flying time of the wave in the sample.

Because of the lower speed of shear waves compared to compression waves [6], the arrival of the shear wave propagating through the material is detected when the received signal reaches it first local minimum. All the process has been

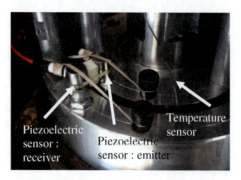

Fig. 2. Picture of ASR and waves propagation system, general view (left), ASR cell equipped with the piezoelectric sensors (right)

carefully evaluated and analyzed in other studies, which gave reasonable results [11; 16; 17].

To find the "real" travel time, it is necessary to subtract the time lag induced by the electronic and mechanic systems. This time lag has been measured on the ASR mold without sample. The shear wave propagation rate C_S is equal to $\Delta t/e$, where Δt is the real travel time and e is the thickness of the sample (Figure 3), which is equal to 5 mm. The value of C_S leads to the determination of shear modulus at 500 kHz (the chosen frequency) $|G^*|_{500\,KHz}$ by applying equation (3):

$$|G^*|_{500KHz} = \rho \left(C_S \cos\left(\frac{\phi_{500KHz}}{2} \right) \right)^2 \qquad (3)$$

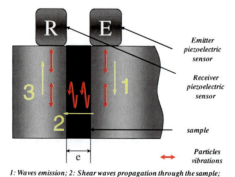

1: Waves emission; 2: Shear waves propagation through the sample; 3: Waves reception

Fig. 3. Schematic representation of the ASR cell during a wave propagations test, emitted and received waves signal during test

2.3 Testing Procedure

The new proposed testing protocol was used in a study of the fatigue behavior of bituminous mastics led at the ENTPE/DGCB laboratory in collaboration with Total company. It is called "advanced fatigue test" [17]. The general testing protocol is presented in Figure 4. Fatigue cyclic loadings in rather "high" strain domain are applied to a sample of bitumen or mastic at a 10 Hz frequency and at a constant regulated temperature (≈10°C). The great originality and improvement of the test consists in the insertion of complex shear modulus measurements periods (in the linear domain) at different time interval during fatigue test. The specimen is then punctually loaded at "low" strain levels. The small strain amplitude cyclic measurements are made at 6 different frequencies (0.03, 0.1, 0.3, 1, 3 and 10 Hz) every 20000 fatigue cycles. Two additional measurements are performed, one before the beginning of the test, which is representative of the undamaged material, and one after the first 10000 fatigue cycles. The testing protocol is associated with the waves propagation system. A dynamic measurement is made every twenty seconds.

The presented results concerns a bitumen composed of a 35/50 penetration grade. The bitumen is called "B3550". Two "advanced fatigue tests" at the same strain level ε_{rz} (= $\gamma/2$) of 0.5%, are presented, B3550D5000 test 1 and B3550D5000 test 2. This strain level is not within the small strain domain (linear domain). Then, the complex modulus measured during the fatigue test G_e^* is a non-linear complex modulus.

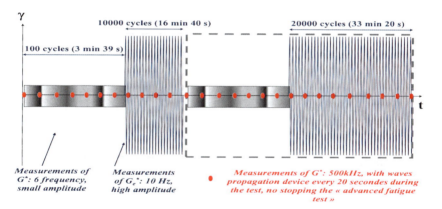

Fig. 4. Schematic representation of the advanced fatigue test protocol with the waves propagation device : G* is measured at 6 frequencies from 0.03 to 10 Hz.

3 Results

3.1 Equivalent Complex Modulus (G_e^*)

Results obtained on G_e^* during the two fatigue tests B3550D5000 test1 and B3550D5000 test2 are both presented Figure 5. The modulus |G_e^*| and the phase lag ϕ_e are plotted as a function of the number of fatigue cycles (N) on graphs a and b respectively. The evolution of G_e^* in Black space ($|G_e^*|$ versus ϕe) is plotted on graph c.

Discontinuities can be seen on the results every 20000 cycles. They are due to the G^* measurement periods (about 3 min 40 seconds), which correspond to quasi-rest periods for the material. The fatigue loading is stopped during those periods, and the material is loaded at low strain levels. So, a recovery process (untreated in this paper) may occur: $|G_e^*|$ slightly increases and ϕ_e decreases. This phenomenon seems partially reversible. Indeed, it can be observed that, after several fatigue cycles following the G^* measurements periods, the measurements points reach a unique curve. It can be noticed that both tests give very close results, which seems to indicate that G^* measurements periods have no influence on the progress of the fatigue process. The good repeatability of the test can also be underlined.

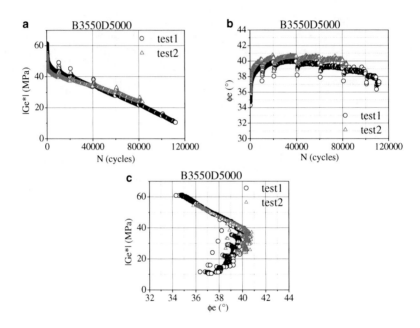

Fig. 5. Equivalent complex modulus $|G_e^*|$ (a) and phase lag ϕ_e (b) as a function of the number of fatigue cycles and representation of G_e^* in Black space (c) Tests B3550D5000 test1 and B3550D5000 test

3.2 LVE Complex Modulus (G*)

As explained in section 2.3, LVE complex modulus G^* is measured at 6 frequencies at the beginning of the test, and then every 20000 fatigue cycles until the end of the test, plus an additional measurement after 10000 cycles. The evolution of G*, at each frequency, is qualitatively the same than G_e^* (Figure 6). At the beginning a decrease of |G*| and an increase of φ can be observed. After |G*| continues to decrease when the phase lag is quasi stable. The main interest of G* measurements during fatigue tests is to allow modeling of the LVE behavior during fatigue process. An analogical model, called 2S2P1D [12; 9; 7; 8], developed at the ENTPE/DGCB laboratory to model LVE behavior of bitumen, mastics and asphalt mixes, is used. Due to the lack of space, this modeling is not developed in this paper and we invite reader to consult given references. G* measurements, at 6 frequencies, give by optimization the constants of the 2S2P1D model. Using the model, it is then possible to plot the whole |G*| master curve (for all frequencies) for each cycle where G* measurements is done (principle is explained in Figure 7). The calculated $|G^*|_{500\ kHz}$, is then compared with the $|G^*|_{500\ kHz}$ value obtained with the waves propagation system at the same fatigue cycle in.

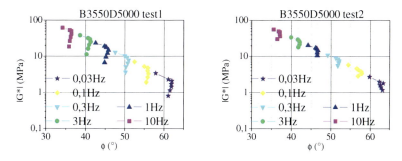

Fig. 6. LVE complex modulus G*, representation of G* in Black space, B3550D5000 test 1 (left) and B3550D5000 test 2 (right)

4 Waves Propagations

As explained in section 2.2, the waves propagation system allows to obtain the value of |G*| at 500 kHz. A special protocol, presented in section 2.3, is used during the two tests. To compare the two linear viscoelastic moduli $|G^*|_{500kHz}$ and $|G_e^*|$ the elapsed time "t", the common data, is used. Cycle equivalent N_{eq} is introduced : $N_{eq}=t*10$ (10Hz = frequency). N_{eq} is different of the number of fatigue cycles N. This difference is the result of small strain measurements period.

The value of the density ρ of the bitumen is equal to 1035 kg.m^{-3}. Modeling back analysis, with 2S2P1D model, shows that, at 500 kHz, the value of the phase lag φ does not vary too much during the considered fatigue tests. Then, it is considered equal to its initial modeled value of 8° during the two tests.

Results obtained on $|G^*|_{500kHz}$ during the two fatigue tests B3550D5000 test1 and B3550D5000 test2 are both presented in figure 9b. The results show a behavior very similar between the two tests. The repeatability observed in part 3.1,

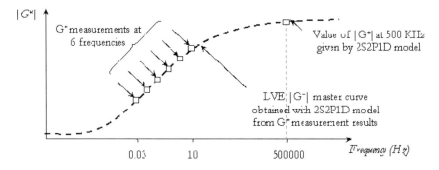

Fig. 7. Schematic |G*| master curve at a given cycle N during the fatigue test, and |G*| at 500 kKHz values obtained from 2S2P1D model. T is fixed at 10°C

with the ASR, can also be underlined for waves propagation measurements. In figure 8, observed evolution of $|G_e^*|$ and $|G^*|_{500kHz}$ are slightly different. At the beginning of the test $|G_e^*|$ decreases quickly. It is not the case for $|G^*|_{500kHz}$. It as been shown in previous studies by the team [10] that the large initial decrease is due to a temperature artifact and to thixotropy. In fact these biased effects do not have the same effect at 10Hz and 500 kHz (on the $|G^*|_{500kHz}$ modulus and $|G_e^*|$ (10Hz modulus)).

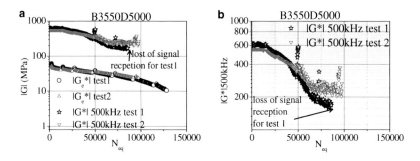

Fig. 8. Complex modulus $|G^*|_{500kHz}$ and equivalent complex modulus $|Ge^*|$ (a) as a function of N_{eq} (=t*10), and zoom on $|G^*|_{500kHz}$ (b)

It can be noticed (figure 9a) that at the beginning of the test, the values of $|G^*|_{500kHz}$ obtained with the waves propagation system (from equation 2) are close to the values given by the 2S2P1D model (calibrated from cyclic quasi-static measurements in the small strain domain). After about 20 000 equivalent cycles (2000 seconds), both procedures give a decrease of $|G^*|_{500\ kHz}$. The gap between the two procedures must be explained by the recovery ,as illustrated with waves propagation measurements given in figure 9b. During the G* measurement period the material is nearly at rest and recovers (mainly because of temperature decrease and thixotropy effect reversibility). This point is not developped in this paper due to the lack of space.

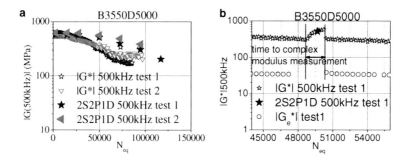

Fig. 9. |G*| at 500 KHz waves propagation and 2S2P1D model as a function of the number of cycles equivalent (a) and a zoom on a G* measurement period with $|G_e^*|$, $|G^*|_{500kHz}$ waves propagation and $|G^*|_{500kHz}$ 2S2P1D model

5 Conclusion

The ASR prototype, developed at the ENTPE/DGCB laboratory, is used to investigate fatigue behaviour of bitumen and mastics and therefore to obtain a better knowledge of fatigue properties of bituminous materials. The addition of a wave propagation system provides new and interesting information about material behavior. The presented results indicate good repeatability of $|G_e^*|$ and $|G^*|_{500kHz}$ measurements, which correspond to complex modulus measured at 10 Hz during fatigue loading and modulus from back analysis of waves propagation tests at 500kHz, respectively. However the evolution of these two moduli with the number of cycles (N) is quite different at the beginning of the fatigue test. Modulus values at 500 kHz, that are much less affected at the beginning of the test by the different biasing effects (temperature and thixotropy), does not show rapid decrease with N. A new protocol consisting in measuring modulus in the small strain domains at different frequencies and different fixed time during fatigue test gives enough information to calibrate a powerful linear viscoelastic model "2S2P1D". A comparison between dynamic modulus at 500 kHz and "2S2P1D" modeling for this frequency gives quite good agreement.

References

[1] Airey, G.D., Behzad, R.: Combined bituminous binder and mixture linear rheological properties. Construction and Building Materials 18, 535–548 (2004)
[2] Baaj, H.: Comportement des matériaux granulaires traités aux liants hydrocarbonés. Mecanique, Energétique, Acoustique et Génie Civil. PhD ENTPE - INSA Lyon. p. 247 (2002) (French)
[3] Bocci, M., Cerni, G., Santagata, E.: Rheological Characterization of the Fatigue Resistance of Asphalt Binders. In: ISAP 2006 - 10th International Conference on Asphalt Pavements, Québec, p. 11 (2006)

[4] Delaporte, B.: Etude de la rhéologie des mastics bitumineux à l'aide d'un rhéomètre à cisaillement annulaire. Mecanique, Energétique, Acoustique et Génie Civil. PhD ENTPE - INSA Lyon, p. 248 (2007) (in french)

[5] Delaporte, B., Van Rompu, J., Di Benedetto, H., Gauthier, G., Chaverot, P.: New procedure to evaluate fatigue of bituminous mastics using an annular shear rheometer prototype. In: 6th RILEM International Conference on Cracking in Pavement, Chicago (2008)

[6] Di Benedetto, H.: Small strain behaviour and viscous effects on sands and sand-clay mixtures. Soil Stress-Strain Behavior: Measurement, Modeling and Analysis. Solid Mechanics and its Applications 146, 159–190 (2006)

[7] Di Benedetto, H., Neifar, M., Sauzéat, C., Olard, F.: Three-dimensional thermo-viscoplastic behaviour of bituminous materials: the DBN model. Road Materials and Pavement Design 8(2), 285–315 (2007)

[8] Di Benedetto, H., Delaporte, B., Sauzéat, C.: Three-dimensional behavior of bituminous materials: experiments and modeling. International Journal of Geomechanics (ASCE) 7, 149–157 (2007)

[9] Di Benedetto, H., Olard, F., Sauzéat, C., Delaporte, B.: Linear viscoelastic behaviour of bituminous materials: from binders to mixes. Road Materials and Pavement Design 5 (Special Issue EATA), 163–202 (2004)

[10] Di Benedetto, H., Nguyen, Q.T., Sauzéat, C.: Nonlinearity, Heating, Fatigue and Thixotropy during Cyclic Loading of Asphalt Mixtures. International Journal of Road Materials and Pavement Design 12(1), 129–158 (2011)

[11] Flohart, L.: Comportement à la fatigue des bitumes et mastics bitumineux (mémoire de Master Recherche). DGCB. ENTPE - Université de Lyon, p. 158 (2008) (French)

[12] Olard, F., Di Benedetto, H.: General "2S2P1D" model and relation between the linear viscoelastic behaviors of bituminous binders and mixes. Road Materials and Pavement Design 4(2) (2003)

[13] Savary, M.: Propagation d'ondes dans les enrobés bitumineux (mémoire de Master Recherche). DGCB. ENTPE - Université de Lyon. p. 159 (2008)

[14] Soenen, H., De La Roche, C., Redelius, P.: Predict Mix Fatigue Tests from Binder Fatigue Properties, measured with DSR. In: 3rd Euraspahlt & Eurobitume Congress, Vienna (2004)

[15] Thom, N.H., Osman, S., Collop, A.C., Airey, G.D.: Fracture and Fatigue of Binder and Binder/filler Mortar. In: 10th International Conference on Asphalt Pavements, ISAP 2006, Québec (2006)

[16] Van Rompu, J., Di Benedetto, H., Gauthier, G., Gallet, T.: New fatigue test on bituminous binders and mastics using an annular shear rheometer prototype and waves propagation. In: 7th International RILEM Symposium ACTBM 2009 on Advanced Testing and Characterization of Bituminous Materials, Rhodes, Greece, pp. 69–79 (2009)

[17] Van Rompu, J.: Etude de la fatigue des liants et mastics bitumineux à l'aide d'un rhéomètre à cisaillement annulaire (PhD thesis). ENTPE - Université de Lyon. Mécanique, Energétique, Génie Civil et Acoustique. Lyon. p. 364 (2010)

Effect of Steel Fibre Content on the Fatigue Behaviour of Steel Fibre Reinforced Concrete

Mofreh F. Saleh, T. Yeow, G. MacRae, and A. Scott

Department of Civil and Natural Resources Engineering, University of Canterbury, Christchurch, New Zealand

Abstract. Rigid pavements are widely used for very heavily trafficked freeways because of their long design period and high performance. Rigid pavements are designed for two modes of failure, namely, fatigue and erosion. Most of the fatigue damage occurs due to very heavy axle loads. In this research, steel fibre was added to Portland cement concrete at 20 kg/m^3 and 60 kg/m^3 to improve fatigue resistance, which could allow for thinner pavements and hence lower construction costs. In addition, the prediction of fatigue life according to the Portland Cement Association and Corps of Engineers models were compared with the measured fatigue of the plain concrete and fibre reinforced concrete. Fatigue tests were carried out using constant stress mode. A range of stresses were applied to cover a range of stress ratios from 0.26 to 0.616. Comparisons between measured fatigue lives and the predicted lives using the Portland Cement Association and Corps of Engineers models have shown that none of these models provided a good match with the measured values. It was found that steel fibres improved fatigue resistance. However, high fibre contents showed detrimental effect on fatigue at high stress ratios.

1 Background

Portland cement concrete is a material widely used in construction, such as pavement and bridge projects. Tensile stresses are developed in the concrete member due to bending under traffic axle loadings. The repeated flexing of the pavement structure causes the development of micro-cracks and these cracks grow over time, resulting in fatigue failure. The use of fibre reinforced concrete (FRC) is steadily increasing. FRC is likely to be a reasonably cost-effective alternative due to its high fatigue performance and strength characteristics. However, the available design methods and guidelines for fatigue in rigid pavements are based on conventional unmodified concrete. Therefore, there is a need to study the effect of fibre reinforcing on the fatigue properties of the concrete and to compare this against existing concrete fatigue models. The subject of fibre reinforcement and fatigue behaviour of concrete has been studied by several researchers. Gao and Hsu [1] found that the mechanism of fatigue could be divided into three stages.

The first stage, termed as flaw initiation, involves flaws forming within weak regions of the concrete. The second stage, known as microcracking, is the slow growth of flaws to a critical size. The final stage occurs when continuous or macrocracks form, eventually leading up to failure.

Cornelissen and Reinhardt [2] showed that the first and last stage makes up approximately 10% of the total curve each, while the second stage accounts for the remaining 80%. It has been found by Hordijk [3] that the slope of the second stage can be correlated with fatigue life. Zhang [4] showed that the internal structural degeneration of concrete may be characterised by the presence of micro-defects, such as gel pores. With increasing number of loading cycles, the development of micro-defects results in a decrease in strength and stiffness of the concrete.

Hsu [5] showed that low-cycle fatigue (temperature variation related fatigue) and high-cycle fatigue (i.e. traffic loadings) had different mechanisms. Low-cycle fatigue mechanism tends to be formed by the formation of mortar cracks which leads to a network of continuous cracks. High-cycle fatigue however tends to produce bond cracks in a slow and gradual process. Saito and Imai [6] found that there was no clear fatigue limit for concrete. This means that there were no stress level which gave an infinite fatigue life for concrete. In fact the results of this research agree with Saito and Imai findings as there was no endurance limit at low stress ratio as indicated by the Portland Cement Association model.

Zhang and Stang [7] found that steel fibre reinforced concrete (SFRC) can sustain significantly more damage and strain than plain concrete, and hence is more ductile. Grzybowski and Meyer [8] deduced that, although fibres retard the growth of cracks, that they can also increase the amount of pores and initial microcracks. Work done by Cachim [9] has shown that it is possible for the flaws introduced by fibre addition to outweigh the benefits. Thus, the size and quantity of fibres are important. Yin and Hsu [10] showed that the presence of fibres enhanced the fatigue behaviour for low cycle fatigue but do not seem to have any effect on high cycle fatigue. This was due to fibres being able to increase fatigue life in part of mortar cracking, but unable to do so for bond cracking.

Overall it was found that the addition of fibre reinforcing can increase the fatigue performance of the concrete under flexural fatigue loading. The explanation for this, explained by Barr and Lee [11], was that, under tensile forces, the fibres would be able to bridge cracks and prolong fatigue life. Falkner et al. [12] conducted extensive work on the fatigue performance of SFRC road pavements. They tested several concrete slabs in bending with a subgrade foundation to model road pavements under repeated axle loading. They found that the load capacity did not increase, however the rupture load capacity increased and the fibres were shown to retard crack growths. The experiment was specifically conducted to investigate the effect the addition of fibre reinforcing on typical road pavements used in Germany. No fatigue models for SFRC were developed from this experiment.

2 Plain Concrete Fatigue Models

There are several fatigue models developed mainly for plain concrete. Examples of these models include the Zero-Maintenance Design Fatigue Model [13], ERES/COE Fatigue Model [14] and the Portland Cement Association (PCA) model [15]. The PCA model is quite popular and it has been adopted in the PCA rigid pavement design procedure which is the basis for the Austroads rigid pavement design method [15]. In the following sections, only the ERES/COE and PCA models are discussed. The ERES/COE model was developed from Corps of Engineers data from 51 full-scale field sections conducted between 1943 and 1973. Darter [14] obtained the relationship shown in Equation 1, where N is the number of load cycles to failure. This model was originally intended for airport pavements only, but has shown good results in other applications.

$$\log N = 2.13\ SR^{-1.2} \tag{1}$$

SR is given by Equation 2 and is termed as the stress ratio which is defined as the ratio of the maximum tensile stress, σ, to the flexural strength (modulus of rupture) of concrete, σ_{MR}.

$$SR = \frac{\sigma}{\sigma_{MR}} \tag{2}$$

The PCA fatigue model assumes that, for stress ratios below 0.45, that the fatigue life of the concrete is infinite. Packard and Tayabji [16] recommended the PCA model shown in Equation 3.

$$N = \text{unlimited} \qquad \text{for } SR \leq 0.45 \tag{3a}$$

$$N = \left[\frac{4.2577}{SR - 0.4325}\right]^{3.268} \qquad \text{for } 0.45 < SR < 0.55 \tag{3b}$$

$$\log N = 11.737 - 12.077\ SR \qquad \text{for } SR \geq 0.55 \tag{3c}$$

3 Objectives of the Research

The objectives of this research is (i) to establish fatigue models for fibre reinforced concrete that best model localized materials that are currently in use in New Zealand and (ii) to compare the models against the fatigue behaviour of the fibre reinforced concrete with plain concrete and the commonly used models such as PCA and ERES/COE models.

4 Specimen Preparation, Material Properties and Experimental Setup

4.1 Mix Design and Fresh Concrete Properties

In this research, three concrete mixes were designed with different dosages of steel fibres. The target 28 day compressive strength of the concrete is 30 MPa. The properties of the fresh concrete such as slump and viscosity were measured to ensure mix workability. The three types of mixes are:

- Control mix; plain concrete with no fibre.
- Concrete mix with steel fibre content of 20 kg/m^3.
- Concrete mix with steel fibre content of 60 kg/m^3.

The steel fibre mixes were compared to the control mix which did not contain any fibre, with all the mixes having a constant water/cement, (w/c) ratio of 0.6. Details of the optimized mix designs are provided in Table 1. In addition to the measured slump, the fresh properties of the concrete were determined using a coaxial cylinder viscometer to obtain the yield shear stress and plastic viscosity of each mix. The results for the slump, yield shear stress and plastic viscosity are provided in Table 2.

Table 1. Concrete mix design

Mix Number	30/0	30/20	30/60
GP Cement	280	287	308
Water	168	172	185
13mm agg	1100	1040	750
Sand	868	906	1129
Fibres	0	20	60
Water reducing admixture (L)	750	1150	1650
Super plasticizer (L)	0.4	0.5	0.8
Retarder (L)	1.12	1.148	1.232
Density	2416	2424	2433

Table 2. Slump, yield shear stress and plastic viscosity

Mix Number	30/0	30/20	30/60
Slump (mm)	167	127	105
τ (Pa)	204	440	324
μ (Pa.s)	44	58	64

The addition of 20 kg steel fibre required only minor modifications to the mix design, through the removal of 60 kg of coarse aggregate, to produce a material with similar fresh properties as shown in Table 2. The addition of 60 kg steel fibre however resulted in the removal of approximately 30% of the coarse aggregate which was replaced with sand and a slight increase in the paste content. Despite attempts to maintain similar workability there was an increase in both the shear stress and plastic viscosity of the mixes containing steel fibre compared to the control.

4.2 Hardened Concrete Properties

The target flexural strength (modulus of rupture) is 4.5 MPa. The steel fibres used in the specimens are the 80/60 Dramix fibre from Bekaert. This means that the length of fibres is 60 mm and the aspect ratio (length to diameter of fibre) is 80. The fibre is low carbon and has a minimum tensile strength of 1050 MPa. There are three properties which are of importance; compressive strength, modulus of rupture and fatigue properties. Concrete cylinders with dimensions of 100 mm diameter and 200 mm height were casted for the compressive strength test. For the modulus of rupture test, rectangular beams with dimensions of approximately 150 mm depth, 150 mm width and 450 mm length were prepared. For the fatigue test, concrete slabs with dimensions of 400 mm length, 245 mm width and 75 mm depth were casted. Each slab was then sawed into three smaller beams with dimensions of 400 mm length, 75 mm width and 70 mm depth. All specimens were cured in humid room for 28 days.

4.3 Compressive and Flexural Strength Results

Three cylinders and two beams were tested for compressive strength and modulus of rupture tests respectively. The results are as shown in Table 3. It can be seen that, while the average compressive strength results were similar, the average rupture of modulus has more variability. This could be due to casting the beams on separate days, leading to variability in environmental/climate conditions.

Table 3. Results of 28 Day Compression and Modulus of Rupture Tests

Steel Fibre Dosage	Compressive Strength (MPa)				Modulus of Rupture (MPa)		
	1	2	3	Average	1	2	Average
0	42.3	38.8	32.4	37.8	3.74	3.72	3.73
20 kg/m^3	43.1	32.5	34.7	36.8	3.61	3.45	3.53
60 kg/m^3	38.0	34.7	37.8	36.8	4.85	4.46	4.65

4.4 Fatigue Test Setup

The concrete beams were loaded under four point loading, subjecting to repeated cyclic loading using sinusoidal load pulses at 10 Hz frequency. The test setup is as shown in Figure 1.

Fig. 1. Fatigue Test Set Up

The method used to determine the fatigue life of the samples was the constant stress method. In this method, the load applied to the beam was kept constant until failure.

The fatigue test using constant stress mode was carried out for 20 concrete beams prepared with different percentages of steel fibre content in addition to control concrete beams. The stress level applied on the beams span a wide range of stress ratios from as low as 0.26 to as high as 0.616. It should be noted that the models developed are valid within the tested range of stress ratios; any extrapolation outside this range will likely to provide inaccurate predictions.

5 Effect of Fibre Content on Fatigue Behaviour

Figure 2 shows comparison between the measured fatigue lives of the control concrete. It is obviously clear both the PCA and the ERES/COE are over estimating fatigue lives at low stress ratio. In fact the PCA predicts unlimited fatigue values at stress ratios lower than 0.45 which is not the case with the tested samples. This finding agrees with Saito and Imai [6]. ERES/COE model provided a better match with the measured values at stress ratios greater than 0.4, however, it highly overestimates the fatigue lives at stress ratios lower than 0.4. There is no model developed for plain concrete beams in this research because of the limited data available.

Effect of Steel Fibre Content on the Fatigue Behaviour

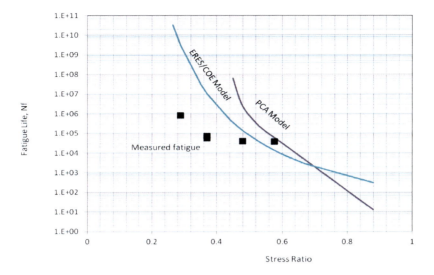

Fig. 2. Comparisons between fatigue behaviour of control concrete, ERES/COE and PCA fatigue models

The fatigue results for the concrete mix reinforced with 20kg/m³ steel fibre were modelled with exponential model shown in Equation 4 and portrayed in Figure 3. The model represented by Equation 4 provided a reasonable match to the measured fatigue values compared to the PCA and ERES/COE models. The coefficient of determination of the models, R^2, is 0.91. The PCA and ERES/COE deviated significantly from the measured fatigue values for the fibre reinforced concrete mix with 20 kg/m³. The PCA and ERES/COE overestimated the fatigue lives at stress ratios lower than 0.6. However, at stress ratios higher than 0.6 the developed fatigue model showed a higher fatigue lives compared to the PCA and ERES/COE fatigue models. The higher fatigue lives at higher stress ratios of the steel fibre reinforced mix could be attributed to the delay in crack propagations offered by the steel fibre matrix. In addition, the measured fatigue values for the 20 kg/m³ reinforced concrete at low stress ratio agrees with the results measured for the plain concrete which are much lower than the PCA and ERES/COE predictions. After reaching the fatigue failure, it was noted that fibre reinforced concrete beams did not snap into pieces such as the plain concrete beams. The fibre reinforcement holds the beam parts together. In the field, fibre reinforcement is expected to keep good interlocks between the rigid pavement slab parts post fatigue cracking.

$$N_f = 283623 * e^{-2.7146 * SR} \qquad R^2 = 0.91 \qquad (4)$$

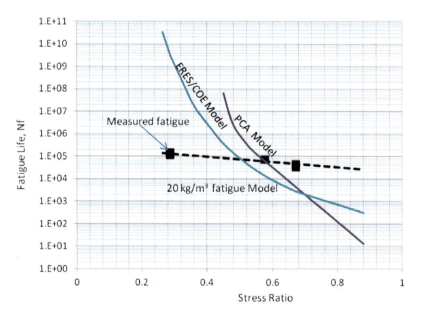

Fig. 3. Comparisons between fatigue behaviour of the 20 kg/m³ fibre reinforced concrete and ERES/COE and PCA

Figure 4 shows the measured fatigue values for the 60 kg/m³ and comparisons with the PCA and ERES/COE models. It is again clear that the PCA and ERES/COE do not provide any good matching with the measured fatigue. The PCA significantly overestimated the fatigue values at stress ratios lower than 0.6 and underestimated fatigue values at stress ratios over 0.6. ERES/COE overestimated the fatigue values at stress ratios lower than 0.55 and underestimated fatigue values at stress ratios greater than 0.55. Equation 5 provided good fit to the measured fatigue values with a coefficient of determination, R^2, equal 0.83. In this model nine fatigues values were used, however, more data will be required to provide validation to the model.

$$N_f = 598300.6 * e^{-3.94069*SR} \qquad R2 = 0.83 \qquad (5)$$

Figure 5 compares the fatigue lives predicted by the two models shown in Equations 4 and 5 for the 20 kg/m³ and 60 kg/m³ fibre reinforced concrete respectively. It appears from Figure 5 that the higher doses of fibre has only tangible effect on the fatigue lives at lower stress ratio (SR<0.6). At stress ratios higher than 0.6 the higher doses of fibre showed detrimental effect on the fatigue lives. This could be explained by the higher content of pores that can be created by the higher fibre content; consequently, this can lead to weaker areas in the concrete and therefore accelerating the crack initiation and propagations leading to the shorter fatigue lives. Therefore, it is clear that there is an optimum fibre content that can maximize the fatigue properties; beyond, this optimum the fatigue life will be compromised.

Effect of Steel Fibre Content on the Fatigue Behaviour

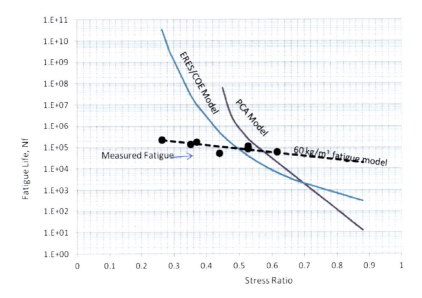

Fig. 4. Comparisons between fatigue behaviour of the 60 kg/m3 fibre reinforced concrete and ERES/COE and PCA

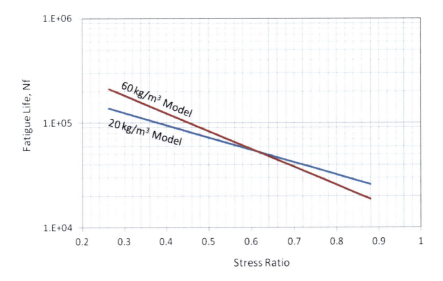

Fig. 5. Comparison between fatigue lives of the 20kg/m^3 and 60kg/m^3 fibre reinforced concrete

6 Conclusions

Fatigue behaviour of the plain concrete and steel fibre reinforced concrete has been studied in this research. Two doses of steel fibre were used in this research, 20 kg/m^3 and 60 kg/m^3. The target 28 days compressive strength for the concrete is 30 MPa. While the compressive strength of the concrete was insensitive to the fibre content, the flexural strength has shown considerable improvement at higher doses of the fibre content. Fatigue behaviour was investigated using four point bending beam fatigue and constant stress model. A wide range of stress ratios were used to examine the relationship between the stress ratio and fatigue life. The PCA and ERES/COE fatigue models showed poor correlations with both the plain and fibre reinforced fatigue values. The PCA model assumes fatigue endurance limit at stress ratio less than 0.45. However, the actual test results showed a limited fatigue life for both plain and fibre reinforced concrete at stress ratio less than 0.45. Both PCA and ERES/COE over estimated fatigue lives at lower stress ratio for plain and fibre reinforced concrete. Both PCA and ERES/COE underestimated fatigue lives at stress ratios higher than 0.55 for fibre reinforced concrete. Comparing the 20 kg/m^3 and the 60 kg/m^3 fibre reinforced concrete, it was clear that the higher fibre content only beneficial at lower stress ratio, however, at stress ratios higher than 0.6, the higher fibre content is detrimental to the fatigue resistance.

Acknowledgement. The authors would like to thank Mr. Alan Ross, the business development manager of Bekaert OneSteel Fibres Australasia, BOSFA, for his support in providing the required steel fibre and providing research articles that helped in this research. The authors are also grateful to Mr. John Kooloos for his technical support in the Transportation laboratory that led to the development of the beam fatigue loading gear.

References

[1] Gao, L., Hsu, T.C.C.: Fatigue of concrete under uniaxial compression cyclic loading. ACI Materials Journal 95(5), 575–581 (1998)
[2] Cornelissen, H.A.W., Reinhardt, H.W.: Uniaxial tensile fatigue failure of concrete under constant-amplitude and programme loading. Magazine of Concrete Research 36(129), 216–226 (1984)
[3] Hordijk, D.A.: Local approach to fatigue of concrete, PhD thesis, Delft University of Technology, p. 210 (1991)
[4] Zhang, B.: Relationship between pore structure and mechanical properties of ordinary concrete under bending fatigue. Cement and Concrete Research 28(5), 699–711 (1998)
[5] Hsu, T.C.C.: Fatigue and microcracking of concrete. Materials and Structures 17(97), 51–54 (1984)
[6] Saito, M., Imai, S.: Direct tensile fatigue of concrete by the use of friction grips. ACI Journal 80, 431–438 (1983)
[7] Zhang, J., Stang, H.: Fatigue performance in flexure of fibre reinforced concrete. ACI Materials Journal 95(1), 58–67 (1998)

[8] Grzybowski, M., Meyer, C.: Damage accumulation in concrete with and without fibre reinforcement. ACI Materials Journal 90(6), 594–604 (1993)
[9] Cachim, P.B.: Experimental and numerical analysis of the behaviour of structural concrete under fatigue loading with applications to concrete pavements. PhD Thesis, Faculty of Engineering of the University of Porto, p. 246 (1999)
[10] Yin, W., Hsu, T.C.C.: Fatigue behaviour of steel fibre reinforced concrete in uniaxial and biaxial compression. ACI Materials Journal 92(1), 71–81 (1995)
[11] Barr, B.I.C., Lee, M.K.: Dynamic Analysis of Cracked Sections (Literature Review), Report from Test and Design Methods for Steel Fibre Reinforced Concrete, Technical report, EU Contract-BRPR-CT98-813, University of Wales, Cardiff, UK (2001)
[12] Falkner, H., Teutsch, M., Klinkert, H.: Load bearing capacity and deformation of dynamically loaded plain and steel fibre reinforced concrete road pavements. iBMB University Brunswick (1997)
[13] Darter, M.I., Barenberg, E.J.: Design of Zero-Maintenance Plain Jointed Concrete Pavement, Volume 2 – Design Manual. Report FHWA-RD-77-112 FHWA, US Department of Transportation (1977)
[14] Darter, M.I.: A Comparison Between Corps of Engineers and ERES Consultants, Inc. Rigid Pavement Design Procedures, Technical Report Prepared for the United States Air Force SAC, Urbana, IL (1988)
[15] Austroads Guidelines, Guide to Pavement Technology, Part2: Pavement Structural Design (2009) ISBN 978-1-921329-51-7
[16] Portland Cement Association, Thickness Design for Cocnrete Highway and Street Pavements, Portland Cement Association, Skokie, Ill (1984)
[17] Packard, R.G., Tayabji, S.D.: New PCA Thickness Design Procedure for Concrete Highway and Street Pavements. In: Concrete Pavement & Rehabilitation Conference, Purdue, USA (1985)

Effect of Specimen Size on Fatigue Behavior of Asphalt Mixture in Laboratory Fatigue Tests

Ning Li[1], A.A.A. Molenaar[1], A.C. Pronk[1], M.F.C. van de Ven[1], and Shaopeng Wu[2]

[1] Road and Railway Engineering, Faculty of Civil Engineering and Geosciences,
Delft University of Technology, Delft, the Netherlands
[2] State Key Laboratory of Silicate Materials for Architectures,
Wuhan University of Technology, Wuhan 430070, Hubei, P. R. China

Abstract. Laboratory fatigue testing has been extensively used to estimate the resistance to fatigue cracking of asphalt mixtures. Researchers developed a number of test methods to estimate the fatigue behavior of an asphalt mixture. However, based on the classical fatigue analysis, fatigue lives obtained from different test devices are not comparable even when they are performed under the same mode of loading and environmental conditions for the same material. The differences in specimen geometry and load configuration will result in different stress-strain distributions inside the tested specimens leading to differences in local fatigue damage.

In this paper, the size effect on the fatigue life is investigated. Uniaxial tension and compression (UT/C) fatigue tests were carried out using cylindrical specimens with three different sizes. Size effect on the fatigue test results are compared and analyzed using the partial healing (PH) material model, which describes the change of the complex modulus for a unit volume due to loading. The fatigue failure points determined by the classic approach, dissipated energy ratio and the PH model were compared. The results show that fatigue behavior of the specimens in the UT/C fatigue tests can be fitted excellently by the PH model, including estimation of the endurance limit. It is concluded that fatigue behavior from the UT/C fatigue test is not dramatically influenced by specimen size.

1 Introduction

Various fatigue test devices are currently used to evaluate the fatigue performance of asphalt concrete and are accepted in the European standard EN 12697-24 [1]. The two-point bending (2PB) test with trapezoidal specimen was adopted by the researchers from Shell [2] and LCPC [3]. The Shell Laboratory at Amsterdam has used the three-point bending loading equipment to estimate the fatigue life [2]. In the USA [4] and the Netherlands [5], the four-point bending test (4PBT) is specified. In the UK and Sweden, the standard fatigue test is the indirect tensile fatigue test (ITFT). The Nottingham Asphalt Tester (NAT) was specially designed for this test [6]. Di Benedetto et al, [7] reported an interlaboratory investigation.

Eleven different test methods, including uniaxial tension/compression (UT/C), bending and indirect-tension tests, were used to evaluate fatigue properties of an asphalt mixture. The results showed that fatigue test results obtained from different test equipments are difficult to compare. Two main reasons for this are the different stress-strain distributions in the samples and the different fatigue analysis approaches, as discussed below.

The (internal) stress-strain distribution in a specimen is determined by the loading conditions, geometry and dimensions. When this stress-strain field is constant everywhere in the specimen, the test is considered to be homogeneous. For non-homogeneous tests, such as the beam bending tests and the indirect tensile test, the stress-strain field is not uniform along the specimen and cross section. During testing, the local stiffness does not decrease at the same rate for every unit of volume in the specimen. The back calculated stiffness using the measured deflection and applied load is not a material property but a specimen stiffness, which depends on geometry and dimension of the specimen. Therefore, with these test results it is not possible to directly evaluate the fatigue performance of materials. In the UT/C fatigue test, the stress-strain field is in theory uniform over the length and the cross area. In that case, the back calculated stiffness is in principle a material property [8].

In the classical fatigue analysis, the point of failure is normally defined as the moment at which the stiffness of the specimen is reduced to 50% of its initial value. In fact this definition is based on an empirical analysis and sensitive to the loading condition and the geometry of specimen. Currently some other fatigue models were proposed. Based on the ratio of dissipated energy theory [9], Shen [10] developed the PV model, which can predict the fatigue life from the material properties and loading conditions. A mechanical damage model was proposed by Bodin [11] based on the non-local damage theory. This damage model has been used to describe the local complex modulus decrease induced by microcrack development. Kim [12] and Lundstrom [13] applied the work potential theory to simulate the damage evolution under cyclic tests. However, most of these fatigue damage models do not take into account the test type and the specimen size. The partial healing (PH) model proposed by Pronk [14] has been proven to be a good material model, making it possible to simulate the evolution of a material property for a unit of volume.

This paper focuses on the influence of the specimen size on the fatigue behavior in the UT/C fatigue test based on the PH model. In addition, the different fatigue life definitions are also compared.

2 Theory of the Partial Healing (PH) Model

The PH model developed and modified by Pronk [8, 14-15], is a material model that describes the evolution of the complex modulus and phase angle for a unit volume during a fatigue test. During a cyclic loading test, energy is dissipated into the device and the specimen due to the visco-elastic behavior. It is assumed that the total energy includes three parts:

(1) System losses ΔW_{sys}: The system losses ΔW_{sys} caused by the test setup can be ignored if the test machine is good enough.

(2) Visco-elastic losses ΔW_{dis}: ΔW_{dis} is the area of the stress-strain loop and generally represented in Eqn. (1). This part of the energy is completely transformed into heat and increases the temperature of specimen.

$$\Delta W_{dis} = \pi \cdot \sigma_i \cdot \varepsilon_i \cdot \sin(\varphi_i) \tag{1}$$

where σ_i is the stress amplitude at cycle i, MPa; ε_i is the strain amplitude at cycle i, m/m; φ_i is the phase angle at cycle i, °.

(3) Fatigue consumption ΔW_{fat}: The fatigue consumption ΔW_{fat} is the main reason for the decrease of stiffness during the fatigue test. In the proposed model, the mathematical formulation of this part is modeled as a very small part of the visco-elastic losses ΔW_{dis}. In a strain controlled mode, ΔW_{fat} is expressed by Eqn. (2):

$$\Delta W_{fat} = \delta \cdot \Delta W_{dis} = \delta \cdot \pi \cdot \sigma_i \cdot \varepsilon_i \cdot \sin(\varphi_i) = \delta \cdot \pi \cdot S_i \cdot \varepsilon_0^2 \cdot \sin(\varphi_i) \tag{2}$$

where δ is the very small value, S_i is the stiffness modulus at cycle i, MPa; ε_0 is the strain amplitude, m/m; φ_i is the phase angle at cycle i, °.

A new parameter, the stiffness damage Q, is introduced, which relates to the fatigue consumption ΔW_{fat}. The damage factor Q reduces the stiffness modulus, including the loss modulus F and the storage modulus G, following by Eqn. (3) and (4), respectively.

$$F\{t\} = S \cdot \sin\varphi = F_0 - \int_0^t \frac{dQ\{\tau\}}{d\tau} \left[\alpha_1^* e^{-\beta(t-\tau)} + \gamma_1^*\right] \cdot d\tau \tag{3}$$

$$G\{t\} = S \cdot \cos\varphi = G_0 - \int_0^t \frac{dQ\{\tau\}}{d\tau} \left[\alpha_2^* e^{-\beta(t-\tau)} + \gamma_2^*\right] \cdot d\tau \tag{4}$$

where F_0 is the initial loss modulus, MPa; G_0 is the initial storage modulus, MPa; t is the testing time, s; $\alpha_{1,2}^*$, β and $\gamma_{1,2}^*$ are the model parameters.

The damage initiated at a certain moment will partially diminish in time; this is expressed by means of the exponential function. One can denote this as Partial Healing (PH). In a strain controlled mode, the fatigue damage rate is given by Eqn. (5):

$$\frac{d}{dt}Q = \frac{d}{dt}W_{fat} \approx \frac{\Delta W_{fat}}{\Delta t} = \delta \cdot \frac{\pi \cdot S_i \cdot \varepsilon_0^2 \cdot \sin(\varphi_i)}{\Delta t} = \delta \cdot \pi \cdot f \cdot \varepsilon_0^2 \cdot F_i \tag{5}$$

where Δt is the time duration in a cycle, s; F_i is the loss modulus at cycle i, MPa; f is the frequency, Hz.

Eqn. (5) is substituted into Equation (3) and (4). The solutions for the UT/C fatigue test are given as follows:

$$\alpha_{1,2} = \delta \cdot \pi \cdot f \cdot \varepsilon_0^2 \cdot \alpha_{1,2}^*; \quad \gamma_{1,2} = \delta \cdot \pi \cdot f \cdot \varepsilon_0^2 \cdot \gamma_{1,2}^* \tag{6}$$

$$B = \frac{\alpha_1 + \beta + \gamma_1}{2}, C = \sqrt{B^2 - \beta\gamma_1}, D = \frac{\beta - B}{C}, E = \frac{B - \gamma_1}{C} \tag{7}$$

$$F\{t\} = S \cdot \sin\phi = F_0 e^{-Bt}\left[Cosh\{Ct\} + DSinh\{Ct\}\right] \quad (8)$$

$$G\{t\} = G_0 - F_0 \begin{bmatrix} \dfrac{\alpha_2}{C} e^{-Bt} \cdot Sinh\{Ct\} + \\ \dfrac{\gamma_2}{\gamma_1}\left(1 - e^{-Bt} \cdot \left[Cosh\{Ct\} + ESinh\{Ct\}\right]\right) \end{bmatrix} \quad (9)$$

3 Experimental Work

3.1 Materials

A dense asphalt concrete DAC 0/8 with a maximum aggregate size of 8mm was used and designed in accordance with the Dutch RAW specifications [16]. The aggregate consists of Scottish crushed granite, Norwegian Bestone and crushed sand. A 40/60 penetration grade bitumen was used with a design binder content by mass of 6.5%. Wigras 40K was used as filler. Three different sizes (0.5, 1 and 1.5) were used, in which size 1 corresponds to the standard size. The dimensions of the specimens with different sizes are presented in Table 1. The target air voids content is 3.5%.

Table 1. Dimensions of the cylindrical specimen

Specimen size	Diameter [mm]	Height [mm]
0.5	25	62.5
1	50	125
1.5	75	175

3.2 Test Procedures

The larger specimens (size 1 and 1.5) and the smaller specimens (size 0.5) were tested in the MTS and UTM-25 test machine, respectively. The UT/C fatigue test set-ups, shown in Figure 1, consist of a rigid frame in a temperature-controlled

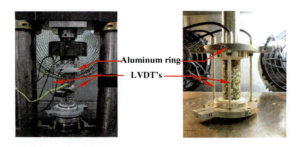

Fig. 1. The UT/C fatigue test setups in MTS (left) and UTM-25 (right)

cabinet. The axial deformation is measured via three LVDT's, which are fixed in an aluminum ring and placed around the specimen. The fatigue test was conducted at the temperature of 20°C and the frequency of 10Hz. During the test, the specimens were subjected to a continuously sinusoidal axial loading in both tension and compression. The tests were performed in strain controlled mode.

Three different fatigue criteria are used in this study.

(1) $N_{f,50}$: the point of failure is defined as the moment at which the back calculated stiffness of the specimen was reduced to 50% of its initial value

(2) N_R: the fatigue life is defined as the point at which the slope of the dissipated energy ratio (DER) versus number of load cycles deviates from a straight line [17]. DER is the ratio of the accumulated dissipated energy up to cycle N and the dissipated energy in cycle N, given as Eqn. 10.

$$DER = \frac{\sum_{n=1}^{n=N} w_i}{w_N} \qquad (10)$$

(3) N_{PH}: the fatigue life is defined as the point where the measured stiffness deviates from the fitted evolution based on the PH model.

4 Results and Discussion

4.1 Determination of the Model Parameters

From Figure 2, three phases of stiffness response can be observed during the test. In the beginning the complex stiffness decreases rapidly. After a short number of cycles, the decline rate becomes constant. At the end, the stiffness drops quickly again. It is assumed that in this phase the material starts to disintegrate. The

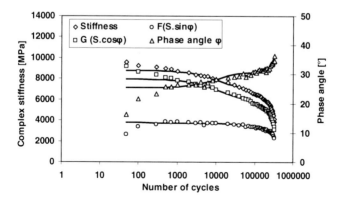

Fig. 2. Measured and predicted evolutions of stiffness and phase angle of the specimen C-3-26

evolution of the phase angle shows an opposite trend. The model parameters are determined, by minimizing the differences between the measured and fitted values for F, G, φ and S in the interval from N=1,000 to N=N$_r$. They are presented in table 2. In the first and the second phase, the fitted line corresponds very well with the measured values.

Table 2. The PH model parameters

	Sample code	Average ε$_0$ [μm/m]	δα$_1$*	δα$_2$*	δγ$_1$*	δγ$_2$*	β [10^{-5} s^{-1}]
Size 0.5	C-3-26	130	0	687	19.6	54.7	104
	C-3-4	146	0	474	36.9	89.7	123
	C-3-7	162	0	1521	42.1	115.7	300
	C-3-14	168	0	1226	47.7	126.7	322
	C-3-13	188	0	1927	53.1	127.3	407
Size 1	C-6-11	112	0	936	18.4	44.1	76.6
	C-8-11	144	0	1829	26.4	78.5	236.4
	C-8-12	174	0	1191	47.3	107.5	172.1
	C-7-5	240	0	2414	90.7	288.9	657.1
	C-7-4	283	0	1230	226.9	514.6	919
Size 1.5	C-5-1	90	0	1014	9.3	20.8	56.5
	C-5-2	120	0	1113	47	78.2	74.3
	C-4-2	143	0	2460	33.7	87.1	244.9
	C-4-1	182	0	2993	113.8	321.9	644.7
	C-9-2	231	0	2251	158	417.9	612.3

4.2 Size Effect on the Endurance Limit

In the model, the parameters δγ$_1$* and δγ$_2$* represent the irreversible damage in the loss and storage modulus, respectively. Figure 3 shows that the irreversible damage is higher at a higher strain level. A linear relationship between the parameters δγ$_{1,2}$* and strain level is assumed. The regression fits for sizes 0.5 and 1 are nearly the same. For size 1.5, the slope of the regression line is much higher, because the specimen with larger volume has a higher chance to create weak spots during testing.

$$\delta\gamma_{1,2}^* = k_2 \cdot (\varepsilon - \varepsilon_{limit}) \tag{11}$$

The value of ε$_{limit}$ represents the intersection point of the straight line with the x axis. The strain range between the two intersection points of δγ$_{1,2}$* indicates the existence of an endurance limit, because the irreversible damage is zero below this strain range. Table 3 presents the regression equations of the parameters δγ$_{1,2}$* and the calculated ranges of the endurance limit. The specimen size does not significantly influence the endurance limit range, which is between 85 to 98 μm/m.

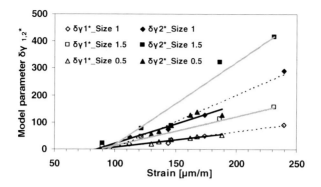

Fig. 3. Relationship between $\delta\gamma_{1,2}{}^*$ and strain level

Table 3. The relationship between $\delta\gamma_{1,2}{}^*$ and strain level

		Regression equation	Endurance limit range
Size 0.5	$\delta\gamma_1{}^*$	$\delta\gamma_1{}^* = 0.6658 \times (\varepsilon - 96)$	85~96 μmm/mm
	$\delta\gamma_2{}^*$	$\delta\gamma_2{}^* = 1.4288 \times (\varepsilon - 85)$	
Size 1	$\delta\gamma_1{}^*$	$\delta\gamma_1{}^* = 0.5928 \times (\varepsilon - 91)$	91~98 μmm/mm
	$\delta\gamma_2{}^*$	$\delta\gamma_2{}^* = 1.9580 \times (\varepsilon - 98)$	
Size 1.5	$\delta\gamma_1{}^*$	$\delta\gamma_1{}^* = 1.0750 \times (\varepsilon - 88)$	88~94 μmm/mm
	$\delta\gamma_2{}^*$	$\delta\gamma_2{}^* = 3.0440 \times (\varepsilon - 94)$	

4.3 Size Effect on the Fatigue Life

As mentioned above, three different fatigue life definitions are used in this study. An example is given in Figure 4. Table 4 presents all three fatigue lives. In nearly all cases the difference between the N_R and N_{PH} is much smaller compared to difference between N_R and the traditional fatigue life, $N_{f,50}$. For an asphalt mixture, the classical Wöhler curve of strain versus fatigue life is generally regarded as a straight line in a double logarithmic coordinate system.

$$N = k \cdot \varepsilon_0^b \quad (12)$$

where N is the fatigue life; ε_0 is the strain amplitude, μmm/mm; k and b are the material coefficients.

All the regression equations and material coefficients of size 0.5, 1.0 and 1.5 are gathered in Table 5. It seems that the fatigue lines do not differ significantly between the different specimen sizes. In Figure 5, the data points from different specimen sizes are combined to one fatigue line. The R^2 value of this fatigue function is 0.98. Therefore, in spite of the differences in $\delta\gamma_1$ and $\delta\gamma_2$ for the size 1.5 with respect to the sizes 0.5 and 1, the size effect on the fatigue life can be ignored for the UT/C fatigue test.

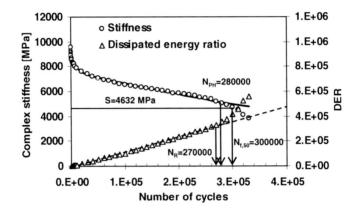

Fig. 4. Different fatigue life definitions for specimen C-3-26

Table 4. Comparison of the different fatigue life definitions

	Sample code	Average ε_0 [μm/m]	N_{PH}	N_R	$N_{f,50}$
Size 0.5	C-3-26	130	280000	270000	300000
	C-3-4	146	115000	100000	136000
	C-3-7	162	78000	74000	80000
	C-3-14	168	70000	64000	78000
	C-3-13	188	66000	70000	70000
Size 1	C-6-11	112	530000	550000	610000
	C-8-11	144	165000	160000	170000
	C-8-12	174	82000	82000	84000
	C-7-5	240	23000	22000	18500
	C-7-4	283	9200	8900	8000
Size 1.5	C-5-1	90	1260000	1230000	1380000
	C-5-2	122	360000	360000	370000
	C-4-2	143	210000	200000	170000
	C-4-1	187	34000	32000	29000
	C-9-2	231	17000	16000	13000

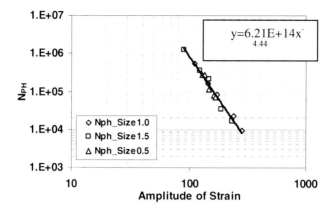

Fig. 5. Fatigue lives N_{PH} of the specimens with different sizes

Table 5. Regression equation and material coefficients of the fatigue lines

Specimen size	Material coefficients		R^2
	k	b	
Size 0.5	5.73E+14	-4.43	0.86
Size 1	2.50E+14	-4.24	1.00
Size 1.5	2.67E+15	-4.74	0.98
Size 0.5+1+1.5	6.21E+14	-4.44	0.98

5 Conclusions

Based on the results presented in this paper the following conclusions can be drawn:

1. The PH model provides a good prediction for the evolutions of the complex stiffness and phase angle. The results from the uniaxial tension and compression fatigue test can be directly used for the determination of model parameters.
2. The model parameter $\delta\gamma_1^*$ and $\delta\gamma_2^*$ can be used to determine the range of endurance limit, and this range does not change significantly with the increase of the specimen size.
3. Compared to the $N_{f,50}$, the fatigue lives determined by DER and the PH model are more close to each other.
4. For the UT/C fatigue test, the size effect is not significant for the fatigue life. It is possible to combine the fatigue results from the different specimen sizes into a unique fatigue line using the fatigue life N_{PH}.

References

1. European committee for standardization, Bituminous Mixtures-Test Methods for Hot Mix Asphalt, BS EN 12697: Part 24: Resistance to Fatigue. CEN, Brussels (2004)
2. van Dijk, W.: Practical fatigue characterization of bituminous mixes. In: Proceedings of the Association of Asphalt Paving Technologists, p. 38 (1975)
3. Bonnot, J.: Asphalt aggregate mixtures, Transportation Research Record 1096, Transportation Research Board, pp. 42–50 (1986)
4. Strategic Highway Research Program (SHRP), Fatigue response of asphalt-aggregate mixes, Executive summary, National Research Council (1992)
5. Pronk, A.C.: The theory of the four point dynamic bending test-Part 1. Report P-DWW-96-008, Delft University of Technology, the Netherlands (1996)
6. Brown, S.F.: Practical test procedures for mechanical properties of bituminous materials. In: Proceeding of ICE Transport, vol. 111, pp. 298–297 (1995)
7. Di Benedetto, H., de la Roche, C., Baaj, H., Pronk, A.: Fatigue of bituminous mixtures. Materials and Structures 37(3), 202–216 (2004)
8. Pronk, A.C.: Partial Healing, A new approach for the damage process during fatigue testing of asphalt specimen. In: Proceedings of the Symposium on Mechanics of Flexible Pavements, ASCE, Baton Rouge (2006)
9. Hopman, P.C., Kunst, P.A., Pronk, A.C.: A renew interpretation method for fatigue mea-surements, verification of Miner's rule. In: Proceedings of the 4th Eurobitume Symposium, Madrid, pp. 557–561 (1989)
10. Shen, S., Carpenter, S.H.: Development of an Asphalt Fatigue Model Based on Energy Principles, vol. 76. Association of Asphalt Paving Technologists (AAPT) (2007)
11. Bodin, D., Pijaudier-Cabot, G., De La Roche, C., Piau, J.M., Chabot, A.: Continuum damage approach to asphalt concrete fatigue modeling. Eng. Mech. 130(6), 700–708 (2004)
12. Kim, Y.R., Lee, H.J., Little, D.: Fatigue characterization of asphalt concrete using viscoelasticity and continuum damage theory, vol. 66, pp. 520–569. Association of Asphalt Paving Technologists (AAPT) (1997)
13. Lundstrom, R.: Characterization of Asphalt Concrete Deterioration Using Monotonic and Cyclic Tests. Pavement Engineering 4(3) (2003)
14. Pronk, A.C.: Partial healing in fatigue tests on asphalt specimen. Road Materials and Pavement Design 4(4) (2001)
15. Pronk, A.C., Molenaar, A.A.A.: The Modified Partial Healing Model used as a Prediction Tool for the Complex Stiffness Modulus Evolutions in Four Point Bending Fatigue Tests based on the Evolutions in Uni-Axial Push-Pull Tests. In: Proceedings of the 11th Int. Conf. on Asphalt Pavements, Nagoya, Japan (2010)
16. CROW, In: Standaard RAW Bepalingen, CROW, Ede (2005) (in Dutch)
17. Pronk, A.C., Hopman, P.C.: Energy dissipation: the leading factor of fatigue. In: Proceedings of the Conference on the United States Strategic Highway Research Program, pp. 255–267 (1991)

Long-Life Overlays by Use of Highly Modified Bituminous Mixtures

D. Simard[1] and François Olard[2]

[1] R&D project manager – Central
 Laboratory – EIFFAGE Travaux Publics, France,
 delphine.simard@eiffage.com
[2] R&D project manager – Research and Development
 Division – Eiffage Travaux Publics, France
 françois.olard@eiffage.com

Abstract. Polymer modified asphalt mixtures have usually been used in wearing courses in order to improve both crack growth resistance and rutting performance where temperatures, vertical stresses and shear strain levels are more severe. Nonetheless, the need for either thinner yet high-performing wearing courses or ever-increasing durability provides the motivation for using higher polymer contents in the wearing course of bituminous pavements.

Therefore, instead of the conventional use of 2-3% styrene-butadiene-styrene (SBS) in polymer modified binders (PMB's), the resort to higher SBS contents in the range 6-7% allows for a phase inversion in the PMB microstructure: the swollen polymer becomes the continuous phase in which asphaltene nodules are dispersed. This significant change in PMB microstructure brings about significantly higher performances. Besides, some other benefits may be related to the possible layer thickness reduction: less natural materials (aggregate, bitumen) used, less resources required for construction (man-hours, emissions during transport and laying) and, overall, cost saving.

Microstructure, both empirical and rheological characteristics were investigated in laboratory for two different PMB's with very high SBS content (referred to as Biprene® or Orthoprene®). In addition, in-situ testing was carried out: a brief follow-up of the highly trafficked Millau Viaduct surfacing (constructed in France in 2004) where this type of PMB was used, is in particular proposed.

The paper illustrates that the proposed innovative highly modified bituminous mixes may be from now on potentially considered as a relevant solution for sustainable long-life and high-performance overlays, needing only rare surface maintenance.

1 Introduction

1.1 Technical Background on Highly Polymer Modified Binders (HPMB's)

The development of highly polymer modified binders (HPMB's) and the use of additives (such as thermoplastics) is very much linked with the development of new mix designs for bituminous mixes for thin surfacings, wich provide improved practical qualities and durability. Indeed, when using HPMB's what is looking for is phase inversion with polymer continuous phase. Fluorescence microscopy is the most frequently used technique for assessing the state of dispersion of the polymer and bitumen phases (Brion and al) [1]. It is based on the principle that the polymers, swollen by some of the constituents of the bitumen to which they have been added, fluoresce in ultraviolet light. They emit yellow-green light while the bitumen phase remains black (Bouldin and al) [2] (Figure 1).

Fig. 1. Microstructure of bitumen / polymer blends: a) Continuous bitumen phase b) Bitumen with inter-twisted phases c) Continuous polymer phase (AIPCR) [3]

Thermoplastics triblock copolymers such as Styrene Butadiene Styrene (SBS) have a styrenic endblock (PS) and a rubbery midblock (PB). In asphalt blends, the PS endblocks are forming spherical micelles acting as physical crosslinks while the PB part is swollen by the maltene part of asphalt (saturates, aromatics and resins) (Adedeji and al) [4] (Kraus and al)[5]. The variation of their chemical structure, specifically total styrene and also the 1,2- or 1,4-butadiene vinyl contents can modify the inversion phase (Hernandez and al) [6]. Indeed, with higher styrene content (between 40 and 50%) systems are too rigid and a homogeneous dispersion is difficult to achieve. Usually, SBS with between 25 and 35% of styrene are used to allow an equilibrium point, improving the compatibility between SBS and asphalt.

This physical network can be enhanced by adding a crosslinker thus creating a chemical network between the PB parts. Indeed, macromolecular chains are linked together thanks to sulfur forming a tri dimensional network [7]. This new sulfur based network, enhances mechanical and thermal properties of PMB, in particularly, crosslinking enables to stabilize polymer in the bitumen giving outstanding storage stability properties.

Usually, phase inversion occurs when adding around 5% of SBS but this rates also depends on the base bitumen (grade an nature) and if the binder is crosslinked or not (Figure 2) (Planche and al) [8]. Hence, as SBS swollen in the maltenes part of the bitumens, in the past, HPMB's were mainly developed using high grade base bitumen.

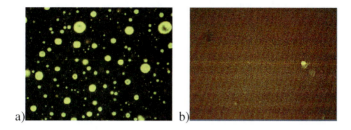

Fig. 2. 5% polymer a) physical blend b) Crosslinked blend

1.2 Technical and Environment Stakes

Polymer Modified Binders (PMB's) started being used extensively about thirty years ago with the quest for improving the mechanical performance of bituminous pavements particularly on the wearing course such as better resistance to rutting and reduction in reflective cracking. This development came in response to traffic increase, to reduce maintenance periods, which are a major source of costly traffic disturbance. In this new area of sustainable development where materials enabling durability enhancement are being asked for, highly polymer modified binders (HPMB's) are being developed. Moreover, the ever-increasing production of porous asphalts and open-graded asphalts revolves around an increase in the use of PMB's.

The principal cause of bituminous binder aging in service is commonly known to be oxidation by the oxygen from the air of certain molecules resulting in the formation of highly polar and strongly interacting oxygen containing functional groups. The chemistry of asphalt oxidation reactions is based on a dual sequential hydrocarbon oxidation mechanism (Petersen) [9] leading to the formation of both ketones and sulfoxides which are the major oxidation products. In the latter stages of oxidation in highly oxidized asphalts, dicarboxylic anhydrides and carboxylic acids are being formed in a lesser amount. However, during road service life, (i) standard binder characterizations showed that plain asphalt is much more affected by aging than the PMB and (ii) crosslinked PMB feature a significantly lower oxidation degree as measured by Fourier Transformed InfraRed Spectroscopy (Dressen and al) [10]. This can be explained by a very homogenous polymer repartition in the binder matrix (Mouillet and al) [11].

Traffic and weather conditions are also affecting binder aging. Nevertheless, the great chemical stability of the SBS polymer is responsible for its reliable long-term performances and great durability (Gallet and al.) [12]. Indeed, Gel Permeation Chromatograpy GPC linked to IRTF studied over 20 years showed that only the really thin surface (~15mm) is affected whereas the inferior layer remains intact giving the great durability to the PMB over years.

In an environmental context, what is mainly looking for is the reduction of the emissions of polycyclic aromatic hydrocarbons (PAHs) and semivolatile organic compounds (SVOCs) during the manufacturing stage. As said in technical background, the SBS swollen in the maltenes part of asphalt (saturates, aromatics and resins) leading to a two phases composed PMB: a first one inflated by bitumen maltenes and a second one containing all the bitumen components which have not been absorbed by polymer. As a result, the more important the SBS content is the more emissions could be limited due to volatile compounds from binder trapping by polymer (Gaudefroy and al) [13].

1.3 Objective

In order to develop long-life overlays, Eiffage Travaux Publics attempts to develop high-performance binders with polymer continuous phase. This can be easily achieved with high grade bitumen base and very high SBS content. The objective of this work is to obtain the inversion phase with all grades of bitumen and with the optimum SBS content.

2 Experimental Laboratory Study

Three highly polymer modified bitumens (HPMB's) with 6% of SBS triblock copolymer and using 3 different base bitumens (hard, medium and soft) were studied at EIFFAGE Travaux Publics central laboratory in Corbas (France). All of them are cross-linked. Their respective compositions are proprietary. These three HPMB's are referred as to PMB.A, PMB.B and PMB.C (cf. Table 1).

2.1 Empirical Tests

Properties of HPMB's were obtained doing empirical tests and following European norms (Table 1): penetration grade (NF EN 1426) [14], ring and ball softening point (NF EN 1427) [15], elastic recovery (NF EN 13398) [16], Brookfield viscosity (MOPL 102) [17], storage stability (NF EN 13399) [18] and Fraass breaking point (NF EN 12593) [19].

Table 1. Empirical tests on HPMBs

Effet of asphalt on HPMBs @ 6% SBS cross-linked			
Properties	PMB.A	PMB.B	PMB.C
Penetration (dmm)	30	39	62
Ring & ball softening point (°)	88	86	92.5
Elastic recovery (%)	88	90	97
Viscosity at 160°C (Po)	5.8	4.5	5.51
Storage stability test (%)	91	100	99
Fraass breaking point (°C)	-13	-13	-20

Adding 6% of SBS in a hard bitumen grade is audacious; penetration, R&B softening point and elastic recovery are acceptable however viscosity is high and storage stability in the vicinity of boundaries. When working with softer bitumens with still a high level of polymer, viscosity is more acceptable and storage stability remains unaffected. Indeed, when the maltene content increase meaning when the grade of bitumen became softer, polymer is more swollen leading to better properties such as outstanding elastic recovery and fraass breaking point with very soft bitumen.

2.2 Microscopy on Binders

Fluorescence measurements of HPMB's were carried out at EIFFAGE Travaux Publics laboratory with a Zeiss Axioskop microscope. The microscope is equipped with a specific filter set composed of: a blue violet excitation filter (395-440nm), a beam splitter (FT 460nm) and an emission filter (LP 470nm). The light source is based on mercury and emits in the all wavelengths (from 350 to 650nm). The principle of fluorescence is based on electronic transitions between an exited singular state and a fundamental state; according to the Jablonski diagram.

With the fluorescence microscopy technique and using this specific filter set, the asphalt rich phase appears dark, while the polymer-rich phase appears light.

The three blends made in laboratory were studied using the microscope and the evolution of the results in function of the bitumen grade is presented in Figure 3.

At a constant polymer rate of 6%, when the maltene content increase, microstructure is changing. With hard bitumen the continuous phase seems to be bitumen (case a). Indeed, there is a black continuous phase and the high level of polymer appears very shiny. When increasing the oil content and working with softer bitumen, the inter-twisted phase is seen (case b); yellow and black parts are slightly melted. However, with very soft bitumen and a highly oil content, the continuous phase become clearly the polymer (case c) as black spots are defined among a large yellow continuous phase. This microstructure evolution highlights the role playing by maltene part in the swollen polymer.

Fig. 3. Microstructure of bitumen / polymer blends

2.3 Rheological Tests on Binders

Complex modulus tests were performed at EIFFAGE Travaux Publics laboratory using a Dynamic Shear Rheometer DSR (Physica 501 Anton Paar) over a frequency range from 1 to 100Hz and a temperature range from -30°C to 70°C. The geometry is an 8mm diameter plate-plate configuration. Thus, thanks to the shift factor, master curves were obtained on modified bitumen (Figure 4). Then, complex modulus and Cole-Cole diagrams were obtained (Figure 5). Some fatigue tests were also carried out using the same DSR at 10°C-25Hz in order to rank HPMB's faced with the base bitumen used and the rate of polymer (Figure 6). This work has been done for PMB.A and further work is needed for the others HPMB's.

The complex modulus of a PMB is mainly influenced by the modulus of the polymer in the high-temperature and low frequency domain; as used in this case, the thermoplastic SBS has a phase angle near 0°. Those three HPMB's have the same level of SBS but not the same asymptotic elastic behaviour illustrating the influence of the maltene content on the SBS. All of the three curves are quite similar however, with a stiffer base binder, the phase angle is lower. This can be explain by the fact that in softer bitumen polymers are better swollen and thus give better elasticity to the binder. As regards fatigue curves it is shown that the increase of polymer improve the fatigue resistance at 1 million cycle.

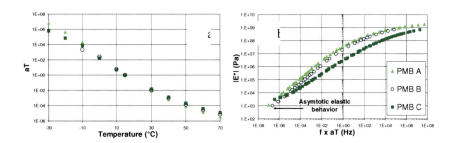

Fig. 4. Rheological results a) Shift Factor; b) Master curves at 15°C

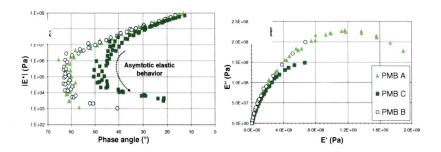

Fig. 5. Rheological results a) Black diagram; b) Cole Cole Diagram

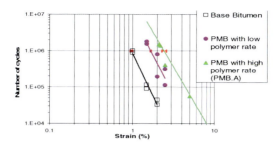

Fig. 6. Fatigue resistance curves

3 Development and Large Scale Roadworks with HPMB's

Polymer modified asphalt mixtures have usually been used in wearing courses in order to improve both crack growth resistance and rutting performance where temperatures, vertical stresses and shear strain levels are more severe. Nonetheless, the need for either thinner yet high-performing wearing courses or ever-increasing durability provides the motivation for using higher polymer contents in the wearing course of bituminous pavements.

The previous laboratory results presented in section 2 were found very encouraging and led to many large scale roadworks with HPMB's. The two following sub-sections 3-1 and 3-2 present two case studies:

- section 3.1 presents the case study of the Millau viaduct surfacing with a soft bitumen modified with more than 7% of cross-linked SBS (proprietary composition);
- section 3.2 deals with the case study of a highly trafficked bus lane in Lyon city center for which a medium pen grade bitumen was modified with more than 5% of cross-linked SBS (proprietary composition).

3.1 Design of a Specific Bituminous Surfacing for Orthotropic Steel Bridge Decks: The Millau Viaduct

Both the geometry of the structure and the very high flexibility of metallic plates make the deformations and stresses very severe in steel bridge surfacings (Figure 7). In particular, the repeated loading make the fatigue strength be an important parameter for the design of such bituminous wearing courses. In addition, these specific surfacings must also have durability over the expected temperature range: it must be resistant to thermal cracking at low temperature and to rutting at high temperature. The technical studies led in parallel to the construction of the Millau Viaduct (France) –the tallest bridge in the world– have provided in particular the opportunity of new progress in the development of appropriate laboratory testing equipment and of original highly polymer modified surfacing [20][21][22]. Indeed, EIFFAGE Travaux Publics pulled the socks up and led a comprehensive research program including a large laboratory testing campaign. Those studies led to an HPMB named Orthoprène® applied successfully on the bridge in December 2004.

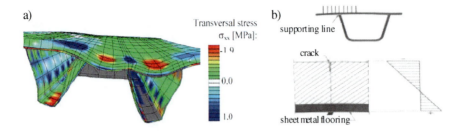

Fig. 7. a) Example of deformed bridge deck (250x) with transversal stress (Huurman 2003); b) Bituminous mix on orthotropic plate (Méhue 1981) [23]

Microstructure tests as explained previously (2.2. Microscopy on binders) were hence performed on Orthoprène® (Figure 8 – a) as well as on the coated aggregate (Figure 8 –b). Photographs were carried out in 2 dimensions even for the aggregate (explaining trouble parts on the photographs).

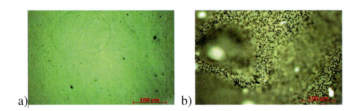

Fig. 8. Microstructure observations a) Soft bitumen modified with Orthoprène® b) Arvieu aggregate coated by Orthoprène®

The microstructure of Orthoprène®, consisting of a soft bitumen modified with more than 7% cross-linked SBS, highlights a polymer continuous phase. Indeed, black spots representing the asphaltenes are surrounded by a homogeneous yellow phase representing the SBS swollen in the maltenes part of the bitumen. This particularly microstructure gives outstanding properties (resistance to rutting and fatigue tests...) to the asphalt concrete. Besides, photographs taken on the coated aggregate show that the modified binder is homogeneously spread at the surface of the aggregate.

Deformation levels are illustrated in Figure 9. According to Pouget et al. [24][25][26], maximum strain magnitudes may reach 500 10^{-6} m/m, obviously depending upon the temperature and loading rate.

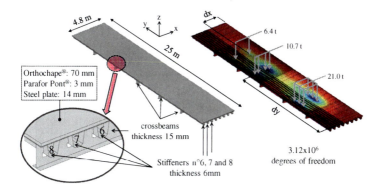

Fig. 9. Geometry, mesh, load and vertical displacement field of the Millau Viaduct structure in the Finite Element Code

Furthermore, the high fatigue resistance of this highly SBS modified surfacing was studied with the two-point bending fatigue test; carried out following the NFEN12657-24 standard at 10°C-25Hz. For strain amplitudes below 350.10^{-6}m/m, there is no failure observed up to 10 million cycles (a plateau is observed), which indicate that 350.10^{-6}m/m is the so called "endurance limit", an outstanding value for such bituminous material.

3.2 Development of a New HPMB for Lyon City Center – France 2011

Making roads in city center such as Lyon is challenging as even by night the traffic can't be stop; buses and trolleys are still working until one in the night. Thus, in order to achieve a sustainable wearing course (less maintenance periods) and resistant to highly trafficked such as bus lane, a medium pen grade bitumen modified with 5% of cross-linked SBS was developed.

In this street, 2000m² (or 280t) of the BBSG 0/10mm with granitic aggregates (Lafarge La Patte at St Laurent de Chamousset 69 France) were implemented with a thickness of 5cm. This asphalt concrete was prepared with 5.1% of the innovative HPMB described previously.

After planing the previous surface course, asphalt concrete was implemented on the paving slab just after the laying off a classic tack coat. The laying off the asphalt concrete was done with a traditional compaction train. It was observed a good behaviour of asphalt concrete during lay off and at compaction. Regarding surface texture, the asphalt concrete was great looking. Thus, this experiment was successfully done and traffic (car, buses and trolleys) started back just after the first way was done in order not to stop traffic.

After one month under traffic, the asphalt concrete behaviour is quite well but a much more accurate follow-up of this work site is planned during the next few years. Indeed, resistance to rutting and observations will be studied over months and years to confirm the great durability of this new HPMB confronted with a very aggressive traffic (particularly from low vehicles such as buses and trolleys).

As previously, microscopy was studied on the HPMB (Figure 10 – a) used to make the experiments presented in this paper. Regarding the aggregates, photographs are also carried out in 2 dimensions (Figure 10 –b).

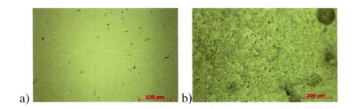

Fig. 10. Microstructure observations a) medium pen grade bitumen modified with 5% cross-linked SBS b) Coated aggregate using this HPMB

The microstructure of the binder used for this experiment and done with medium pen grade bitumen modified with 5% of cross-linked SBS, highlight a polymer continuous phase. Indeed, black spots representing the asphaltenes are surrounding by a homogeneous yellow phase representing the SBS swollen in the maltenes part of the bitumen. This particularly microstructure gives outstanding properties to the asphalt concrete. Besides, photographs taken on the aggregate show that the modified binder is homogeneously spread at the surface of the aggregate.

4 Conclusions and Perspectives

Nowadays, the need for ether thinner yet high-performing wearing courses or ever-increasing durability provides the motivation for using higher polymer contents in the wearing course of bituminous pavements. Besides, it has been proven that working with HPMB helps reducing environmental impact and also has longer term durability than usual PMB with less oxidation and less aging of the binder. As only the really thin surface of the road is the most affected to aging, using HPMB is very useful.

HPMB's developed at Eiffage Travaux Publics (Biprene® and Orthoprene®) enabled to work on two particularly outstanding cases studies: the Millau Viaduct realised in 2004 and Lyon city center in 2011. In the first case, studies were made with a soft bitumen with more than 7% of polymer in order to have good fatigue resistance test. New HPMB developed in 2011 with a hard bitumen and more than 5% of polymer, gives both rigidity and good fatigue test resistance in the same binder; this is very encouraging. In both cases, the inversion phase is obtained and highlights a polymer continuous phase. Moreover, as hard bitumens are cheaper than soft bitumens, HPMB in hard bitumen is not only technically outstanding but also economically interesting.

To answer and respect sustainable development demands, HPMB are still being improved in Eiffage Travaux Public laboratories, in order to reach more performing and long lasting binders.

References

[1] Brion, Y., Brule, B.: Etude des mélanges bitumes-polymères. Composition, structure et propriétés. Bulletin LCPC, Paris, p. 123, rapport PC-6 (1986)
[2] Bouldin, M.G., Collins, J.H., Berker, A.: Rheology and microstructure of polymer / asphalt blends. Rubber Chemistry and Technology 64, 577 (1990)
[3] AIPCR Association mondiale de la route - Used of modified bituminous binders, special bitumens and bitumens with additives in road pavements (1999)
[4] Adedeji, A., Grunefelder, T., Bates, F.S., Macosko, C.W.: Asphalt modified by SBS triblock copolymer: structure and properties. Polymer Engineering and Sci. (12), 1707–1733 (1996)
[5] Kraus, G.: Modification of asphalt by block polymers of butadiene and styrene. Rubber Chemistry and Technology 55, 1389–1402 (1982)
[6] Hernandez, G., Medina, E., Sanchez, R., Mendoza, A.: Thermomechanical and rheological asphalt modification using styrene butadiene triblock copolymers with different microstructure. Energy & Fuels 20, 2623–2626 (2006)
[7] Société Anonyme d'Application des Dérivés de l'Asphalte SAADA. Composition de vulcanisation, procédé pour sa préparation, et son utilisation dans les liants routiers. Trinh Cu Cuong et Million Denis. EP 0 299 820 (June 21, 1988)
[8] Planche, J.-P.: Special features of polymer modified binders. In: Petersen Asphalt Research Conference Symposium on Additives (2004), http://www.petersenasphaltconference.org/download/2004/14Planche.pdf
[9] Petersen, J.C.: A dual sequential mechanism for the oxidation of petroleum asphalts. Petroleum Science and Technology 16 (9-10), 1023 (1998)
[10] Dressen, S., Ponsardin, M., Planche, J., et al.: Durability study: field aging of conventional and polymer modified binders. TRB, Annual Meeting (2010)
[11] Mouillet, V., Lamontagne, J., Durrieur, F., Planche, J.-P., Lapalu, L.: Infrared microscopy investigation of oxidation and phase evolution in asphalt modified with polymers. Fuel 87, 1270 (2008)
[12] Gallet, T., Dressen, S., Dumont, A.-G., Pittet, M.: Evolution à long terme de la structure chimique d'un bitume modifié SBS. RGRA (890) (December 2010, January 2011)
[13] Gaudefroy, V., Olard, F., Beduneau, E., De La Roche, C.: Influence of the low-emission asphalt LEA® composition on total organic compounds emissions using the factorial experimental design approach. In: Enviroad Congress (2009)
[14] NF EN 1426. Détermination de la pénétrabilité à l'aiguille (Septembre 2009)
[15] NF EN 1427. Détermination de la température de ramollissement – Méthode Bille et Anneau. Février (June 9, 2011)
[16] NF EN 13398. Détermination du retour élastique des bitumes modifiés (Aout 2010) (in French)
[17] MOPL 102. Notice d'utilisation d'un viscosimètre Brookfield. Eiffage Travaux Publics (Mars 2006) (in french)

[18] NF EN 13399. Détermination de la stabilité au stockage des bitumes modifiés (Aout 2010)
[19] NF EN 12593. Détermination du point de fragilité Fraass (Mars 2010)
[20] Héritier, B., Olard, F., Saubot, M., Krafft, S.: Bituminous wearing course on steel deck – Orthochape®: outstanding technical solution for the Millau Viaduct surfacing. RGRA (2004)
[21] Héritier, B., Olard, F., Loup, F., Krafft, S.: Design of a specific bituminous surfacing for the world's highest orthotropic steel deck bridge. Transportation Research Record: Journal of the Transportation Research Board, No. 1929
[22] Olard, F., Héritier, B., Loup, F., Krafft, S.: New French standard test method for the design of surfacing on steel deck bridges: case study of the Millau Viaduct. Road Materials and Pavements Design 6 (2005)
[23] Méhue, P.: Platelages métalliques et revêtements de chaussées. Bull. de liaison des Laboratoires des Ponts et Chaussées (111) (1981)
[24] Pouget, S., Sauzéat, C., Di Benedetto, H., Olard, F.: Modeling of viscous bituminous wearing course materials on orthotropic steel deck. Materials and Structures (2011)
[25] Huurman, M., Medani, T.O., Scarpas, A., Kasbergen, C.: Development of a 3D-FEM for Surfacings on Steel Deck Bridges. In: International Conference on Computational & Experimental Engineering (2003)
[26] Pouget, S.: Influence des propriétés élastiques ou viscoélastiques des revêtements sur le comportement des ponts à dalle orthotrope PhD ENTPE-INSA, p. 254 (2011)

Investigation into Tensile Properties of Polymer Modified Bitumen (PMB) and Mixture Performance

E.T. Hagos[1], M.F.C. van de Ven[2], and G.M. Merine[1]

[1] Gebr. van der Lee V.O.F., Lelystad, The Netherlands
[2] Faculty of Civil Engineering, Delft University of Technology, Delft, The Netherlands

Abstract. Pavement performance in relation to cracking and durability of asphalt mixtures is largely dependant on the low temperature characteristics of the bitumen. For this reason, the use of modified binders is usually adopted to improve the low temperature performance and in this way to extend the pavement life. In this study the low temperature tensile properties of two types of modified bitumen were investigated with the Force-ductility test method. In addition, other rheological properties of the PMBs, such as the complex modulus G* and phase angle δ, were determined using the Dynamic Shear Rheometer (DSR). The results indicate that there is a trade-off between the elasticity and stiffness properties of the modified binders. Tests conducted on an asphalt mixture with the PMB binders show that the effect of the modification resulted in a lower initial stiffness but higher fatigue life compared to a reference mixture in a strain controlled fatigue test.

1 Introduction

Improved low temperature properties of a bitumen will extend the service life of pavements. In other words, the use of polymer modified bitumen (PMB) improves the crack resistance of asphalt mixtures which usually is a problem at low pavement temperatures. It is also recognized that the use of PMBs in asphalt mixtures improves the resistance to permanent deformation (rutting) at high temperatures and pavement fatigue life at intermediate temperatures [1-3]. Moreover, the improved elastic properties of modified bitumen remain in general constant during the service life of the pavement [4].

In this study the enhanced performance of asphalt mixtures with modified bitumen is investigated by studying the low temperature elastic behaviour (toughness) and other fundamental characteristics of the binder together with mixture properties. In this case, laboratory prepared modified binders were tested along with commercially available PMBs to examine their properties. The Force-ductility test method and the Dynamic Shear Rheometer (DSR) results showed enhanced binder characteristics. The results indicate a trade-off between the elasticity/ductility and stiffness properties of the modified binders. Tests conducted on an asphalt mixture with the PMB binder show that the effect of the modification resulted in a lower initial stiffness but higher fatigue life compared to a reference mixture in a strain controlled fatigue test. In other studies it is reported that not only the fatigue properties but also the permanent deformation characteristics are improved [2]. Hence, possible increase of the overall pavement thickness with PMB mixtures will depend on the need to increase the overall fatigue resistance of the pavement structure, but not to overcome permanent deformation in the asphalt layers at high temperatures.

2 Materials and Testing Plan

2.1 Material Preparation

Two types of Polymer Modified Bitumen (PMB1 and PMB2) each with 6% polymer content were produced in the laboratory. PMB1 was produced with linear SBS polymer and PMB2 with a combination of linear SBS and EVA polymers. An additive was used to improve the storage stability of the PMBs. In addition to the laboratory prepared modified binders, commercially available PMB samples were also tested. The two types of modified binders in the laboratory were made from a 70/100 pen bitumen. An overview is given in Table 1.

Table 1. Materials used in the research

Material	Type	Description
Bitumen	70/100 pen	Penetration grade bitumen
Polymer	SBS (Kraton D1101*)	Styrene-Butadiene-Styrene * Styrene 31%
	EVA (Polybilt)	Ethylene-Vinyl-Acetate
Additive	Code name: Z	Added to improve storage stability

Laboratory prepared materials:
1. PMB-L1: Modified bitumen with SBS polymer
2. PMB-L2: Modified bitumen with SBS+EVA polymers

Commercially available PMB products:
1. PMB-C1: SBS modified commercial PMB
2. PMB-C2: SBS+EVA modified commercial PMB

The procedure adopted to produce the PMBs in the laboratory was as follows:
- The base bitumen was heated to its EVT temperature and kept at that temperature for about 3 hours.
- Then a high shear mixer was used to mix the polymer in the base bitumen.
- The polymer was added at low shear rate and the shearing was gradually increased to 5000 rpm.

- The mixing temperature was maintained at 170-190°C during the blending of the polymer in the base bitumen. The shearing was conducted for 1 hour.
- An additive was added after the polymer was thoroughly blended in the base bitumen and was stirred at a lower shearing rate.

Standard AC 22 base asphalt mixtures were prepared to carry-out fundamental performance tests on the mixtures. A mixture with 50% recycled material (RAP) and 70/100 pen bitumen was used as refernce mixture. The other two mixtures were produced without RAP. One mixture was made with PMB-L1 and the other mixture with PMB-C1 binder.

2.2 Testing Plan

The testing plan for the PMBs included the determination of empirical and fundamental properties. The tests performed are described below.

1. Empirical tests
    a. Penetration at 25°C (NEN-EN 1426)
    b. Softening point (NEN-EN 1427)
    c. Elastic recovery at 25°C (NEN-EN 13398)
    d. Force-ductility at 5°C, 50 mm/min (NEN-EN 13589)
2. Fundamental tests
    a. Complex modulus G^* and phase angle δ (DSR)
        - Temperature
            - With 8 mm parallel plates: -10, 0, 10, 20, 30°C
            - With 25 mm parallel plates: 30, 40, 46, 58, 70, 82°C
        - Frequency 0.1 – 100 rad/s

Tests on asphalt mixtures were conducted in accordance to the test methods specified in the "Standard RAW bepalingen" [5]. The test conditions for stiffness and fatigue tests are described below.

1. Stiffness test
    - Four-point-bending (4PB) test
        - Inner clamp 120 mm, Outer clamp 420 mm
    - Beam dimension 450 x 50 x 50 mm (L x W x H)
    - Temperature 20°C
    - Strain 50 µm/m
    - Frequency 0.1, 0.2, 0.5, 1, 2, 5, 8, 10, 20, 30 Hz

2. Fatigue test
    - Frequency 30 Hz
    - Temperature 20°C
    - Strain – three strain levels 120, 150, 180 µm/m

3 Test Results

The test results of the PMBs are shown in Table 2.

Table 2. Test results of the PMBs

	Norm NEN-EN	Laboratory PMBs		Commercial PMBs	
		PMB-L1	PMB-L2	PMB-C1	PMB-C2
Base bitumen		70/100 pen	70/100 pen		
Polymer (Lab PMBs) 6%		SBS	SBS+EVA	SBS	SBS+EVA
Penetration (0.1 mm)	1426	53	47	48	52
Softening point, $T_{R\&B}$ (°C)	1427	101.1	96.4	94.6	76.5
Elastic recovery at 25°C	13398	98	94	91	90
Force ductility at 5°C - Emax (J/cm^2) - E (200 – 400 mm) - Elongation (mm)	13589	17.5 7.4 310	16.2 8.0 371	13.1 7.3 407	14.6 7.3 360
Storage stability, ΔT_{RB} (°C)	13399	2.4	4.6	0.1	1.3
Viscosity at 135°C (Pa.s)		3.50	2.54	1.97	2.19
Viscosity at 185°C (mPa.s)	13702-2	596	904	335	382
G*/Sin(delta) 70°C	--	5.24	3.67	1.33	2.0
82°C	--	2.42	1.34	0.53	0.8

In Table 3, mixture performance test results are shown for the reference mixture (AC 22 base with 50% recycled asphalt) and the mixtures with the SBS modified binders PMB-L1 and PMB-C1. The mixtures had voids content in the range 4.5 – 5.3 %.

Table 3. Mixture performance test results

Test type	Reference Mixture	PMB-L1 mixture	PMB-C1 Mixture
1. Stiffness @ 8 Hz (MPa)	8142	4939	4522
2. Fatigue at 10^6 cycles (µm/m)	109	152.6	142.0
3. Permanent deformation f_c	0.35	--	--

In Figure 1 and Figure 2 respectively the high temperature frequency sweep test results of SBS modified binders and binders modified with combined SBS+EVA polymers are shown.

As can be seen in Figure 1, the laboratory made PMB has a higher stiffness in the high temperature region than the commercial binder. The same is true with the binders modified with SBS+EVA. The phase angles of all PMBs are relatively

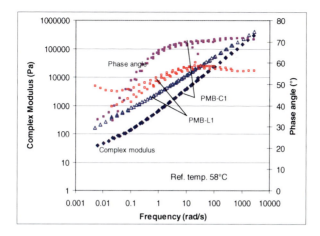

Fig. 1. Complex modulus and phase angle master curves - SBS modified binders at 58°C

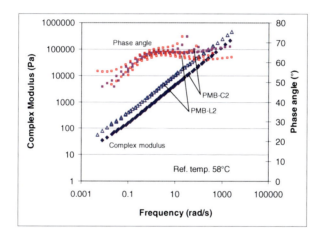

Fig. 2. Complex modulus and phase angle master curves - SBS+EVA modified binders at 58°C

very low in this temperature region and decrease at low frequencies even to 50° or lower, indicating still large elastic part of the complex modulus. This behaviour is very good for permanent deformation resistance.

In Figure 3, the frequency sweep test results at 58°C are shown for a reference binder, i.e. binder recovered from AC 22 base asphalt mixture with 50% recycling, PMB-L1 and PMB-C1. Although the reference binder has a higher stiffness than the PMBs, it shows a sharp increase in phase angle at lower frequencies. This trend is an indication of susceptibility to permanent deformation at higher pavement temperatures.

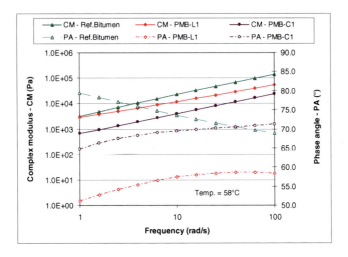

Fig. 3. Complex modulus and phase angle of binders at a temperature of 58°C

Results of Force-Ductility test for binder samples prepared in the lab are shown in figure 4 and the results for the commercial binders are shown in Figure 5.

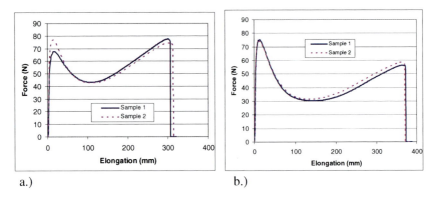

Fig. 4. Force-ductility test results for a.) PMB-L1 and b.) PMB-L2 binders

Two peaks can be observed in the force-ductility test as shown in Figure 4 and Figure 5. The second peak is a characteristic peak for modified binders and will not be found with standard pen grade bitumens. With regards to the laboratory prepared PMBs, it can be seen that the addition of EVA has resulted in a reduction in the second peak value. The commercial binders did not show a reduction in the second peak, but even a slight increase in both peaks as shown in Figure 5. In general it is expected that addition of EVA in the binder will result in a reduction of the maximum elongation at failure. However, this is not true for the binder PMB-L2 and the reason is not clear.

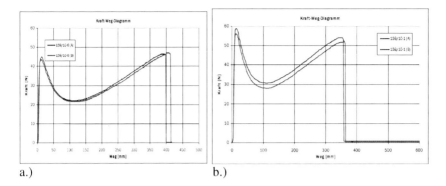

Fig. 5. Force-ductility test results for a.) PMB-C1 and b.) PMB-C2 binders

The toughness, i.e. the area under the 200 to 400 mm elongation, is usually described as a requirement in PMB specifications [6]. The flexibility of the binder, which is related to the resistance to cracking at low temperatures, is measured both by the toughness and elongation to failure of the binder.

4 Discussion

Force-ductility tests were performed to evaluate the low temperature performance of the PMBs. From Figure 4 and Figure 5, it is difficult to see if the PMBs with EVA+SBS polymer show an increase in stiffness and a reduction in elastic/ductile property. It is also not clear from Table 2. A typical increase in rigidity of PMBs due to the effect of EVA modification is however well documented [7]. It could also be confirmed by the DSR frequency sweep test results shown in Figure 6 at a reference temperatures of 5°C. The effect of EVA in the PMB binder increased the stiffness and decreased the viscous component as shown in Figure 6. The trade-off between stiffness and elastic characteristics of the PMBs is an important consideration for the use of binders in different types of asphalt mixtures. It can also be noticed from Figure 6 that the complex modulus and phase angle master curves of the PMBs show a higher viscous component compared to the reference binder at 5°C. This is an indication that the PMBs are more flexible at lower temperatures. Such characteristic improves the resistance to brittle cracking of asphalt mixtures at lower temperatures.

It is apparent from Table 2 that all the SBS modified binders have comparable toughness but show different ductility properties. Especially in the elongation range between 200 – 400 mm, i.e. the specification requirement, the PMBs seem to exhibit similar toughness characteristics. But that doesn't fully show the ductile properties of the binders. For this reason it is important to consider both the toughness of the materials and the maximum elongation to failure as essential characteristics to indicate the low temperature elastic properties of PMBs.

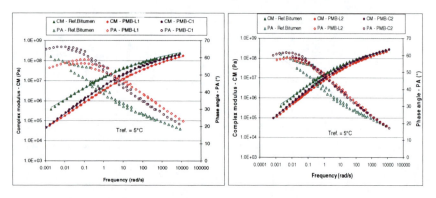

Fig. 6. Complex modulus and phase angle mastercurves of binders at a reference temperature of 5°C

In Figure 7 and 8 the stiffness and fatigue characteristics of a standard/reference AC 22 base mixture used as a base layer in pavements, and two other mixtures with the same aggregate gradation but with modified bitumen, i.e. PMB-L1 and PMB-C1, are shown. From the figures it can be observed that the mixtures with the modified bitumen show comparable stiffness values and fatigue, but they are very different from the performance of the reference mixture.

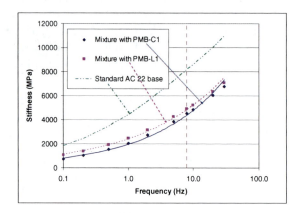

Fig. 7. Stiffness of a standard AC 22 base mixture with 50% reclaimed asphalt (RAP) and mixtures with modified binders without RAP

In the reference mixture, the combined bitumen from the reclaimed asphalt and new 70/100 pen bitumen results in a bitumen penetration value between 40-60. This value is equivalent compared to the penetration values of the PMBs. However, the effect of the polymer modification has resulted in a considerably lower stiffness and higher fatigue life of the mixture. The lower stiffness, nevertheless, does not

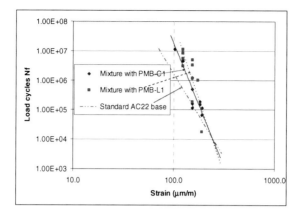

Fig. 8. Fatigue life of a standard AC 22 base mixture with 50% reclaimed asphalt (RAP) and mixtures with the modified binders without RAP

necessarily mean that the susceptibility to permanent deformation of the mixture is higher than the reference mixture. Studies indicate that the resistance to permanent deformation of mixtures with PMB at high temperatures is excellent due to the delayed elastic behaviour of the modified binders [2]. The binder stiffness and phase angle values shown in Figure 1 and Figure 2 also indicate this fact.

The PMBs improved the fatigue performance of the asphalt mixture at a strain level of 100 µstrain with a factor of about 5. This has an advantage of delaying the bottom-up crack initiation and propagation problems often encountered in the base layers of pavement structures. Figure 8 also gives the indication that the endurance limit (strain level at which fatigue is not a point anymore) of the PMB mixtures is considerably higher compared to the reference mixture. In addition to enhanced fatigue performance of the mixture at intermediate temperatures, the resistance to cracking at lower temperatures (ductility test) and the resistance to permanent deformation at higher temperatures plays an important role in prolonging the pavement service life. This in turn implies an overall reduction in maintenance cost. Future research will focus on aging tests of the modified binders to validate the effect of aging on the long term performance of the mixtures.

5 Conclusions

The following conclusions were drawn from the study into the binder and mixture performance of PMBs.

1. A relatively steady phase angle of PMB over a range of frequencies at higher temperatures is an indication of the stable functional characteristics of the modified binders.
2. Lower stiffness as a result of the modification does not imply reduction in mixture performance. Polymer modification enhances not only the fatigue performance of the asphalt mixtures but also the resistance to deformation.

3. To evaluate low temperature stiffness and flexibility/ductility of PMBs and make a choice for a particulate type of PMB in asphalt mixtures, both the total toughness, in addition to the toughness between 200 – 400 mm elongation, and the maximum elongation at failure are important parameters.
4. Modified bitumen showed improved low temperature elastic properties. Improved low temperature elastic behaviour of the PMBs implies improved resistance to crack initiation and growth which considerably prolongs pavement service life.

References

[1] Wen, H., Kutay, M.E., Shen, S.: In: Proceedings of the 89th TRB annual meeting, paper number 10-3971, Washington, DC, USA (2010)
[2] Schunselaar, R., Stigter, J.: Asfalt blad, vol. 2, p. 16. VBW Asfalt, The Netherlands (2011)
[3] Sivpatham, P., Beckedahl, H.J., Janssen, S.: International conference on asphalt pavements. In: Proceedings of the 11th ISAP Conference, Nagoya, Aichi, Japan (2010)
[4] Read, J., Whiteoak, D.: The Shell Bitumen Handbook, 5th edn. Thomas Telford Publishing, London (2003)
[5] CROW. Standard RAW Bepalingen 2005, wijziging mei (2008)
[6] NEN-EN 14023, Specification framework for polymer modified bitumens. Nederlandse norm (2010)
[7] Sengoz, B., Isikyakar, G.: Const. and Build. Mat. 22(9). Elsevier bv. Publishing (2008)

Effect of Polymer Dispersion on the Rheology and Morphology of Polymer Modified Bituminous Blend

Ibrahim Kamaruddin[1], Noor Zainab Habib[2], Isa Mohd Tan[1], Masaharu Komiyama[3], and Madzlan Napiah[1]

[1] Associate Professor Universiti Teknologi PETRONAS, Malaysia
[2] PhD Student Universiti Teknologi PETRONAS, Malaysia
[3] Professor, Graduate School of Medicine and Engineering, University of Yamanashi, Japan

Abstract. Increase in axle wheel load and traffic volume has led to the use of polymer modified bitumen (PMB) on roads as it offers better rutting, thermal and fatigue performances. In this paper the viscosity function of PMB obtained at 135°C was studied in the context of polymer dispersion in the bitumen blend. Polypropylene (PP) was used as the modifier for 80/100 pen bitumen. The morphological analysis using Scanning Electron Microscopy (SEM), Atomic force Microscopy (AFM) and Field Emission Electron Microscopy (FESEM) were presented. It was found that although the polymer resin was not fully digested by the virgin bitumen, there was evidence of a significant alteration of the Newtonian behavior of virgin bitumen to non-Newtonian behavior by the addition of the polymer.Presence of thixotropic behavior in the blend can be considered benefical in recovery of stress related deformation. SEM examination of PP resin revealed that partial breakage of the periphery of the resin was sufficient to enhance the viscosity of the PMB significantly. AFM phase images revealed that up to 2% polymer concentration in bitumen significantly enhances the viscoelastic property of the final PMB blend.The phase separated layer in PMB blend with sufficient stiffnesss and viscoelastic property of PP also acts as stress relaxant surface. Thus the PMB benefited by the incorporation of the polymer as it induces phase separated layer in the blend that can potentialy offer better fatigue and cracking properties to the resulting mix.

1 Introduction

In order to predict the engineering properties of PMB, an understanding of its rheological behavior is necessary. It is essential that the bituminous mix possess a high degree of compatibility in order for it to demonstrate high resistance against stress related degradation, thermal stability, load spreading capability and chemical stability [1]. The flow behavior or rheology of PMB strongly depends on the dispersion of the polymer in the blend. The concentrations of polymer, mixing technique and blending temperature also have a profound effect on the morphology and thus rheology of the PMB blend. The chemical structure of bitumen which is discussed as colloidal model introduces three types of bitumen namely sol (Newtonian behavior), gel (non-Newtonian behavior) and sol-gel or "elastic –sol" [2] depending on the presence of asphaltene micelle in the maltene phase. The asphaltene micelle affects the viscosity of the bitumen. Thus the behavior which is proposed on the basis of agglomeration ability of micelles [3] shows significant effect on the rheology and morphology of the bitumen. In general polymers always have the tendency for phase separation in the blend as high molecular weight polymers are immiscible with lower molecular weight bitumen [2].

Polypropylene which belongs to the thermoplastic group showed a lower tendency of dispersion but enhances the viscosity of the blend. Although the mixing was achieved at lower shear rates and at higher temperature, complete digestion of polymer in bitumen was not attained. Thus the bitumen- polymer blend shows only physical interaction, with polymer being dispersed in the bitumen, where dispersion in the blend depends on the polymer structure besides mixing technique [4]. Usually the polymer in the blend absorbs oil from the lighter component of the bitumen i.e resin and eventually swells up. However, with increase of polymer a phase inversion is observed with the flocculation of agglomerated particles, which leads to the instability of the blend [5].PMB blend will thus get profited from this optimum level of polymer concentration which incorporates for enhanced PMB performance.

Optical microscopy is used as an effective tool to investigate the morphology of immiscible, partially dispersed polymer in the blend as the flow behavior is strongly affected by its local morphology [6] as most polymers are thermodynamically incompatible with the bitumen [7].

The objective of this paper is to discuss the influence of polymer dispersion in PMB blend focusing on the flow behavior in context with the thixotropic behavior of PMB. Morphological analysis was done to know the effect of blend composition on the rheology of the blend. AFM results indicate the presence of phase separated layer, where difference in stiffness in this phase separated layer would play a positive role in preventing pavement crack propagation.

1.1 Materials and Methods

Materials used in this study includes

- 80/100 penetration grade base bitumen obtained from the PETRONAS Refinery, Malaysia.
- The polymer Polypropylene (PP) powder used for modification was supplied by PETRONAS Polypropylene Sdn Bhd Malaysia, with a melt flow index of 8g/10min and density of 0.887g/cm3.

1.2 Preparation of PMB Sample

PMB blend was prepared by mixing bitumen with polymer using Silverson laboratory mixer at 120 rpm. Slow rate of mixing was adopted in order to make sure the polymer get dispersed into the bitumen without agglomeration. Blending continued for one hour at temperature of 170°C. The concentration of PP was kept between 1 - 3% by weight of bitumen, after concluding that blend produced with 5% to 7% polymer concentration, induces the excessive agglomeration of polymer particles in the blend.

1.3 Conventional and Morphological Test Methods

Conventional tests performed on virgin and polymer modified bitumen includes, penetration test at 25°C (ASTM D-5), softening point test (ASTM D-36). Viscosity test was conducted at 135°C (ASTM D-4402). Morphological analysis was accomplished by SEM using LEO 1430 VPSEM and FESEM with high resolution Zeiss Supra 55VP, while topographical and surface information was obtained by using AFM model SII NANO NAVI E- Sweep.

2 Results and Discussion

2.1 Penetration and Softening Point Test Results

Table 1 presents conventional test results. It can be observed that there is a sharp decrease in the penetration value of 84 dmm for base bitumen to 34 dmm with the addition of only 1.0 % polymer in bitumen. The decrease in penetration value was observed for all concentration of polymer in blend. It reflects that increase in the hardness of the PMB is associated with polymer loading. The use of the high molecular weight PP having a melt flow index of 8g/10min increases the viscosity of the PMB with the increase in polymer content. It is obvious from the results that PP belonging to thermoplastics family influences more on the penetration with the increase in the viscosity of the bitumen [8]. Although PP has a melting temperature between 160-166°C,it does not completely dissolved into the bitumen but it absorbs some oil and release low molecular weight fractions into the bitumen which increases the viscosity of the PMB[9]. Thus the incompatible polymer blend with phase separated polymer layer at top act as a sheath against the penetration needle where the stiffness of the top separated layer increases with the increase in polymer concentration. The softening point of PMB shows an insignificant variation as one hour mixing time at a temperature of 170°C is insufficient to chemically break the bond in isotactic PP. The insignificant difference in softening temperature of virgin and PMB was due to the phase separated layer in the brass ring. The lower phase separated layer consisting of bitumen portion with minimum amount of partially dispersed polymer will deform at slightly higher temperature than the softening temperature of virgin bitumen while the upper separated polymer layer stays there. Thus lower phase separated layer with partially dispersed polymer was considered to be responsible for minimum increase in softening temperature till 5% polymer concentration in PMB.

Table 1. Properties of Virgin Bitumen and PP PMB

	Pent. (dmm)	Soft Pt. (°C)	Visc (Pa s) at 135°C
Bitumen 80/100 pen	84	53	0.44
1% PP+ Bitumen	34	54	0.78
2% PP +Bitumen	30	55	0.81
3% PP +Bitumen	28	55	0.83
5% PP+ Bitumen	15	59	1.25

2.2 Viscosity Test Results

The flow behavior of the material described in terms of viscosity exhibits Newtonian as well as non- Newtonian characteristics depending on the composition and source of the crude. The temperature, loading levels and internal structure of bitumen also affect the viscoelastic properties of the blend [10]. With reference to Fig. 1 base bitumen which has viscosity of 0.44Pa s at 135°C shows an increase in viscosity with the increase in the polymer concentration. The non-Newtonian characteristics as seen by a decrease in viscosity with the increase in shear rate was observed for all polymer concentrations. The non-Newtonian phenomenon is dependent on the shear rate as it influences the internal structure of the PMB [11]. Viscosity test results show that as the shear rate is increased the viscosity of the blend reduces, but still all the values of PMB blend are well below 3 Pa s, as mentioned by ASSHTO MP1 [12] for a workable mix. This behavior of bitumen from the viscosity stand point shows that it is neither Newtonian nor non-Newtonian as there was a mild change in the viscosity observed with an increase of shear rate. It may be due to the presence of asphaltene component in the bitumen with an accompanying increase in polarity and increase in molecular mass as well as decrease in the aromatics content of the bitumen. Thus collectively they cause the formation of gel type bitumen behaving more or less like polymeric solution [13]. While mixing either using a mechanical or chemical method the differences in molecular weight and polarity of base bitumen, polymer dispersion has a crucial effect on compatibility [14]. Shear thinning phenomenon was observed for all PP concentration up to 3%.However it is very difficult to say that polymer modified bitumen purely exhibits shear thinning phenomenon with the increase in shear rate as shear thickening phenomenon was also observed for 5% polymer concentration in the blend between 2000- 4000 sec-^{1} shear rate, as polymeric blend containing dispersed particle usually exhibit pseudoplasticity and thixotropic behavior when being sheared.[15] This pseudo plastic behaviour of PMB at higher concentration of polymer was observed because of the breakdown of structure existing in equilibrium state become more aligned thus offering lesser resistance to flow. However with the increase in shear rate, a higher resistance due to agglomeration or flocculation of particle in multiphase system becomes prominent due to inter particle forces like Brownian, van der Waals and electrostatic forces[15].Thus this instability in colloidal or microstructure of PMB blend would be considered beneficial to fatigue related phenomenon of pavement material (binder) where during rest period of load application on pavement, chances of healing and recovery can't be neglected. Self healing process which addresses the recovery of binder during fatigue cracking [16], based on the rearrangement of these dispersed macromolecules which rearrange themselves during rest period of repeated load application. Thus blended PMB would show promising behavior in overcoming fatigue related distress in pavement, although this instability in the rheological behavior of PMB blend would leads to morphological phase separation [17].

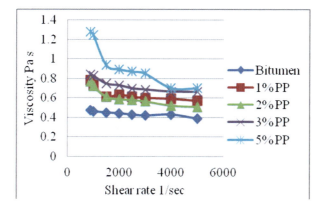

Fig. 1. Viscosity of Bitumen & PP PMB blend at 135°C

2.3 Morphological *Analysis*

FESEM was used to study the internal morphology of the PMB. The analyzed samples describe the extent of the continuous phase and compatibility of the blend. From Fig. 2(a) & (b) it was observed that PP tends to swell in base bitumen till 2% concentration. As the concentration was increased from 2%, more creaming effect was observed (refer Fig.2c) due to the agglomeration of partially dissolved polymer. The main cause of this phase separation of PMB was due to Brownanian coalescence followed by gravitational flocculation ending with creaming [17]. At higher temperature of mixing, partially dispersed fine particles which are present among large polymer particles coalesce after meeting each other splitting the bitumen film inducing phase separation. The movement of these tiny particles due to Brownian motion increases the volume of the resulting particle due to agglomeration, which causes an increase in the buoyancy force acting on the resulting large particle which when moves upward, it captures other slow upward moving particles in its way to the surface of the blend [18].Thus the mechanism of this phase separation in PMB is also defined as coalescence followed by creaming [19], which is observed more for 3% PP in blend.

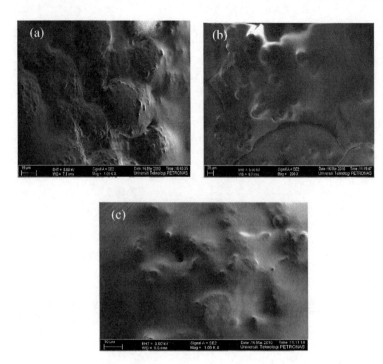

Fig. 2. FESEM image (a) 1% PP PMB (b) 2% PP PMB (c) 3% PP PMB

SEM was used to examine the toluene washed polymer resin particles from the blended bitumen to determine the extent of polymer miscibility in blend. From Fig. 3 (a) it was observed that the average nominal diameter of PP resin varies from 11.83μm to 16.75μm. From Fig 3(b) & Fig 4 it was inferred that blending at temperature of 170°C for one hour, no major change in peripheral diameter of the polymer resin was observed. Only a slight breakage of periphery and fractured area was observed for 2% PP resin. As the concentration was increased to 3%, more breakage and fracture was observed (refer Fig. 4). Thus it was inferred that these fractured dispersed polymer particles induces creaming effect due to the coalescence of broken particles and thus becomes the major cause of phase separation.

AFM images were taken in tapping mode, which provides information in three dimensional topography and phase shift, as shown in the images of virgin and modified bitumen in Fig.5-10. Phase shift contrast in tapping mode of AFM reveals different surface compositions on a surface. Fig.5 shows a two dimensional phase image of virgin bitumen in which "bee" structure can be observed, which was believed to be asphaltene micelles [20][2]. The three dimensional phase image of the same bee structure is shown in Fig 6, in which protruding tubers from the surface of maltene fraction can be observed representing asphaltene[2][21]. Within this protruding structure two different phases can be observed with altering dark and bright region [22]. These alternating light and dark regions in bee represents portion of different relative

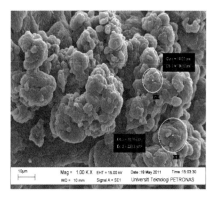 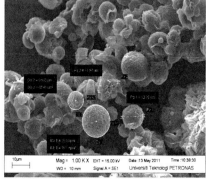

Fig. 3. (a) SEM image of PP resin   (b) SEM image of 2% PP washed resin

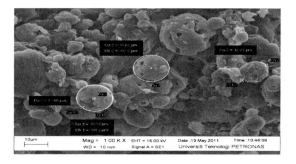

Fig. 4. SEM image of 3% PP washed resin

stiffness (different viscoelastic properties) having contrasting mechanical properties. According to Loeber et al [21], catanaphase or bee can be assigned to the most polar fraction of the bitumen which is asphaltene, where the surrounding phase around bee structure formed by lighter or less polar fraction maltene composed mainly of resin and aromatics. The surrounding flat surface may be concluded as non polar saturates or alkanes. The phases observed within protruding structure has varying stiffness offering different mechanical properties depending upon the amount of protruds, commonly considered as asphaltene[23], the most aromatic structure composed of fused aromatic rings stacked together forming heaviest molecular weight fraction of the bitumen. Thus the dark and light protruding tubers emerging from the base as observed in the topographic image (Fig.7) needs further investigation although it is being confirmed that it has varying viscoelastic properties. Fig.8-10 shows the evolution of phase change with the addition of PP in bitumen. Phase evolution was observed with the formation of phase separated polymer layer on the surface of bitumen whose thickness seems to be increasing with concentration of polymer in blend as higher molecular weight polymer seems to be incompatible with the lower molecular weight bitumen thus causing the phase separation [2]. The 3D phase images of PP modified bitumen shows variation in surface roughness which decreases with the increase in polymer concentration in the blend. From Fig.8, 1%PP PMB phase image shows phase separated polymer layer totally covering the underneath bitumen layer, as evidenced by the disappearance of the "bee" structures. It is noted in Fig.8 that the phase image of 1% PP PMB is very rough, indicating that the surface is micro-mixture of two materials having different viscoelastic properties. It is possible to interpret this image that PP segregated to the sample surface but has not covered it completely, and forming a surface that is a micro-mixture of bitumen and PP. Fig.9 of 2% PP PMB also shows surface roughness, but to much less degree compared to 1% PP PMB, indicating that the cover- up of the surface by segregated PP is more complete when PP concentration is increased to 2%. 3% PP PMB (Fig.10) phase image expose a complete flat phase separated layer of polymer above the bitumen. These AFM images unfold the surface evolution by the segregated PP. Polymer phase segregation in the blend is considered as one of the main cause of the drastic decrease in penetration value of 84 dmm for virgin bitumen to 34dmm for 1% PP modified bitumen. Besides acting as sheath, this phase separated layer of different relative stiffness potentially benefits pavement mechanical characteristics as at higher service temperature the stiffness modulus of segregated polymer layer would be higher than surrounding medium as observed by enhanced viscosity of the PP PMB while at lower service temperature the stiffness of the dispersed polymer medium is lower than surrounding medium which thus reduces its brittleness and would accommodate large stresses induced in pavement due to repeated loading.

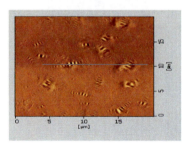

Fig. 5. 2D Phase image of Bitumen   **Fig. 6.** 3D Phase image of Bitumen

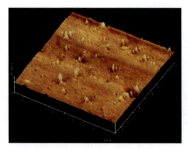
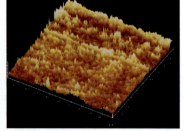

Fig. 7. Topo image of Bitumen   **Fig. 8.** Phase image of 1% PP PMB

 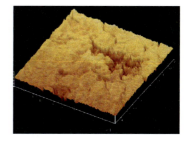

Fig. 9. Phase image of 2% PP PMB   **Fig. 10.** Phase image of 3% PP PMB

3 Conclusions

The addition of polypropylene in base bitumen changes the Newtonian behavior of base bitumen to non-Newtonian. The PP induces stiffness in the blend as observed by drastic change in penetration value but at the same time shows both thixotropy and viscoelastic characteristics due to change in rheological behavior of the blend as observed by viscosity test results. The immiscible partially dispersed polymer in PMB blend would thus benefit in self healing of pavement cracks during rest periods as partially miscible dispersed polymer would retreat to its original shape on removal of stress. The phase segregated polymer layer as observed by AFM results confirms that PP PMB blend is composed of two segregated layers of different stiffness, higher stiffness PP layer at the top while the lower stiffness partially miscible polymer - bitumen layer at the bottom, which would help in inhibiting crack propagation thereby enhancing the fatigue life of in-service polypropylene modified bituminous pavement.

References

[1] Lucena, M.D., Cavalcante, C., Jorge, B.S.: Mater. Resh. 7, 529–534 (2004)
[2] Lesueur, D.: Advan. in Colloid and Interface Sci. 145, 42–82 (2009)
[3] Guern, M., Le, Chaillex, E., Frarca, F., Dreesen, S., Mabille, I.: Fuel 89, 3330–3339 (2010)
[4] Luo, W.Q., Chen, J.-C.: Const. and Build. Mat. 25, 1830–1835 (2011)
[5] Baginska, K., Gawel, I.: Fuel Proc. Tech. 85, 1453–1462 (2004)
[6] Grizzuti, N., Bifulco, O.: Rheol. Acta 36, 406–415 (1997)
[7] Kranse, S.: In: Paul, D.R., Newman, S. (eds.) Polymer Blends, pp. 16–113. Academic Press, New York (1978)
[8] Whieoak, D., Read, J.: The Shell Bitumen Handbook, 5th edn. Thomas Telford Publishing (2003)
[9] Yousefi, A.A., Kadi, A.: J. of Mat. in Civil Eng. 12, 113–123 (2000)
[10] Sybilski, D.: Mat. and Struct. 30, 182–187 (1997)
[11] Drozdov, A.D., Yuan, Q.: Inter. J. of Solids and Struct. 40, 2321–2342 (2003)
[12] Fuentes-Audén, C., Sandoval, J.A., Jerez, A., Navarro, F.J., Martinez, B.F., Partal, P., Gallegos, C.: Poly. Test 27, 1005–1012 (2008)

[13] Vinogradov, G.V., Isayev, A.I., Zolotarev, V.A., Verebskaya, E.A.: Rheol. Acta 16, 266–281 (1977)
[14] Isacsson, U., Lu, X.: In: Francken, L. (ed.) Bituminous Binders and Mixe, pp. 1–38 (1998)
[15] Bhattacharya, S.N.: Rheology Fundamentals and Measurement, pp. 1–32. Royal Melbourne Institute of Technology, Australia (1997)
[16] Shan, L., Tan, Y., Underwood, S., Kim, R.Y.: J. of the Transport. Resh. Board 2179, 85–92 (2010)
[17] González, O., Munoz, M.E., Santamari, A., Morales, G.M., Navarro, F.J., Partal, P.: Europ. Poly. J. 40, 2365–2372 (2004)
[18] Yousefi, A.A.: Prog. Color Colorants Coat 2, 53–59 (2009)
[19] Hesp, S.A.M.: PhD dissertation, University of Toronto, Canada (1991)
[20] Loeber, L.: J. of Micro. 182, 32–39 (1996)
[21] Wu, S.-P., Ling-Tong, Yong-Chun, M.C., Guo-Jun, Z.: Const. and Build. Mat. 23, 1005–1010 (2009)
[22] Dourado, E.R., Simao, R.A., Leite, L.F.: J. of Micro 245, 119–128 (2011)
[23] Jäger, A., Lackner, R., Eisenmenger-Sittner, C., Blab, R.: Road Mat. and Pav. Design, 9–24 (2004)

Effect of Organoclay Modified Binders on Fatigue Performance

Nader Tabatabaee and Mohammad Hossein Shafiee

Sharif University of Technology, Tehran, Iran
nader@sharif.edu, shafiee87@gmail.com

Abstract. Organoclay modification is receiving attention as a nano-modifier for asphalt binders. Nanoparticles are able to effectively mend the damaged sites of the nanocomposite without external intervention. Particle-polymer interaction results in packing particles into developed cracks, thereby mending the cracks during patch formation. This study investigates the rheological properties of organoclay modified binders with a focus on their fatigue properties. To this end, the dissipated energy concept, the ratio of dissipated energy change and plateau value (PV) energy were examined during fatigue tests using a dynamic shear rheometer. The effects of strain level, frequency and temperature on unaged and aged neat PG64-22 asphalt binder modified with organoclay were evaluated in strain-controlled time sweep fatigue tests. Also evaluated was the effect of rest periods introduced at cycles corresponding to different damage levels over the course of the time sweep tests on the fatigue life of neat and modified binders. It was found that increased amounts of organoclay and decreased levels of strain led to lower PV values and higher fatigue resistance. It was also found that organoclay modified binders heal more effectively during the rest periods, which results in higher fatigue life.

1 Introduction

Fatigue resistant and self-healing asphalt materials using nanoparticles is still in its early stages. Studies have shown that nanoparticles can be driven toward damaged zones in nonocomposite materials and repair cracks as a result of polymer-induced depletion attraction localizing particles in the cracks to form patches [1]. The positive effects of organo modified montmorillonite (OMMT) on the physical and rheological properties of clay-asphalt composites, such as increased stiffness, resistance to aging, and thermal stability, have been investigated by several researchers [2-4]. The focus of this study is on the fatigue resistance evaluation of OMMT modified binders based on the ratio of dissipated energy change (RDEC) and the healing properties obtained from changes in the speed of microcrack growth after introduction of rest periods (RPs) during time sweep (TS) tests.

2 Background

Studies have shown that the fatigue response of asphalt binders is better predicted using repeated cyclic loading tests than by the evaluation of binder fatigue resistance based on G*sinδ during the initial few cycles [5]. This stems from accurate simulation of damage accumulation in an asphalt binder during repeated cyclic loading TS tests. Since the amount of energy dissipated at each loading cycle in a TS test is a good indicator of damage accumulation in asphalt, dissipated energy (DE) is a fundamental factor used to study the fatigue behaviour of asphalt [6].

Research has revealed that only a portion of DE is responsible for fatigue damage and that the energy dissipated through passive behaviors, such as plastic dissipated energy and thermal energy, should be excluded from the calculations [6, 7]. In response to this, an approach called the ratio of dissipated energy change (RDEC) has been developed in which the relative amount of energy dissipation coming from each additional cycle is taken into account. According to this definition, Eqn (1) shows the ratio of dissipated energy change [7]:

$$RDEC_m = |DE_m - DE_n|/DE_n(m-n) \qquad (1)$$

where $RDEC_m$ is the ratio of dissipated energy change at cycle m, DE_m and DE_n are dissipated energies at loading cycles m and n, respectively, and $m>n$.

The dissipated energy versus the number of loading cycles under strain-controlled loading mode is divided into three distinct stages. The reduction rate of DE increases in the first stage, then DE reduces at an almost constant rate in the second stage and, in the third stage, the DE reduces at a decreasing rate. This corresponds to the three stages in the RDEC vs. loading cycle curve for HMA mixtures and asphalt binders (Figure 1). The second stage, where the RDEC value is almost constant, is considered the plateau value (PV).

The effect of introducing several rest periods at predetermined levels of damage on increasing the fatigue life or healing the asphalt has been reported by different investigators [8, 9]. These studies have recognized the healing effects and improved fatigue properties of neat and polymer modified asphalt. However, limited research has been conducted to characterize the effect of OMMT modified binders on fatigue and healing properties.

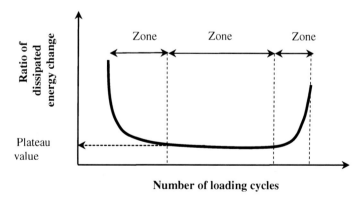

Fig. 1. Schematic RDEC plot with three behaviour zones [7]

3 Research Approach

Fatigue characteristics are of great importance in flexible pavement design and there is a need to enhance the fatigue behaviour of asphalt using new and more cost-effective asphalt modifiers. OMMT has been widely used as a nonometer filler to improve the fatigue resistance of polymer materials [10]. It is expected that better fatigue resistance can be achieved by the OMMT modification of asphalt due to lipophilicity and the enlargement of interlayer spacing in OMMT. The present research studied the capability of nanoclay to improve the fatigue resistance of clay-asphalt nanocomposite under various testing conditions on the basis of changes in the PV value and fatigue life calculated using a traditional TS test. The healing potential of OMMT modified samples was also investigated by introducing rest periods over the course of the TS tests. The effect of rest periods and OMMT modification on healing was quantified by defining three healing indices to reflect the changes in microcracking speed. These indices were defined to separately identify the effect of rest periods and OMMT on healing.

4 Experimental Design

4.1 Materials

A neat PG 64-22 binder was selected as the base binder to be modified with 1%, 4% and 7% OMMT by weight of the base binder. The nanolin DK1 series of organoclay with 95% to 98% montmorillonite and a cation exchange capacity (CEC) of 110 meq/100 g was used. The modified binders, labled B+M1%, B+M4% and B+M7%, were prepared using a high-shear melt blending process at 150°C and 4000 rpm rotation speed for 60 min. Aging simulation of asphalt samples were carried out according to AASHTO T 240 (RTFOT).

4.2 Testing Methods

The fatigue characteristics of asphalt binders were assessed using TS binder fatigue testing in a dynamic shear rheometer (DSR). Specimens 8 mm in diameter and 2 mm thick were subjected to continious shear loading in the strain-controlled mode at specific strain levels, frequencies and temperatures. Unaged samples were subjected to TS tests at 25°C, 10 Hz in frequency and a 3% strain level. A limited sensitivity study was conducted to account for the effect of control parameters on binder fatigue. To investigate the effect of rest periods (to simulate healing) on fatigue, two TS tests (with one and 10 rest periods) were performed on unaged samples at 3% and 5% strain levels to represent the state of strain of the binder in the compacted surface course. The corresponding healing indices were then calculated for each condition.

4.3 X-Ray Diffraction

Two types of structures exist in layered silicate modified asphalt: intercalated and exfoliated. An intercalated structure corresponds to a well organized multi-layer silicate in the binder in which asphalt chains expand the clay galleries. In an exfoliated structure the clay layers have been completely seperated such that they are no longer close enough to interact and silicate layers disperse throughout the binder randomly [11]. The extent of nanoclay particle intercalation and exfoliation were examined using x-ray diffraction (XRD). XRD patterns of pure organoclay and three OMMT-modified binders were investigated using an X'Pert MPD instrument with Co Kα radiation (λ = 1.78897 Å, 40 kV, 30 mA) at a scanning rate of 0.02 °/s from 1° to 50° in the 2θ range. The XRD patterns of pristine organoclay and three modified binders are shown in Figure 2. As shown in Table 1, different crystalline peaks were found for the three modified binder samples corresponding to different interlayer spacings calculated using the Bragg equation (λ = 2d sinθ). It was observed that the peaks shift to lower angles after binder modification.

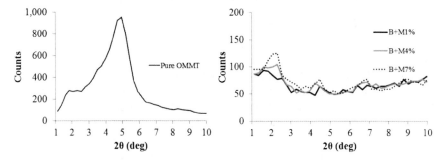

Fig. 2. XRD patterns of organoclay and modified binders

Table 1. Interlayer spacing of organoclay and modified asphalt binders.

Sample	2θ (deg)	Intensity (AU)	d-spacing (Å)
Pure OMMT	4.8759	973.44	21.04724
B+M1%	1.3549	98.01	75.65871
B+M4%	2.3581	100.97	43.50388
B+M7%	2.2980	125.56	44.64092

This suggests the widening of the interlayer, formation of an exfoliated structure and a good dispersion of silicate layers in binders under the aforementioned blending conditions.

5 Results and Discussion

The addition of 1% to 7% organoclay increased the original asphalt rotational viscosity from 7.7% to 166.8% from the exfoliation of the high aspect ratio silicate layers. This behaviour is a result of the silicate layers preventing movement of asphalt molecules. The higher viscosity is indicative of the higher degree of network build-up associated with the higher degree of exfoliation.

5.1 Rutting Resistance

The rutting potential of neat and modified binders were evaluated using a DSR based on the Superpave high temperature binder test for unaged and RTFO-aged samples and the multiple stress creep and recovery (MSCR) test for RTFO-aged samples according to AASHTO M-320 and ASTM D7405-08, respectively. Using the MSCR tests, average non-recoverable creep compliance (J_{nr}) and average percent recovery (R) for neat and modified binders were determined. Loading consisted of 10 cycles of 1 sec creep plus 9 sec of recovery time at a 100 Pa stress followed by another 10 cycles at 3200 Pa. MSCR test results at 64°C and the Superpave rutting factors are tabulated in Table 2. Both $G^*/\sin\delta$ and J_{nr} suggest higher rut resistance with increased amounts of organoclay in asphalt. Although changes in R do not reflect a significant increase in this parameter at either stress level, the addition of higher amounts of OMMT improved the resistance of materials to rutting. The percent difference in non-recoverable creep compliance, $J_{nr\text{-}diff}$, showed low sensitivity of all tested binders to stress level.

5.2 Fatigue Characteristics Testing

Evaluation of asphalt fatigue life was performed using strain-controlled oscillation testing at constant shear strain amplitude with no rest periods. Strain-controlled testing was used to achieve a zero mean displacement during the test. The cycle corresponding to 50% reduction in the initial complex modulus (N_{f50}) was used as the criteria for fatigue failure. PVs were then calculated for the plateau stage between N_{f15} and N_{f50}. Table 2 shows a summary of fatigue testing results

Table 2. Permanent deformation parameters for neat and modified binders at 64°C

Binder type	$G^*/\sin\delta$ (kPa)		J_{nr} 100 (1/kPa)	R 100 (%)	J_{nr} 3200 (1/kPa)	R 3200 (%)	$J_{nr\text{-}diff}$ (%)
	Unaged	RTFO aged					
Neat	1.11	2.26	2.45	6.32	3.48	1.34	29.6
B+M1%	1.21	2.53	2.18	6.94	2.52	2.08	13.5
B+M4%	2.22	3.63	1.67	9.67	2.00	2.41	16.8
B+M7%	2.34	4.06	0.97	20.53	1.33	6.48	27.1

conducted under five testing conditions. The effect of loading frequency on fatigue was evaluated by conducting TS tests at 10 Hz and 15 Hz loading at 25°C and 3% strain level. As shown in Table 3, increasing loading frequency increased the corresponding PVs, indicating higher fatigue resistance at lower frequencies from the decreased damage accumulation rate. At both loading frequencies, a higher OMMT content improved the fatigue resistance of the binders, however, no linear trend was observed between OMMT content and increase in N_f caused by lower frequencies.

In a similar manner, changes in PV for unaged samples subjected to 3% and 5% strain level TS tests at 10 Hz loading and 25°C reflect the influence of strain level on fatigue. It was evident that lower PVs at 3% strain represented better fatigue performance in all of the tested binders. Higher amounts of OMMT led to more fatigue resistance at both strain levels, while the positive effect of OMMT was more pronounced at 5% strain. In comparison with the neat binder, at 3% and 5% strain level, the B+M7% modified binder reduced the PVs up to 63% and 67%, respectively. This shows the significant fatigue resistance caused by the addition of nanoparticles to asphalt.

Also as shown in Table 3, an increase in temperature from 25°C to 30°C under 3% strain-controlled TS at 10 Hz loading decreased PVs for neat and B+M1% modified binders. Nonetheless, this temperature change resulted in a marginal decrease in PVs for 4% and 7% modified binders. Hence, fatigue performance of samples with higher amounts of organoclay was less susceptible to temperature. A comparison between PVs before and after RTFO aging indicated that it caused a reduction in the fatigue life. However, aged samples containing higher amounts of nanoclay were more fatigue resistant, based on PV and N_{f50}. After RTFO aging, the stiffness of the tested asphalt samples increased and the RTFO-aged OMMT modified samples were found to be stiffer than the neat sample. This is also in accordance with improved aging resistance of OMMT modified binders. It should be noted that, particularly for the B+M7% binder, the slight difference in PVs before

Table 3. Effect of testing conditions on fatigue parameters

Time sweep test condition	PV (10^{-4})				N_f (10^4)			
	Neat	B+M 1%	B+M 4%	B+M 7%	Neat	B+M 1%	B+M 4%	B+M 7%
25°C, 3% 10 Hz, unaged	0.495	0.385	0.214	0.175	4.20	4.70	6.09	6.42
25°C, 3% 15 Hz, unaged	0.598	0.395	0.338	0.210	3.02	2.70	3.51	4.56
25°C, 5% 10 Hz, unaged	1.239	1.156	0.924	0.404	1.37	1.58	2.07	3.37
30°C, 3% 10 Hz, unaged	0.455	0.340	0.217	0.164	3.26	3.32	5.41	9.37
30°C, 3% 10 Hz RTFO aged	0.875	0.588	0.500	0.261	1.66	1.98	2.60	3.43

and after aging showed the effectiveness of nanoclay in improving the aging resistance of asphalt. This may be the result of the barrier properties of the OMMT layers to heat and oxygen [4]. PVs were plotted against N_f for all binders under different testing conditions in Figure 3, where R^2 equaled 0.85, providing a good PV-N_f correlation.

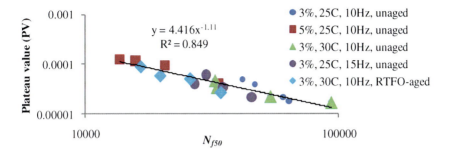

Fig. 3. PV versus N_{f50} for all binders

5.3 Effect of Rest Periods on Fatigue Life

Strain-controlled interrupted loading was used to investigate the effect of rest periods on asphalt fatigue and healing. The unaged samples were subjected to interrupted TS tests with ten 5-min equally-spaced rest periods or one 50-min rest period at 3% and 5% strain levels. During the TS test with 10 rest periods (10RPs) each RP was inserted at the damage level corresponding to a 5% reduction in the initial complex modulus. During the fatigue test with one rest period (1RP), a 50-min rest period was inserted at the damage level corresponding to a 50% reduction in the initial complex modulus in the TS fatigue test without RP.

The introduction of RPs at specific damage levels can effectively bring the material healing potential into consideration [9]. It was observed that, at both strain levels, N_f increased more effectively during the 10RPs tests than the 1RP; this may have been due to the shorter intervals between RPs inserted at lower damage levels (Figure 4). Nevertheless, for 4% or more nanoclay, there was a noticeable difference between the trends of fatigue life at 3% and 5% strain indicating that the the nanoclay in the binder increased the fatigue resistance at the higher strain level, particularly for 10 RPs.

In consideration of the factors influencing the healing potential of binders, three healing indices were defined in this research to characterize the role of organoclay and rest periods in the healing, as shown in Eqns. (3) to (5):

$$\text{Net healing index (NHI)} = \left| 100 \times (S(i,10) - S(0,0))/S(0,0) \right| \quad (3)$$

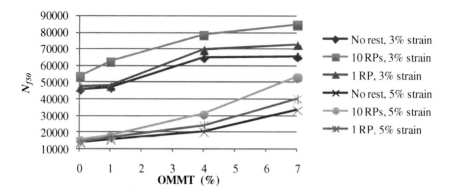

Fig. 4. Effect of rest periods on N_{f50} at 25°C and 10 Hz

Healing agent effect index (HAI) = $\left| 100 \times (S(i,10) - S(0,10))/S(0,10) \right|$ (4)

Rest period effect index (RPI) = $\left| 100 \times (S(i,10) - S(i,0))/S(i,0) \right|$ (5)

where $S(i,j)$ is the absolute slope representing the microcracking speed for the $i\%$ OMMT modified asphalt binder during a healing test with j number of RPs. After normalizing the complex modulus at each loading cycle to the initial complex modulus, $S(i,0)$ was calculated for data points corresponding to 15% to 50% modulus reduction during a fatigue test and $S(i,10)$ was calculated for data points corresponding to the third RP to tenth RP that best fit the straight line regression equation during a healing test. The NHI parameter presents the effectiveness of both RP insertion and OMMT addition on the microcracking speed test relative to damage growth speed during a fatigue test of the neat binder. HAI and RPI show the influence of the addition of OMMT on damage growth speed reduction relative to neat binder and the effect of rest time on speed reduction relative to the no rest condition.

Healing indices for unaged neat and modified binders at two strain levels using 10 Hz are plotted in Figure 5. Organoclay modification showed a considerable impact on the improvement of healing indices at both strains levels. This is generally consistent with the well-established concept that stiffer asphalt binders are more prone to microcracks and, thus, have more noticeable healing potential [12]. Furthuremore, the higher strain level inversely affected NHI for neat and B+M1% samples, probably because of higher damage development at 5% strain, while the addition of higher amounts of OMMT appears to have prevented the opening of microcracks at the higher strain level.

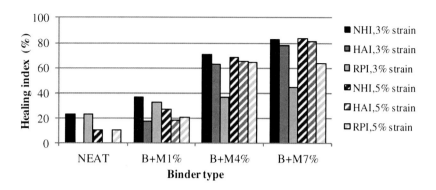

Fig. 5. Healing indices at 3% and 5% strain-controlled healing tests at 25°C

6 Conclusions

Rheological characteristics of different OMMT and bitumen composites were studied with a focus on fatigue properties. This research has shown that melt blending can effectively form an exfoliated structure in OMMT modified binders and OMMT modification can improve the rutting factors of asphalt binders. Using RDEC to evaluate the fatigue properties for neat and organoclay modified binders under several testing conditions revealed that nanoclay effectively increased the fatigue life and the aging resistance in accordance with changes in the mechanical properties. The good PV-N_f correlation in the results of the time sweep tests show the applicability of RDEC in fatigue life prediction of neat and OMMT modified binders. Finally, the calculation of three newly defined healing indices based on rest time and the OMMT effect on healing showed that a higher strain level and loading frequency decreased the healing potential of neat binder. However, samples containing organoclay were more prone to microcracks and more healable.

References

[1] Lee, J.Y., Buxton, G.A., Balazs, A.C.: Using nanoparticles to create self-healing composites. J. Chem. Physics 121, 5531–5540 (2004)
[2] You, Z., Mills, J., Foley, J., Roy, S., Odegard, G., Dai, Q., Goh, S.: Nano-clay modified asphalt materials: Preparation and characterization. Constr. Bldg. Mat. 25, 1072–1078 (2010)
[3] van de Ven, M.F.C., Molenaar, A.A.A., Besamusca, J.: Nanoclay for binder modification of asphalt mixtures. In: Proceedings of the 7th International RILEM Symposium on Advanced Testing and. Characterization of Bituminous Materials, Rhodes, Greece, vol. 1, pp. 133–142 (2009)
[4] Zhang, H.L., Wang, H.C., Yu, J.Y.: Effect of aging on morphology of organo-montmorillonite modified bitumen by atomic force microscopy. J. Microscopy 242, 37–45 (2011)

5. Tabatabaee, N., Tabatabaee, H.A.: Multiple stress creep and recovery and time sweep fatigue tests: crumb rubber modified binder and mixture performance. Transp. Res. Rec. 2180, 67–74 (2010)
6. Carpenter, S.H., Shen, S.: A dissipated energy approach to study hot-mix asphalt healing in fatigue. Transp. Res. Rec. 1970, 178–185 (2006)
7. Shen, S., Chiu, H.M., Huang, H.: Characterization of fatigue and healing in asphalt binders. ASCE J. Mat. in Civil Eng. 22(19), 846–852 (2010)
8. Kim, Y.R., Little, D.N., Lytton, R.L.: Fatigue and healing characterization of asphalt mixtures. ASCE J. Mat. in Civil Eng. 15(1), 75–83 (2003)
9. Johnson, C.M.: Evaluate relationship between healing and endurance limit of asphalt binders, Asphalt Research Contortium, quarterly technical progress report, pp. 39–49 (2008)
10. Utracki, L.A.: Clay-containing polymeric nanocomposites, Smithers Rapra Technology Ltd., Shropshire (2004)
11. Yu, J., Zeng, X., Wu, S., Wang, L., Liu, G.: Preparation and properties of montmorillonite modified asphalts. Mat. Sci. Eng. A 447(2), 233–238 (2007)
12. Si, Z., Little, D.N., Lytton, R.L.: Characterization of microdamage and healing of asphalt concrete mixtures. ASCE J. Mat. in Civil Eng. 14(6), 461–470 (2002)

Effects of Polymer Modified Asphalt Emulsion (PMAE) on Pavement Reflective Cracking Performance

Yu Chen[1], Gabriele Tebaldi[2], Reynaldo Roque[1], and George Lopp[1]

[1] University of Florida, Civil & Coastal Engineering, United States
[2] University of Parma, Civil & Environmental Engineering and Architecture, Italy

Abstract. Hot mix asphalt (HMA) overlay is widely used to restore functional and structural capacity of existing asphalt pavements. In order to ensure that the overlay and the underlying layer act as a uniform composite layer and more effectively transfer and distribute the external load over a large area, a good bond between the overlay and underlying layer is a necessity. Overlay performance greatly depends on both the bond strength along the interface and the cracking resistance across the interface provided by the interface materials. For example, one way to potentially enhance HMA overlay cracking performance is by using a highly polymer modified asphalt emulsion (PMAE) at the interface to help relieve stress transferred across the interface, as well as to enhance bonding between layers. Consequently, it is necessary to evaluate the effects of the interface conditions on overlay cracking performance. The effects of two types of interface conditions on reflective cracking were evaluated: conventional tack coat and PMAE. Tests were performed on composite specimens with these two different interface conditions using a newly developed Composite Specimen Interface Cracking (CSIC) test. Tests were performed under repeated tensile loading while monitoring the rate of damage development. Results clearly indicated that the PMAE interface can significantly improve reflective cracking resistance. However, these test results need further experimental road test evaluation and verification.

Keywords: Polymer Modified Asphalt Emulsion, Interface, Cracking, Bond, Asphalt Overlay.

1 Introduction

Hot-mix asphalt (HMA) overlays are used to restore safety and ride quality and increase structural capacity for existing pavements as a preventive maintenance and/or rehabilitation technique. One of the concerns associated with HMA overlay is reflective cracking, which is initiated by discrete discontinuities such as cracks on existing pavement and propagates upward through the HMA overlay. The most recognized driving force of reflective cracking is the horizontal movement concentrated at cracks and differential vertical movement across the cracks in existing pavement [1]. The resistance of HMA overlay to reflective cracking relies on the quality of overlay mixture, the bond between the overlay and existing pavement, and the conditions of the existing pavement. It is important to ensure a good bond between HMA overlay and existing pavement since poor bonding or debonding can reduce load transfer capability and lead to pavement distress such as slippage [2]. Poor tack coat between pavement layers has been reported to be among the causes of debonding [3]. Despite the potential stress relieving benefits of polymer modified asphalt emulsion (PMAE), little research has been done to evaluate the effect of PMAE on overlay reflective cracking resistance when it is applied between overlay and existing pavement.

Meanwhile, various methods have been used to reduce reflective cracking in HMA overlay including reinforcing the overlay, stress relieving interlayers, and restrengthening original pavement prior to overlaying [4]. However, field performance of interlayers in reflective cracking reduction has ranged from clear successes to total failures [5, 6, 7, 8]. In order to evaluate the effect of interlayers on pavement reflective cracking, laboratory test methods must allow reflection cracks to initiate in existing pavement and propagate across the interlayer and into the overlay without slipping along the interlayer [9, 10]. Test methods commonly used to evaluate reflective cracking resistance of interlayers require large specimens, which are relatively difficult to fabricate in the laboratory and more difficult to obtain from the field [11, 12, 13, 14, 15].

This paper presents the results of a research work that investigated the phenomenon of crack propagation through pavement layer interface and into overlay(s) using a newly developed test method, the Composite Specimen Interface Cracking (CSIC) test. Reflection cracks were simulated by installing teflon spacer in existing pavement layer of composite specimen, which can be prepared through Superpave^{TM} gyratory compaction.

2 Objectives

The objectives of this study are as follows:

- To present a detailed test method development process for the evaluation of tack coat interface effect on overlay reflective cracking performance;
- To present specimen preparation, testing and data interpretation methods;
- To analyze the effect of PMAE on overlay reflective cracking performance.

3 Scope

This study primarily focused on evaluating effects of two different bonded interfaces on pavement reflective cracking resistance. Dense-graded mixture was used for both composite specimen layers with two types of interface bonding agents, i.e. conventional tack coat and PMAE. Tests were conducted at one temperature (10°C), which has been determined in prior fracture research on the same material at the University of Florida to correlate well with cracking performance of pavements in the field.

4 Materials

A dense-graded mixture commonly used by the Florida Department of Transportation (FDOT) as a structural layer, identified as Dense-GA-Granite, was used to produce composite specimens. Its aggregate was made up of four components: coarse aggregate, fine aggregate, screenings, and sand. Its gradation is shown in Table 1. The mixture was designed according to the Superpave^{TM} volumetric mix design method. The binder used for the mixture was PG 67-22 at the rate of 4.8%.

Table 1. Aggregate gradation

Sieve Size	19.00 mm	12.50 mm	9.50 mm	4.75 mm	2.36 mm	1.18 mm	600 μm	300 μm	150 μm	75 μm
% Passing	100.0	99.0	86.0	65.0	47.0	32.0	23.0	14.0	7.0	4.2

Two types of tack coats, conventional unmodified asphalt emulsion and polymer modified asphalt emulsion (PMAE, a plant-produced, anionic emulsion), were evaluated in this study. The properties of conventional tack coat (an anionic slow setting asphalt emulsion, ASTM type SS-1) and PMAE (The polymer family was of the styrene-butadiene or styrene-butadiene-styrene block copolymer type) are presented in Table 2.

5 Test Method and Sample Preparation

The CSIC test system included the environmental chamber cooling system, MTS loading system, measurement and data acquisition system. The testing composite specimen geometry and loading configuration are shown in Figure 1. The test was performed by applying a repeated haversine waveform load to the specimen for a period of 0.1 second followed by a rest period of 0.9 seconds (See Figure 1-B). The distinctive features of this test are specimen symmetry and application of load inside the stress concentrator. The hole at the center of the specimen serves both as a stress concentrator and a platform for load application.

In the laboratory, composite specimens can be prepared by compacting loose dense-graded mixture on top of the pre-compacted dense-graded existing pavement

Table 2. Properties of conventional tack coat and PMAE

Tests on Conventional Tack Coat Residue	AASHTO/ASTM	Specification	Test Result
Penetration (dmm), 25°C(77°F), 100g, 5s	T 49 / D 5	100.0 – 200.0	129.0
Solubility in Trichloroethylene, %	T 44 / D 2042	97.5 min	99.6
Ductility, 25°C (77°F), 5cm/min	T 59 / D 113	40.0 min	58.0 +
Tests on PMAE Residue	AASHTO/ASTM	Specification	Test Result
Penetration (dmm), 25°C (77°F), 100g, 5s	AASHTO T49	90.0 – 150.0	115.0
Elastic recovery at 10°C (50°F), %	AASHTO T301	58.0 min	75.0

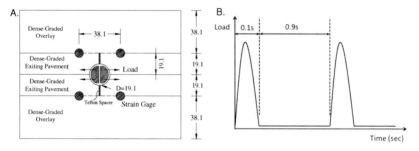

Fig. 1. A. Composite specimen geometry (Unit: mm) and B. loading mode

layer using Superpave^{TM} Gyratory Compactor (SGC). By half-slicing, two dense-graded specimens for lower layer were obtained from each of the Superpave^{TM} gyrator compacted specimen pill. Dense-graded lower layers were cut in half along the diameter to install the telfon spacer. The diamond saw blade was barely thinner than the telfon sheet. Interface bonding agents, conventional tack coat and PMAE, were applied on the cut surface of dense-graded specimen using a silicon rubber mold. Loose dense-graded overlay mixture was then compacted on top of the lower dense-graded layer after it was reinserted into the gyratory compaction mold. The overlay mixture was compacted to the desired thickness to achieve the design air voids. Through a series of cutting, gluing, and grooving operations, the completed composite specimen was obtained. The composite specimen preparation process is illustrated in Figure 2. Teflon spacer (See Figure 2-F) was introduced in composite specimens to represent an existing crack.

Fig. 2. Composite specimen preparation A. Half slicing; B. Cut along diameter; C. Layered compaction; D. Half-composite specimen; E. Cutting; F. Final epoxying, coring stress concentrator and carbon fiber reinforcement of the ends.

6 Data Collection and Interpretation Method

As reported elsewhere [16], extensometer data was acquired for calculation of the specimen's total recoverable deformation if a sudden deformation change occurred, or whenever desired. The number of loading cycles required to break the composite specimen (See Figure 3) and the damage rate were used to compare reflective cracking resistance for specimens with different interface conditions subjected to the same loading conditions. The damage rate was defined as the slope of the steady state response portion of total recoverable deformation progression curve as shown by the line in Figure 3.

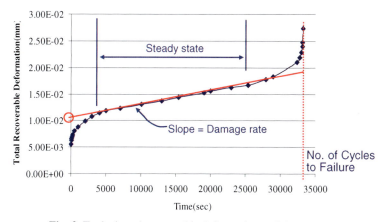

Fig. 3. Typical total recoverable deformation and damage rate

7 Test Results

As part of the test method development, three replicate specimens without teflon spacer for each of the two types of interfaces, 0.453 l/m^2 diluted conventional tack coat and 0.905 l/m^2 PMAE, were first prepared for reflective cracking tests. These two application rates were selected according to manufacturer's recommendation. The composite specimen was prepared using the sample preparation approach presented earlier without teflon spacer installation. The geometry and strain gauge distribution of testing specimens were the same as presented in Figure 1-A. Tests were performed under the loading mode stated in Figure 1-B with 2535N peak load. Tests results are presented in Figure 4 for number of loading cycles to failure (See Figure 3 for failure criterion) and Figure 5 for damage rate.

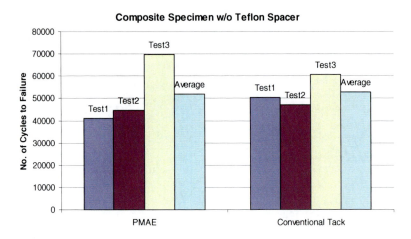

Fig. 4. Number of cycles to failure of PMAE and conventional tack

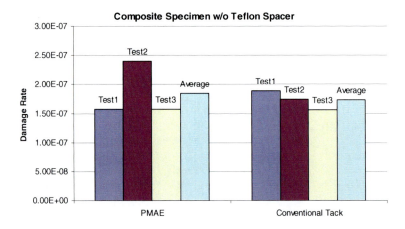

Fig. 5. Damage rate of PMAE and conventional tack

Specimens with PMAE and diluted conventional tack interface exhibited almost the same reflective cracking resistance. This might be explained by the fact that composite specimens with both types of interface conditions have the same crack length (See Zone 1+Zone 2 in Figure 6). Interface bonding agents started to dissipate stresses only after cracks propagated to the interface and considerable amount of damage was accumulated in zone 2 during zone 1 crack propagation. For both types of bonding agents, zone 2 crack propagation entered unstable/final crack propagation stage. As compared with zone 1, zone 2 crack propagation time is relatively short, which is the time allowed for PMAE to dissipate stresses. This led to the same reflective cracking resistance for both types of bonding conditions.

Three replicate specimens with teflon spacer for each of the two types of interfaces, 0.453 l/m^2 diluted conventional tack coat and 0.905 l/m^2 PMAE, were prepared. Half the load used in composite specimens without teflon spacer testing, 1245N, was applied because of the newly introduced teflon spacer stress concentration. Typical composite specimen with teflon spacer failure mode is shown in Figure 6; this appears to correspond nicely with the crack propagation phenomenon in the field. Careful examination of the strain gauge deformations indicated that the specimens were not uniformly loaded, which made the results unreliable. Therefore, the results with 1245N peak load were not included in the following analysis.

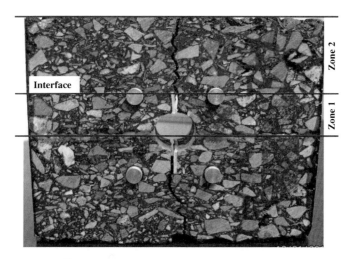

Fig. 6. Typical composite specimen failure mode

Peak load was increased to 1912 and 2313N to reduce the testing time. Tests results were presented in Figure 7 for number of loading cycles to failure and Figure 8 for damage rate. Results presented in Figures 7 and 8 indicate that specimens with PMAE interface exhibited higher fracture resistance than specimens with conventional tack coat interface. These results also indicate that PMAE applied at the interface took effect right from the moment of loading with the introduction of teflon spacer as stress concentrator, which leads to better cracking performance for specimens with PMAE interface.

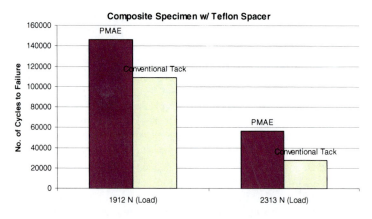

Fig. 7. Number of cycles to failure of PMAE and conventional tack

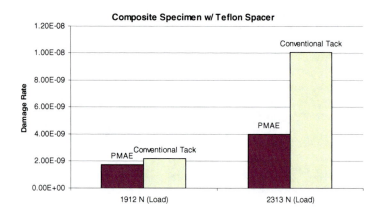

Fig. 8. Damage rate of PMAE and conventional tack

8 Discussion

The test results presented above indicate that composite specimens with telfon spacer successfully simulated reflective cracking in existing pavement layer and its propagation through interface. For composite specimens without telfon spacer, both reflective cracking initiation in existing pavement layer and propagation through interface and into overlay was simulated. Interface bonding agents were not engaged in stress dissipation and/or reducing stress transmitted through interface until cracks propagated to interface (Zone 1 in Figure 6). Because of high fracture energy density of dense-graded mixture in zone 1, it took large number of loading cycles for the crack to initiate and propagate through zone 1; considerable amount of damage was accumulated in zone 2 during this crack initiation and propagation throughout zone 1.

However, it took considerably less number of loading cycles for the crack to initiate and propagate through zone 1 when dense-graded mixture in zone 1 was replaced with open-graded mixture because of its low fracture energy resistance [17]. This explains the reason why composite specimen composed of open-graded friction course and dense-graded structural layer successfully identified stress dissipating benefits of interface bonding agents for top-down cracking evaluation without the introduction of teflon spacer [16]. This analysis concludes that reflection cracks represented as teflon spacer installed in existing pavement layer is a necessity for the evaluation of interface/interlayer effects on reflective cracking resistance.

9 Summary and Conclusions

Symmetric composite specimen with teflon spacer was successfully used to simulate reflective cracking in existing pavement for interface cracking resistance evaluation. This composite specimen testing method identified the enhancement of PMAE on reflective cracking resistance when it was applied between dense-graded overlay and existing pavement layer. The overlay reflective cracking resistance depends on both tack coat characteristics and its application rate. The PMAE helps to dissipate the stresses accumulated near the interface and/or reduce the stress transmitted through the interface.

References

[1] Von Quintus, Mallela, Lytton: Techniques for mitigation of reflective cracks. In: FAA Worldwide Airport Technology Transfer Conference, P10067, Atlantic City, New Jersey (2010)

[2] Romanoschi, S.A.: Ph.D. Dissertation: Characterization of Pavement Layer Interfaces. Louisiana State University, Baton Rouge, LA (1999)

[3] Muench, S.T., Moomaw, T.: De-bonding of hot mix asphalt pavements in washington state: an initial investigation, WA-RD 712.1 and TNW 2008-10. Washington State Department of Transportation (WSDOT) and Transportation Northwest Regional Center X (TransNow), Seattle, WA (2009)

[4] Button, J.W., Lytton, R.L.: Guidelines for Using Geosynthetics with HMA Overlays to Reduce Reflective Cracking. Report 1777-P2, Project Number 0-1777, Texas Department of Transportation, Austin, TX (2003)

[5] Barksdale, R.D.: Fabrics in Asphalt Overlays and Pavement Maintenance. NCHRP Synthesis 171, National Cooperative Highway Research Program, National Research Council, Washington, DC (1991)

[6] Blankenship, P., Iker, N., Drbohlav, J.: Interlayer and Design Considerations to Retard Reflective Cracking, pp. 177–186. Transportation Research Record, Washington, DC (1896)

[7] Amini, F.: Potential Applications of Paving Fabrics to Reduce Reflective Crackin, Final Report, Report No. FHWA/MS-DOT-RD-05-174, Jackson State University Department of Civil & Environmental Engineering, Jackson MS, pp.1–32 (2005)

[8] Verspa, J.W.: An Evaluation of Interlayer Stress Absorbing Composite (ISAC) Reflective Crack Relief System. Final Report, Report No. FHWA/IL/PRR150. Illinois Department of Transportation (2005)
[9] Pickett, Lytton: Laboratory Evaluation of Selected Fabrics For Reinforcement of Asphaltic Concrete Overlays. Research Report, Report No. FHWA/TX-84+261-1. State Department of Highways and Public Transportation (August 1983)
[10] Mukhtar, M.T.: Interlayer Stress Absorbing Composite (ISAC) for Mitigating Reflection Cracking in Asphalt Concrete Overlays, Ph.D., University of Illinois at Urbana-Champaign, Urbana, IL (1994)
[11] Mukhtar, M.T., Dempsey, B.J.: Interlayer stress absorbing composite (ISAC) for mitigating reflection cracking in asphalt concrete overlays, Final Report Project IHR-533, Illinois Cooperative Highway Research Program (1996)
[12] Kim, K.W., Doh, Y.S., Lim, S.: Mode I reflection cracking resistance of strengthened asphalt concretes. Construction and Building Materials 13(5), 243–251 (1999)
[13] Brown, S.F., Thom, N.H., Sanders, P.J.: A study of grid reinforced asphalt to combat reflection cracking. In: Annual Meeting of Association of Asphalt Paving Technologists, pp. 543–569 (2001)
[14] Zhou, F., and Scullion, T. (2004). Overlay tester: a rapid performance related crack resistance test, Report No. FHWA/TX-05/0-4467-2, Texas Transportation Institute
[15] Khodaii, A., Fallah, S., Nejad, F.M.: Effects of geosynthetics on reduction of reflection cracking in asphalt overlays. Geotextiles and Geomembranes 27, 1–8 (2009)
[16] Chen, Y., Lopp, G., Roque, R.: Test Method to Evaluate the Effect of Interface Bond Condition on Top-down and Reflective Cracking. International Conference on Road and Airfield Pavement Technology (ICPT), Thailand (2011)
[17] Koh, C.: Tensile Properties of Open Graded Friction Course (OFGC) Mixture to Evaluate Top-Down Cracking Performance, Ph.D. Dissertation, University of Florida, Gainesville, FL (2009)

Characterization of Long Term Field Aging of Polymer Modified Bitumen in Porous Asphalt

D. van Vliet[1], S. Erkens[2], and G.A. Leegwater[1]

[1] TNO
[2] Rijkswaterstaat, Dutch Ministry of Infrastructure and Environment

Abstract. The effect of long term field aging on different types of polymer modified binders used in two-layer porous asphalt is studied using different test methods. Chemical and rheological tests are performed on samples taken from road sections at different moments in time in search of trends in long term field aging.

The results show that chemical analyses performed with GPC and infrared can be used to study the effect of short term aging. However, these tests are not accurate enough to establish the more subtle trends that play a role in long term field aging. The tests proved to be very valuable to determine the type of modification in the binder.

Master curves showing the complex modulus and the phase angle at different temperatures determined with the Dynamic Shear Rheometer (DSR) show linear trends for long term aging when an aging period of 8 years is observed. As expected the complex modulus increases with aging, while the phase angle decreases. There is one exception to this, in polymer modified binders the phase angle decrease with age at low loading frequencies which can be related to degradation of the polymer modification.

Long term field aging, polymer modified binder, GPC, FTIR, DSR.

1 Introduction

The performance of two layered porous asphalt surfaces is superior with respect to the reduction of the noise produced by road-tyre interaction and the amount of splash and spray during rain, compared to dense surface layers. However these open surface roads have a drawback with respect to durability, namely their relative short service life compared to dense asphalt layers. The governing damage mechanism for two layered porous asphalt is ravelling, the loss of stones due to cracking of the connecting mastic. As aging processes are accelerated by the open structure of this mixes, it is assumed that the failure mode ravelling is related to the fast aging process of open mixtures. Two-layered porous asphalt in the Netherlands is usually made with a polymer modified binder. Therefore the aging of polymer modified binder will be investigated within this research.

While aging is a well known phenomenon in asphalt [1], the long term aging behaviour in practise of porous asphalt and the mechanisms behind this are relatively unknown. To obtain a broad view of the characteristics of aging of two layered porous asphalt an extensive research program has been set-up to monitor several parameters that are related to aging. On the one hand tests are performed to describe the change in mechanical behaviour of the material trough time. On the other hand chemical tests are done to track changes in the composition of the material in order to look for driving mechanism behind aging.

In order to monitor actual aging in practice, the measurements are performed on samples obtained from field sections. The samples are taken from the so called ZEBRA sections; these are test sections of two-layered porous asphalt roads realized by Rijkswaterstaat for research purposes. At four different locations in the Dutch highway network (A15, A28, A30, A59), the same eight types of two layered porous asphalt surfaces were realized between 2002 and 2004. For this research three of the eight asphalt mixtures were selected for further research on aging of polymer modified binders. At different moments in time cores were drilled at the four locations of the three selected mixtures. Performance of the virgin binder is also incorporated within the research program.

The aim of the research is to look for trends in the material behaviour of polymer modified bitumen over time in practice. Chemical and mechanical tests are performed in order to look for trends in behaviour and the possible relation between these trends. In order to asses if the type of binder influences the aging, three different types of polymer binders are compared. In order to see if the loading influences aging four road locations and different positions in the road are evaluated. This article describes the most important observations, however as the results cover over a hundred samples, not all results can be described in detail, full results can be found in [2, 3].

2 Aging of Bitumen

The material properties of asphalt are closely related to the behaviour of bitumen [4]. The rheological behaviour of bitumen changes during production of asphalt and continues to change over time. This phenomenon of changing behaviour is called aging [5]. In asphalt a distinction is made between aging that occurs while producing and applying asphalt, this is referred to as short term aging and aging that occurs during the service life of the road, so called long term aging.

Aging is characterized by an increase in stiffness and an increase in viscosity of the binder. To quantify this change in mechanical behavior, a DSR test can be performed. In this test a sinusoidal loading is applied at a certain frequency and the deformation under this loading is measured. Due to the fact that bitumen is a visco-elastic material there is time delay between the loading and the deformation, this is called a phase change. Therefore this test results in a complex modulus (G^*) and a phase angle (δ) which together give an impression of the visco-elastic behavior. In general for bitumen the complex modulus will increase and the phase angle will decrease due to aging.

As aging is not fully understood at the moment there are several processes that are associated with aging [5]. The processes that were thought relevant with respect to long term aging, the main topic of this research, are introduced in short below. The test methods that are chosen in this research to monitor the described processes are also introduced. In the next paragraph more details about the used methods will be given.

With time molecules within the bitumen interconnect with each other, which increases the average molecule size. This chemical change to larger molecules leads to an increase in stiffness of the bitumen [6]. The change in molecule size can be determined with Gel Permeation Chromatography (GPC).

The aging process of bitumen is further characterized by oxidation of the bitumen. During oxidation mainly ketones (C=O) and sulfoxides (S=O) are formed. Due to the fact that this reaction turns non-polar fractions into polar fractions the viscosity of the binder increases with aging. The sum of the ketones and sulfoxides indicate the relative oxidation of the binder [1]. The amount of ketones and sulfoxides can be determined with help of Infra-red Spectroscopy (FTIR).

Steric hardening is the process that the individual molecules in bitumen tend to slowly form structures that are more energy efficient [1]. This type of hardening can be reversed by adding energy to the material true reheating or by applying a load. There are no tests performed to quantify this type of aging.

The processes where lighter constituents of the bitumen either evaporate, segregate or are absorbed by aggregates, won't be considered separately in this article as they are closer related to the short term aging which is less of interest in this research [5]. However it is possible that part of the change in behaviour measured with DSR or changes in the molecule size measured by the GPC are caused by these effects.

The previous remarks all discus the general effects of aging of unmodified binders. However in case of this research the focus is on polymer modified binders. Therefore in analyzing the results the following has to be considered on aging of polymer modified binders. Aging of polymer modified binders will result in the degradation or break-up of the polymer [7, 8, 9]. Therefore the effectiveness of the modification could be reduced as aging progresses. However the reduction of molecule sizes will also decrease the stiffness of the binder and will therefore counteract the effects of aging.

3 Test Program

3.1 Samples

Two-layer porous asphalt has a bottom layer with coarse aggregates (11/16 mm) and a top layer with fine aggregates (2/6 or 4/8 mm). The fine top layer always contains a polymer modified binder in the Netherlands. The bottom layer can be made with normal binders or with modified binders. In this research the behaviour of three different types of two-layer porous asphalt are analyzed, A, B and C. In section A and B the same polymer modified binder has been used for the top and the bottom layer. In section C different binders are used for the layers, the binder used for the top layer is SBS modified and straight run bitumen is used for the bottom layer. An overview of modifications per mixture is given in table 1. All three mixtures have 4/8 mm graded top layers. The binder was recovered from laboratory and field specimens following RAW 2005 test 110, which refers to a Dutch norm (NEN-3917). The method uses dichloromethane as a solvent to extract the binder from the mixes, after this the solvent is evaporated from the binder by vacuum distillation.

Table 1. Modification type of the bitumen samples per road section

Section code	Modification type toplayer	Modification type bottumlayer
A	SBS +EVA	SBS +EVA
B	EVA	EVA
C	SBS	-

Cores were drilled over a period of four years (2006, 2007, 2008 and 2009), as the roads were constructed from 2002 to 2004, the moment of sampling results in data points showing the performance from 2 to 7 years after construction. At each road location, of each of the three sections, seven cores were taken from the right wheel track of the right lane and seven cores from the emergency lane. The two-layer PA is cut along the interface of the top and the bottom layer, parallel to the pavement surface, in order to separate the layers. From all the top layers the bitumen was recovered. In order to reduce the amount of specimens, the bitumen was only recovered from the bottom layers of mixture B and C for three of the four sites A15, A28 and A30. Six of the seven asphalt cores are processed together in order to be able to extract enough bitumen for all tests and to obtain representative average that is representative for the road. The seventh core is kept as a back-up sample.

From two of the four locations, the A59 and the A15, samples were taken when the road was constructed, at the age of 0 years. This means that these samples show the effect of short term aging caused by construction, but long term aging effects are not yet present. As these measurements at t=0 are an important reference with respect to long term aging, they are named startingpoint value within this research.

3.2 Test Methods

Gel Permeation Chromatography (GPC) is used to analyze the Molecular Weight Distribution of the (polymer modified) bitumen samples. GPC is a chromatographic method in which particles are separated based on their molecular size. In the standard testing procedure [7], the sample is solved in THF (tetra hydro furan). This method is used to analyse the 70/100 pen grade bitumen samples and SBS polymer modified samples. EVA doesn't dissolve well in THF, therefore toluene that is heated to 60°C is used as solvent with a solvation period of 30 minutes. The changes in molecular weight for the bitumen component and the modification component are evaluated separately.

The Attenuated Total Refraction Fourier Transform Infrared (ATR/FTIR) spectroscopy is a technique used to identify functional groups in organic compounds, which is an effective method to investigate the chemical composition of materials [10]. The apparatus used to conduct the IR test was a Galaxy Series FTIR 3000. To minimize absorption of moisture, which could influence the results of the IR spectrum, all bitumen samples were put in a desiccator. In addition, a device was connected to the IR apparatus to introduce nitrogen into the system to help decrease the presence of moisture inside the ATR/FTIR apparatus.

Dynamic shear tests were conducted using a Rheometrics RAA asphalt analyzer – Dynamic Shear Rheometer (DSR). With help of the DSR test the complex modulus and phase angle at different temperatures and loading frequencies were determined. The specimen is placed between two circular parallel plates. The upper plate is fixed; while the lower part oscillates applying the shear strain during testing. The test is carried out in a temperature controlled mini-oven (chamber). The temperature of the sample is controlled with air. The temperature control has an accuracy of ± 0.1°C when adequate time (usually 10 min) is provided to stabilize the temperature. In the DSR test, the bituminous materials were subjected to a sinusoidal loading of constant strain at different loading frequencies (frequency sweep). The frequency sweep test was conducted at eight different temperatures ranging between -10 °C and 60 °C. Every test was carried out at frequencies ranging between 0.1 –400 rad/s. Two parallel plate geometries with a diameter of 8 mm and 25 mm were used.

4 Test Results

4.1 General Remarks

During the analyses of the data, the importance of having a startinpoint value became very clear. As the amount of short term aging caused by production is a very important value as a reference for the long term aging. The quality of the trends found in long term aging is less if these startingpoint values are missing. In future research on long term aging in practice the effect of short term aging should be documented more extensively.

4.2 Molecular Size

Tests with GPC showed that some bitumen samples of later constructed road sections turned out the have another "recipe". On confrontation with this observation contractors admitted that they had tried to improve their binder from the first constructed section. These deviating samples were left out for further analyses.

The GPC results show a clear and significant reduction in the molecular size of the SBS polymer modification from virgin binder to the mixture on the road at t=0. The average reduction of the molecule size is less than half of the original molecule size. This phenomenon is already observed in other research [7, 9]. Changes in the molecular size over time from 2 to 7 years do not lead to any trends as the differences are not statistically significant due to scatter in the data. The degradation in the first three years after construction seems to be most significant, but as we have limited data of t=0 and short after, these changes are also not statistically significant. The degradation of EVA was not retraceable as the concentration of the EVA was extremely low in all retracted samples. It is assumed that this is caused due to the applied binder recovery process, where cold extraction with methylene chloride is performed.

The aging of the bitumen component is also studied. Here a slight increase in molecular size can be seen between the virgin binder and the mixture on the road at t=0, which indicates a polymerisation and/or an evaporation. A further increase in the molecular size over time is visible, but no trend in time can be deduced as the scatter data is large and not many data points are available from the first three years where the most changes are observed. However, one conclusion can be based on the data, there is a difference in aging behaviour of the bitumen component between the top and the bottom layer of mixture B, the bottom layer ages slower. This is shown in figure 1, by the fact that M_w, a parameter that indicates the amount of larger molecules, increases faster for the top layer.

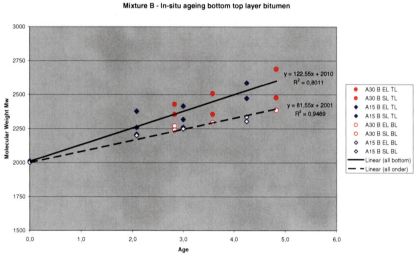

Fig. 1. Different aging trends between bottom (open points (BL)) and top layer (closed data points (TL)) with respect to molecule size M_w

4.3 Oxidation

With help of quantitative analysis of the IR spectrum changes in oxidation levels and modification levels are examined. Aging of the bitumen component is assessed by looking for changes in the absorption peaks that correspond to the aging products ketones (C=O) at 1700 cm-1 and sulfoxides (S=O) at 1030 cm-1. For the modifications also specific peaks can be distinguished. EVA, poly(ethylene-vinylacetate), is a combination of polyethylene and polyvinylacetate. The infrared spectrum is a combination of the peaks of these two components. Characteristic peaks are the carbonyl stretch (C=O) at 1739 cm-1 and the C-O single bond stretch of the acetate group at 1242 cm-1. SBS, poly(styrene-butadiene), is a combination of polystyrene and polybutadiene. The infrared spectrum is a combination of the peaks of the two components as listed above. Characteristic peaks are the C=C stretch of butadiene at 966 cm-1 and the C-H bend of styrene at 699 cm-1.

The FITR results showed an increase in C=O and S=O peaks after short term aging for all mixtures. Trends with respect to long term aging (2 to 7 years) were not visible as the scatter in the data was large, as is demonstrated by figure 2. The degradation of EVA was not retraceable as the concentration of the EVA was extremely low in all retracted samples, probably due to the used extraction method. The degradation of the SBS was not clearly visible for short and long term aging in practise based on the chosen peaks, again large scatter in data is observed.

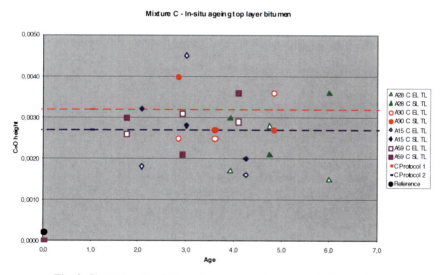

Fig. 2. Changes in the C=O peak (1700 cm-1) over time in field aging

4.4 Rheological Behaviour

The master curves measured with the DSR show that as the loading is applied faster the material responds stiffer (higher complex modulus) and more elastic (lower phase angle). Using the Time-Temperature Superposition (TTS) principle master curves of the complex modulus and phase angle were constructed for a reference temperature of 20°C. Polymer modifications influence the master curve quite significantly especially at lower loading rates, as can be seen in figure 3, where the master curve of an unmodified binder and of three modified binders is shown.

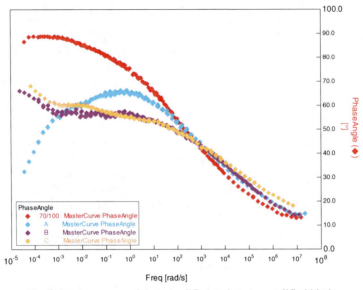

Fig. 3. Master curves of three modified and one unmodified binder

The effect of ageing for conventional bitumen is an increase in the complex modulus and a decrease in the phase angle over the whole frequency range [11]. To quantify the rate of ageing of the binders, the phase angle values and the complex modulus values were determined at four different frequencies, 0.001, 0.1, 10 and 1000 rad/sec for the different samples.

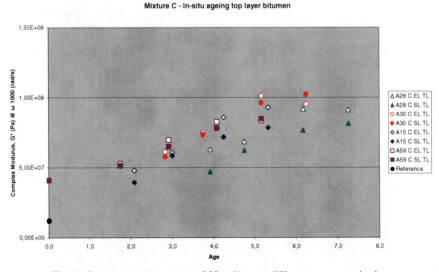

Fig. 4. Complex modulus at ω 1000 rad/sec at different moments in time

The analyses for the complex modulus and the phase angle show a trend in time, an example of a trend can be seen in figure 4. The figure shows that the binder becomes increasingly stiffer with time. Short term aging can be seen as the black spot at t=0 corresponds to the virgin binder and the square at t=0 with the measured value after construction. The linear trends are statistically significant both for the complex modulus as the phase angle.

At all frequencies the complex modulus is increasingly stiff as time passes. The phase angle however becomes lower for frequencies 0.1, 10 and 1000, while it increases for the frequency 0.001, see figure 5. The increase in phase angle at low frequencies might be explained by degradation of the polymer, which means that the polymer network degrades. There are two arguments that support this assumption. Looking at the GPC results trends were not statistically significant, however it seems likely that more measurements will show that the modification degrades over time. A second argument is the fact that this increase in phase angle at low frequencies isn't seen in mixture B, where the EVA modification is assumed to be lost in the extraction process. The effect of degradation of SBS polymer trough aging and its impact in the phase angle is also described in [12].

Furthermore the faster degradation of the top layers compared to the bottom layers can also been found in the complex modulus and the phase angle.

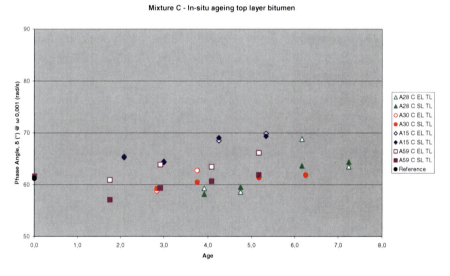

Fig. 5. Phase Angle at ω 0.001 rad/sec at different moments in time

5 Conclusions and Recommendations

Molecular size determination performed with GPC can be used to determine the effects of short term aging in polymer modified bitumen. However for long term aging in the field the method is less useful as the scatter is large compared to the changes over time. Nonetheless GPC very useful to determine the type of binder. As the method shows degradation of polymers after short term aging clearly it might be interesting to check the presence of modifications in a really old polymer modified porous mixture as these differences are expected to be large compared to the scatter.

Infrared analyses results in significant changes in short term aging of polymer modified bitumen, however looking for trends in long term field aging with this method is less interesting due to a large scatter in results compared to expected changes over time. It would be interesting to investigate the absolute amount of oxidation in long term aging in general, as the oxidation observed in this research is limited.

Rheological measurements performed with the DSR show a clear trend in long term aging of polymer modified bitumen. Results measured up to 8 years show a linear trend in between the complex modulus and the phase angle and service life in the field. As expected the complex modulus increases with aging, while the phase angle decreases. There is one exception to this, the phase angle of polymer modified binders decrease with aging at low loading frequencies. This could be related to the degradation of the polymer. A next step in this research could be to match the phase angle and the complex modulus to the performance of the mixture on the road with respect to ravelling as this is the main degradation mechanism for porous asphalt.

References

[1] Petersen, J.C.: Chemical composition of asphalt as related to asphalt durability: state of the art. Transportation Research Record 999, 13–30 (1984)
[2] Leegwater, G.A., van Vliet, D.: Aging of polymer modified bitumen in practise and under laboratory conditions, TNO-034-DTM-2009-03538, Delft (2010)
[3] van Vliet, D., Telman, J.: Aanvullend DSR bitumenonderzoek 2-laags ZOAB, TNO-060-DTM-2011-01919, Delft (2011)
[4] Bahia, H.U., Anderson, D.A.: The new proposed rheological properties of asphalt binders: Why are they required and how do they compare to conventional properties, Physical properties of asphalt cement binders, ASTM STP 1241, Philadelphia (1995)
[5] Roberts, F.L., Kandhal, P.S., Ray Brown, E., Lee, D.-Y., Kennedy, T.W.: Hot mix asphalt materials, mixture design, and construction, NAPA Research and Education Foundation, Lanham, Maryland (1996)
[6] Noureldin, A.S.: Oxidation of asphalt binders and its effect on the molecular size distribution and consistency, Physical properties of asphalt cement binders. ASTM STP 1241, Philadelphia (1995)
[7] Sanches, F.: Veroudering asfalt - GPC-bepaling van polymeermodificaties (methode ontwikkeling), Ministerie van Verkeer en Waterstaat, Rijkswaterstaat, Dienst Weg-en Waterbouwkunde, Delft (2002)
[8] Sanches, F.: Veroudering van polymeergemodificeerde bitumen voor ZOAB, Ministerie van Verkeer en Waterstaat, Rijkswaterstaat, Dienst Wegen Waterbouwkunde, Delft (2004)

[9] Wua, S.-P., Pang, L., Mo, L.-T., Chen, Y.-C., Zhu, G.-J.: Influence of aging on the evolution of structure, morphology and rheology of base and SBS modified bitumen. Construction and Building Materials 23, 1005–1010 (2009)
[10] Jemison, H.B., et al.: Application and use of the ATR, FT-IR Method to asphalt aging studies. Fuel Science and Technology 10, 795–808 (1992)
[11] Bahia, H.U., Anderson, D.A.: The Pressure Aging Vessel (PAV): A Test to Simulate Rheological Changes Due to Field Aging. In: Physical Properties of Asphalt Cement Binders, ASTM STP 1241, Philadelphia (1995)
[12] Lu, X., Isacsson, U.: Chemical and rheological evaluation of ageing properties of SBS polymer modified bitumens. Fuel 77(9,10), 961–972 (1998)

Bending Beam Rheological Evaluation of Wax Modified Asphalt Binders

Gaylon L. Baumgardner[1], Geoffrey M. Rowe[2], and Gerald H. Reinke[3]

[1] Paragon Technical Services, Inc.
[2] Abatech, Inc.
[3] Mathy Technology, Inc.,

Abstract. A simple Bending Beam Rheometer (BBR) test to determine binder low temperature properties from asphalt mixtures was recently developed by the University of Minnesota. The mixture BBR test was performed concurrent with binder BBR testing to evaluate the effect wax addition has on stiffness/physical hardening and potential for low temperature cracking in asphalt mixtures. Asphalt mixture and binder BBR tests were performed at low temperatures typical of binder grading, after equivalent low temperature conditioning in air for incremental extended periods up to 32 days. Results from temperature saturation are compared to conditioning corresponding to normal 20 hours of PAV aging and testing in accordance with parameters specified in AASHTO M320. Data produced by the BBR suggests that at temperatures close to or below the glass transition temperature mixture beams became less stiff with time when held at a constant temperature. This effect appears to be reversible if a heating/annealing cycle is applied to the beams. This paper reports further investigation of the observed phenomenon and the potential that observations may be indicative of low temperature micro-cracking and subsequent healing in asphalt mixture.

1 Introduction

Wax-like additives which melt in a temperature range between the highest pavement temperature and the desired compaction temperature have been used as an asphalt additive for warm mix and asphalt compaction aide applications [1]. Waxes have long been viewed as a problematic component within an asphalt binder, largely due to their negative impact on bitumen temperature susceptibility. With this in consideration, the primary concern in this study was how addition of wax to asphalt to reduce construction temperatures can be beneficial with respect to overall binder performance. More importantly, can current specifications distinguish between beneficial additives versus those that might have a negative impact on the performance of hot mix asphalt (HMA).

Typical waxes melt within the pavement service temperature range. When even a small fraction of the asphalt undergoes a phase change from solid to liquid over a short temperature range, the Shell bitumen test data charts exhibit a unique behaviour as defined by "W" type asphalts. With added wax, the resulting binder is both harder at low pavement temperatures and softer at high pavement temperatures. Both of these characteristics are considered as detrimental performance characteristics. When hot candle-wax is poured on a surface, it quickly solidifies to a soft, pliable mass. Over time it crystallizes into to a hard, non-ductile chip which occupies significantly less volume. This volume change also causes the well- known indention of the candle wax around the wick as the ductile amorphous wax continues to crystallize.

More recent asphalt research studies suggest that waxes also exist in bitumen as two different physical states corresponding to amorphous and microcrystalline wax. As pavements cool to low temperatures, the solid-solid phase transition between the two states is accompanied by a significant decrease in volume, which yields a corresponding increase in binder density. This phenomenon, called reversible physical hardening (RPH), was first identified by Bahia and Anderson during Strategic Highway Research Program (SHRP) studies of the Bending Beam Rheometer [2]. They noted continuous stiffening of certain asphalt beams as they were held at -15°C for up to four days. Dilatometric studies confirmed that an increase in stiffness was directly correlated to an increase in density under the corresponding storage conditions. Brule et al. [1] used analytical tools such as Differential Scanning Calorimetry (DSC), Phase Contrast Microscopy, Polarized Light Microscopy, Dilatometric measurements, Nuclear Magnetic Resonance (NMR), and Dynamic Shear rheology to conclusively tie RPH to the wax solid-solid phase transition from amorphous to microcrystalline states. The amount of hardening is significant, and detrimental to asphalt quality. Asphalt AAM, the SHRP core asphalt highest in wax, changes from a PG 64-22 to PG 64-10 after being stored at -15°C for four days. Upon reheating to 60°C, the wax crystals melt, and the binder is again PG 64-22. Two research teams led by Planche and Turner separately identified the crystallizable fraction as measured by DSC to be directly related to the physical hardening effect as measured by DSR [3, 4, 5].

For the purposes of the study of this paper, waxes were defined to be Paraffin and Non-paraffin wax. Paraffin waxes are those waxes which have molecular size less than C45 and have melting points less than 70°C (158°F). Non-paraffin waxes are those waxes that have molecular size greater than C45 and have melting points greater than 70°C (158°F). Paraffin waxes are, or are related to, refined/de-oiled microcrystalline waxes derived from crude oil. Non-paraffin waxes include, but are not limited to, natural waxes (animal and vegetable waxes), modified natural waxes (brown coal derived wax), partial synthetic waxes (ester and amid waxes) and synthetic waxes (Fischer Tropsch (FT) and polyethylene (PE) waxes) [6, 7].

The objective of this work was to evaluate the effect of non-paraffin wax additives on physical properties and characteristics of asphalt binders. Testing to include binder master curve development, binder true-grading, rotational viscosity profile, bending beam rheometer (BBR), direct tension (DTT), was used to evaluate changes in mechanical properties, other analytical methods were employed to offer effective means to evaluate the potential for waxy materials as warm-mix additives such as modulated differential scanning calorimetry (MDSC) to provide the glass transition temperature, change in heat capacity on melting, amount of crystallizable fraction, and melting point range of the wax in asphalt.

Further characterization of wax stereochemistry, Infrared Spectroscopy (IR) and/or Nuclear Magnetic Resonance (NMR) were used to determine the relative degree of branching in the wax molecules. Atomic Force Microscopy (AFM) was also used to evaluate the degree of crystallization of wax additives in asphalt [8].

This paper reports on the evaluation of BBR mixtures tests when the beams have been subjected to standard and extended temperature saturation.

2 Materials

A single source of asphalt binder was used which was selected as a PG64-22 Lion Oil produced at El Dorado, Arkansas.

Nine waxes were selected for the study. The products selected cover the range of waxes discussed earlier to include; paraffin, natural, partial synthetic and synthetic materials. In addition, the selection considered specific synthetic materials in common usage for asphalt modification (for example Sasobit). A paraffin wax (Astra Wax) that was anticipated to give properties resulting in inferior performance was also selected. Materials selected are presented in Table 1.

Asphalt binders were prepared which consisted of the one (1) neat binder and twelve (12) wax modified binders using low shear blending at 160°C. The wax modified binders were made with 3% wax additive and (for three additional modified binder blends) with 1% wax additive. The data with 1% wax blends has been reported elsewhere [8]. The control binder is referenced by a "0" in the various tables and figures of this paper whereas the wax modified binders are represented by the modifier number – 1 to 9.

Table 1. Waxes selected for study

Ref.	Category	Material	Notes
1	Natural	Romanta Normal Montan	
2	Natural/Synthetic	Romanta Asphaltan A	Blend of Montan normal and amide wax
3	Natural	Romanta Asphaltan B	Refined normal Montan
4	Partial synthetic	Licomont BS 100	N,N'-bisstearamide, stearic acid pitch
5	Synthetic	Sasobit	Fischer-Tropsch Wax
6	Partial synthetic	Luxco Pitch # 2	N,N'-bisstearamide, stearic acid pitch
7	Synthetic	Alphamin GHP	Also referenced as THP
8	Wax Ester	Strohmeyer and Arpe Montan LGE	
9	Paraffin	Astra Wax 3816D Microcrystalline	Refined microcrystalline wax

The aggregate source selected for mixture analysis was a Vulcan Barin Quarry Granite (9205 Fortson Rd., Fortson, GA 31808). This material is described as Granite Gneiss/Amphibolite and has been used extensively in research projects such as the NCAT test track. The quarry produces both stone fractions and manufactured sand. Typical properties for this material [9] are presented in Table 2.

Asphalt mixtures were prepared with the various binders using a 12.5mm (½-inch) nominal maximum aggregate dense graded SuperPave® gyratory designed mixture, as presented in the project report [8]. This mix design contained 5.15% binder and was compacted to a nominal 7.0 % air voids for testing.

Table 2. Typical properties for Barin Aggregate [9]

Property	Value	Source of Information
G_{sb} bulk specific gravity of aggregate	2.707	GDOT - 2009
G_{sa} apparent specific gravity of aggregate	2.732	
Water Absorption	0.34	
Sand Equivalent Value	86	

3 Superpave® Binder Testing

All testing other than true grading and master curve development was performed on pressure aging vessel (PAV) aged binders.

Superpave true grade was performed in accordance with AASHTO M320 Tables 1. One of the noted issues with the Superpave specifications has been that the high temperature specification parameter in Table 1 of AASHTO M320 ($G^*/\sin\delta$) has been shown to relate poorly to rutting for many "premium grade," modified asphalt binders. This has led to the development of the multiple stress creep-recovery (MSCR) (AASHTO TP70) test as the replacement for the conventional $G^*/\sin\delta$ parameter in the specification. From the MSCR test, the new high temperature specification parameter is determined by dividing the non-recoverable (or permanent) shear strain by the applied shear stress. The result is called the non-recoverable creep compliance, or Jnr. In addition, the percent recovery (% recovery MSCR) is also computed which provides more efficient method of characterizing the elasticity of a binder than that currently done with the elastic recovery test (AASHTO T301). These parameters were determined for the materials considered in this project.

PG grades can be considered within the AASHTO M320 specification using either Table 1 or 2. In addition a new table has been introduced which evaluates the performance by the Multi-Stress Creep and Recovery (MSCR) test (ASSHTO MP19) and this data has been presented elsewhere [10]. In addition to grade evaluation the data from testing can also be shown as "true grades" by evaluating the pass/fail temperature for any given criteria. Data of this format has been evaluated for the various products and this is illustrated in Figure 1.

It can be observed from this data that considerable differences exist in the different products. The Astra Wax which was selected as a product unlikely to perform well has the poorest performance with a temperature spread of a mere 60.9°C. Six of the other waxes improved the performance range while two had reduced ranges. Most of the products reduced the low temperature grade by a few degrees but with careful design of modified products with the possible selection of softer products this aspect can be considered in the formulation stage of an asphalt binder.

Brookfield viscosity data obtained from the M320 specification evaluation is presented in Figure 2 which illustrates that all of the waxes reduce the viscosity within a range of 15 to 32%. However, it should be noted that the largest viscosity reduction was with the Astra wax which was selected as the "poor" performing product. This means that the range of viscosity production for the 3% wax addition is in the 15 to 23% range for possible effective products. The data with 1% wax showed smaller changes but an overall comment that could be applied is that the viscosity reduction appears to be linearly related to the percentage of wax used. It should be noted that 1% data was only obtained with 3-waxes so this comment is based on a limited data set.

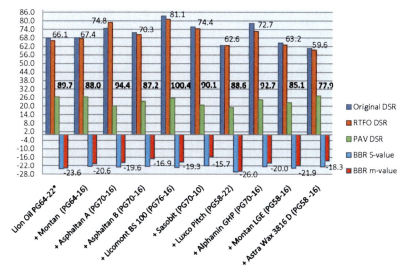

Notes: Numbers at top of bars indicates high grade passing temperature. Numbers at bottom of bars indicates low grade temperature. Numbers in middle of figure indicates PG grade range using M320 Table 1 criterion.

Fig. 1. PG true grades (AASHTO M320 Table 1) for control and 3%wax modified products

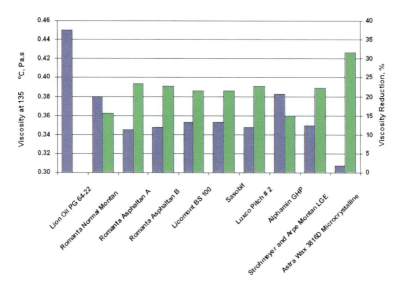

Fig. 2. Brookfield viscosity

3.1 BBR Binder and Mixture Testing

A simple BBR test to determine binder low temperature properties from asphalt mixtures was developed by Zofka et al. [11] at the University of Minnesota. This mixture BBR test was performed concurrent with binder BBR testing, BBR stiffness/physical hardening testing was performed on mixture beams. In this test the BBR measures the mid-point deflection of a beam of asphalt mixtures subjected to a constant load applied to the mid-point of the beam. The BBR operates only in the loading mode; recovery measurements were not obtained. Conditioned test beams were placed in the controlled temperature fluid bath at -12°C and -18°C (for both binder and mixture), temperature saturation testing was performed at 0, 1, 2, 4, 8, 16 and 32 days. Mixture specimens were loaded with a constant load (1961 ±50 mN or 4413 ±50 mN) for 1000s. The test load and the midpoint of deflection of the beam are monitored versus time.

The maximum bending stress at the midpoint of the beam is calculated from the dimensions of the beam, the span length, and the load applied to the beam for loading times of 8, 15, 30, 60, 120, and 240 seconds. The maximum bending strain in the beam is calculated for the same loading times from the dimensions of the beam and the deflection of the beam. The stiffness of the beam for the loading times specified is calculated by dividing the maximum stress by the maximum strain.

Data from the BBR binder and mixture testing is presented by Baumgardner et al. [8].

4 Discussion

Initially it was planned to construct master curves from both binder and mixture BBR data. However, when the data was inspected it became apparent that in many cases the stiffness of the BBR data collected at the colder test temperature (-18°C) was less stiff than the data at -12°C at the extended conditioning times. This can be seen by inspection of the data in Table 3 and 4. The stiffness drops as isothermal conditioning time is increased at the -18°C temperature. An example of this is shown in illustrated in Figure 3 by the variation in BBR stiffness at 60 seconds for the materials considered. It should be noted that this phenomena also occurred for the control binder as well as the wax modified products. At -12°C the data was generally as expected, with increasing stiffness with time and the large drop in stiffness at the intermediate conditioning times did not occur.

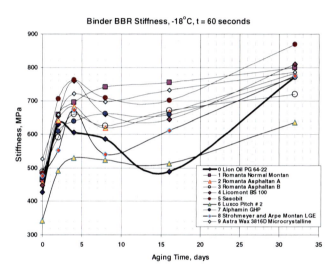

Fig. 3. BBR Binder Stiffness at t=60 seconds for -18°C data versus conditioning time (days)

An example of the results obtained for the mixture BBR tests is illustrated in Figure 4. The data in this figure shows the three replicates and average results for the initial testing conducted. It can be clearly seen that the -18°C data has stiffness isotherms that are considerably lower than the data collected for -12°C for the initial set of testing. After this phenomenon was observed it was decided to retest the beams two times, the first after sitting in a lab environment for a few weeks and then again after an annealing period. The annealing temperature of 64°C was selected as it was believed to represent what may actually occur in actual pavement performance with 64°C representing pavement temperature during a reasonably warm day. The test data from the annealed results are also shown in Figure 4 and these results rank in the manner expected with the data from the -18°C isotherm being higher than the -12°C isotherm. The data from the retesting before the annealing also showed the correct order of results but with the -18°C data lying between that obtained for the -12°C data and that representing the -18°C data after annealing.

In all cases what is believed to be healing of the specimen occurred after both laboratory ambient conditioning and overnight annealing at 64°C resulting in increased stiffness of the -18°C isotherm. Most significantly, it resulted in the BBR stiffness of the -18°C isotherm being greater than the -12°C isotherm as was expected. This stiffness reduction is believed to be due to micro-crack formation over extended low temperature saturation and is significant at -18°C. The result supports the research reported by Eshan et al. [12] who showed that cracking of asphalt mixtures, as monitored by acoustic emission events, started to occur at a temperatures 2 to 11°C higher that the PG low temperature grade of the mix.

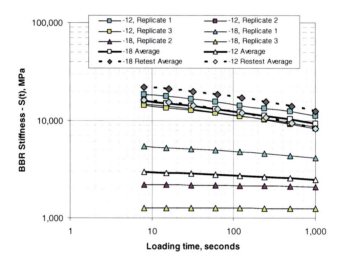

Fig. 4. BBR Mix Stiffness Isotherms of Sasobit (#5) at 16 days conditioning showing replicates, average of initial testing and average testing of annealed specimens

5 Conclusions

Addition of select waxes (Fischer-Tropsch and Fatty Acid Amides) to asphalt is an accepted practice in warm-mix asphalt production. Additionally Fischer-Tropsch, Montan wax and Montan wax blends have been used in Europe for several years as compaction aids for bituminous mixtures. Addition of waxes to binders prompts concern as to detrimental effects they may have on asphalt binder performance, especially fatigue and low temperature performance. In this study one base asphalt and nine wax additives, for possible use in warm-mix asphalt binders, were used to evaluate the effect binders modified with wax additives have on asphalt mixture properties in mixture bending beam rheology testing.

Results of binder testing reported by [8] revealed that the low temperature binder Tg values predicted by the peak in the G" data using the DSR are comparable to the predicted binder low PG grade values from the BBR. Low temperature binder Tg values predicted by the peak in the G" data using the DSR were also generally similar to the Tg values predicted by Modulated DSC. Some significant exceptions may have been related to wax chemical type, solubility of some wax fraction in the binder, differences in heating and cooling rates or some other source. More comparative work needs to be performed to account for these results and to determine which Tg value is more closely related to mixture performance. Physical hardening of binders does not appear to be a major problem at the lower addition rates and the loss of temperature range appears to be a good indication if that is a problem.

Data produced by the mixture BBR testing suggested that at temperatures close to or below the glass transition temperature the mixture beams were becoming less stiff with time when held at a constant temperature. This effect appears to be reversible if a heating/annealing cycle was applied to the beams. General assumption is that the phenomenon of loss of stiffness during long low temperature saturation and recovery of stiffness upon annealing is due to microcracking and healing of the asphalt binder in the mixture. This seems to be supported by the work of [12] which reports that micro-cracking may actually occur at a temperature 2-11 degrees higher than the binder low temperature PG grade temperature. It should be noted that even the control binder with no wax exhibited this behaviour.

The theory of micro cracking and healing upon extended low temperature saturation is a phenomenon that merits continued research. The acoustic emission testing performed by University of Illinois [12] could provide a useful avenue for extending the research work conducted with these binders and mixes. The variability in performance due to micro-cracking that occurs above the low temperature PG grade is not currently evaluated or understood.

Acknowledgements. The authors thank the support and assistance of Dr. Isaac Howard, Mississippi State University, Department of Civil and Environmental Engineering, during the development of this work and the production of this paper.

References

[1] Brule, B., Planche, J.-P., King, G., Claudy, P., Letoffe, J.M.: Relationships Between Characterization of Asphalt Cements by Differential Scanning Calorimetry and Their Physical Properties. In: Proceedings of the American Chemical Society Symposium on Chemistry and Characterization of Asphalts, Washington, DC (1990)

[2] Anderson, D.A., Christensen, D.W., Bahia, H.U., Dongre, R., Sharma, M.G., Antle, C.E., Button, J.: Binder Characterization and Evaluation, Physical Characterization, Strategic Highway Research Program. In: Report ref. SHRP-A-369, vol. 3. National Research Council, DC (1994)

[3] Planche, J.-P., Claudy, P.M., Letoffe, J.M.: Using Thermal Analysis Methods to Better Understand Asphalt Rheology. Thermochimica Acta 324, 223–227 (1998)

[4] Robertson, R. E., Thomas, K.P., Harnsberger, P.M., Miknis, F.P., Turner, T.F. Branthaver, J.F., Huang, S-C., Pauli, A. T., Netzel, D. A., Bomstad, T. M., Farrar, M. J., Rovani, Jr., J. F., McKay, J. F., McCann, M., Sanchez, D., Wiser, W. G., Miller, J.: Fundamental Properties of Asphalts and Modified Asphalts II, Final Report. Volume II: New/Improved Test Methods, Federal Highway Administration, Contract No. DTFH61-99C-00022 (2005) (submitted for publication)
[5] Michon, L.C., Netzel, D.A., Turner, T.F., Martin, D., Planche, J.-P.: A 13C NMR and DSC Study of the Amorphous and Crystalline Phases in Asphalts. Energy & Fuels 13(3), 602–610 (1999)
[6] Edwards, Y.: Influence of Waxes on Bitumen and Asphalt Concrete Mixture Performance, Doctorial Thesis in Highway Engineering, Stockholm Sweden (2005)
[7] Radenburg, M.: Temperature Reduced Asphalts - Basics and Experiences, Presentation at German Federal Highway Research Institute, BASt (2007)
[8] Baumgardner, G.L., Reinke, G., Anderson, D.A., Rowe, G.M.: Laboratory Evaluation: Wax Additives in Warm-Mix Asphalt Binders, Report from FHWA Binder Expert Task Group and Warm Mix Asphalt Technical Working Group, Report submitted to Engineering & Software Consultants, Inc. and Federal Highways Administration (2009)
[9] GDOT, Qualified Products List, Georgia Department of Transportation, Office of Materials and Research (2009)
[10] Rowe, G.M., Baumgardner, G.L., Reinke, G., Anderson, D.A.: Wax Additives in Warm-Mix Asphalt Binders and Performance in the Multi-Stress Creep and Recovery Test (MSCR). Paper submitted to the 5th Eurasphalt & Eurobitume Congress, Istanbul (2012)
[11] Zofka, A., Marasteanu, M.O., Xinjun, L., Clyne, T.R., McGraw, J.: Simple Method to Obtain Asphalt Binders Low Temperature Properties from Asphalt Mixture Properties. Journal of the Association of Asphalt Paving Technologists 74, 255–282 (2005)
[12] Eshan, V.D., Behnia, B., Ahmed, S., Buttlar, W.G., Reis, H.: Low Temperature Fracture Evaluation of Asphalt Mixtures using Mechanical Testing and Acoustic Emissions Techniques. Journal of the Association of Asphalt Paving Technologists (2011)

Reducing Asphalt's Low Temperature Cracking by Disturbing Its Crystallization

Ellie H. Fini[1] and Markus J. Buehler[2]

[1] North Carolina A&T State University
[2] Massachusetts Institute of Technology

Abstract. This paper investigate effect of a new bio-based modifier, "bio-binder" on disturbing bituminous asphalt crystallization using molecular dynamics simulation and X-Ray powder diffraction while studying effects of introduction of bio-binder on bituminous asphalt low temperature properties and low temperature cracking resistance. The proposed bio-binder is produced from the thermochemical conversion of swine manure. Bio-binder is then blended with virgin binder to produce bio-modified binder (BMB). Bio-binder can be used as a renewable partial replacement for petroleum-based asphalt. The production and application of bio-binder can facilitate swine waste management. In addition, the bio-binder resources (swine manure) is renewable and is not competing with food supplies. This paper argues that the improved low temperature rheological properties in BMB can further enhance pavement low-temperature cracking resistance.

1 Introduction

The U.S. asphalt binder market is valued at approximately $11.7 billion/year [1]. Asphalt binder supplies are shrinking, while the demand for it is increasing rapidly [2, 3]. When the price of asphalt binder increased from $235/ton in 2004 to $520/ton in 2007, it represented an increase of 53% in price [4]. As the price of asphalt binder increases, the demand for alternative and renewable binder resources increases. This motivated several unsuccessful attempts by researchers to produce bio-asphalt from various materials (sugar, molasses, potato starches, vegetable oils, lignin, cellulose, palm oil waste, coconut waste, and dried sewage). However, those bio-asphalts either found not to be feasible or never reached the asphalt market due to low performance or high production cost [5, 6]. To the best knowledge of the authors of this article , no one has developed a bio-asphalt from swine manure, and this is the first attempt in this area. Therefore, the main application of bio-binder is envisioned to be as a modifier for the petroleum-based asphalt binder used in pavement construction. Although other applications such as roofing, soil stabilization, crack and joint sealing and carpeting are recognized for bio-binder, considering the huge extent of the paving application, this paper focuses only on the application of bio-binder as a modifier for the asphalt binder used in pavement construction. It is estimated that the 40.2 million tons of swine manure (solid weight) produced in the U.S. annually [7] can supply about 28 million tons of bio-binder to be used in pavement (calculation based on 70% conversion efficiency [1]. Pork production is a major agricultural enterprise in the U.S., involving over 75,000 swine producers, creating 35,000 full-time equivalent jobs directly and an additional 515,000 jobs indirectly. With the sale of 116 million pigs in 2008, the U.S. hog industry generated gross income of $16 billion [8]. Generally, growing and finishing pigs weighing 21 to 100 kg can be expected to generate 0.39 to 0.45 kg of waste per day per pig on a dry matter basis [9, 10]. Derived from swine manure, bio-binder is composed mostly of carbon (about 72%), which is sequestered from the manure [11]. Swine manure is disposed of by storing it in lagoons. This process has significant negative environmental impacts, particularly with respect to surface water and groundwater quality and to air quality as affected by odors and gaseous emissions from large-scale swine production operations [10]. To make use of manure, researchers developed various methods to convert manure to gas and oil [11,12]. Fini and her coworker further utilized the oil to produce bio-binder [13-16]. This paper discusses how introduction of bio-binder can enhance low temperature rheological properties of asphalt binder, which in turn can facilitate application of RAP and RAS in new pavement construction.

2 Chemical Characterization of Bio-binder

The chemical characterization of bio-binder is needed to better understand the material physical characteristics and its effect on petroleum-based asphalt when used as modifier. This in turn, can help predict the performance characteristics of the bio-modified binder before its actual placement in the field. Previous works demonstrated our initial efforts in determining the chemical characteristics of the bio-binder material and how it compares directly with petroleum asphalt binders [13, 16].

An elemental analysis was first performed on the bio-binder to determine the carbon, hydrogen, nitrogen, and oxygen content. The pig waste-derived bio-binder has much less aromatic character than conventional binders but is more polar, with roughly four times the average nitrogen level as well as 10 times the average oxygen level.

Though useful in understanding the chemical makeup, the elemental analysis does not provide enough information to determine the potential compatibility of the biomaterial with conventional binders. Several other analytical techniques were employed to improve the understanding of the material. First, a chromatographic separation of the bio-binder into the solvent-defined saturates, aromatics, resins, and asphaltenes fractions (SARA) showed major differences between the bio-binder and the typical petroleum asphalt. Our analysis showed that bio-binder mainly consists of resins and asphaltenes with almost no evidence of saturates and aromatics. The elemental and SARA results (Table 1) indicate the potential for the bio-binder material to improve mixture moisture damage resistance due to the higher concentration of polar nitrogen and oxygen containing functional groups [17-19]. These same polar components are an indication that the bio-binder is a promising candidate for use in crack and joint sealants and roofing shingles..

Table 1. SARA Components of Bio-adhesive and AAD-1

Adhesive Type	Saturate (aliphatic) (wt %)	Naphthene Aromatics, (wt %)	Polar aromatic (Resin), (wt %)	Asphaltenes, Percentage, (wt %)
Bio-Adhesive	2.48	1.67	45.87	43.39
AAD-1	8.6	41.3	25.1	20.5

Further investigation of the bio-binder's chemical composition was conducted utilizing nuclear magnetic resonance (NMR) spectroscopy, gas chromatography-mass spectrometry (GC-MS) and Fourier transform infrared spectroscopy (FT-IR). Proton NMR showed the presence of olefinic carbons as well as alcohols or esters, none of which are found in petroleum asphalt and are susceptible to oxidation chemistry [13]. The carbon-13 NMR spectrum indicated that the bio-binder is comprised mainly of carbons in straight chain aliphatic compounds. The 1H NMR spectrum showed presence of olefins and alcohols. The GC-MS data agreed well with results from other techniques, indicating presence of olefinic carbons as well as a variety of oxygen and nitrogen moieties. An approximate molecular weight distribution of between 250 g/mol to 450 g/mol, also found by GC-MS, found to be much lower than the 700 g/ml estimated for petroleum binders which is comprised of a mixture of multi-ring systems and linear, aliphatics. Finally, FTIR data was in agreement with the above data that the amount and type of aliphatic hydrocarbons in bio-binder is different from petroleum asphalt with indications of olefins, amines, alcohols, and aromatics. the analysis showed that bio-binder has a high level of nitrogen relative to typical asphalt binder but a very close ratio of carbon to hydrogen content. The specific gravity of bio-binder (1.01) was found to be close to that of asphalt binder (1.03). It was also shown that both materials have similar solubility characteristics in common asphalt solvents.

3 Rheological Characterization

To investigate the effect of incorporation of bio-binder on pavement performance, bio-binder was added to PG 64-22. To modify the binder, bio-binder was added to the base binder (PG 64-22) at 2, 5, and 10 percent by weight of the base binder to produce bio-modified binder. Bio-binder and base binder were heated to 60°C and 120°C, respectively. The base binder (Table 2) and bio-binder were mixed thoroughly at shear rate of 3000 rpm for 30 minutes while the temperature was kept at 120°C. Specimens were prepared and tested to study low temperature rheological properties of modified and non-modified binder.

Table 2. Properties of Base Binder PG 64-22

Specific Gravity @15.6 °C	Flash Point, Cleveland Open Cup, °C	Change in Mass RTFO	Absolute Viscosity at 60 °C, Pa.s
1.039	335	-0.0129	202

3.1 Creep Stiffness and M-Value

To examine the effect of bio-modification on low temperature properties of binder, creep compliance, stiffness, m-value and cracking temperature were measured using Bending Beam Rheometer (BBR) according to ASTM D6648.

The BBR evaluates the binder efficiency at low temperature and its propensity to crack. Thermal cracking is caused by stresses build up during pavement contraction when the temperature drops rapidly. The accumulated stress may exceed the stress relaxation capability of the material, resulting in crack initiation. The temperature at which pavements show specific stiffness is defined as the limiting stiffness temperature. Figure 1 shows master curves at low temperature for modified and non-modified specimens. It can be seen addition of bio-binder decreases stiffness. Figure 2 shows the m-value increases due to the addition of bio-binder, improving binders' stress relaxation capability which results in less stress accumulation. At 5% and 10% modification the specimens were too soft to be tested and their deflections were above the equipment range. It is expected that the improvement in low temperature properties of the binder results in reduced low temperature cracking due to the general reduction in binder stiffness and increase in m-value.

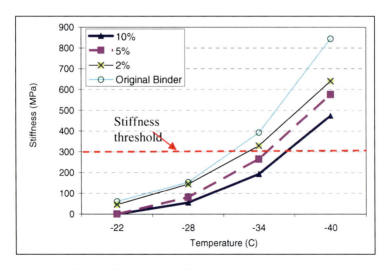

Fig. 1. Stiffness for modified and non modified binder

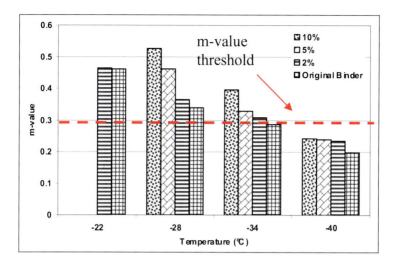

Fig. 2. M-value at various temperature for modified and non-modified binder

3.2 Asphalt Cracking Temperature

The cracking temperature, Tcr, is usually calculated from Bending Beam Rheomter (BBR) test using the AASHTO T 313 procedure. The higher value between the temperature where the stiffness at a loading time of 60s is 300 MPa and the temperature where the m-value at a loading time of 60s is 0.3 is considered as the Tcr. Using this method, the cracking temperatures for modified and non-modified binders were calculated. In addition to BBR test, the Asphalt Binder Cracking Device (ABCD) was used to determine the cracking temperature of the modified and non-modified asphalt binder after a progressive temperature drop from room temperature through 0°C to -60°C. In ABCD test, the micro-strains developed in the binder specimen at the cracking temperature are used as a measure of the thermal cracking resistance. Table 3 shows the cracking temperature of the tested specimen using BBR and ABCD test. As can be seen, the cracking temperature decreases as the amount of bio-binder increases with BMB-10 showing cracking temperature of -37.3 °C which is 5.6 degree lower that that of base binder which is -31.7 °`C.

Table 3. Cracking Temperature of Asphalt Specimens

Binder	BBR	ABCD
PG 64-22	-31.7	
2% BMB	-33.1	-32.4
5% BMB	-34.7	-37.9
10% BMB	-36.3	-37.3

As can be seen, the cracking temperature decreases as the amount of bio-binder increases, which in return, may improve pavements' low temperature cracking resistance. To understand the mechanism at the molecular level, following molecular dynamics simulation and X-Ray analysis were conducted.

4 Molecular Dynamics Simulation and X-Ray Powder Diffraction

The enhancement in low temperature cracking of BMB can be attributed to disturbed pi-pi stacking and reduced crystallization of asphalt molecules due to presence of bio-binder. This in turn, increases flexibility of alkyl chains. Therefore, reduction of pi-pi stacking can facilitate segmental motion in asphalt molecules and increase BMB's amorphous components. To study this phenomenon in molecular level, X-ray powder diffraction analysis was conducted on BMB samples at both room temperature and after being refrigerated for 1 hour (Figure 3). Powder XRD can be used to determine the crystallinity by comparing the integrated intensity of the background pattern to that of the sharp peaks. XRD results showed significantly broader diffraction peak for BMB (2% BMB) compared to the base binder (PG 64-22). Also the peaks were broader in both cases (modified and non-modified binder) at room temperature compared to those at lower temperature when specimens were chilled to 5°C. It should be noted that in contrast to a crystalline pattern consisting of a series of sharp peaks, amorphous materials produce a broad peak. This observation further confirms that crystallized structure was increased by reduction in temperature. However, the level of crystallization in modified binder was less than non-modified binder.

To further examine the experimental observations, molecular interactions within BMB was studied using molecular dynamic simulation (MD) for base asphalt and BMB MD simulations represent a numerical implementation to solve the equations of motion of a system of atoms or molecules, to predict the motion of each atom in the material characterized by the atomic position, velocity, and acceleration. The collective behavior of atoms then is used to understand how a material undergoes deformation, phase change, or other phenomena by relating the atomic scale to meso or macroscale phenomena [21].

In this study molecular structures of BMB and base binder were constructed using the geometric parameters and the graphic resources of the Materials Studio (version 5) program [22]. The structures were initially minimized by Reax-FF reactive force field in an existing MD code called LAMMPS [23]. After constructing the atomistic models, computation experiments was conducted under NVT ensemble at 300 K. It was found that presence of bio-binder molecules changes asphalt molecules' chain conformation, hindering the pi-pi stacking of asphalt molecules. This in turn can cause BMB to remain more flexible at low temperatures compared to non-modified asphalt.

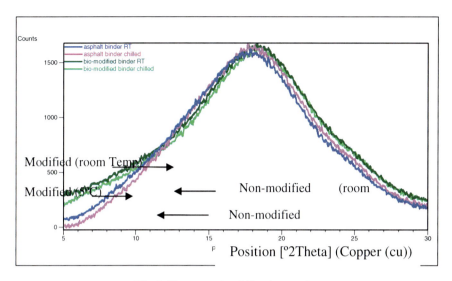

Fig. 3. X-ray powder diffraction results

5 Discussion

This paper discusses effect of application of bio-binder on bituminous asphalt's low temperature rheological properties. In addition, production and application of bio-binder including its impact of waste management has been discussed. The proposed bio-binder is produced from the thermochemical conversion of swine manure. Bio-binder is then blended with virgin binder to produce bio-modified binder (BMB). Bio-modified binder chemical and rheological characterization was used to determine effect of bio-modification on bituminous asphalt performance. It is shown that bio-binder, upon its introduction to bituminous asphalt binder; improves low temperature rheological properties of bituminous binder by disturbing asphalt molecules stacking resulting in delayed crystallization at low temperature. The low temperature cracking prediction using the BBR and ABCD test methods showed that cracking temperature decreases as the amount of bio-binder increases indicating that BMB is more flexible than base binder at low temperature. This was further attributed to the extended amorphous region in BMB. Extension of amorphous region due to the introduction of bio-binder was further confirmed by XRD analysis and was examined by MD simulation.

Acknowledgements. The materials in this paper are based upon work supported by the National Science Foundation (grants number 0955001 and 1150695) as well as partial support of the Materials Research Science and Engineering Center at the Massachusetts Institute of Technology. The author would like to acknowledge the guidance and assistance of Dr. S. Speakman with MIT, Dr. E. Kalberer with NuVention Solutions Inc. and Dr. Z. You with Michigan Tech. The contents of this paper reflect the view of the authors, who are responsible for the facts and the accuracy of the data presented herein. This paper does not constitute a standard, specification, or regulation.

References

1. Specialists in Business Information (SBI), Asphalt Manufacturing in the U.S. Market, Research Report, R460-201(2009), http://www.sbireports.com/about/release.asp?id=1289 (accessed July 15, October 2011)
2. National Asphalt Pavement Association, http://www.hotmix.org (accessed November 22, 2011)
3. US Department of Transportation, National Transportation Statistics. Bureau of Transportation Statistics. Washington, DC (2005), http://www.transtats.bts.gov
4. Hassan, M.M.: Life-Cycle Assessment of Warm-Mix Asphalt: An Environmental and Economical Perspective. Paper No. 09-0506. In: The 88th Transportation Research Board Annual Meetings, Washington, DC (2009)
5. Abdel Raouf, M., Williams, R.C.: General Physical and Chemical Properties of Bio-Binders Derived from Fast Pyrolysis Bio-oils. In: Proceedings of the 2010 Mid-Continent Transportation Research Forum, Madison, WI (2010)
6. Airey, G., Mohammed, M.H., Fichter, C.: Rheological Characteristics of Synthetic Road Binders. Fuel 87(10), 1763–1775 (2008)
7. Blue Marble Energy - AGATE Conversion and Biochemical Recovery Technology (2009), http://bluemarbleenergy.net/technology
8. Ecopave Australia - Ecopave Australia Bio-Bitumen Asphalt Concrete Research (2009), http://www.ecopave.com.au/bio_bitumen_asphalt_concrete_research_ecopave_australia_005
9. National Research Council Canada, New NRC Initiative on Bioproducts for Construction (2009), http://www.nrc-cnrc.gc.ca/eng/ibp/irc/ci/v14no2/7.html
10. USDA, Agricultural Statistics. USDA National Agricultural Statistics Service, Washington, DC, http://www.usda.gov (accessed August 31, 2005)
11. Ocfemia, K.: Hydrothermal Process of Swine Manure to Oil Using a Continuous Reactor System AAT 3202149. University of Illinois at Urbana-Champaign, Dissertation (2005)
12. Xiu, S.N., Rojanala, H.K., Shahbazi, A., Fini, E.H., Wang, L.: Pyrolysis and Combustion Characteristics of Bio-Oil from Swine Manure. Journal of Thermal Analysis and Calorimetry (2011) (in Press), http://www.springerlink.com/content/d784007tx7551501, doi:10.1007/s10973-011-1604-8
13. Fini, E.H., Kalberer, E.W., Shahbazi, G., Basti, M., You, Z., Ozer, H., Aurangzeb, Q.: Chemical Characterization of Bio-Binder from Swine Manure: A Sustainable Modifier for Asphalt Binder. ASCE Journal of Materials, American Society of Civil Engineering (ASCE) 23(11), 1506–1513 (2010, 2011), http://dx.doi.org/10.1061/ASCEMT.1943-5533.0000237
14. Fini, E., Shahbazi, G.H., Xiu, S., Zada, B.: Bio-Binder from Swine Manure: Production of a Sustainable Binder for Asphalt Pavement Construction. In: 1st International Conference on Green and Sustainable Technology, Greensboro, NC (2010)
15. Mogawer, W.S., Fini, E., Austerman, A.J., Booshehrian, A., Zada, B.: Performance Characteristics of High RAP Biomodified Asphalt Mixtures, Paper No. 12-2411. In: The 91st Transportation Research Board Annual Meetings, Washington, DC (2012)

16. Fini, E.H., Al-Qadi, I.L., Xiu, S., Mills-Beale, J., You, Z.: Partial Replacement of Binder with Bio-binder: Characterization and Modification to Meet Grading System. Submitted to International Journal of Pavement Engineering (2011), http://www.tandfonline.com/doi/abs/10.1080/10298436.2011.596937
17. Petersen, J.C., Plancher, H., Ensley, E.K., Miyake, G., Venable, R.L.: Chemistry of the Asphalt–Aggregate Interaction: Relationships with Pavement Moisture Damage Predication Tests. In: Transportation Research Record 483, pp. 95–104. TRB, National Research Council, Washington, D.C (1982)
18. Petersen, J.C., Branthaver, J.F., Robertson, R.E., Harnsberger, P.M., Duvall, J.J., Ensley, E.K.: Effects of Physicochemical Factors on Asphalt Oxidation Kinetics. In: Transportation Research Record, vol. (1391), pp. 1–10. TRB, National Research Council, Washington, D.C (1993)
19. Rostler, F.S., White, R.M.: Influence of Chemical Composition of Asphalts on Performance, Particularly Durability. American Society for Testing Materials, Special Technical Publication 277, 64–88 (1959)
20. Schabron, J.F., Rovani Jr., J.F.: On-column precipitation and re-dissolution of asphaltenes in petroleum residua. Fuel 87(2), 165–176 (2007)
21. Buehler, M.J.: Atomistic Modeling of Materials Failure. Springer, New York (2008)
22. Materials Studio, http://accelrys.com/products/materials-studio/index.html (accessed January 17, 2011)
23. Plimpton, S.: Fast Parallel Algorithms for Short-Range Molecular Dynamics. Journal of Computational Physics 117, 1–19 (1995)

Mechanistic Evaluation of Lime-Modified Asphalt Concrete Mixtures

Amjad H. Albayati

Civil Engineering Department, University of Baghdad, Iraq

Abstract. Frequently, Load associated mode of failure (rutting and fatigue) as well as, occasionally, moisture damage in some sections poorly drained are the main failure types found in some of the newly constructed road within Baghdad as well as other cities in Iraq. The use of hydrated lime in pavement construction could be one of the possible steps taken in the direction of improving pavement performance and meeting the required standards.

In this study, the mechanistic properties of asphalt concrete mixes modified with hydrated lime as a partial replacement of limestone dust mineral filler were evaluated. Seven replacement rates were used; 0, 0.5, 1, 1.5, 2, 2.5 and 3 percent by weight of aggregate. Asphalt concrete mixes were prepared at their optimum asphalt content and then tested to evaluate their engineering properties which include moisture damage, resilient modulus, permanent deformation and fatigue characteristics. These properties have been evaluated using indirect tensile strength, uniaxial repeated loading and repeated flexural beam tests. Mixes modified with hydrated lime were found to have improved fatigue and permanent deformation characteristics, also showed lower moisture susceptibility and high resilient modulus. The use of 2 percent hydrated lime as a partial replacement of mineral filler has added to local knowledge the ability to produce more durable asphalt concrete mixtures with better serviceability.

1 Introduction

In the recent five years, some of the newly constructed asphalt concrete pavements in Baghdad as well as other cities across Iraq have shown premature failures with consequential negative impact on both roadway safety and economy. Frequently, Load associated mode of failure (rutting and fatigue) as well as, occasionally, moisture damage in some sections poorly drained are the main failure types found in those newly constructed roads. Investigations on the reasons beyond these failure showed that it can be grouped into two categories, extrinsic and intrinsic, the first one due to the heavy axle loading coupled with relentless high summer temperatures (ambient air temperature for nearly three months can reach 50 degree Celsius and pavement surface temperature can reach up to 60 degree Celsius) , whereas the second category is limited to the mixture itself , improper gradients, excess use of natural sand and lack of mineral filler all of these factors acts either in collect or in single manner for the deterioration in the mix strength and also loss of durability of asphalt concrete pavement.

Based on the preceding it is clear that there is a real need to the development of modified asphalt concrete mixtures to improve the overall performance of pavements. The use of hydrated lime in pavement construction could be one of the possible steps taken in this direction. In the United States of America, Hydrated lime has been added to hot mix asphalt pavements for over 30 years, improving the mixtures in many ways and increasing the life of highways. Extensive experimental studies have revealed that the use of hydrated lime in Hot-Mix Asphalt (HMA) mixtures can reduce permanent deformation, long-term aging, and moisture susceptibility of mixtures. In addition, it increases the stiffness and fatigue resistance of mixtures. The structure of hydrated lime consists of different size fractions. The larger size fraction performs as a filler and increases the stiffness of the bituminous mixture. The smaller size fraction increases binder film thickness, enhancing viscosity of the binder, and improving the binder cohesion and stiffness.

In view of this , the primary objective of this study is to evaluate the mechanical properties of asphalt concrete mixtures containing hydrated lime (as a partial replacement of limestone filler) based on the following tests, Marshall properties (Mix Design), Indirect tensile test (Moisture susceptibility) , uniaxial repeated load test (Resilient Modulus and permanent deformation) and repeated flexural beam test (fatigue characteristics).

2 Background

Hydrated lime which is also known as calcium hydrate ($Ca(OH)_2$) has been used in asphalt mixes for a long time, both as mineral filler and as an antistripping additive. Researchers observed that when hydrated lime coats an aggregate particle, it induces polar components in asphalt cement to bond to the aggregate surface. This effect also inhibits hydrophilic polar groups in the asphalt from congregating on the aggregate surface (McGennis et al. 1984). In addition, lime can neutralize acidic aggregate surfaces by replacing or coating acidic compounds and water-soluble salts on the aggregates and can react pozzolanically to remove deleterious materials (Epps et al. 2003).

Al-Suhaibani [1992] evaluated the mineral filler properties of hydrated lime and other local fillers available in Saudi Arabia. The mechanical properties of the mixes were studied using tests such as the resilient modulus test, the indirect tensile strength test, Hveem stability, and Marshall criteria. The research results revealed that the amount and characteristics of the mineral fillers can have an effect on the rutting susceptibility of flexible pavements, and that the use of hydrated lime can improve resistance of the mixes to rutting. The lime showed improved stiffening properties when incorporated into the mixture.

Shahrour and Saloukeh [1992] conducted a research study to evaluate the influence of ten types of different fillers (including hydrated lime) on the physical properties of filler-bitumen mixtures and two types of asphalt mixtures namely- Asphaltic concrete (AC) and Dense Bitumen Macadam (DBM) commonly used in Dubai, U.A.E. The mixtures were designed using Marshall mix design method, and the fillers were incorporated in various ratios to the mixtures. Marshal parameters (% VFB, % VIM, % VMA, and Bulk Specific Gravity) for asphalt mixtures were reported to be not significantly affected by changing the type of filler at specific filler contents. On the other hand, Penetration, Ring and Ball Softening point, absolute Viscosity, and Kinematic Viscosity tests on the filler-binder mixtures (mastics) showed that all types of mineral fillers acted as an extender to the binder with minimal stiffening effect. But comparing to others, hydrated lime showed superior stiffening performance. The authors also recommended to use hydrated lime as a mineral filler in a ratio of 0.5 to 0.8 of the bitumen content in the asphalt mixtures. Afterwards Paul, 1995; Khosla et al., 2000; Mohammad et. al., 2000; Little et al., 2005; Atud et al., 2007; Khattak and Kyatham, 2008 conducted numerous researches to evaluate the influence of hydrated lime on the moisture damage of HMA pavements. In those studies, hydrated lime was reported to improve the resistance against the moisture induced damages of HMA mixtures. By maintaining a good adhesion between the aggregate and the asphalt cement in the presence of water, hydrated lime worked successfully as an antistripping agent. Its ability to reduce viscosity building polar components in the asphalt binder enabled hydrated lime to show effect as an oxidation reducing agent. Also, its ability to increase mixture stiffness by filling air voids in the mixture with its tiny particles makes it effective mineral filler.

Lime-treated mixtures also showed cost efficiency in terms of pavement life. Sebaaly et al. [2003] conducted a research to quantify the improvements of pavement performance that contained lime. Performances of HMA mixtures from the northwestern part of Nevada were evaluated both in the laboratory and in the field. In the laboratory evaluation, both lime treated and untreated sections were sampled and then evaluated through laboratory test. On the other hand, pavement performance data from pavement management system (PMS) were used to assess field performance of lime treated and untreated sections. The study showed that lime treatment on HMA mixtures significantly improved their moisture resistance and resistance to multiple freeze-thaw cycles than that of untreated HMA mixtures. From the long-term pavement performance data it was also evident that under similar environmental and traffic conditions, lime treated mixtures provided better performance with lesser maintenance and rehabilitation activities. Again, the analysis of the impact of lime on pavement life indicated that lime treatment extended the performance life of HMA pavements by an average of 3 years which represented an average increase of 38% in the expected pavement life.

Hydrated lime can be introduced into asphalt mixes by several methods: lime slurry to dry or wet aggregate, dry lime to wet aggregate, dry lime to dry aggregate and dry lime to asphalt. Although little researches have been done to quantify the difference in effects of these methods, it is sufficient to say that asphalt mixes benefit from the addition of hydrated lime, no matter how it is introduced into the mix (Epps et al. 2003). Typically, the amount of hydrated lime added is 1 to 2 percent by weight of the mix, or 10 to 20 percent by weight of the liquid asphalt binder but If an aggregate has more fines percent, it may be necessary to use more lime additive due to the increased surface area of the aggregate.

3 Material Characterazation

The materials used in this work, namely asphalt cement, aggregate, and fillers were characterized using routine type of tests and results were compared with State Corporation for Roads and Bridges specifications (SCRB, R/9 2003).

3.1 Asphalt Cement

The asphalt cement used in this work is a 40-50 penetration grade. It was obtained from the Dora refinery, south-west of Baghdad. The asphalt properties are shown in Table (1) below.

Table 1. Properties of Asphalt Cement

Property	ASTM designation	Penetration grade 40-50	
		Test results	SCRB specification
1-Penetration at 25C,100 gm,5 sec. (0.1mm)	D-5	47	40-50
2- Rotational viscosity at 135°C (cP.s)	D4402	519	......
2- Softening Point. (°C)	D-36	47	......
3-Ductility at 25 C, 5cm/min,(cm)	D-113	>100	>100
4-Flash Point, (°C)	D-92	289	Min.232
5-Specific Gravity	D-70	1.041	......
6- Residue from thin film oven test	D-1754		
- Retained penetration,% of original	D-5	59.5	>55
- Ductility at 25 C, 5cm/min,(cm)	D-113	80	>25

3.2 Aggregate

The aggregate used in this work was crushed quartz obtained from Amanat Baghdad asphalt concrete mix plant located in Taji, north of Baghdad, its source is Al-Nibaie quarry. This aggregate is widely used in Baghdad city for asphaltic mixes. The coarse and fine aggregates used in this work were sieved and recombined in the proper proportions to meet the wearing course gradation as required by SCRB specification (SCRB, R/9 2003). The gradation curve for the aggregate is shown in Figure (1).

Routine tests were performed on the aggregate to evaluate their physical properties. The results together with the specification limits as set by the SCRB are summarized in Table (2). Tests results show that the chosen aggregate met the SCRB specifications.

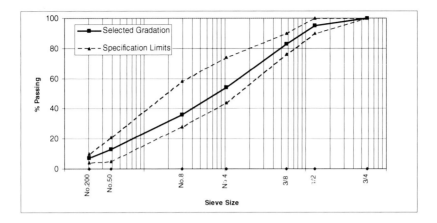

Fig. 1. Aggregate Gradation

Table 2. Physical Properties of Aggregates

Property	ASTM designation	Test results	SCRB specification
Coarse aggregate			
1. Bulk specific gravity	C-127	2.614	......
2. Apparent specific gravity		2.686	......
3. Water absorption,%		0.441	......
4. Percent wear by Los Angeles abrasion ,%	C-131	17.5	30 Max
5. Soundness loss by sodium sulfate solution,%	C-88	3.4	10 Max
6. Fractured pieces, %		98	90 Min
Fine aggregate			
1. Bulk specific gravity	C-127	2.664	......
2. Apparent specific gravity		2.696	......
3. Water absorption,%		0.724	......
4. Sand equivalent,%	D-2419	57	45 Min.

3.3 Filler

The filler is a non plastic material that passing sieve No.200 (0.075mm). In this work, the control mixes were prepared using limestone dust as a mineral filler at a content of 7 percent, this content represent the mid-range set by the SCRB specification for the type IIIA mixes of wearing course. Mixes in which the limestone dust was partially replaced by a hydrated lime were also prepared. The replacement percentages were 0, 0.5, 1.0, 1.5, 2, 2.5 and 3% by total weight of aggregate. The limestone dust and hydrated lime were obtained from lime factory in Karbala governorate, south east of Baghdad. The chemical composition and physical properties of the fillers are presented in Table (3) below:

Table 3. Properties of Fillers

Filler type	Chemical Composition ,%							Physical Properties		
	CaO	SiO_2	Al_2O_3	MgO	Fe_2O_3	So_3	L.O.I	Specific gravity	Surface area* (m^2/kg)	% Passing sieve No. 200(0.075)
Limestone Dust	68.3	2.23	-	0.32	-	1.20	27.3	2.41	244	94
Hydrated Lime	56.1	1.38	0.72	0.13	0.12	0.21	40.6	2.78	398	98

* Blain air permeability method (ASTM C204).

4 Experimental Work

The experimental work was started by determining the optimum asphalt content for all the asphalt concrete mixes using the Marshall mix design method. To investigate the stiffening effect of hydrated lime on the filler-asphalt mortar, filler-asphalt mixes were then prepared and tested using the conventional binder tests, penetration and softening point. Also, asphalt concrete mixes were made at their optimum asphalt content and tested to evaluate the engineering properties which include moisture damage, resilient modulus, permanent deformation and fatigue characteristics. These properties have been evaluated using indirect tensile strength, uniaxial repeated loading and repeated flexural beam tests.

4.1 Marshall Mix Design

A complete mix design was conducted using the Marshall method as outlined in AI's manual series No.2 (AI, 1981) using 75 blows of the automatic Marshall compactor on each side of specimen. Based upon this method, the optimum asphalt content is determined by averaging the three values shown below:

Asphalt content at maximum unit weight
Asphalt content at maximum stability
Asphalt content at 4% air voids

For each percentage of hydrated lime content, six Marshall specimens were prepared with a constant increments rate in asphalt cement content of 0.2 percent. The selected asphalt cement content starts from 4.2 percent for the control and 0.5 percent hydrated lime mixes and increased 0.2 percent for each 1 percent increase in hydrated lime content, so for the mixes with 3 percent hydrated lime, the starting value for asphalt cement content was 4.8 percent. This procedure is followed since it was found earlier in this work that the use of low asphalt content was not sufficient to provide proper coating for the aggregate with high content of hydrated lime.

To investigate the stiffening effect of hydrated lime upon the filler- asphalt mixture, the penetration as well as softening point tests was conducted according the ASTM –D5 and ASTM D 36, respectively for the mixes prepared using 7 percent filler but with different hydrated lime contents as a partial replace of limestone filler and corresponding optimum asphalt cement content.

4.3   Indirect Tensile Test

The moisture susceptibility of the asphalt concrete mixtures was evaluated using ASTM D 4867. The result of this test is the indirect tensile strength (ITS) and tensile strength ratio (TSR). In this test, a set of specimens were prepared for each mix according to Marshall procedure and compacted to 7±1 % air voids using different numbers of blows per face that varies from (34 to 49) according to the hydrated lime replacement rate. The set consists of six specimens and divided into two subsets, one set (control) was tested at 25°C and the other set (conditioned) was subjected to one cycle of freezing and thawing then tested at 25°C. The test involved loading the specimens with compressive load at a rate of (50.8mm/min) acting parallel to and along the vertical diametrical plane through 0.5 in. wide steel strips which are curved at the interface with specimens. These specimens failed by splitting along the vertical diameter. The indirect tensile strength which is calculated according to Eqn. (1) of the conditioned specimens (ITSc) is divided by the control specimens (ITSd), which gives the tensile strength ratio (TSR) as the following Eqn. (2).

$$ITS = \frac{2P}{\pi t D} \qquad (1)$$

$$TSR = \frac{ITS_c}{ITS_d} \qquad (2)$$

where
    ITS= Indirect tensile strength
    P = Ultimate applied load
    t   = Thickness of specimen
    D = Diameter of specimen
    Other parameters are defined previously

4.4   Uniaxial Repeated Loading Test

The uniaxial repeated loading tests were conducted for cylindrical specimens, 101.6 mm (4 inch) in diameter and 203.2 mm (8 inch) in height, using the pneumatic repeated load system (shown below in fig.(2)). In these tests, repetitive compressive loading with a stress level of 0.137 mPa (20 psi) was applied in the form of rectangular wave with a constant loading frequency of 1 Hz (0.1 sec. load duration and 0.9 sec. rest period) and the axial permanent deformation was measured under the different loading repetitions. All the uniaxial repeated loading tests were conducted at 40°C (104°F). The specimen preparation method for this test can be found elsewhere (Albayati, 2006).

The permanent strain (εp) is calculated by applying the following equation:

$$\varepsilon_p = \frac{p_d \times 10^6}{h} \quad (3)$$

where
εp= axial permanent microstrain
pd= axial permanent deformation
h= specimen height

Also, throughout this test the resilient deflection is measured at the load repetition of 50 to 100, and the resilient strain (εr) and resilient modulus (Mr) are calculated as follows:

$$\varepsilon_r = \frac{r_d \times 10^6}{h} \quad (4)$$

$$M_r = \frac{\sigma}{\varepsilon_r} \quad (5)$$

where
εr= axial resilient microstrain
rd= axial resilient deflection
h= specimen height
Mr= Resilient modulus
σ = repeated axial stress

The permanent deformation test results for this study are represented by the linear log-log relationship between the number of load repetitions and the permanent microstrain with the form shown in Eqn. (6) below which is originally suggested by Monismith et. al., (1975) and Barksdale (1972).

$$\varepsilon_p = aN^b \quad (6)$$

where
εp= permanent strain
N=number of stress applications
a= intercept coefficient
b= slope coefficient

Fig. 2. Photograph for the PRLS

4.5 Flexural Beam Fatigue Test

Within this study, third-point flexural fatigue bending test was adopted to evaluate the fatigue performance of asphalt concrete mixtures using the pneumatic repeated load system, this test was performed in stress controlled mode with flexural stress level varying from 5 to 30 percent of ultimate indirect tensile strength applied at the frequency of 2 Hz with 0.1 s loading and 0.4 s unloading times and in rectangular waveform shape. All tests were conducted as specified in SHRP standards at 20°C (68°F) on beam specimens 76 mm (3 in) x 76 mm (3 in) x 381 mm (15 in) prepared according to the method described in (Al-khashaab, 2009). In the fatigue test, the initial tensile strain of each test has been determined at the 50th repetition by using Eqn. (7) shown below and the initial strain was plotted versus the number of repetition to failure on log scales, collapse of the beam was defined as failure, the plot can be approximated by a straight line and has the form shown below in Eqn. (8).

$$\varepsilon_t = \frac{\sigma}{Es} = \frac{12h\Delta}{3L^2 - 4a^2} \tag{7}$$

$$N_f = k_1(\varepsilon_t)^{-k_2} \tag{8}$$

where
ε_t = Initial tensile strain
σ = Extreme flexural stress

Es = Stiffness modulus based on center deflection.
h = Height of the beam
Δ = Dynamic deflection at the center of the beam.
L = Length of span between supports.
a = Distance from support to the load point (L/3)
N_f = Number of repetitions to failure
k_1 = fatigue constant, value of Nf when ε_t = 1
k_2 = inverse slope of the straight line in the logarithmic relationship

5 Test Results and Discussion

5.1 Effects of Hydrated Lime on Filler - Asphalt Mixes

The consistency of filler-asphalt mixes with different percentage contents of hydrated lime as partial replacement of limestone dust was determined using the penetration and softening point tests, the result of tests are presented in table (4) and shown graphically in figures (3) and (4).

Table 4. Penetration and softening point tests result

	Hydrated Lime Content*, %	0	0.5	1.0	1.5	2.0	2.5	3.0
	Optimum Asphalt Content, %	4.73	4.75	4.88	4.92	5.07	5.13	5.34
Test Results	Penetration at 25C,100 gm,5 sec. (0.1mm) (original=47)	34	32	29	28	26	22	19
	Softening Point, (°C) (original=47°C)	51.0	54.2	58.0	59.6	61.3	64.4	69.0

* As partial replacement of limestone dust, 7 percent filler content is constant.

As can be seen from the presented data, hydrated lime content has a substantial influence on the the consistency of filler-asphalt mixes. With respect to the penetration test, the penetration decreases with increasing hydrated lime content, for example, the penetration value for 1.5 percent hydrated lime content is 0.875 times the value of 0.5 percent, the constant of proportionality which can be driven from figure (3) is approximately -4.857 (1/10 mm) for each 1 percent increase in hydrated lime content. By contrast, the softening point increases with increasing the hydrated lime content and the constant of proportionality is +5.55 (°C) for each 1 percent increase in hydrated lime content. From the above results, it may be possible to argue that the higher the replacement rate of limestone dust with hydrated lime, the higher the stiffness of the resulted filler-asphalt mixes.

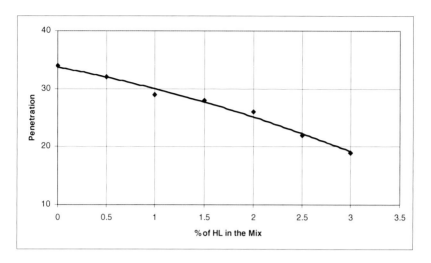

Fig. 3. Effect of hydrated lime content on penetration

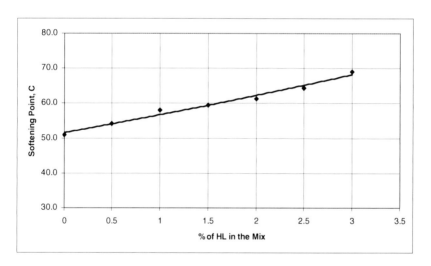

Fig. 4. Effect of hydrated lime content on softening point

5.2 Effects of Hydrated Lime on Marshall Properties

The variation of Marshall properties with hydrated lime content is shown in figure (5) which is based on the data presented in Table (5). Examinations of the presented data suggest that the mixes with higher hydrated lime content possess higher optimum asphalt cement content, the highest value of optimum asphalt content (5.34%) was obtained with 3% hydrated lime, while the lowest value (4.73%) was obtained with 0% hydrated lime which is the case that the mineral filler entirely consists of limestone dust.

Table 5. Summary of the Marshall properties of asphalt concrete mixes at optimum asphalt content

	Hydrated Lime Content*, %	0	0.5	1.0	1.5	2.0	2.5	3.0
	Optimum Asphalt Content, %	4.73	4.75	4.88	4.92	5.07	5.13	5.34
Marshall Properties	Stability, kN	8.65	8.87	9.76	10.3	11.05	11.14	11.2
	Flow, mm	3.23	3.5	3.62	3.73	3.8	3.53	3.41
	Density, gm/cm^3	2.320	2.329	2.333	2.336	2.348	2.337	2.331
	Air Voids, %	4.3	4.14	4.03	3.94	3.71	4.21	4.7
	VMA, %	15.83	15.45	15.49	15.42	15.12	15.57	15.97

* As partial replacement of limestone dust, 7 percent filler content is constant.

These differences can be attributed to the higher surface area of hydrated lime as compared to that of limestone dust. As shown earlier in this study, the surface area of hydrated lime is 1.63 times that of limestone dust, and hence the demand for asphalt has increased with increasing the replacement rate of limestone dust with hydrated lime. With respect to stability, the results indicate that the stability increases with increasing hydrated lime content, also the increment rate varies with hydrated lime content, the maximum rate obtained is 1.09 kN/1 percent for the hydrated lime content ranged from 0.5 to 2 percent, whereas for the hydrated lime content ranged from 0 to 0.5 percent and from 2 to 3 percent the rate was 0.44 and 0.15 kN/1 percent, respectively. From the stability plot, it may be possible to argue that the maximum benefit can be obtained with the use of 2 percent hydrated lime since further increase in hydrated lime content associated with just slight increases in stability value and require more asphalt cement content as compared to mixes with 2 percent hydrated lime.

The results of flow as a function of varying the hydrated lime content is shown in plot "c", its obvious that the flow value increases as the hydrated lime content increases from 0 to 2 percent, and then decreases as the hydrated lime content increases. This is due to the fact that air voids are too low at 2 percent hydrated lime content, addition of hydrated lime higher than this value tend to increase air voids due to insufficient compaction effort so the flow value decrease. The relationship between hydrated lime content and density which is shown in plot "d" follow the same trend of that between the hydrated lime content and Marshall flow, an optimum hydrated lime content which yields the highest Marshall density is 2 percent, further increases in hydrated lime content tend to decrease the Marshall density. As demonstrated in plot "e", the trend observed for the effect of hydrated lime content on air voids values is exactly opposite to that observed between hydrated lime content and flow, for a hydrated lime content from 0 to 2 percent, the air voids decreases with a rate of -0.295 percent for each 1 percent change in hydrated lime content, beyond 2 percent, the air voids content increases rapidly with a rate of +0.99 percent for each 1 percent change in hydrated lime content, this can be easily explained by the fact that the hydrated lime is finer than limestone dust so it can efficiently fill the voids pockets and stiffens the mixes for a certain amount beyond which there will be a lack in the compaction effort resulting in high air voids content. Plot "f" demonstrates the effect of hydrated lime content on voids in mineral aggregate (VMA), as its clear from the plot until a 2 percent of hydrated lime content the VMA decreases as the hydrated lime content increases, the minimum VMA value corresponding to 2 percent hydrated lime is 15.12 percent which means less spaces to be accommodated by asphalt cement, after 2 percent hydrated lime content, an addition of hydrated lime result in increasing the VMA values.

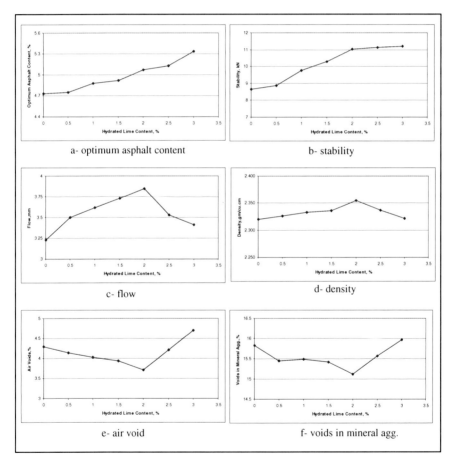

Fig. 5. Effect of hydrated lime content on Marshall properties

5.3 Effects of Hydrated Lime on Moisture Susceptibility

Based on the data shown in Table (6) and Figure (6), it appears that the examined hydrated lime contents have influence on the moisture susceptibility of the asphalt concrete mixes. The indirect tensile strength results for both control and conditioned mixes approximately linearly proportional to the hydrated lime content with constants of proportionality of +92.5 for the former and +150.5 kPa per 1 percent change in hydrated lime content for the latter. It is interesting to note that the improvement rate in the indirect tensile strength for the mixes with hydrated lime, added as part of the mineral filler, is higher in the case of conditioned mixes than that of control mixes. These findings beside that related to tensile strength ratio shown in figure (6) confirm that the resistance to moisture induced damage is enhanced in asphalt concrete pavement modified with hydrated lime.

Table 6. Moisture susceptibility test results

Hydrated Lime Content, %	ITS, kPa		TSR, %
	Control	Conditioned	
0	1290	1051	81.5
0.5	1342	1053	78.4
1.0	1373	1124	81.9
1.5	1441	1200	83.3
2.0	1496	1281	85.6
2.5	1532	1405	91.7
3.0	1554	1467	94.4

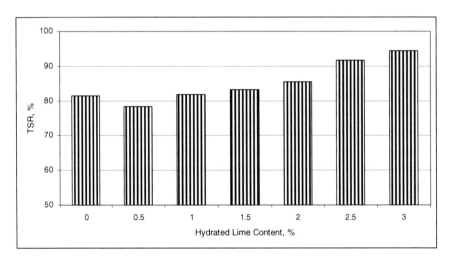

Fig. 6. Effect of hydrated lime content on tensile strength ratio

5.4 Effects of Hydrated Lime on Resilient Modulus

Table (7) as well as figure (7) exhibits the variation of the resilient modulus values with the hydrated lime content. The relation is in reverse order up to 1 percent content of hydrated lime (i.e., as the hydrated lime content increases the resilient modulus decreases), but further increase in hydrated lime content reflects this relation, the resilient modulus of the mixes with 3 percent hydrated lime (1098 mPa) is 1.4 times the value for mixes with 1 percent hydrated lime which was 779 mPa, these results can be explained as follow; since the test was conducted under relatively high temperature (40°C (104°F)), so the low level of hydrated lime content (below 1 percent) is insufficient to stiffening the asphalt concrete mixes whereas the higher values of resilient modulus resulted from the high level of hydrated lime content (above 1 percent) indicate that the hydrated lime did increase the stiffness of the asphalt concrete mix.

Table 7. Resilient modulus test results

Hydrated Lime Content, %	0	0.5	1.0	1.5	2.0	2.5	3.0
Resilient Modulus, mPa	838	796	779	871	989	1047	1098

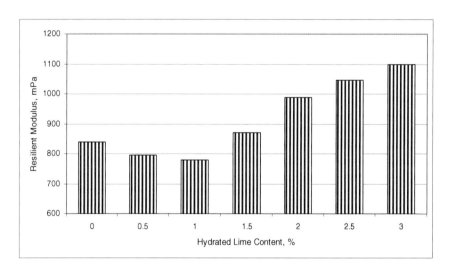

Fig. 7. Effect of hydrated lime content on resilient modulus

5.5 Effects of Hydrated Lime on Permanent Deformation

The result of permanent deformation tests is shown in figure (8) which is based on the data presented in table (8), Examinations of the presented data suggests that the permanent deformation parameters intercept and slope generally improved with the use of hydrated lime, for mixes containing 0 percent hydrated lime, the slope value which reflects the accumulation rate of permanent deformation is approximately 20 percent higher than that of mixes with 3 percent hydrated lime. For the intercept, the value is slightly increase as the hydrated lime content increases from the 0 to 0.5 percent, but then the addition of extra amount of hydrated lime tend to decrease the intercept value in a rate of 19.7 microstrain per each 1 percent change in hydrated lime content. This finding confirms that the rutting mode of failure in asphalt concrete pavement which is enhanced at hot summer temperature can be reduced into large extent with the introduction of hydrated lime to asphalt concrete mixtures.

Table 8. Permanent deformation test results

Hydrated Lime Content, %	0	0.5	1.0	1.5	2.0	2.5	3.0
Intercept	108	113	102	95	80	70	66
Slope	0.372	0.366	0.355	0.341	0.324	0.312	0.300

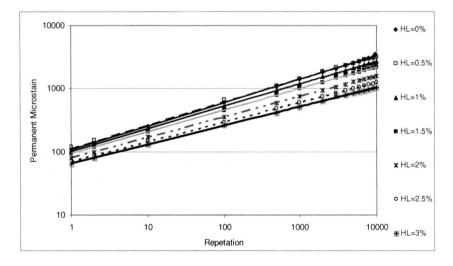

Fig. 8. Effect of hydrated lime content on permanent deformation

5.6 Effects of Hydrated Lime on Fatigue Performance

The fatigue characteristic curves for all mixtures are presented in Fig. 9. The fatigue parameters k1 and k2 are shown in Table 9. Values of k1 and k2 can be used as indicators of the effects of hydrated lime on the fatigue characteristics of a paving mixture. The flatter the slope of the fatigue curve, the larger the value of. k2 If two materials have the same k1 value, then a large value of k2 indicates a potential for longer fatigue life. On the other hand, a lower k1 value represents a shorter fatigue life when the fatigue curves are parallel, that is, k2 is constant. Test results indicate that the use of hydrated lime with a rate of content ranged from 0 to 1 percent does not have significant effect on fatigue life but the mixes with more than 1 percent hydrated lime showed better fatigue performance, the k2 value for mixes with 2 and 3 percent hydrated lime was more than that of 1 percent hydrated lime by 30.6 and 54.4 percent, respectively. Considering k1, it can be concluded from the data shown in table (9) that there is an agreement between the results of k1 and k2 in the field of fatigue resistance, k1 has the smallest value (9.561x E-12) when the hydrated lime content was 3 percent and it

Table 9. Fatigue parameters result

Hydrated Lime Content, %	0	0.5	1.0	1.5	2.0	2.5	3.0
k_1	1.339 x E-7	4.162 x E-7	1.780 x E-7	1.555 x E-8	1.822 x E-9	5.499 x E-11	9.561 x E-12
k_2	2.74	2.56	2.61	3.08	3.41	3.83	4.03
Crack Index	5.34	5.47	5.38	3.19	2.63	2.01	1.45

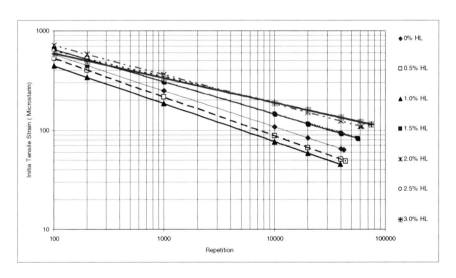

Fig. 9. Effect of hydrated lime content on fatigue performance

was increase as the hydrated lime content decreased from 3 to 1 percent, but for the mixes with 0.5 percent hydrated lime content, k1 value was more than that of 0 percent hydrated lime. Using vesys5w software for analyzing pavement section consisted of a 150 mm asphalt concrete layer over a 400 mm base course layer with 1 million ESALs application during 10 years service life, the crack index value which is a dimensionless parameter providing an estimate for the amount of fatigue cracking is obtained and shown in table 9 below. 0 percent hydrated lime result in crack index value of 5.34 (severe cracking) whereas the use of 2 and 3 percent hydrated lime result in crack index value of 2.63 (moderate cracking) and 1.45 (light cracking), respectively.

6 Conclusions and Recommendations

The following conclusions and recommendations are based on the results of the laboratory tests and analysis presented in this study:

1. In the filler-asphalt mixes, hydrated lime has shown significant stiffening properties when mixed with asphalt cement as partial replacement for limestone dust. For each 1 percent increase in hydrated lime, the penetration value decreases at a rate of 4.857 (1/10 mm) whereas the softening point increases at a rate of +5.55 (°C).
2. The addition of different percentages of hydrated lime as a filler substitute has a significant effect on volumetric mixture properties, some of the obtained results can be summarized as follow:
    - The mixes with higher hydrated lime content possess higher optimum asphalt content, the highest value of optimum asphalt content (5.34%) was obtained with 3 percent hydrated lime, while the lowest value (4.73%) was obtained with 0 percent hydrated lime
    - An optimum hydrated lime content which yields the highest density is 2 percent, further increases in hydrated lime content tend to decrease the density
    - For mixes with a hydrated lime content ranged from 0 to 2 percent, the air voids decreases with a rate of -0.295 percent for each 1 percent change in hydrated lime content. Beyond 2 percent, the air voids content increases rapidly with a rate of +0.99 percent for each 1 percent change in hydrated lime content
3. The addition of hydrated lime has improved the indirect tensile strength for both control and conditioned mixes with a rate of +92.5 and +150.5 kPa per 1 percent increase in hydrated lime content, respectively. The resistance to moisture induced damage is enhanced in asphalt concrete pavement modified with hydrated lime.
4. The addition of hydrated lime as a filler substitute with a rate ranged from 1.5 to 3 percent has shown an increase in resilient modulus. The resilient modulus for mixes with 3 percent hydrated lime was 1.31 times that for mixes with 0 percent hydrated lime.

5. The permanent deformation parameters, slope and intercept, was significantly effected with the addition of deferent percentages of hydrated lime. The modified mixes show higher resistance to permanent deformation when the percentage of hydrated lime is increased as a filler substitute.
6. The use of hydrated lime as a filler substitute within a range of 1.5-3 percent has improved the fatigue property of the asphalt concrete mixes as determined by flexural test. The k_2 value (inverse slope of fatigue line) for mixes with 2 and 3 percent hydrated lime was more than that of 0 percent hydrated lime by 24.4 and 47.0 percent, respectively.
7. The use of 2 percent hydrated lime has shown a significant improvement of asphalt concrete behavior, and has added to the local knowledge the possibility of producing more durable mixtures with higher resistance to distresses.

References

AI: Thickness Design-Asphalt Pavements for Highways and Streets. Asphalt Institute, Manual Series No.1,College Park, Maryland, USA (1981)

Albayati, A.H.: Permanent Deformation Prediction of Asphalt Concrete Under Repeated Loading. Ph.D. Thesis, Baghdad University (2006)

Alkhashab, Y.Y.: Development of Fatigue Prediction Model for Local Asphalt Paving Materials. Ph.D. Thesis, Baghdad University (2009)

Al-Suhaibani, A., Al-Mudaiheem, J., Al-Fozan, F.: Effect of Filler Type and Content on Properties of Asphalt Concrete Mixes. In: Meininger, R.C. (ed.) Effects of Aggregates and Mineral Fillers on Asphalt Mixture Performance, ASTM STP 1147, pp. 107–130. American Society for Testing and Materials, Philadelphia

Atud, T.J., Kanitpong, K., Martono, W.: Laboratory Evaluation of Hydrated Lime Application Process in Asphalt Mixture for Moisture Damage and Rutting Resistance. In: Transportation Research Board 86th Annual Meeting CD-ROM, Paper No. 1508, Washington, D.C. (2007)

Barksdale, R.: Laboratory Evaluation of Rutting in Base Course Materials. In: Proceedings of Third International Conference on the Structural Design of Asphalt Pavements, London (1972)

Epps, J., Berger, E., Anagnos, J.N.: Treatments. In: Moisture Sensitivity of Asphalt Pavements, A National Seminar, Transportation Research Board, Miscellaneous Report, pp. 117–186. Transportation Research Board, Washington D.C (2003)

Khattak, M.J., Kyatham, V.: Visco-Elastic Behavior of Asphalt Matrix & HMA under Moisture Damage Condition. Transportation Research Board 87th Annual Meeting CD-ROM, Washington, D.C (2008)

Khosla, N.P., Birdshall, B.G., Kawaguchi, S.: Evaluation of Moisture Susceptibility of Asphalt Mixtures, Conventional and New Methods. Transportation Research Record: Journal of the Transportation Research Board, No. 1728, 43–51 (2000)

Little, D.N., Petersen, J.C.: Unique Effects of Hydrated Lime Filler on the Performance-Related Properties of Asphalt Cements: Physical and Chemical Interactions Revisited. Journal of Materials in Civil Engineering 17(2), 207–218 (2005)

McGennis, R.B., Kennedy, T.W., Machemehl, R.B.: Stripping and moisture damage in asphalt mixtures. Center for Transportation Research, Bureau of Engineering Research, The University of Texas at Austin (1984)

Mohammad, L.N., Abadie, C., Gokmen, R., Puppala, A.J.: Mechanistic Evaluation of Hydrated Lime in Hot-Mix Asphalt mixtures. Transportation Research Record: Journal of the Transportation Research Board, No. 1723, 26–36 (2000)

Monismith, C., Ogawa, N., Freeme, C.: Permanent Deformation Characteristics of Subgrade Soils due to Repeated Loadings, TRR 537 (1975)

Paul, H.R.: Compatibility of Aggregate, Asphalt Cement and Antistrip Materilas. Louisiana Transportation Research Center, Report No. FHWA/LA-95-292 (1995)

SCRB/R9, General Specification for Roads and Bridges, Section R/9, Hot-Mix Asphalt Concrete Pavement, Revised Edition. State Corporation of Roads and Bridges, Ministry of Housing and Construction, Republic of Iraq (2003)

Sebaaly, P.E., Hitti, E., Weitzel, D.: Effectiveness of Lime in Hot Mix Asphalt Pavements. Transportation Research Record: Journal of the Transportation Research Board, No. 1832, 34–41 (2003)

Shahrour, M.A., Saloukeh, B.G.: Effect of Quality and Quantity of Locally Produced Filler (Passing Sieve No. 200) on Asphaltic Mixtures in Dubai. In: Meininger, R.C. (ed.) Effects of Aggregates and Mineral Fillers on Asphalt Mixture Performance, ASTM STP 1147, pp. 187–208. American Society for Testing and Materials, Philadelphia (1992)

Determination of Crack Growth Parameters of Asphalt Mixtures

Maarten M.J. Jacobs[1], Arian H. De Bondt[2], Piet. C. Hopman[3], and Radjan Khedoe[2]

[1] BAM Wegen, Utrecht, the Netherlands
 m.jacobs@bamwegen.nl
[2] Ooms Civiel, Scharwoude, the Netherlands
 {adebondt,rkhedoe}@ooms.nl
[3] KOAC•NPC, Apeldoorn, the Netherlands
 hopman@koac-npc.com

Abstract. A test procedure has developed where the crack growth parameters of an asphalt concrete mixture can be determined easily and uniformly. The semi circular bending (SCB) test is the basis of this procedure. However, in this procedure SCB-specimens with a base length of 225 mm are used instead of 150 mm as is indicated in EN 12697-44. In this way the total crack length increases and this is beneficial for the repeatability and reproducibility of the test results. The tests are force controlled with a haversine as loading signal. The length of the crack is measured indirectly by using a crack opening displacement (COD) gauge over the notch, which is situated in the centre of de base of the SCB-specimen. With finite element simulations, the relationship between the COD and the crack length and the stress intensity factor K are determined.

All the information is used to determine the Paris' parameters A and n of the tested mixture. The (draft) procedure has been used by two laboratories on two asphalt concrete mixtures. It is concluded that with the test procedure appropriate crack growth parameters can be determined with acceptable values for the repeatability and reproducibility of the test results.

In the paper the test procedure will be presented. Also information on the tested materials and test results will be incorporated.

1 Introduction

Although reflective cracking is an important damage phenomenon of asphalt concrete pavements on a cement treated subbase or a cracked concrete base, there is not much practical information available on the design procedure of these types of pavements. Also the determination of crack growth parameters of the used materials is often a problem. For these reasons the use of cement treated subbase layers or the reuse of cracked concrete layers is often not an acceptable option for road authorities and contractors. From the viewpoint of sustainability and environmental friendly reuse of materials, this is not acceptable.

In the Netherlands, the use of the bearing capacity of the cracked pavement is an important criterion in the tender procedure for the rehabilitation of existing roads. To demonstrate the value of the cracked pavement, the contractor has to prove how reflective cracking can be prevented using various kinds of solutions (e.g. SAMI's, reinforcement, highly modified mixtures). In this procedure the determination of accurate crack growth parameters of the applied asphalt concrete mixtures is one of the critical points. Especially the poor repeatability and reproducibility of the crack growth tests with small specimens (e.g. SCB-specimens with a diameter of 150 mm) is a big issue.

For this reason, a procedure for the determination of reliable crack growth parameters is developed. The main starting point for this procedure is the fact that the test can be performed in laboratories which have test facilities to perform dynamic tests to determine properties like stiffness, fatigue and/or permanent deformation. In the Netherlands all the major road contractors own this kind of equipment.

In this paper the test procedure is presented. This procedure is tested by two laboratories, performing tests on two kinds of asphalt mixtures. The compositions of these mixes are described and the test results are presented. Finally the conclusions and recommendations are given.

2 Test Procedure

2.1 General

The test procedure is based on experiences from [1]. To determine the crack growth parameters, the Semi Circular Bending (SCB) test is used.

During the test, a sinusoidal load is applied on an SCB-shaped specimen. The load is always compressive to prevent contact loss between loading strip and specimen. During the test, the crack growth is measured using a crack opening displacement (COD) gauge, which is located over an artificial crack in the middle of the basis of the SCB specimen. During the test the force, the displacement and the COD-gauge are measured and recorded. These data are analysed using Paris' law [2] and the results of finite element simulations of the test.

The SCB specimens are tested in a bending frame. The characteristics of this loading frame are mentioned in EN 12697-44. In the following chapters the requirements with respect to the specimen, the test configuration and the calculations of Paris' parameters are described.

2.2 Specimens

The specimen shall have a diameter of (225±3) mm. The specimen shall be cut from slabs or cored from a pavement. If the thickness is large enough, compaction cracks shall be removed. From the disk shaped core, two SCB-specimens can be cut. In case the base of the SCB specimen is not fully flat, this can be eliminated by polishing.

At mid length of the SCB-specimen, a notch is cut using a saw blade. The final width of the notch is (0,35± 0,10) mm. The notch shall be perpendicular to the base of the SCB-specimen. The length of the notch shall be (15,0±1,0) mm.

The properties of the SCB-specimen are determined using a dry specimen. According to EN 12697-6, a specimen is dry when the change in mass between two weightings, which are performed with a time interval of at least one hour, is less than 0,1%. The dimensions of the specimen are determined using the following procedure (based on EN 12697-29):

- Determine on two locations the diameter $D_{i=1,2}$ of the specimen with an accuracy of 0,1 mm. The diameter D is the average of the two measurements;
- Determine on both sides of the specimen the height $H_{i=1,2}$ of the specimen with an accuracy of 0,1 mm. The height H is the average of the two measurements;
- Determine on 4 locations (three measurements on the basis and one on the top) the thickness $T_{i=1,4}$ of the specimen with an accuracy of 0,1 mm. The thickness T is the average of the four measurements;
- The width and length of the artificial crack shall be determined with an accuracy of 0,1 mm.

In Table 1 the requirements with respect to the dimensions of the SCB-specimen are gathered.

Table 1. Required dimensions of the SCB specimen

Diameter D	Heigth H	Thickness T	Notch length a
(225,0 ± 10,0) mm	(112,0±3,0) mm	(50±1) mm	(15,0±1,0) mm

The upper sieve size of the tested asphalt concrete mixture shall not exceed 25% of the height H of the specimen; the thickness T shall be at least 2 times the upper sieve size of the mixture.

To determine the crack growth parameters of an asphalt concrete mixture, at least 6 SCB-specimens are necessary:

- 2 specimens are used to determine the load level of the dynamic tests. To determine this level, 2 static tests according to EN 12697-44 are carried out on specimens with a diameter of 225 mm at the test temperature of the crack growth measurements;
- The crack growth parameters are determined by testing 4 specimens.

2.3 Loading Facility

The loading frame is schematically given in Figure 1. The characteristics of the loading frame are given in EN 12697-44.

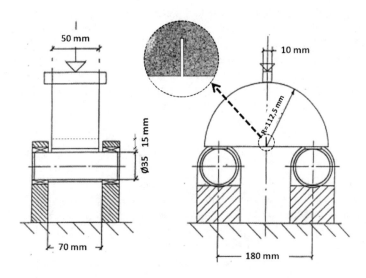

Fig. 1. The SCB loading frame with specimen

2.4 Test Performance

All tests are carried out in a temperature controlled room with a temperature of $(5\pm1)°C$.

Before starting the actual crack growth measurements, first the load level F_{dyn} has to be determined. The magnitude of this dynamic load level F_{dyn} is such that the life span of a notched specimen in the dynamic test is about 200.000 load cycles. In practice this implies that a dynamic load level of 35% of the mean maximum load of the static tests, performed at a temperature of 5°C and a deformation speed of the piston of 50 mm/minute, gives a reasonable load level. So $F_{dyn,max}=0{,}35 F_{stat,mean}$.

After the static tests, the dynamic tests are performed at a loading frequency of 30 ± 1 Hz. The chosen load signal is a haversine, varying between $F_{dyn,min}$ and $F_{dyn,max}$. The $F_{dyn,min}$ is 10% of the $F_{dyn,max}$, so the specimen is always under compression. The R-value of the test is +0,1.

During the cyclic crack growth test, the crack length is measured indirectly using a COD-gauge. This gauge is mounted over the notch at mid length of the SCB-specimen. The accuracy of the COD-gauge shall be chosen appropriately. Deformations in the range of 10 μm are usual values found during the dynamic tests. A gauge with a measuring length of 12 mm and a dynamic measuring range of 9 mm able to measure at a frequency of 100 Hz with a non-linearity of 0,15% and an accuracy class of 0,2 can comply to these requirements.

During the dynamic crack growth test, the measurements of the force, deformation and COD-value must be carried out at least each time the stiffness of the specimen has decreased 1% with respect to its initial stiffness. For a load controlled test, this implies that data have to be recorded when the dynamic deformation of the specimen has increased by 1% with respect to the initial deformation level.

The test will be finished when the specimen is broken into two pieces. Because the specimen will break progressively, at least 60 to 70 data points will be available after the test.

2.5 Determination Paris' Parameters

Based on the test conditions and the recorded data, the Paris' parameters A and n can be determined:

$$\frac{dc}{dN} = A \cdot K_I^n \qquad (1)$$

where:
- c = crack length (mm);
- N = number of load repetitions (-);
- A,n = Paris' parameters;
- K_I = stress intensity factor (Nmm$^{1.5}$ or MPa√mm)

The aim of the total procedure is to determine the value of the Paris' parameters A and n. With these parameters, each set of A and n are characteristics for a specific mixture, asphalt mixtures can be compared or the life span of a pavement structure with crack growth as design procedure can be determined. The total procedure has 6 steps (see Table 2), where:

- COD = measured COD-amplitude in the dynamic crack growth test;
- COD = f(N) means that the COD-value is a function of N.

Table 2. Procedure to determine Paris' parameters from COD measurements

Step	Action	Result
1	Perform the dynamic SCB-test	COD = f(N)
2a	Simulate the crack growth process with	K_I = f(COD)
2b	finite element calculations	c = f(COD)
3	Combine step (1) with step (2a)	K_I = f(N)
4	Combine step (1) with step (2b)	c = f(N)
5	Combine step (3) with step (4)	c = f(K_I)
6	Combine step (4) with step (5)	dc/dN = f(K_I)

In the determination of A and n the following equations are used:

$$K_I = \sigma_{SCB} \cdot f\left(\frac{c}{H}\right) \qquad (2)$$

The relation between K and c can be determined using linear elastic finite element method (FEM-)calculations. It is found that:

$$f\left(\frac{c}{H}\right) = -4,9965 + 155,58\left(\frac{c}{H}\right) - 799,94\left(\frac{c}{H}\right)^2 +$$
$$+ 2141,9\left(\frac{c}{H}\right)^3 - 2709,1\left(\frac{c}{H}\right)^4 + 1398,6\left(\frac{c}{H}\right)^5 \tag{3}$$

where:

H = the maximum crack length. This is the height of the SCB-specimen.

In the SCB-specimen the magnitude of the stress σ_{SCB} is determined using equation (4):

$$\sigma_{SCB} = \frac{4,263 \cdot F_{SCB,N=0}}{T_{SCB} \cdot L_{SCB}} \tag{4}$$

where:

$F_{SCB,N=0}$ = external applied force (=F_{max}-F_{min}) at the start of the test (N). The value of F_{max} is equal to $F_{dyn,max}$;
L_{SCB} = distance between the rollers in the SCB-test = 180 mm;
T_{SCB} = thickness of the SCB specimen (mm).

In equation (3) and (4) the results of the experiment are taken into account. The effect of the stiffness modulus E and Poisson's ratio υ on the various parameters is investigated. From this it is concluded that the K_I-value does not depend on the stiffness and the COD-value has a linear relationship with the stiffness. Poisson's ratio has no effect on both K_I and COD.

From the FEM-calculations also a relation between the COD-value and the crack length can be determined. It is found that:

$$\left(\frac{c}{H}\right) = 0,1642 \cdot \ln\left(\frac{COD_N}{COD_{N=1}}\right) + 0,1413 \tag{5}$$

where:

COD_N = crack opening displacement value at the Nth load cycle (mm);
$COD_{N=1}$ = crack opening displacement value at the first load cycle (mm).

From equations (1) to (5) the following relations can be determined:

- The relation between crack length and the number of load repetitions (using the COD-measurements from the test) and
- The relation between the K_I-value and the crack length (from the FEM-calculations).

This implies that there is a relationship between the crack length c and the stress intensity factor K_I as a function of the number of load repetitions N. From these relationships the Paris' parameters can be determined.

Finally a set of data is available with the relationship between the number of load repetitions N and the crack length c. Now two different approaches can be followed:

- The available data can be fit using the following relationship:

$$c = C_0 + C_1(1 + C_2 N)^{C_3} + C_4 N^{C_5} \qquad (6)$$

  The coefficients C_1 to C_5 can be determined using a curve fitting procedure. The value of (C_0+C_1) is the length of the notch at the start of the test (N=0). After determination of the coefficients C_1 to C_5 the first derivative of equation (6) can be determined and dc/dN as a function of N can be defined.
- From the available dataset between c and N the value $\Delta c/\Delta N$ can be determined where Δc is the increment in crack length during an interval of load repetitions ΔN. The $\Delta c/\Delta N$-values as a function of N can be used as the left term in Paris' law.

At the end the parameters dc/dN and K_I as a function of the number of load repetitions N are available. Using regression analysis, the Paris' parameters A and n can be determined using equation (1).

3 Test Program

To check the procedure a test program has been carried out by two laboratories using two different kinds of asphalt mixtures. Characteristics of the mixtures are presented in Table 3.

Table 3. Characteristics of the tested mixtures

	AC 5 surf 40/60	AC 16 surf 40/60
Passing sieve:		
C22,4		100
C16		98,8
C11,2		87,5
C8	100	75,0
C5,6	98,4	62,5
2 mm	50,0	40,0
180 µm	14,4	10,3
63 µm	7,6	6,2
Bitumen content (%m/m)	6,4	5,7
Density (kg/m^3)	2280	2361
Air voids (%V/V)	5,2	2,4
Stiffness @8 Hz, 20°C (MPa)	3797	6532
ε_6 @30 Hz, 20°C (µm/m)	154	102

In the test program, first the load level in the dynamic tests is determined. This load level was set at 30% of the failure load in a static test.

4 Test Results

In this section the results of the tests and the determination of the Paris' parameters will be presented. The average diameter D of all specimens is 224,7 mm; the average thickness T is 50,0 mm and the average height H is 109,5 mm. All tests are performed at a test temperature of 5°C.

In Figure 2 one of the results of the measurements is shown. On the left of Figure 2, the results of the COD-measurements during the test are illustrated.

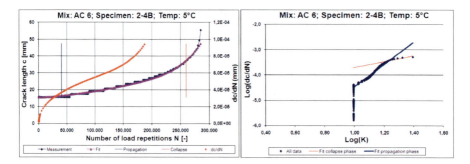

Fig. 2. Results of the crack growth measurements using the curve fit procedure

From Figure 2 it can be concluded that the curve fit procedure can describe the measured data accurate. After fitting the data the dc/dN-values can be determined. On the right of Figure 2 the data points are presented in the Paris' plot.

The main disadvantage of the curve fit procedure is that sometimes no solution can be found automatically. Another disadvantage is the fact that the final magnitude of the crack growth parameters sometimes depends on the start values of the curve fitting procedures.

In Figure 2 three processes can be distinguished:

- The crack initiation phase, which is represented by the data points with a log(dc/dN)-value smaller than -4,4;
- The crack propagation phase, represented by the blue line in the right of Figure 2. On the left of Figure 2 this phase occurs in the period between the vertical blue and orange line (between 40.000 and 260.000 load repetitions);
- The progressive crack growth or collapse phase, represented by the orange line on the right in Figure 2. This phase occurs in the period between 260.000 and complete failure of the specimen. However, the slope of the progressive crack growth process is unexpected: an almost vertical line was expected (dc/dN becomes very large) and a almost horizontal line is found. An explanation of this can be the fact that the high K-values during this phase are compensated by lower dc/dN-values.

Looking closer to the cracked planes of specimen 2-4B (see Figure 3), these three phases can be recognized again. On the left of Figure 3, the notch is visible. In the middle part (up to 55% of the height of the specimen), the crack propagation can be seen where the cracked surface has a matt appearance. The progressive collapse phase can be recognized by the shining surface. At this moment only differences in colour and light reflection are recognised in the crack surfaces. In future high accurate texture scans will be used to recognise differences in crack growth phases.

Fig. 3. The cracked surface of specimen 2-4B

In Figure 4 the results of the same crack growth test are presented as in Figure 2, but now using the $\Delta c/\Delta N$-approach to determine the Paris' parameters, so no curve fitting has been carried out on the crack length data. The main disadvantage of this procedure is that a substantial part of the data points cannot be used: because in the $\log(\Delta c/\Delta N)$-$\log(K_I)$-relation only increasing crack length data can be used in the analyses.

In Table 4 the Paris' parameters of both procedures are presented.

Table 4. Results of the 2 procedures to determine the Paris' parameters

Crack phase	Paris' parameter	Curve fitting procedure	$\Delta c/\Delta N$-approach
Propagation	Log(A)	-8,04	
	n	3,61	
Collapse	Log(A)	-6,52	
	n	2,45	
All data	Log(A)	-8,36	-9,47
	n	3,96	4,90

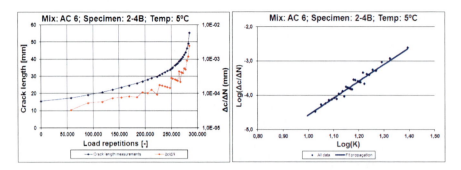

Fig. 4. Results of the crack growth measurements using the Δc/ΔN-approach

So the procedure to determine Paris' parameters from the data points influences the final values of the Paris' parameters substantially. Especially the data points at the end of the crack growth process influences the final Paris' parameters.

In Table 5 the Paris' parameters of the tested specimens of both laboratories are presented. The Paris' parameters determined by Lab1 are calculated using the curve fit procedure for the crack propagation phase; Lab2 determined the Paris' parameters using the Δc/ΔN-approach.

From these results the following conclusions can be formulated:

- The repeatability of the crack growth test using SCB-specimens with a base length of 225 mm is rather good. In the n-value an average variation coefficient of 2,7% is found;
- The reproducibility of the crack growth process is rather poor. This is mainly caused by the interpretation of the measured data points using a curve fitting procedure or the Δc/ΔN approach;
- The results of the procedure depend on the way the data are analysed. For this reason the Δc/ΔN approach is the preferred procedure and not the curve fitting procedure. The restriction of the Δc/ΔN approach is that a substantial amount of data point should be available, but this is nowadays with the current data recording instruments not an issue anymore;
- The determined values of the Paris' parameters using the Δc/ΔN approach comply with values found in literature (e.g. in [3]). Due to the absence of larger aggregate parts, which act as crack retarders, the n-value of the AC 6 mixture is expected to be larger than for the AC 16 mixture;
- In Figure 5 the relation between the Paris parameters n and log(A) is given. Again it is found that there is a linear relationship between these two parameters. This implies that the log(A)-value can be calculated if the n-value is known or vice versa.

Determination of Crack Growth Parameters of Asphalt Mixtures

Table 5. The Paris' parameters determined by the 2 laboratories on 5 specimens for each laboratory

Mix	Lab1: curve fitting procedure		Lab2: Δc/ΔN procedure	
	Log(A)	n	Log(A)	n
AC 5 surf 40/60	-7,56	3,34	-9,30	5,08
	-7,67	3,36	-9,32	5,20
	-7,85	3,31	-9,18	5,06
	-8,04	3,61	-9,79	5,29
	-7,58	3,37	-9,46	5,00
Average	-7,74	3,40	-9,41	5,13
Standard deviation	0,20	0,12	0,24	0,12
AC 16 surf 40/60	-8,32	3,85	-8,72	4,37
	-8,11	3,68	-8,65	4,30
	-8,32	3,96	-8,29	3,67
	-8,42	3,77	-9,13	4,47
	-8,26	3,75	-8,41	4,30
Average	-8,29	3,80	-8,64	4,22[*]
Standard deviation	0,11	0,11	0,30	0,08[*]

[*]) The results of the third test are not taken into account.

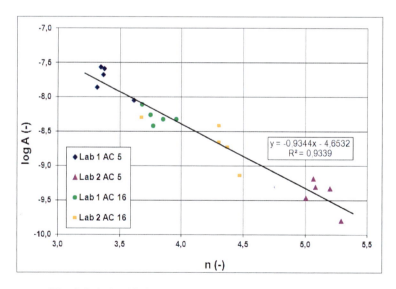

Fig. 5. Relationship between the Paris' parameters log(A) and n

5 Conclusions and Recommendations

Based on the experiences from the test program, the following conclusions and recommendations can be formulated:

1. The cyclic SCB-test (the static version of the test is described in EN 12697-44), produces reliable crack growth parameters if the following starting-points are taken into account:
   - The base of the SCB-specimen shall be 225 mm;
   - The accuracy class of the COD-gauge shall be 0.2;
   - During the crack growth test, the COD-value shall be recorded regularly. An appropriate number of COD-measurements during a test is 1000;
   - In the data interpretation procedure, the $\Delta c/\Delta N$ approach is preferred and not a curve fitting procedure;
2. There seems to be a relationship between the Paris' parameters n and log(A). If one of these parameters can be determined indirectly (e.g. by using the approach in [4]), maybe appropriate Paris' parameters can be determined using a simple test;
3. In the cracked surface, the three crack growth processes (crack initiation, crack propagation and progressive collapse) can be distinguished. By looking in detail to these cracked surfaces (e.g. by using highly accurate texture scans), the actual crack growth process can be coupled with the data from the crack growth process.

References

[1] CROW, COMPASS+, the next step in the functional specification of asphalt concrete mixtures, Final report CROW working group FEA, CROW-report 06-09 (2006) (in Dutch)
[2] Paris, P.C., Erdogan, F.: A critical analysis of crack propagation law. Transactions of the ASME, Journal of Basic Engineering, Series D 85(3) (1963)
[3] Jacobs, M.M.J.: Crack growth in asphaltic mixes. PhD-dissertation Delft University of Technology (1995)
[4] Schapery, R.A.: A theory of crack growth in visco-elastic media, Report MM 2764-73-1. Mechanics and Materials Research Centre, Texas A&M University (1973)
[5] Van Rooijen, R.C., De Bondt, A.H.: Crack propagation performance evaluation of asphaltic mixes using a new procedure based on cyclic semi-circular bending tests. In: Proceedings Sixth RILEM International Conference on Cracking in Pavements, Chicago, Illinois, USA, pp. 437–446 (2008)

Differential Thermal Contraction of Asphalt Components

Ignacio Artamendi, Bob Allen, Chris Ward, and Paul Phillips

Research & Development Department, Aggregate Industries, UK

Abstract. Large differences between the coefficient of thermal contraction of mineral aggregate and binder has been associated with localised damage at the aggregate-binder interface at low temperatures. In this work, the coefficients of thermal contraction of different binders, aggregates and asphalt mixtures have been determined. Binder specimens were first produced by pouring hot bitumen into 200 x 50 x 50 mm^3 moulds. The specimens were conditioned at various temperatures ranging from 10 to -20 ^{0}C. The change in length was then measured to determine thermal strains as a result of cooling. It was found that linear coefficients of thermal contraction varied between 115 and 175 x 10^{-6} mm/mm/^{0}C depending on grade and type of binder. Coefficient of thermal contraction of different aggregates was also determined. Rock specimens of the same dimension as the binder specimens were cut from large rock cores. The specimens were then conditioned at different temperatures and their change in length was measured. Three types of rocks namely limestone, granite and greywacke typically used in asphalt mixtures were employed. It was found that CTC of the aggregates varied between 7 and 10 x 10^{-6} mm/mm/^{0}C, thus, 10 to 25 times lower than those of the binders. The coefficient of thermal contraction of various asphalt mixtures was determined using a volumetric and a composite model. Furthermore, predicted values were compared with those determined experimentally using beam shaped asphalt specimens cut from roller compacted slabs manufactured in the laboratory.

1 Introduction

Pavements are subjected to a wide range of temperatures as a result of daily and seasonal temperature variation. Most materials including mineral aggregates and bituminous binders experience changes in volume due to changes in temperature, thus, they expand when the temperature rises and contract when the temperature drops. This thermovolumetric dependency is characterised in terms of the coefficient of thermal expansion (CTE) and contraction (CTC).

The thermal behaviour of an asphalt mixture depends primarily on that of its individual components i.e. mineral aggregate and bituminous binder. Coefficients of thermal contraction of these materials are, however, markedly different. Thus, at microscopic scale, thermal stresses may develop on the bituminous binder film surrounding an aggregate particle as result of differential thermal contraction

(DTC). Under extreme low temperatures these thermal stresses may cause localised damage in the form of hairline cracks that develop at the interface and propagate into the binder film. Furthermore, the presence of these cracks may accelerate moisture associated damage and reduce durability [1].

As aggregates represent about 85 % of the total volume of a typical asphalt mixture, the CTC of asphalt is greatly influence by that of the aggregate. Moreover, the CTC of aggregates depends on the mineralogy of the rock. In general, siliceous aggregates with high quartz content exhibit high CTC whereas limestone which consists mainly of calcite exhibit low CTC [2]. Comparing to the CTC of aggregates, the CTC of a bituminous binder is between 10 to 20 times higher than those of the mineral aggregate. Typical value used for thermal stress analysis is 170×10^{-6} mm/mm/^{0}C [3]. Moreover, the CTC of an asphalt mixture can be obtained using a simple model based on the CTC of the asphalt components and the volumetric characteristics of the mixture. More complex composite models which include mixture parameters such as binder and aggregate modulus have also been developed [4].

The aim of this work was to determine the CTC of various binders and aggregates typically used in asphalt mixtures in the UK. It is believed that differential thermal contraction (DTC) due to large differences between the CTC of aggregates and binders could be the cause of localised damage at low temperatures leading to early pavement failures. CTC of aggregates and binders were also employed to calculate the CTC of various asphalt mixtures using a volumetric and a composite model. Furthermore, calculated values were compared with those determined experimentally using asphalt specimens.

2 Materials

2.1 Aggregates

Three types of aggregates namely, greywacke, granite and limestone, were used in the study. Greywacke is a type of sedimentary rock belonging to the sandstone group. Petrographic examination showed that the greywacke comprised of quartz (45 %), feldspars (15 %), chlorite (20 %) and biotite (9 %). Quartz is composed of pure silica whereas feldspars are aluminosilicates containing potassium, sodium and calcium. Chemistry of the greywacke indicated relatively high silica content of 66 %. Some of this silica combines with the main oxides to form the silicates.

Granite, on the other hand, is an intrusive igneous rock composed of interlocking crystals. Petrographic examination showed that the granite aggregate comprised mainly of quartz (25 %), orthoclase feldspars (46 %), amphibole (18 %) and biotite (5 %). Orthoclase feldspars are aluminosilicates containing potassium and are the main component in the granite. Chemistry of the granite indicated a silica content of 64 %.

Limestones are sedimentary rocks formed in a marine environment from the precipitation of calcium carbonate and compressed to form a solid rock. They are composed primarily of calcium carbonate ($CaCO_3$) in the form of calcite.

Petrographic examination of the limestone used in the study showed an almost single mineral phase nature of the aggregate.

2.2 Binders

Three binders were used in the present study, two penetration grade binders, a 10/20 pen and a 40/60 pen, and an elastomeric polymer modified binder (PMB 50). The 10/20 pen binder is primarily used in EME2 base and binder course mixtures whereas the 40/60 and PMB 50 are used in more generic type of mixtures.

2.3 Asphalt Mixtures

Three types of materials, asphalt concrete (AC), stone mastic asphalt (SMA) and porous asphalt (PA), were used in the study. Maximum aggregate nominal size for all the mixtures was 10 mm. Design binder contents were 5.8 % (AC), 6.4 % (SMA) and 5.3 % (PA). In order to assess the effect of aggregate type on CTC, a first set of AC mixtures was produced with the same binder (PMB 50) but with three different types of aggregates. A second set was produced with the same aggregate (granite) but with three different types of binders to assess the effect of the binder type on CTC. Finally the same binder and aggregate were used to produce a set of AC, SMA and PA mixtures in order to evaluate the effect of the type of mixture on CTC.

3 Experimental

3.1 Specimen Preparation

Binder beam specimens were produced by pouring hot bitumen into 200 x 50 x 50 mm^3 steel moulds. Gang moulds suitable to produce three prisms with recessed inserts fixed to the centre of the inside faces of the moulds were used (see Figure 1). A release agent of talc and glycerol was first applied with a paint brush to the surface of the moulds. The moulds were then covered with degrease paper and sprayed with silicon.

Rock specimens were obtained by cutting 200 x 50 x 50 mm^3 beams from large rock cores. Three specimens per rock type were employed. Steel pegs were then glued on the centre of both sides of the prisms using epoxy resin. Rock specimens are also shown in Figure 1.

Finally, asphalt specimens were produced by cutting 200 x 50 x 50 mm^3 beams from 300 x 300 x 50 mm^3 asphalt slabs. The slabs were manufactured in the laboratory using a roller compactor. Bulk density and air voids of the beam specimens were determined before testing commenced. Three asphalt beam specimens were used for testing. Asphalt specimens are shown in Figure 1.

Fig. 1. Binder, rock and asphalt specimens

3.2 Testing

The measuring apparatus consisted of a digital dial gauge (micrometer) mounted in a measuring frame with a recessed end which could be located on the inserts attached to the beam specimens. The apparatus is used for the determination of drying shrinkage of aggregates.

After manufacturing, the specimens were first conditioned at the initial test temperature for 24 h. After 24 h, the first measurements of the length of the beams were taken. Immediately after this the specimens were conditioned at the second test temperature for further 24 h after which the second measurement was taken. This sequence was maintained until the specimens were tested at the lowest temperature, i.e. -20 ^{0}C approximately. Due to the viscoelastic character of the binders and the asphalt mixtures, the initial test temperatures selected were 10 ^{0}C for the binders and 20 ^{0}C for the asphalt mixtures. At higher temperatures specimen deformation due to self weight as a result of creep was observed after 24 h conditioning. The initial test temperature for rock specimens, on the other hand, was 60 ^{0}C.

4 Results and Discussions

4.1 Thermal Contraction of Asphalt Components

Thermal strain of binder specimens was determined by dividing the change in length as a result of cooling by the initial length measured at 10 ^{0}C. Thermally induced strains at temperatures ranging from 10 ^{0}C to -15 ^{0}C are presented in Figure 2. The linear coefficient of thermal contraction (CTC) was then obtained from the slope of the straight line fitted through the experimental data.

CTC of the three binders investigated are presented in Table 1. It can be seen that CTC of the hard 10/20 pen binder was the highest, followed by the 40/60 pen and the PMB 50. This might suggest that the CTC increases as the grade of the binder decreases. Also, it indicates that the CTC of a binder can be reduced by using an elastomeric polymer modifier.

CTC values obtained in this work shown in Table 1 have been found similar to those found in the literature. For instance, a typical value used for thermal stress analysis is 170 x 10^{-6} mm/mm/^{0}C (510 ml/ml/^{0}C) [3]. Also, Ojo et al. [5] reported values between 152 and 194 mm/mm/^{0}C (456 and 583 ml/ml/^{0}C) for various modified and unmodified binders above the glass transition temperature using a dilatometer. Work by Nam and Bahia [6] showed there were no clear trends between the CTC of different binder grades and modified binders. Furthermore, they reported a wide range of CTC values ranging from 175 to 242 mm/mm/^{0}C (525 to 725 ml/ml/^{0}C) for various modified binders above the glass transition temperature using a dilatometer glass transition test.

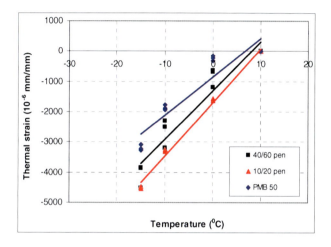

Fig. 2. Thermal strain vs temperature (binders)

Thermal strain of aggregate specimens was determined by dividing the change in length as a result of cooling by the initial length measured at 60 ^{0}C. Thermally induced strains at temperatures ranging from 60 ^{0}C to -20 ^{0}C are presented in Figure 3. The linear coefficient of thermal contraction (CTC) was then obtained from the slope of the line fitted through the experimental data.

CTC of the three aggregates investigated are presented in Table 1. It can be seen that CTC of the greywacke was the highest, followed by the granite and the limestone. CTC of an aggregate is related to the mineralogical and chemical composition. It can be seen that aggregates with relatively high silica (SiO_2) content such as greywacke (66 %) and granite (64 %) have higher CTC than those with low or not silica at all such as limestone. Also, although the silica content of the greywacke and the granite were very similar, higher content of pure silica in the form of quartz crystals in greywacke (48 %) compared to that in the granite (25 %) gave higher CTC values.

CTC of aggregates reported in the literature have been compared with those obtained in this study. For instance, Mukhopadhyay et al. [2] reported values of 6.45 x 10^{-6} mm/mm/^{0}C for limestone, 8.90 x 10^{-6} mm/mm/^{0}C for granite and 11.15 x 10^{-6} mm/mm/^{0}C for sandstone using a dilatometer. It can be seen that these values are similar to those obtained in the current study. Also, Kim et al. [4] reported values between 4.0 and 11.4 x 10^{-6} mm/mm/^{0}C for a wide range of aggregate sources using a strain gauge technique.

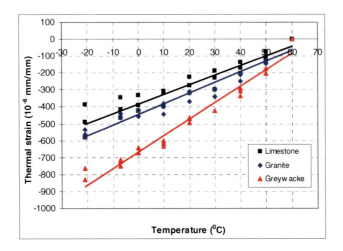

Fig. 3. Thermal strain vs temperature (aggregates)

Table 1. Coefficient of thermal contraction of binders and aggregates

Material	Type	CTC $10^{-6}/^0C$	R^2
Binder	PMB 50	126.60	0.876
	40/60	161.21	0.923
	10/20	176.62	0.994
Aggregate	Limestone	5.69	0.903
	Granite	6.28	0.944
	Greywacke	9.74	0.964

Thermally induced damage in asphalt can be originated as a result of large differences in the CTC of the aggregate and the binder. During cooling, both the binder film surrounding and aggregate particle and the aggregate itself contract. Thermovolumetric contraction of the binder is, however, much larger than that of the aggregate as a result of higher CTC. As the binder contracts more than the aggregate, compressive stresses develop on the aggregate particles leading to tangential stresses in the bitumen film. At intermediate temperatures, these stresses may relax due to the viscoelastic character of the binder. At very low temperatures, however, the behaviour of the binder becomes practically elastic and thermally induced stresses might build up to a point where they exceed the tensile strength of the binder. When this happens, localized damage at the aggregate-binder interface in the form of micro-cracking will develop.

Once these cracks have been formed, they might propagate into the binder film. Water then may be able to penetrate and fill the spaces between the aggregate and the bitumen film causing de-bonding and loss of adhesion. So, although the failure mechanism might be associated to moisture damage it might have been originated as a result of thermal induced damage due to differential thermal contraction.

4.2 Thermal Contraction of Asphalt Mixtures

Thermal strain of asphalt specimens was determined by dividing the change in length as a result of cooling by the initial length measured at 20 ^{0}C. The linear coefficient of thermal contraction (CTC) was then obtained from the slope of the line fitted through the experimental data. Experimental CTC values for the mixtures investigated are presented in Table 2. It can be seen that for the same material (AC) and binder (PMB 50), the mixture produced with the greywacke aggregate had the highest CTC, followed by the granite and the limestone. Thus, as expected, the higher the CTC of the aggregate, the higher the CTC of the mixture. Similarly, for the same material (AC) and aggregate (granite), the mixture produced with the 10/20 pen binder had the highest CTC. Thus, the higher the CTC of the binder, the higher the CTC of the mixture. This, however, was not observed for the mixtures produced with 40/60 and PMB 50.

Table 2. CTC of asphalt mixtures

Material	Binder	Aggregate	CTC_{Exp} $10^{-6}/^0C$	R^2	CTC_{Theo} (1) $10^{-6}/^0C$	CTC_{Theo} (2) $10^{-6}/^0C$
AC	PMB 50	Limestone	20.67	0.966	22.98	29.98
AC	PMB 50	Granite	21.41	0.981	22.88	24.35
AC	PMB 50	Greywacke	23.83	0.979	26.57	28.38
AC	PMB 50	Granite	21.41	0.981	22.88	24.35
AC	40/60	Granite	20.98	0.981	27.66	27.96
AC	10/20	Granite	22.28	0.997	29.79	29.57
PA	PMB 50	Greywacke	21.24	0.983	25.28	27.53
AC	PMB 50	Greywacke	23.83	0.988	26.57	28.38
SMA	PMB 50	Greywacke	24.59	0.979	28.20	29.47

CTC of different mixtures produced with then same binder (PMB 50) and aggregate (greywacke) are also presented in Table 2. It can be seen that the SMA had the highest CTC followed by the AC and the PA. This could be attributed to the relatively high binder content in the SMA compared to the AC and the PA. It is anticipated that, in general, the higher the binder content, the higher the CTC of the mixture. Also, relatively low CTC values of the PA have been attributed to lower binder content and much high air void content in the PA (19.6 %) compared to the air void content in the AC (4.3 %) and the SMA (4.3 %).

Dilatometric results for laboratory prepared asphalt beam specimens reported by Marasteanu et al. [7] showed that, during cooling, the Glass Transition Temperature (T_g) ranged between -27.9 °C and -47.98 °C. Also, CTC values below T_g ranged between 7.18 and 14.30 x 10^{-6} mm/mm/°C whereas those above T_g ranged between 30.48 and 40.74 x 10^{-6} mm/mm/°C. It can be seen that, at temperatures above T_g, CTC values of these mixtures were found higher than those determined in the present study.

4.3 Thermal Models

The CTC of an asphalt mixture can also be predicted using a simple volumetric model where the CTC of the mixture components, i.e. aggregate and binder, and the volumetrics of the mixture are the inputs, as follows [4],

$$\alpha_{mix} = \alpha_a V_a + \alpha_b V_b \qquad (1)$$

where α_{mix}, α_a, α_b are the CTC of the mixture, the aggregate and the binder, and V_a and V_b are the volume fraction of the aggregate and the binder in the mixture, respectively. CTC values calculated using Equation 1 are presented in Table 2. It can be seen that the values predicted using this simple volumetric model are only slightly higher than those obtained experimentally using asphalt specimens. Furthermore, the ranking of the mixtures follows the same trend as those observed using asphalt specimens.

Alternatively, more complex models can be use to predict CTC of an asphalt mixture. The Hirsch's model [8] is a combination of a parallel model and a series model and it is used to predict the properties of a composite from those of its components. It has been used to predict, for instance, the elastic modulus of concrete based on the modulus of the cement mortar and the aggregate. In this work, a model based on Hirsch's composite model was adopted to predict CTC of asphalt mixtures. The model has been successfully used to predict the CTC of aggregates and concrete [2].

The following equation was used to calculate CTC of an asphalt mixture,

$$\alpha_{mix} = X(\alpha_a V_a + \alpha_b V_b) + (1-X)\frac{\alpha_a V_a E_a + \alpha_b V_b E_b}{V_a E_a + V_b E_b} \qquad (2)$$

where α_{mix}, α_a, α_b are the CTC of the mixture, the aggregate and the binder, V_a and V_b are the volume fraction of the aggregate and the binder in the mixture, E_a and E_b are elastic modulus of the aggregate and the binder, and X and $(1-X)$ are the relative proportions of material conforming with the upper and lower bound solutions.

It can be seen that The Hirsch's model (Equation 2) becomes the series model when $X = 0$ and the parallel model when $X = 1$ (Equation 1). In this work a value of 0.5 has been adopted which indicates that the chances of occurrence of either parallel or series arrangements of the constituents components in the mixture are equal. Furthermore, the elastic modulus values of the aggregates used in the present study were 25.2 GPa for the greywacke, 18.1 GPa for the granite and 29.1 for the limestone [9]. Also, the elastic modulus of the binder used in the calculations was 3 GPa [4].

Predicted CTC values of the asphalt mixtures using the composite model are presented in Table 2. It can be seen that the predicted values are in general higher than those obtained using the simple volumetric model and the experimental method. Main difference was observed for the AC PMB 50 Limestone mixture. The relatively high CTC value predicted for this mixture was attributed to the modulus of the limestone aggregate. Equation 2 indicates that an increase in the modulus of the aggregate and/or the binder results in an increase on the CTC value of the mixture. Nevertheless, the ranking of the mixtures in terms of CTC values follows the same trend as those observed using the experimental and volumetric approaches.

5 Conclusions

- CTC of the binders depended on the grade and type of binder.
- CTC of the aggregates were 10 to 25 times lower than those of the binders. Also, aggregates with relatively high silica content in the form of quartz crystals had higher CTC than those with low or not silica at all such as limestone.
- Thermally induced damage in asphalt at low temperatures can be originated as a result of large differences in the CTC of the aggregate and the binder.
- Experimental CTC values of the mixtures varied between 21 and 24 x 10^{-6} mm/mm/^{0}C. Also, the higher the CTC of the aggregate and/or the binder the higher the CTC of the mixture.
- CTC of the different types of asphalt mixtures suggest that mixtures with relatively low binder content and high air voids such as PA are less susceptible to thermal contraction than those with high binder content and low air voids, such as SMA.
- Predicted CTC values of the asphalt mixtures using a volumetric and a composite model were in reasonable good agreement with those obtained experimentally.

References

[1] El Hussein, H.M., Kim, K.W., Ponniah, J.: J. Mater. Civil Eng. 10(4), 269–274 (1998)
[2] Mukhopadhyay, A.K., Neekhra, S., Zollinger, D.G.: FHWA/TX-05/0-1700-5, Texas Transportation Institute. The Texas A&M University System, Texas (2007)
[3] Bouldin, M.G., Dongre, R., Rowe, G.M., Sharrock, M.J., Anderson, D.A.: J. Assoc. Asphalt Pav. 69, 497–539 (2000)
[4] Kim, S.S., Sargand, S., Wargo, A.: FHWA/OH-2009/5, Ohio Research Institute for Transportation and Environment. Ohio University, Ohio (2009)
[5] Ojo, J.O., Fratta, D., Bahia, H.U., Daranga, C., Marasteanu, M.: Pavement Cracking. In: Al-Qadi, Scarpas, Loizos (eds.) Proceedings of the 6th RILEM International Conference on Cracking in Pavements, Chicago, USA, pp. 469–479. Taylor and Francis Group, London (2008)

6. Nam, K., Bahia, H.U.: J. Mater. Civil Eng. 21(5), 198–209 (2009)
7. Marasteanu, M., et al.: MN/RC 2007-43, National Pooled Fund Study 776, Department of Civil Engineering. University of Minnesota, Minneapolis (2007)
8. Hirsch, T.J.: ACI J. 59(3), 427–451 (1962)
9. Artamendi, I., Ward, C., Allen, B., Phillips, P.: In: Paper accepted 7th RILEM International Conference on Cracking in Pavements, The Netherlands, Delft (2012)

Mechanistic Pavement Design Considering Bottom-Up and Top-Down-Cracking

A. Walther and M. Wistuba

Braunschweig Pavement Engineering Centre,
Technische Universität Braunschweig, Germany

Abstract. Pavement design and pavement life-time estimation are generally realized by using mechanistic-empiric pavement design tools. In this paper an improved design procedure is presented that integrates bottom-up-fatigue-cracking and low-temperature-induced top-down-fatigue-cracking in a narrow time-scale. Considering a design period of decades (e. g. 30 years), traffic data and pavement temperature data are considered in every single hour. As time-variation curve of material stiffness is displayed, traffic loads are realistically superimposed in time by temperature-induced loads. Thus, inaccurate time-independent consideration of traffic and temperature loads is avoided, as usually realized in conventional mechanistic design procedures where crucial effects of extreme loading situations are more or less neglected because of simplification by statistic clustering and averaging of load input data. For the assessment of temperature input data, time variation of climate data – i. e. air temperature, humidity, wind velocity and solar radiation – may be obtained from routine meteorological observation, or from road data observation using sensor technology. These data are used in order to assess temperature profiles occurring in the pavement on an hourly time scale. For this purpose, the finite difference method is used, which is a simple numerical procedure based on the Fourier heat equation and which describes unsteady thermal processes in an iterative way. Consequently, temperature profiles considered in short time intervals can be calculated and used in the pavement design process. Using data of hourly pavement temperatures, critical temperature gradients are incorporated in the design process.

1 Background

Routine mechanistic pavement design is usually based on linear elastic multilayer theory. Input data needed for design analysis are related to weather and traffic. Material properties are described in terms of parameters assessed in laboratory tests. In regard to temperature input, data are often summarized in temperature classes, each class representing a characteristic temperature distribution within the pavement with a specific annual distribution of frequency (see, e. g. [1]). For each temperature class, Young's Modulus is calculated in function of temperature distribution over pavement layers. Based on Young's Modulus, the horizontal bending tensile strain ε at the bottom of the asphalt base course layer is calculated.

Fatigue strength evaluation of the asphalt layer is based on cyclic stress tests on asphalt mix samples. According to the European Standard for fatigue testing (EN 12697-24), the classical fatigue criterion is used, and determination of the number of load applications at failure $N_{f/50}$ is undertaken. The results of the tests end in a fatigue line (called Wöhler line) which is drawn by executing a linear regression between $N_{f/50}$ and ε_i, indicating the fatigue life duration in function of the applied load amplitude. As the Wöhler line can be expressed in the general form

$$N = \alpha_1 \cdot \varepsilon^{-\beta_1}, \qquad (1)$$

the linear regression function in the log-log-diagram reads

$$lnN_{f/50} = \alpha + \beta \cdot ln\varepsilon_i, \qquad (2)$$

where α and β are experimentally derived material constants. Finally, the slope of the fatigue line and the initial strain amplitude ε_i corresponding with a fatigue life of 10^6 load cycles are determined, as required for CE-declaration of conformity by the European Standards (EN 13108). According to (EN 12697-46), Resistance to low temperature cracking the uniaxial cyclic tensile stress test (UCTST) is to be performed. Here a specimen is loaded with a cyclic tensile stress, which is characterized by a sinusoidal waveform for the simulation of the dynamic loading due to traffic, exposed in combination with a constant stress, which symbolizes the cryogenic stress derived from thermal stress restrained specimen tests (TSRST). The resulting Wöhler line can be expressed as follows where k_1 and the exponent k_2 are experimentally derived material constants [2].

$$N_{f/50} = k_1 \cdot \Delta\sigma^{k_2}. \qquad (3)$$

Analysis of fatigue evolution requires a cumulative damage hypothesis. This is traditionally realized by linear summation of cyclic ratios applying Miner's law [3]; taken over from fatigue of metals to asphalt materials by Peattie in 1960 [4]. The cumulative fatigue damage D due to load repetitions reads

$$D = \sum_{i=1}^{n} \frac{n_i}{N_i} \leq 1 \qquad (4)$$

where n_i is the number of actual traffic load application at strain/stress level i, and N_i is the number of allowable traffic load application to failure at strain/stress level i. Eqn. (4) allows predicting fatigue life in terms of the number of allowable load applications (due to traffic and thermal load cycles). Sometimes, the theoretical residual fatigue life is expressed in the unit of years. For this purpose, the number of actual traffic load applications used for design considerations is formulated by a mean value for one design year (based on traffic counts and extrapolative estimations). Fatigue life is then calculated by the ratio of the number of allowed load applications to the mean number of traffic load applications of the design year, where the number of allowed load applications is derived from fatigue testing on the asphalt base- and wearing course material by superposition of the results that were obtained from tests at various strain/stress levels.

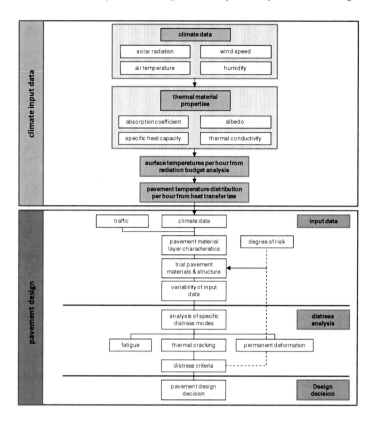

Fig. 1. Flow chart of classical design process with enhanced input approach

The calculation procedure is usually organized in an iterative mechanistic pavement design approach as illustrated in the lower part of Figure 1. In this study such an approach is followed up and climate input data are improved based on hourly calculation of temperature distribution in the pavement structure.

2 Assessment of Temperature Time Series

In this paper, design calculation is based on traffic and temperature data that are superimposed on an hourly time scale. Hence, temperature input data are needed in a rather narrow time scale.

Pavement surface temperature can be determined by heat balance equation. Pavement surface is regarded as a closed thermodynamic system. Accordingly, the radiation budget $[Q]$, soil heat flux $[B]$, convective heat exchange to air $[L]$, and heat flux due to evaporation and condensation of water $[V]$ equals zero, reading

$$Q + B + L + V = 0 \tag{5}$$

In order to satisfy energy conservation law, all thermal effects need to be taken into account. Usually, heat flux V is neglected, as heat gain due to condensation over night balances with heat loss through evaporation at sunrise [5]. Figure 2 illustrates heat flux budget schematically.

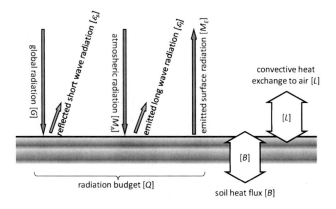

Fig. 2. Heat flux on paving surface at sunny weather conditions

As to radiation, input parameter are global radiation G [W/m²], reflected short wave radiation ε_s [-], atmospheric radiation M_A [W/m²], emitted long wave radiation ε_l [-], and emitted surface radiation M_E [W/m²], reading

$$Q = G \cdot \varepsilon_s - M_E + M_A \cdot \varepsilon_l . \tag{6}$$

Atmospheric radiation M_A can be expressed by

$$M_A = T_{air}^4 \cdot \sigma \cdot (a - b \cdot 10^{-c \cdot h_{rel} \cdot E_{air}}) , \tag{7}$$

where T_{air} [K] is air temperature, σ [W/(K⁴m²)] is Stefan-Boltzmann's-constant, h_{rel} [%] is relative air humidity, E_{air} [Pa] is air saturation vapour pressure, $a = 0.79$, $b = 0.174$, and $c = 9.5\text{E-}04$ [1/Pa]. Soil heat flux B [W/m²] in an isotropic layer of dimension dx is determined in function of temperature from

$$B = \lambda \cdot \frac{dT}{dx} , \tag{8}$$

where λ [W/(mK)] is heat conductivity. Convective heat transfer L between air and road surface is given by

$$L = \alpha_{convection} \cdot (T_{air} - T_{road\ surface}) , \tag{9}$$

with

$$\alpha_{convection} = 10 \cdot (0,174 + 0,941 \cdot v_w^{0,366}) , \tag{10}$$

where the heat transfer coefficient ($\alpha_{convection}$) depends on the wind velocity v_w [6]. By using the expressions mentioned above and by introducing virtual air temperature T^*_{air}, which is given by

$$T^*_{air} = T_{air} + \frac{Q}{\alpha_{convection}}, \qquad (11)$$

surface temperature $T_{Surface,\ k+1}$ at a given time $k+1$ and for interval Δx, that represents the distance of knots where temperatures are assessed, can be calculated from

$$T_{Surface,k+1} = -\frac{T^*_{air,k+1} - T_{1,k+1}}{\left(\frac{\lambda}{\alpha_{k+1}} + \frac{\Delta x}{2}\right)} \cdot \frac{\lambda}{\alpha_{k+1}} + T^*_{air,k+1}. \qquad (12)$$

Extension of surface temperature calculation into pavement depth, is realized by means of

$$T_{1,k+1} = a \cdot \frac{\Delta t}{\Delta x^2} \cdot \left[T_{2,k} - 2 \cdot T_{1,k} - \frac{T^*_{air,k} - T_{surface,k}}{\frac{\lambda}{\alpha_k}} \cdot \left(\frac{\lambda}{\alpha_k} - \frac{\Delta x}{2}\right) + T^*_{air,k} \right] + T_{1,k}, \qquad (13)$$

where $T_{1,k+1}$ represents approximation of pavement temperature at depth 1 at time $k+1$. In order to describe temperature flows within the pavement heat transfer law is applied, reading

$$\Delta T_t = a \cdot \frac{\Delta t}{\Delta x^2} \cdot \Delta^2 T_x, \qquad (14)$$

where Δt is the time shift, a [cm²/h] is thermal diffusivity, and $\Delta^2 T_X$ is the difference between two temperature differences at a certain time in function of pavement depth. Any temperature $T_{n,k+1}$ within the superstructure at time $k+1$ and depth n is calculated from

$$T_{n,k+1} = a \cdot \frac{\Delta t}{\Delta x^2} \cdot \left(T_{n+1,k} - 2T_{n,k} + T_{n-1,k}\right) + T_{n,k}, \qquad (15)$$

with

$$a = \frac{3600 \cdot \lambda}{c \cdot \rho}. \qquad (16)$$

Thermal diffusivity a [cm²/h] represents the ratio between thermal conductivity coefficient λ [W/(mK)] and specific heat capacity c [J/(gK)] multiplied by bulk density ρ [g/m³]. Adiabatic temperature conditions are assumed for a depth of 2.5 m [6]. For validation of Eqn. (5 to 16) in regard to real pavement temperatures see Wistuba in 2002 [7] and Yavuzturk et al. in 2005 [8].

2.1 Input Parameter

Weather files used in this study were derived from The National Meteorological Service of Germany (DWD). The data sets contain hourly values concerning global radiation and values captured in 10 minute intervals concerning air temperature, wind velocity and relative air humidity. The corresponding weather station is located in central Germany (52°18' N, 10°27' E). Figures 4 and 5 exemplarily depict measured values for a time span of 1 year.

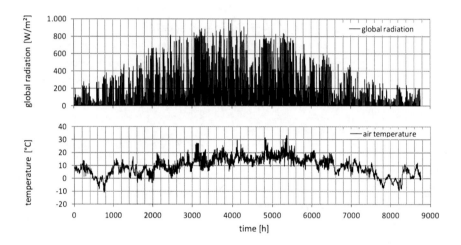

Fig. 3. Global radiation and air temperature data from selected weather station

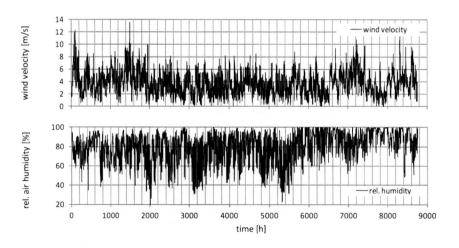

Fig. 4. Wind velocity and relative humidity data from selected weather station

Using asphalt values from literature concerning thermal conductivity coefficient λ, specific heat capacity c and bulk density ρ and a standard German highway pavement construction lead to the following pavement surface temperatures (Figure 5).

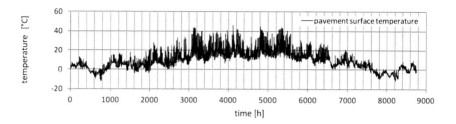

Fig. 5. Example for derived pavement surface temperature per hour

Figure 6 (left) shows temperatures on an hourly time scale within the pavement. Figure 6 (right) shows the corresponding distribution of derived Young's Modulus by Indirect Tension Test (IDT).

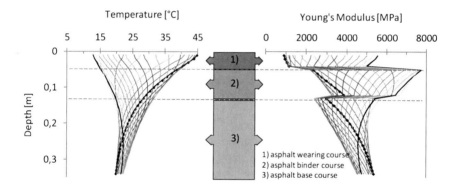

Fig. 6. Pavement temperature distribution (left) and corresponding Young's modulus (right) for 24 hours

Such processed temperature data considering a period of several years enable most detailed design analysis since the steady change of Young's modulus with temperature distribution, is considered for calculation of stresses and strains in the superstructure.

3 Thermal Stresses

The estimation of thermal stresses at the surface of the asphalt wearing course is realized by Maxwell-model. Resulting stress $\sigma(t)$ is assessed in function of initial stress σ_0 relaxation behaviour λ (T), cooling rate ($\dot{T}$) and temperature dependent Young's modulus, reading

$$\sigma(t) = \sigma_0 \cdot e^{-\frac{t}{t_R}} - \beta \cdot \lambda(T) \cdot \dot{T} \cdot \left(1 - e^{-\frac{t}{t_R}}\right), \quad (17)$$

$$\text{with} \quad t_R = \frac{\lambda(T)}{E(T)}. \quad (18)$$

Since the relaxation behaviour λ (T) of asphalt correlates with the softening point ring and ball of the respective bitumen the following assumption is used [9]:

$$log\ log\ \lambda(T) = a + b \cdot log\ T, \quad (19)$$

where a represents the constant, b the slope and T the temperature in Kelvin [K].

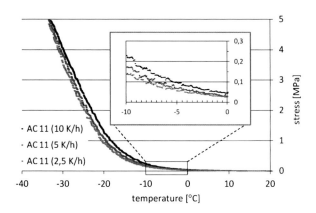

Fig. 7. TSRST data (AC 11) at different cooling rates

In order to calculate cryogenic stress from Eqn. (17), information regarding the relaxation behaviour of the deployed HMA is needed, since an increasing viscosity of the binder results in decreasing relaxation time. By approximation of parameters a and b from Eqn. (19) test results from thermal stress restrained specimen tests (TSRST) shown in Figure 7 can be modelled. The precision of the thermal stress model shows good correlation to measured stress values from TSRST (see Figure 8).

It is coherent that with a fast cooling rate the tensile stress development is increasing per time unit. Concerning the amplitude of the resulting stress level, the cooling rate is less important (see Figure 7) which complies with [10].

As pavement surface temperatures are known per hour, the cooling rate per time unit and the coherent thermal stresses can be derived.

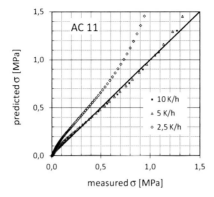

Fig. 8. Precision of thermal stress model

4 Application of Temperature Data for Stress-Strain-Detection

Through determination of temperature distribution within the pavement structure and its detailed development over time, input data for pavement design are advanced. Considering temperature dependency of asphalt Young's modulus (e. g. derived from IDT) at a narrow time scale, the information of resulting stresses and strains represent conditions where single crucial effects of extreme loading situations are incorporated into the design procedure. Figure 9 (bottom) shows resulting strains per hour at the bottom of the asphalt base course due to an 11 t axle load for a period of one year (8760 h). The strains are obtained from linear elastic theory with individual layer thicknesses of 1 cm [11] where a circular tire footprint with uniform pressure serves as input parameter into the calculations.

The application of stress assessment given in previous section lead to thermal tensile stresses at the surface of the asphalt wearing course (Figure 9, top).

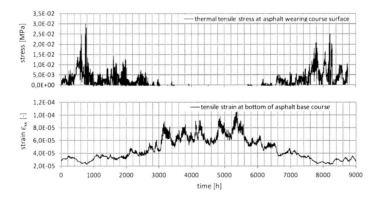

Fig. 9. Resulting thermal stresses per hour at asphalt wearing course surface (top), and resulting strains per hour at bottom of asphalt base course (bottom)

The stress-strain-distribution depicted in Figure 9 can be considered as realistic, since thermal stresses at the surface of the asphalt wearing course evolve only at low temperatures (e. g. wintertime) whereas the corresponding strains at the bottom of the asphalt base course show minor tensile strain values. This is in best agreement with an increasing stiffness of the superstructure. In summertime the strain values show controversial behaviour as only minor stress values develop at summertime.

Use of Eqn. (2, 3 and 4) respectively, results in detailed calculation of pavement design life.

5 Summary and Outlook

This paper presents an approach to enhance the theoretical consideration of the asphalt failure modes Bottom-Up-Fatigue-Cracking and Top-Down-Fatigue-Cracking at low temperatures. The theoretical procedure is described.

By use of real input data as regards air temperature, humidity, global radiation and air velocity, time series of temperature are calculated occurring in the pavement structure.

Through knowledge of temperature distribution within the superstructure and the incorporation of traffic information (axle distribution) on hourly time scale, classic approaches of mechanistic pavement design procedures using linear elastic theory may be advanced. Shear and tensile stresses/strains at the tire edges may be regarded in future within the top-down failure mode since they are most likely to be dominant on main highways and secondary roads [12].

Moreover, as regards future pavement design, climate prognoses (as done e. g. in projects assigned by Intergovernmental Panel on Climate Change (IPCC)) reflect future temperature distributions that may be used as input data in the design procedure.

References

[1] FGSV: Richtlinien für die rechnerische Dimensionierung des Oberbaues von Verkehrsflächen mit Asphaltdecke (RDO Asphalt 2009). Forschungs-gesellschaft für Straßen- und Verkehrswesen, Germany (2009)
[2] Pell, P.S.: Fatigue Characteristics of Bitumen and Bituminous Mixes. In: Proceedings, International Conference on the Structural Design of Asphalt Pavements, Ann Arbor, MI, pp. 310–323 (1962)
[3] Miner, M.A.: Cumulative damage of fatigue. J. of Applied Mechanics (1955)
[4] Monismith, C.L.: Evolution of long-lasting asphalt pavement design methodology: a perspective. In: Lecture presented in June 2004 at Auburn University to the ISAP-sponsored International Symposium on Design and Construction of Long Lasting Asphalt Pavements (2004)
[5] Krebs, H.G., Böllinger, G.: Temperaturberechnungen am bituminösen Straßenkörper. Forschung Straßenbau und Straßenverkehrstechnik, Heft 347, Bonn-Bad Godesberg (1981)

[6] Pohlmann, P.: Simulation von Temperaturverteilungen und thermisch induzierten Zugspannungen in Asphaltstraßen. Institut für Straßenwesen, TU Braunschweig (1989)
[7] Wistuba, M.: Klimaeinflüsse auf Asphaltstraßen – Maßgebende Temperaturen für die analytische Oberbaubemessung in Österreich, Dissertation. Technische Universität Wien, Fakultät Bauingenieurwesen, Wien (2002)
[8] Yavuzturk, C., Ksaibati, K., Chiasson, A.D.: Assessment of Temperature Fluctuations in Asphalt Pavements Due to Thermal Environmental Conditions Using a Two-Dimensional, Transient Finite-Difference Approach. Journal of Materials in Civil Engineering 17, 465–475 (2005)
[9] Arand, W., Dörschlag, S., Pohlmann, P.: Einfluss der Bitumenhärte auf das Ermüdungsverhalten von Asphaltbefestigungen unterschiedlicher Dicke in Abhängigkeit von der Tragfähigkeit der Unterlage, der Verkehrsbelastung und der Temperatur. In: Forschung Straßenbau und Straßenverkehrstechnik, Heft 558, Bonn-Bad Godesberg (1989)
[10] National Pooled Fund Study 776, Investigation of Low Temperature Cracking in Asphalt Pavements, Prepared for: Minnesota Department of Transportation Research Services Section (2007)
[11] ISBS-LEA: Layer Elastic Analysis Computer Program, Braunschweig Pavement Engineering Centre. Technische Universität Braunschweig, Germany (2010)
[12] Molenaar, A.A.A.: Prediction of fatigue cracking in asphalt pavements – do we follow the right approach. In: Transportation Research Record, Washington D.C., pp. 155–162 (2007)

Strength and Fracture Properties of Aggregates

Ignacio Artamendi, Chris Ward, Bob Allen, and Paul Phillips

Research & Development Department, Aggregate Industries, UK

Abstract. This paper presents a study of the mechanical and fracture properties of various types of aggregates used in asphalt mixtures. Three types of rocks namely, greywacke, granite and limestone, were evaluated. Compressive and tensile characteristics of the rocks were determined by means of uniaxial compressive and indirect tensile tests, respectively. Resistance to fracture was determined by means of semi-circular bending tests. Results showed that the greywacke had the highest strength both in compression and in tension. The granite, on the other hand, had high compressive strength but the tensile strength was relatively low. Compressive tests also showed that the response of rock specimens under loading was linear elastic. It was found that the compressive elastic modulus of the limestone was the highest followed by the greywacke and the granite. Similarly, indirect tensile tests indicated that the response of the greywacke and the limestone rocks was linear elastic whereas that of the granite was non-linear. Furthermore, tensile elastic modulus values of the greywacke and the limestone were similar and about five times higher than that of the granite. As regards fracture, load-deflection curves for semi-circular bending tests indicated linear elastic behaviour of the three types of rocks. Thus, linear elastic fracture mechanics theory was applied to determine fracture toughness. Results showed that the greywacke had the highest resistance to fracture.

1 Introduction

Asphalt materials typically comprise a large proportion of graded aggregates bound with a bituminous binder. As a result, the performance of an asphalt mixture is greatly influence by aggregate properties such as grading, shape and strength. As regards strength, aggregates should have the necessary strength to resist degradation during handling, construction and trafficking.

Adequate aggregate strength is required during storage, transportation, drying and mixing activities to avoid fragmentation due to impact between aggregate particles. Fragmentation during these stages reduces particle size and degrades particle shape altering the material grading. Aggregate breakdown also increases the dust content which may affect the adhesion of bitumen to the aggregate. Aggregate crushing, on the other hand, is most likely to occur when the material is initially compacted. Once the asphalt has been installed, the aggregates should also have sufficient strength to bear the load imposed on them by increased traffic loads. When the pavement is loaded as a result of traffic, high stresses develop at the contact points between aggregates. Depending on the aggregate strength these stresses can lead to the fracture of the aggregate [1].

In order to assess aggregate strength, a range of standard tests are commonly used [2]. The majority of these tests, however, are indicator tests, rather than tests that measure fundamental mechanical and strength properties. Nevertheless, some of these tests are widely used by road authorities and suppliers to characterise aggregates.

Alternatively, compressive and tensile tests are typically employed in rock mechanics to determine the strength of rock masses [3, 4]. Fracture mechanics are also applied to describe how a crack initiates and propagates in rocks under loading. Linear elastic fracture mechanics (LEFM) principles are applied to determine fracture parameters such as fracture toughness. Different methods and test geometries are used to determine fracture resistance of rocks. A commonly used method for rock fracture is the semi-circular bending (SCB) test [5].

The aim of this work is to characterise the strength and fracture properties of various types of aggregates used in asphalt mixtures. Both compressive and indirect tensile tests were carried out on rock specimens in order to determine strength and elastic modulus. The semi-circular bending (SCB) test was also used to determine fracture toughness. It is believed that these types of tests could provide valuable information for the characterisation of crushed rock aggregate.

2 Materials and Specimen Preparation

Three types of rocks namely, greywacke, granite and limestone, were used in the study. Greywacke rock was sourced from Haughmond Hill Quarry in Shropshire, England. Geologically, these rocks date from the Upper Pre-Cambrian. They were formed from sediments laid down in shallow waters along the margins of a continental shelf, over 600 million years ago. Granite rock was sourced from Croft Quarry near Leicester, England. These rocks were formed about 500 million years ago from molten igneous materials deep in the earth's crust. Limestone was sourced from Topley Pike quarry, Derbyshire, England. Topley Pike is found on the outcrop of the Carboniferous Limestone series referred to locally as the Derbyshire Dome. The central part of the dome is characterised by massive and chemically pure limestone of several hundred metres total thickness.

For the preparation of specimens for testing, cores of 150 mm and 100 mm diameter were taken from large rocks (1 m^3 approximately) collected from the quarries. Rock cores of 100 mm diameter were then cut to a thickness of 200 mm and used for uniaxial compressive testing. Cores of 150 mm diameter, on the other hand, were cut to a thickness of 150 mm and used for indirect tensile tests. SCB specimens were obtained by cutting 150 mm diameter core specimens to a thickness of 50 mm. These cylinders were cut perpendicular to the axis to obtain the semi-circular specimens, and then notched at the mid-point along their radius.

3 Experimental

3.1 Aggregate Tests

Standard tests were first carried out to characterise the different types of aggregates. These included Los Angeles (LA), aggregate crushing value (ACV), ten per cent fines (TPF) and aggregate impact value (AIV).

The LA test is a measure of the resistance of coarse aggregate to fragmentation resulting from a combination of actions including abrasion or attrition, impact and grinding. The test method is described in EN 1097-2 and is widely used to characterise aggregate. The ACV gives a relative measure of the resistance of an aggregate to crushing when subjected to a compressive load gradually applied to a specific maximum value in a prescribed time. The test method is described in BS 812-110. The TPF value gives a relative measure of the resistance of an aggregate to crushing under a gradually applied compressive load. The test method is described in BS 812-111. The AIV test gives a relative measure of the resistance of an aggregate to sudden shock or impact. The test procedure is described in BS 812-112.

3.2 Uniaxial Compressive Test

Uniaxial compressive tests were carried out using a hydraulic compression machine 3000 kN maximum load. Rock core specimens of 100 mm diameter and 200 mm height approximately were employed. The specimens were instrumented with strain gauges in order to measure both axial and radial strains during loading. The axial strain was measured by means of two strain gauges glued to the specimen and positioned in the middle of it in the direction of the applied load. The radial strain was measured with only one strain gauge situated in the middle of the specimen in a direction perpendicular to the applied load. The length of the gauges, $2L$, was 30 mm. The strain gauges were connected to a compensation device connected to the control system. The test set-up is presented in Figure 1.

Fig. 1. Compressive and tensile tests set-up

Uniaxial compression tests were carried out in load control mode at a rate of 0.6 MPa/s. Five specimens of each rock type were tested. Two of these specimens were tested to failure in order to determine the compressive strength, σ_c. The remaining three specimens were loaded up to 100 MPa in order to determine the compressive elastic modulus, E_c, and the Poisson's ratio, v. The compressive elastic modulus, E_c, was defined as the slope of the linear part of the stress-axial strain relationship. The compressive stress was determined by dividing the applied load by the surface area. The Poisson's ratio was defined as the ratio of the radial strain to the axial strain and was determined from the slope of the linear relationship between radial and axial strain.

3.3 Indirect Tensile Test

In the indirect tensile test, a cylindrical specimen is subjected to a compressive load across its vertical diametral axis. This load originates a relatively uniform constant tensile stress at the centre of the specimen which causes the specimen to fail and crack along the vertical diameter. Indirect tensile test were carried out using rock core specimens of 150 mm diameter and 150 mm thickness. The specimens were instrumented with two strain gauges situated at the centre of each side and along the direction perpendicular to the applied load. The length of the gauges, $2L$, was 30 mm. The test set-up is presented in Figure 1.

The tensile elastic modulus, E_t, was determined using the following expression [6]:

$$E_t = E_s \left\{ \left(1 - \frac{D}{L}\arctan\frac{2L}{D}\right)(1-v) + \frac{2D^2(1+v)}{4L^2 + D^2} \right\} \quad (1)$$

where E_s is defined as the splitting elastic modulus, D is the specimen diameter, L is the strain gauge half-length and v is the Poisson's ratio. The splitting elastic modulus, E_s, can be determined from the slope of the linear part of the stress-strain relationship obtained in the indirect tensile test. Sometimes, however, the evolution of the stress with the strain is non-linear. In this case, E_s, can be determined approximately using the following equation [6].

$$E_s = \frac{\sigma_{t/2}}{\varepsilon_{t/2}} \quad (2)$$

where $\sigma_{t/2}$ is half of the tensile strength and $\varepsilon_{t/2}$ is the tensile strain corresponding to $\sigma_{t/2}$.

3.4 Semi Circular Bending Test

For Mode I, referred to as opening or tensile mode, LEFM theory establishes that, for a material containing a flaw (micro-crack), fracture occurs when the stress intensity factor, K_I, exceeds a critical value, K_{IC}, referred to as the fracture toughness. In general, the fracture toughness, K_{IC}, can be expressed as follows:

$$K_{IC} = Y_I \sigma_0 \sqrt{\pi a} \quad (3)$$

where σ_0 is the critical stress for crack propagation, a is the notch or crack length, and Y_1 is a geometrical factor that depends on the specimen size and geometry.

For a SCB specimen loaded in a three point bending configuration and notched at the mid-point along the radius and in the direction of the load, as illustrated in Figure 2, σ_0 and Y_1 are given by the following expressions [7]:

$$\sigma_0 = \frac{P_0}{2rt} \tag{4}$$

$$Y_1 = 4.782 - 1.219\left(\frac{a}{r}\right) + 0.063\exp\left(7.045\left(\frac{a}{r}\right)\right) \tag{5}$$

where P_0 is the critical load, and r and t are the specimen radius and thickness respectively. It should be noted that Equation 5 is only valid when the span to diameter ratio ($2s/2r$) is 0.8.

Fracture tests were performed using a compression machine. SCB specimens were loaded monotonically at a loading rate of 5 mm/min. The span ($2s$) was 120 mm which gave span to diameter ratio ($2s/2r$) of 0.8. Notch length (a) was 10 mm approximately. Specimen radius (r) and thickness (t) were 70 and 50 mm, respectively. During the test, the crosshead displacement and the load were recorded. Fracture test were conducted at room temperature. Two specimens per material were tested. A rock SCB specimen after testing is presented in Figure 2.

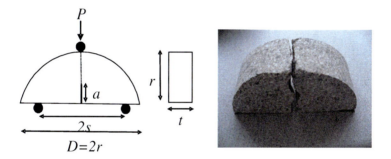

Fig. 2. Schematic SCB test and specimen after testing

4 Results and Discussions

4.1 Standard Aggregate Tests

Results of the standard aggregate tests are presented in Table 1. Results indicated that the granite and the limestone had similar strength characteristics. These tests also show that the greywacke was by far the strongest material in terms of resistance to fragmentation, crushing and impact.

Table 1. Standard aggregate tests results

Aggregate type	LA	ACV	TPF	AIV
		%	kN	%
Greywacke	12	10	410	9
Granite	28	20	180	27
Limestone	31	24	180	25

4.2 Strength and Elastic Modulus

Results from uniaxial compressive tests are presented in Table 2. It can be seen that the compressive strength (σ_c) of the greywacke rock was the highest followed by the granite and the limestone. Similar values have been reported by Bearman [8] in a previous study using very similar type of rocks.

Table 2. Compressive and indirect tensile tests results

Rock type	σ_c	E_c	v	σ_t	E_t	σ_c/σ_t	E_c/E_t
	MPa	GPa		MPa	GPa		
Greywacke	196.8	25.2	0.195	10.5	14.6	18.7	1.7
Granite	150.6	18.1	0.171	3.9	3.1	38.6	5.8
Limestone	90.6	29.1	0.253	5.8	13.4	15.6	2.2

Figure 3 shows typical stress-strain curves obtained during a compressive test. The figure indicates that the stress-strain relationship for rock specimens is practically linear and therefore the behaviour can be considered as linear elastic. Thus, the compressive elastic modulus (E_c) can be determined by fitting a straight line through the stress-axial strain data. The slope of this line is defined as E_c.

Table 2 shows the compressive elastic modulus values of the three types of rocks determined as before. It can be seen that the limestone had the highest modulus followed by the greywacke and the granite. Jianhong et al. [6] reported compressive elastic modulus of 20 GPa for granite, similar to the value of 18 GPa obtained in this study, and 58 GPa for limestone which it is higher than the value of 29 GPa reported here. Visual observation of the limestone rock cores showed that they had numerous defects and cracks and this might one of the reasons why the modulus was relatively low. Thuro et al. [9] also reported values of around 50 GPa for limestone. Furthermore, they showed that the size (diameter) and the shape (length/diameter ratio) of the specimens had an influence on strength and elastic modulus values.

Poisson's ratio values were determined by fitting a straight line through the radial and axial strain data and are shown in Table 2. It should be noted that due to the defects and cracks present in the limestone just one of the tests gave reliable results. Thus, the validity of this value is limited. Nevertheless, Poisson's ratios determined in this work were all within the ranges given in Gercek [10].

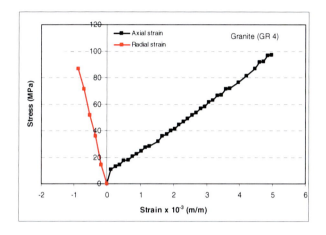

Fig. 3. Stress-strain curves for a compressive test (granite)

Results from indirect tensile tests are also presented in Table 2. Data shows that the indirect tensile strength of the greywacke was the highest followed by the limestone and the granite. Bearman [8] reported values of 3.8 MPa and 15.4 MPa for similar limestone and greywacke rocks, respectively. He also, reported a value of 10.6 MPa for a particular type of granite which is significantly higher than the value of 3.8 MPa obtained in this work.

As regards tensile elastic modulus, the stress-strain curves indicated that the behaviour of the greywacke and the limestone rocks was linear elastic. Thus, the splitting elastic modulus (E_s) could be determined from the slope of the linear part of the stress-strain curve. The stress-strain curve of the granite was, however, non-linear, as seen in Figure 4. Equation 2 was, therefore, used to determine the splitting elastic modulus (E_s). The tensile elastic modulus (E_t) was then calculated using Equation 1. Results presented in Table 2 showed that tensile elastic modulus of the greywacke was the highest followed by the limestone. The elastic modulus of the granite was, however, relatively low.

Data presented in Table 2 shows that the compressive strength and elastic modulus values were higher than the tensile strength and elastic modulus. Results indicated that, for the greywacke and the limestone, the compressive elastic modulus is about twice the tensile elastic modulus. For the granite, however, the compressive modulus is about six times the tensile modulus. Rock materials are in general non-homogenous and contain defects and microcraks. As a consequence they show different behaviour under tensile and compression conditions. Furthermore, anisotropy has also an effect when determining the mechanical properties of rocks under loading.

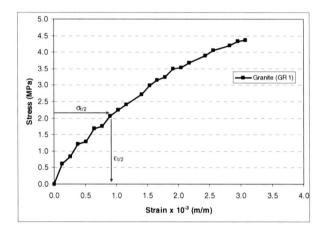

Fig. 4. Stress-strain curve for an indirect tensile test (granite)

4.3 Fracture Toughness

Results from SCB tests are presented in Table 3. Fracture toughness values indicated that the greywacke was more resistance to fracture in the presence of a crack (notch) than the granite or the limestone. Also, small differences were found between the fracture toughness values of granite and limestone.

Fracture toughness values reported in the literature using the SCB method varied from 0.28 MPa m$^{1/2}$ for sandstone to 1.72 MPa m$^{1/2}$ for granite [11]. Values obtained with the chevron bend (CB) test method on similar rocks to the ones tested in this study are also presented in Table 3 [8]. It can be seen that the values obtained with the CB method are higher than those obtained with the SCB method. Nevertheless, the ranking of the rocks from these two methods was the same.

Table 3. Fracture toughness values

Rock type	$K_{I(SCB)}$	$K_{I(SCB)}$ [11]	$K_{I(CB)}$ [8]
	MPa m$^{1/2}$	MPa m$^{1/2}$	MPa m$^{1/2}$
Greywacke	1.45		3.15
Granite	0.58	0.68, 0.88, 1.72	1.83
Limestone	0.57	0.68, 0.85, 1.33	0.73

5 Conclusions

From the experimental work carried out in this study the following conclusions can be drawn:

- Based on standard aggregate tests, the greywacke was the strongest material in terms of resistance to fragmentation, crushing and impact.
- The greywacke rock had the highest compressive and tensile strength. The granite, on the other hand, had high compressive strength but the tensile strength was relatively low.
- Compressive test indicated that the limestone was very resistance to deformation as seen by the compressive elastic modulus value.
- The behaviour of the rocks under tension depended on the mineralogy of the rock. Thus, the response of the greywacke and the limestone rocks was linear elastic whereas that of the granite was non-linear.
- Tensile elastic modulus values of the greywacke and the limestone were similar and about five times higher than that of the granite.
- Fracture tests indicated that the greywacke had the highest resistance to fracture.

References

[1] Mahmound, E., Masad, E., Nazarian, S.: J. Mater. Civil Eng. 22(1), 10–20 (2010)
[2] Pike, D.C.: Standards for aggregates. Ellis Horwood Ltd., Chichester (1990)
[3] ISRM: Int. J. Rock Mech. Min. Sci. Geomech. Abstr. 16, 135–140 (1978)
[4] ISRM: Int. J. Rock Mech. Min. Sci. Geomech. Abstr. 15, 99–103 (1978)
[5] Chong, K.P., Kuruppu, M.D.: Int. J. Fracture 26, R59–R62 (1984)
[6] Jianhong, Y., Wu, F.Q., Sun, J.Z.: Int. J. Rock Mech. Min. Sci. 46, 568–576 (2009)
[7] Lim, I.L., Johnson, I.W., Choi, S.K.: Eng. Fract. Mech. 44(3), 363–382 (1993)
[8] Bearman, R.A.: Int. J. Rock Mech. Min. Sci., Technical Note 36, 257–263 (1999)
[9] Thuro, K., Plinninger, R.J., Zäh, S., Schütz, S.: Rock Mechanics a Challenge for Society. In: Särkkä, Eloranta (eds.) ISRM Regional Symposium Eurorock 2001, pp. 169–174 (2001) (finland)
[10] Gercek, H.: Int. J. Rock Mech. Min. Sci. 44, 1–13 (2007)
[11] Alkiliçgġl, Ç.: Development of specimen geometries for Mode I fracture toughness testing with disc type rock specimens. PhD. Middle East Technical University (METU), Ankara Turkey (2010)

Cracks Characteristics and Damage Mechanism of Asphalt Pavement with Semi-rigid Base

Ai-min Sha and Shuai Tu

School of Highway, Chang'an University, Xi'an, China
aiminsha@263.net, iwtjmh@163.com

Abstract. As the semi-rigid base has such advantages as high bearing capacity, good plate-forming properties, excellent frost resistance and could take full advantage of the local materials, asphalt pavement with semi-rigid base has become a primary pavement structure of highway in China since 80s of the 20th century[1]. However, semi-rigid base has some inherent problems like shrinkage deformation cracking and low erosion resistance[2][3]. With the rapid growth of traffic and loading capacity, some distresses such as cracking, rutting, and pothole appear gradually, in which the cracking is the main part. In this context, the distress types, distribution characteristics and no-disease proportion of the asphalt pavement with semi-rigid base should be further explicated, and further concerns should be taken to whether the structure is fitting for the expressway. In this paper, outdoor investigations including damage observe, core sample examination, deflection test and indoor tests containing extraction test of asphalt mixture, strength test of cores samples are used to investigate the pavement using situation of 11 high-grade highways. The distribution, characteristics, cause, scale, proportion of the cracks in different zones, different climatic features and different structures are analyzed systematically to have a further understanding about the characteristics of the semi-rigid base, which provide the guidance for reducing the pavement distress, improving the performance, promoting the serviceability and its application.

1 Roads and Methods of Investigation

1.1 Roads Situations and Investigation Contents

Outdoor investigations and indoor tests are used in 11 high-grade highways to investigate, analyse and evaluate the using situation of asphalt pavement with semi-rigid base. The main pavement structure of the investigation roads is shown in figure 1.

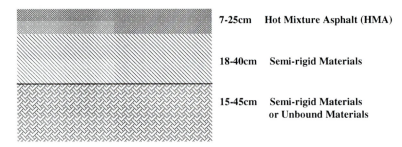

Fig. 1. The main pavement structure of investigation roads

1.2 Investigate Methods

The investigation methods and main contents are shown in table1.

Table 1. The methods and main contents of investigation

Investigation Methods	Damage Observe	Core Sample Examination	Pavement Deflection	CBR of Subgrade	Permeability Test	Extraction of Asphalt Core	Strength of Base Core
Main Contents	Types, Scale and Distribution of Damages	Damages on Surface、Drill Hole Condition、Core Sample Condition	Dynamic Rebound Deflection	CBR Strength in Different Depth of Subgrade	Permeability Coefficients of Pavement with or Without Diseases	Change of Asphalt-aggregate Ratio and Aggregate Gradation	Unconfined Compressive Strength

On the analysis of the cracks distribution and the cracking mechanism, the main ways are damage observe and core sample examination, with reference of geography, climate and traffic.

(1) Damage observes. According to observing the pavement service condition of investigate roads, the types, scale and distribution of damages were analysed. The pavement conditions are shown as figure 2, figure 3 and figure 4.

Fig. 2. Transverse cracks   **Fig. 3.** Massive cracks   **Fig. 4.** Massive cracks

(2) Core sample examination on the investigate roads. The core sampling positions cover running lane, overtaking lane and hard shoulder. Both external and internal situations were noticed. The external situations included surface damages and drainage development; The internal situations included coring time, difficulty and colour of running water in the coring process. After coring operation, the drilling depth, structural condition in the drill hole were scrutinized, including the thickness, uniformity of structural layer, the colour and luster of asphalt, and the bonding between layers. The core sample examinations are shown as figure 5, figure 6 and figure 7.

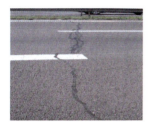  

Fig. 5. Surface cracks   **Fig. 6.** Drill hole   **Fig. 7.** Core sample

2   Damage Types and Proportion of Asphalt Pavement with Semi-rigid Base

From the investigation, common damages on the asphalt pavement with semi-rigid base can be divided into crack, rutting, loosening, pothole, settlement, oil and other type, in which the crack is the primary damage type and the rutting is secondary. [4][5]The damage types and proportions are shown as figure8.

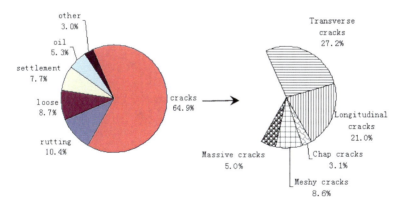

Fig. 8. Damage types and proportions of asphalt pavement with semi-rigid base

Investigation also shows that the service condition of asphalt pavement with semi-rigid base is adequate. Diseases like loosening, pothole and rutting mainly appear on the running lane where the semi-rigid base is in good condition. Damages have obvious regional characteristic, also have direct connection with traffic and vehicle load.[3] Two expressways in the north areas have serious crack diseases, but the first-class highway in the south of the Yangtze River has loosening and potholes as the primary diseases. Overall, disease sections take a proportion of 16% in asphalt pavement with semi-rigid base, the other 84% is in good condition.

3 Influence and Distribution Characteristics of Cracks

3.1 Influence on Pavement Performance from Cracks

The emergence of the cracks makes the pavement lose continuity and provide ways for water. The water will decrease the structural capacity, shorten the pavement service life, lessen the binding power between courses, and induce other damages under vehicle loads later. When the transverse cracks and longitudinal cracks crossed, the carrying mode of asphalt pavement turns into irregular limited size plates from continuum, which will accelerate the diseases.

3.2 Distribution Characteristics of Cracks

According to the direction and appearance, pavement cracks can be divided into transverse cracks, longitudinal cracks, chap cracks, meshy cracks and massive cracks. To investigate the cracks distribution characteristics, external conditions included length, width, area and other parameters were measured or calculated.

Transverse crack is the primary damage form and appear largely in various research sections, usually most cracks are on the running lane; Longitudinal crack is also a main damage form, which principally appear on the wheel tracks of running lane and ultra-lane. Most longitudinal cracks appear in cut-fill sections, high embankment and long vertical slope; Chap crack is often along with other damages like transverse cracks, longitudinal cracks and subsidence; Meshy crack is not the main damage form, only 3 of 11 investigate highways have large area of meshy cracks; Massive cracks rarely appear, which mainly occur in the wheel tracks of running lane. The length, width, interval and area and other distribution characteristics are as table 2.

Table 2. The distribution characteristics of cracks on asphalt pavement with semi-rigid base

Crack types	Transverse cracks	Longitudinal cracks	Chap cracks	Meshy cracks	Massive cracks
Crack length (m)		5~50			
Crack width (mm)	5~15	5~20			
Crack interval (m)	5~50				
Cracks area (m^2)			0.5~6	1~12	3~20
Distribution characteristics	Primary crack, most on running lane	Mainly on running lane, ultra-lane, cut-fill sections	Along with Transverse cracks, longitudinal cracks and subsidence, without regular area	On the wheel tracks of running lane and the cracks intersections, along with subsidence	Rarely appear, in the wheel tracks of running lane, transverse cracks

4 Internal Characteristics and Causes of Pavement Cracks

To analyse the internal characteristics and causes of cracks on asphalt pavement with semi-rigid base, core samples were drilled at the disease place. Tests of pavement deflection, subgrade CBR, strength of base core sample were all important methods.

4.1 Transverse Crack

The causes of transverse cracks from internal characteristics in asphalt pavement with semi-rigid base can be divided into the following kinds.

(1) Temperature shrinkage of surface pavement. Cracks run through the asphalt pavement and the crack widths diminish gradually from top to bottom; (2) Temperature shrinkage of top base course. Cracks run through the asphalt pavement and the crack widths diminish gradually from base surface to upward and downward, serious fracture and loosening emerge in the top base; (3) Bending breakage of bottom base course. Cracks widths diminish from bottom base to road surface, transverse cracks and longitudinal cracks intersect in middle and bottom asphalt layer. The bonding conditions between surface and base are unsatisfactory, serious fracture and loosening emerge in the bottom base; (4) Unequal settlement between structures and roads. When the roads connect with rigid structure like bridge, transverse cracks will emerge due to the unequal settlement between structure and road.

4.2 Longitudinal Crack

The causes of longitudinal cracks from internal characteristics in asphalt pavement with semi-rigid base can be divided into the following kinds.

(1) Bending breakage of bottom surface course. Cracks run through the asphalt pavement, and the crack widths diminish from bottom asphalt layer to the surface, the top base or whole base is broken. The generation process of such longitudinal cracks is that the base course has low strength and breaks under vehicle load and the bending breakages of surface course occur because of losing support; (2) Bending breakage of bottom base course. Structural bonding conditions between surface layer and base layer are fine, cracks widths diminish gradually from base bottom to asphalt surface, serious fracture and loosening emerge in the bottom base; (3) Shear tensile failure of surface course. Cracks are limited to the top surface, the base layer is in good condition. This kind of longitudinal crack is caused by the combined action of horizontal shear stresses and transverse tensile stresses on the surface asphalt layer;

(4) Unequal settlement between new and old subgrade. After road widening project, there exist settlement rate difference between new subgrade and original subgrade. With the road service time increasing, vertical displacement appears in the joint of new base and original base, which will cause the longitudinal cracks; (5) Improper construction methods. Cracks are straight in the central road, serious loosening occurs around. Bonding courses between asphalt layers and the base courses are in good conditions.

When the paving size of asphalt concrete can not satisfy the whole road width, there will exist some joints. If the joints treatment did not strictly according to the demands, the asphalt concrete around the joints will be lack of bonding strength, and longitudinal cracks will emerge under long-term load.

4.3 Chap Crack

The causes of chap cracks from internal characteristics can be divided into the following kinds.

(1) Water in the pavement structure. Separation appears between asphalt layers, and asphalt peels off seriously around cracks. At the same time, severe loosening occurs in bottom surface and top base. The generation process of this type meshy crack is: water entered the pavement structure and accumulates in the adhesive layer, the water damage finally happened under vehicle load; (2) Low strength of base course. Bonding courses between asphalt layers are in good condition, serious cracks occur in the bottom surface, and the whole base course is loose; Because of strength shortage, the base course become loose under the vehicle load. Excessive tensile stress appeared in bottom surface due to the shortage of bearing capacity, which leaded to the meshy cracks finally.

4.4 Meshy Crack

The causes of meshy cracks from internal characteristics can be divided into the following kinds.

(1) Fatigue failure of asphalt layers. There is no obvious deformation in road surface, but abundant tiny cracks appear around massive cracks. The bonding layer and the base course are all in good conditions.

Such cracks were caused by the fatigue failure of asphalt layers from long time vehicle load, and overload would exacerbate these meshy cracks; (2) Bending breakage of bottom surface course. Cracks run through the asphalt pavement, and crack widths diminish from bottom asphalt layer to the surface. The bonding courses between asphalt layers are in good conditions, while the top base is broken. Water caused the strength shortage and vehicle load leaded to the base loosening. Then the excessive tensile stress appears in bottom surface, which caused the meshy cracks; (3) Bending breakage of bottom base course. Cracks run through the asphalt layers and base layer, and the crack widths diminish gradually from bottom base to asphalt surface. The base course becomes loose under the vehicle load because of its strength shortage and tensile stress concentration, and cracks reflect upward to the surface road; (4) Extension of transverse cracks. Most transverse cracks in the meshy cracks are out of the meshy cracks range, while long longitudinal cracks are less. The crack widths diminish gradually from bottom to top, and usually the transverse cracks are wider than longitudinal cracks. The base course is in good condition. Transverse cracks occured when temperature changes rapidly, then the cracks extended to the longitudinal direction, ultimately the criss-cross meshy cracks appeared.

4.5 Massive Cracks

The massive cracks on asphalt pavement with semi-rigid base are chiefly caused by the bending breakage of bottom base course.

Cracks run through the whole pavement layers and their widths diminish gradually from bottom to top, the bottom base is in loose. By the long time effect from load and temperature, load fatigue and temperature shrinkage fatigue had lessened the ultimate bending strength of semi-rigid base. The low strength caused the bonding cracks in the bottom base, and then cracks reflected to the surface road.

5 Damage Mechanism of Pavement Cracks

Research shows that crack is the chief damage of asphalt pavement with semi-rigid base, transverse crack and longitudinal crack are the most common cracks, which are abundant in different zones and different structures. In the cracks, transverse cracks take a proportion of 42%, longitudinal cracks 32.3%, chap cracks 4.8%, meshy cracks 13.3% and massive cracks 7.7%.

The causes of cracks can be divided into temperature, load, water effects, unequal settlement and construction wrongs, in which the temperature and load are primary causes, as shown in figure 9.

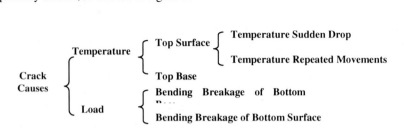

Fig. 9. Causes and classification of cracks on asphalt pavement with semi-rigid base

5.1 Mechanism of Temperature Shrinkage Crack

Temperature shrinkage cracks can be classified into two forms: the first type cracks caused by temperature shrinkage in the surface asphalt layer; the second type cracks reflected from the temperature shrinkage cracks in the surface base course. Temperature shrinkage cracks are mainly characterized by transverse cracks. The two types cracks are shown as figure 10 and figure 11.

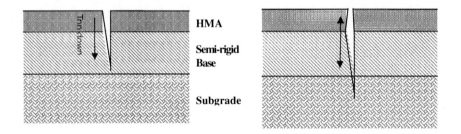

Fig. 10. Shrinkage crack from surface   **Fig. 11.** Shrinkage crack from top base

Most of the first type cracks are slight, with the spacing of 10m ~ 15m and the width of 3mm ~ 10mm, the cracks run through the asphalt layers and their widths diminish gradually from top to bottom; the second type cracks have the spacing of 15m ~ 25m and the width of 5mm ~ 15mm, and the crack widths diminish gradually from top base to up and down. Temperature shrinkage cracks appear not only in the running lane, but also in the ultra-lane and hard shoulder.

(1) Mechanism of temperature shrinkage crack in the surface road. Asphalt temperature in pavement structure changes more slowly than external, and there exist temperature difference in various depths due to the temperature change and transmission. As the asphalt is temperature susceptibility material, the asphalt

concrete also has intensive temperature susceptibility. In high-temperature, the asphalt concrete has excellent stress relaxation, which can effectively avoid the stress concentration under vehicle load. But in low-temperature, stiffness modulus of asphalt concrete increases sharply, and the asphalt concrete shrinkage can easily lead to stress concentration on the road surface.

Reasons of temperature shrinkage cracks can be divided into two forms: when the temperature sudden drop, temperature stress of asphalt concrete exceeded its tensile strength, cracks appeared; temperature repeated movements caused temperature stress fatigue, which leaded to the degradation of ultimate tensile strain and stress relaxation of concrete, then cracks occurred in a modest temperature stress.

Usually, the first kind of cracks occur in regions with chilly winter, the second kind occur in both cold regions and warm regions where temperature varies frequently. After cracks appeared, the crack tip had become the stress focus under vehicle load and caused the cracks extension to top and bottom road.

(2) Mechanism of temperature shrinkage crack in top base course. Compared with flexible base, the semi-rigid base has less thermal capacity, lower adhesive performance with asphalt layer and larger autogenous shrinkage.

When the temperature dropped sharply, the tensile stress in base course increased quickly but the bonding strength between asphalt layer and base layer is limited. Once the tensile stress exceeded the ultimate tensile strength, cracks happened in the base course.

Most base courses are constructed in hot season, during the early molding period the base course can not be covered by the asphalt layers. The large temperature difference between day and night resulted in large thermal stress in base, which would cause cracks. Transverse cracks caused by the shrinkage in surface layer and top base are the main form of cracks, which take a proportion of 80% in the transverse cracks.

5.2 Mechanism of Cracks Caused by Load

Cracks caused by load can be divided into two types: bending breakage of bottom base course caused by load and bending breakage of bottom asphalt surface course.

Mechanisms of the two types cracks are shown as figure 12 and figure13.

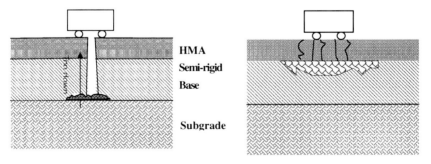

Fig. 12. Bending breakage of bottom base

Fig. 13. Bending breakage of bottom surface

(1) Bending breakage of bottom base course caused by load. Cracks caused by this reason include transverse cracks, longitudinal cracks, meshy cracks and massive cracks. The bottom base course is in serious fracture and loosening, the cracks reflect through the whole pavement structure and crack widths diminish gradually from bottom to top.

Due to the material properties, semi-rigid base has high unconfined compressive strength and stiffness, but these in sub base are much lower. This difference makes the stiffness transition between base courses more difficult, also enlarges the tensile stress in bottom base from load. If the tensile stress exceeded the ultimate bending strength of base course, cracks occur. After base cracked, vertical shear stress increased and become the chief factor causing cracks on the road surface. The large tensile stress in bottom base will make the top sub base cracks easily.

(2) Bending breakage of bottom asphalt surface course caused by load. Cracks caused by bending breakage of bottom asphalt surface course under load include chap cracks and meshy cracks.

When there existed strength shortage or weak layer in base course, they could not provide enough bearing force to surface layers. In this case, compressive stresses in the bottom surface transited to tensile stress. Once the tensile stress exceeded its ultimate bending strength, cracks appear and reflect to road surface.

6 Proportion of Cracks Caused by Base Course

Serious transverse cracks have two types: caused by temperature shrinkage in top base and bending breakage of bottom base, which take a proportion of 70% in transverse cracks; longitudinal cracks caused by bending breakage of bottom base accounts for 40%; chap cracks caused by broken of base account for 50%; meshy cracks caused by semi-rigid base account for 30% and more than 90% of massive cracks are caused by semi-rigid base.

Based on above, 36% of damages on asphalt pavement with semi-rigid base relate to semi-rigid base. The proportion of cracks related to semi-rigid base is shown as table 3.

Table 3. Proportion of cracks related to semi-rigid base

Damage type	Transverse cracks	Longitudinal cracks	Chap cracks	Meshy cracks	Massive cracks	All cracks
Proportion of cracks related to semi-rigid base (%)	70	40	50	30	90	36

7 Conclusion

(1) Damages on asphalt pavement with semi-rigid base show strong regional characteristics, and their occurrences are related directly to traffic and vehicle load.

(2) The main damage forms on the asphalt pavement with semi-rigid base include crack, rutting, loosening, pothole, subsidence and oil. Cracks are the primary damage form and can be divided into transverse cracks, longitudinal cracks, chap cracks, meshy cracks and massive cracks, in which the transverse cracks and longitudinal cracks are dominant.

(3) Disease sections take a proportion of 16% in the asphalt pavement with semi-rigid base, the other 84% is in good condition.

(4) The causes of cracks on asphalt pavement with semi-rigid base can be divided into temperature, load, water effects, unequal settlement and construction wrongs, in which the temperature and load are primary.

(5) 36% damages on asphalt pavement with semi-rigid base are related to semi-rigid base. Analysis results demonstrate that asphalt pavement with semi-rigid base has strong adaptability in kinds of regions, climates and loads. The structure is suitable for current transport level in China, and can be used as a main structure in highway construction for a long period.

References

[1] National Standard of The People's Republic of China. Specifications for Design of Highway Asphalt Pavement. China Communications Press, Peking (2006)
[2] National Standard of The People's Republic of China. Technical Specifications for Maintenance of Highway Asphalt Pavement. China Communications Press, Peking (2001)
[3] Shen, J.A., Li, F.J., Chen, J.: Analysis and Preventive Techniques of Premature Damage of Asphalt Pavement in Expressway. China Communications Press, Peking (2004)
[4] Sha, Q.L.: Phenomenon and Prevention of Premature Damage in Expressway. China Communications Press, Peking (2001)
[5] Yao, Z.K.: Pavement, 3rd edn. China Communications Press, Peking (2006)

Comparing the Slope of Load/Displacement Fracture Curves of Asphalt Concrete

Andrew F. Braham and Caleb J. Mudford

University of Arkansas

Abstract. In order to quantify cracking characteristics of asphalt concrete, the use of fracture testing is becoming more common. When running fracture tests, often the load and displacement are recorded, and from these two values, the fracture energy can be calculated. Unfortunately, fracture energy is only a single value, so it is difficult to differentiate between two different mixtures that have the same fracture energy, but very different Load/Displacement curves. This research examined four sets of mixtures with similar fracture energies, but different material characteristics, including air voids, asphalt cement content, asphalt cement type, and polymer modification. After analyzing these four sets of mixtures, it was found that adding extra asphalt cement or polymer modification to a mixture increased the complaint behavior of the fracture curves for these specific mixtures. In addition, increasing the air voids decreased the compliance, while increasing the high temperature binder grade slightly increased the compliance. This indicated that important information of cracking characteristics of asphalt concrete can be overlooked when only using the fracture energy.

1 Introduction

When looking at asphalt concrete pavements (which constitute 94% of the United State's pavement surfaces), there are fifteen distresses according to the Distress Identification Manual published by the Federal Highway Administration [1]. Of these fifteen, six are related to cracking, including fatigue, block, edge, longitudinal, reflection, and transverse cracking. However, after surveying the current tests performed in the laboratory on asphalt concrete, only two tests that are commonly run measure cracking, or separation, properties of asphalt concrete: the indirect tension test [2] and bending beam fatigue test [3]. Recently, the field of fracture mechanics has been explored with great success in order to quantify and categorize the performance of asphalt concrete materials. When reviewing the literature, three fracture testing configurations have been used: the Disk-Shaped Compact Tension [DC(T)] [4], the Single-Edge Notch Beam [SE(B)] [5], and the Semi-Circular Bend [SC(B)] [6] tests, as seen in Figure 1.

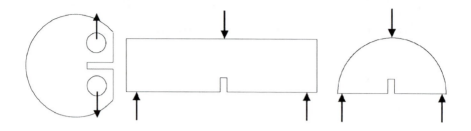

Fig. 1. From left to right: the Disk-Shaped Compact Tension [DC(T)], Single-Edge Notch Beam [SE(B)], and Semi-Circular Bend [SC(B)] fracture tests for asphalt concrete

A common quantification taken from these fracture tests is fracture energy. The fracture energy is found by taking the area under a load/displacement curve, and dividing this number by the area of the crack face. Figure 2 shows some typical fracture energy curves of two types of asphalt concrete.

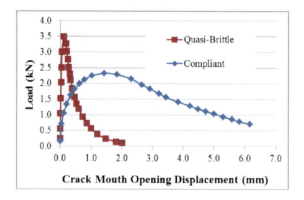

Fig. 2. Fracture curves of asphalt concrete

Unfortunately, fracture energy is a single number. After testing twenty eight asphalt concrete mixtures at three testing temperatures [4], the same fracture energy was calculated for several different asphalt concrete mixtures. This research examined an innovative method of analyzing load/displacement curves in an effort to better understand the material behavior of asphalt concrete.

2 Objectives

The load/displacement curve data from four pairs of asphalt concrete mixtures with similar fracture energy will be examined to further distinguish cracking properties of asphalt concrete mixtures. Differences in the four mixtures include air voids, asphalt cement content, asphalt cement type, and polymer modification. The objective of this research is to develop a new analysis technique that will allow for a deeper understanding of the cracking characteristics of asphalt concrete that can differentiate between asphalt concrete mixtures with similar fracture energies.

3 Background

As part of the Low-Temperature Cracking Pooled-Fund Study TPF-5(080), referred to as LTC study, approximately 252 DC(T) samples were tested [5]. Fracture energy, calculated from the load/displacement curve, was collected for each of these tests. Several parameters were examined, including ten asphalt cement binders, two aggregate types, three testing temperatures, two asphalt cement contents, and two air void levels, in a non-factorial design. Interestingly, several sets of fracture energy between two mixtures were essentially the same. Table 1 summarizes four mixtures that captured this trend.

Table 1. Mixtures with Similar Fracture Energies but Different Properties

Set	Air Voids	Asphalt Cement			Fracture energy (J/m^2)
		Content	Type	Polymer Modification	
1	4%	Optimal	PG58-28	No	316.1
	7%				314.7
2	4%	Optimal	PG58-34	Yes	646.9
		Optimal + 0.5%			632.4
3	4%	Optimal	PG58-28	No	422.2
			PG64-28		425.5
4	4%	Optimal	PG64-28	No	425.5
				Yes	429.7

In Set 1, the testing temperature was low, and the aggregate was limestone. In Set 2, the testing temperature was medium, and the aggregate was granite. In Set 3 and Set 4, the testing temperature was medium, and the aggregate was limestone. The "Low" and "Medium" temperatures were based on the low value of the PG grade in the following format:

- 2°C below the low temperature grade ("Low" testing temperature)
- 10°C above the low temperature grade ("Medium" testing temperature).

For example, for a PG58-28 asphalt cement type, the asphalt concrete mixtures were tested at −30°C and −18°C. In addition to allowing a wide range of testing temperatures, these two temperatures matched PG binder testing temperatures used in the LTC study. The two aggregates studied, limestone and granite, were chosen because they have different coefficients of thermal expansion (granite ≈ 8.2 × 10–6 and limestone ≈ 3.8 × 10–6). This difference emphasizes the effect of the differential contraction between the asphalt binder and aggregate. However, both aggregate gradations had a nominal maximum aggregate size of 12.5mm. Samples were compacted to both 4% and 7% air voids, at either the Superpave Optimal Binder Content, or the Superpave Optimal Binder Content plus an additional one-half percent of asphalt cement. Finally, some of the base asphalt cements were polymer modified, while some were neat.

4 Construction of Slope Comparisons

The two mixtures from Set 1 in Table 1 were chosen to demonstrate the process of analyzing the fracture data for this research. In the first round of data analysis, the full fracture curves were split into twenty data points with an equal Crack Mouth Opening Displacement (CMOD) spacing. However, when these points were plotted, informative trends of pre-peak behavior were being lost, since the majority of data collected occurs after the peak load is reached. Therefore, the data was reanalyzed using eighteen points, eight of which occurred pre-peak load and ten after peak load. Figure 3 shows the two mixtures from Set 1 in Table 1 plotted in this fashion. Note, the two mixtures in Figure 3 had the same testing temperature, aggregate, asphalt cement content, asphalt cement type, but different levels of air voids.

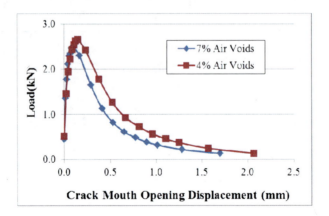

Fig. 3. Load/CMOD Curves of Mixtures with Identical Properties Except for Air Voids

Figure 3 shows that although the two mixtures had very similar fracture energy values (316.1 J/m^2 for 4% air voids, 314.7 J/m^2 for 7% air voids), the curves themselves were not the same. It appeared that the pre-peak and post-peak curves had different slopes, and the two peak loads were different. In order to understand the different slopes, the slope of the Load/Displacement curve was calculated (with units of change in kN/mm) and graphed for the data in Figure 3, and is shown in Figure 4.

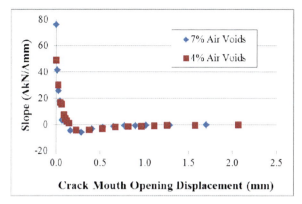

Fig. 4. Load/CMOD Curves of Mixtures with Identical Properties Except for Air Voids

Figure 4 clearly demonstrates that the initial slope for the 7% air voids was greater than the 4% air void, but the test ended sooner than the 4% air voids. This balance of the shape of the initial curve (pre-peak load), the peak load, the shape of the softening curve (post-peak load), and the length of the softening tail determine the fracture energy. When one or more of these components balance out, similar fracture energies with different curves occur. A second format to present a comparison of slopes is to simply plot the two slope values against each other with a line of equality. This technique is shown in Figure 5.

The difference in slopes between 4% and 7% air voids of the fracture curves is readily apparent in Figure 5. There was a small cluster of points around zero slope that appeared to be similar, but there were large differences of slope at the higher values. A higher slope at these larger values indicated a less compliant mixture; a less compliant mixture was an indication of a less crack resistant mixture.

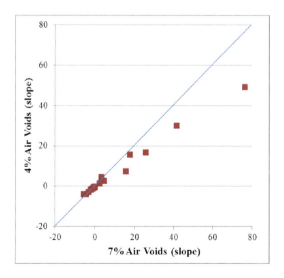

Fig. 5. Slope Comparison of 4% and 7% Air Voids

Therefore, when comparing the slopes of the two mixtures in Figure 5, it appeared that 7% air voids mixture was less compliant, thus less crack resistant, than the 4% air voids mixture. This was intuitive, as a higher level of air voids would indicate lower cohesion of the mixture, and thus lower cracking resistance. This analysis technique was repeated for the next three sets of mixtures.

5 Results

Similar to the process outlined in Figure 3 through Figure 5, the mixtures from Set 2 in Table 1 were analyzed. In set two, the testing temperature, aggregate type, air voids, and asphalt cement type were identical, including Elvaloy polymer modification. The only difference was the asphalt cement content. The mixture compacted at Superpave Optimal Asphalt Content had a fracture energy of 646.9 J/m^2, while the mixture compacted with an additional 0.5% asphalt cement had a fracture energy of 632.4 J/m^2. Figure 6 shows the slope comparison between the two mixtures.

The variation from the line of equality was not as great as the difference in air voids, but the slope was slightly higher for the optimal asphalt content mixture versus the mixture with an additional 0.5% asphalt cement. Therefore, the mixture with lower asphalt cement content was less crack resistant. This was reasonable, as the addition of extra asphalt cement should theoretically increase the cracking resistance of asphalt concrete.

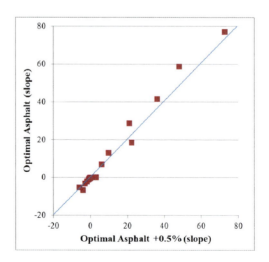

Fig. 6. Slope Comparison of Optimal Asphalt Content and Optimal Asphalt Content +0.5%

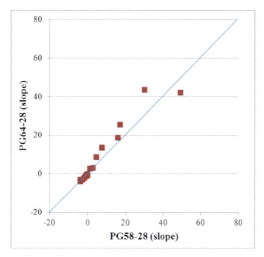

Fig. 7. Slope Comparison of PG64-28 and PG58-28

The next set of mixtures, Set 3 in Table 1, compared two mixtures that had the same testing temperature, the same aggregate type, the same air voids, and the same asphalt cement content. However, one mixture was a PG58-28 (with a fracture energy of 422.2 J/m2) and the second mixture was a PG64-28 (with a fracture energy of 425.5 J/m2). Neither asphalt cement was polymer modified. Figure 7 shows the slope comparison between the two mixtures. The analysis of set of mixtures in Figure 7 was not as straight forward as the first two sets of mixtures. While the slope from the PG64-28 was higher for a portion of the graph, at the highest slope value, the PG58-28 actually had a higher slope. A potential cause for this ambiguity is that the testing temperatures were tied to the low grade of the asphalt, and although these were two different asphalt cement grades, their low grade was the same (-28, with a -18C testing temperature). Therefore, the low temperature properties may be quite similar, as the difference between the asphalt cement may only affect the high temperature properties.

The final set of mixtures, Set 4 in Table 1, compared the effect of polymer modification. All other properties, including testing temperature, aggregate type, air voids, asphalt cement content and type were the same, but one of the asphalt cements was modified with Styrene-Butadiene-Styrene (SBS) polymer. Figure 8 shows the comparison of the two mixes.

The trends in Figure 8 were similar to the trends in Figure 7. The higher slope values seen with the no modification asphalt cement indicates less compliance, which indicates lower cracking resistance. With similar fracture energy values (425.5 J/m^2 for no polymer versus 429.7 J/m^2 for SBS modification), the further analysis of examining the slopes of the fracture curves indicates the power of analyzing the entire curve and not just the fracture energy. The entire curve shows that polymer modification increases crack resistance, while the single fracture energy value did not. However, only two mixtures were analyzed with this data, so this is an area that needs more exploring in order to better understand the ability of polymer to reduce cracking characteristics through fracture testing.

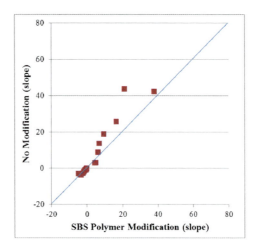

Fig. 8. Slope Comparison of No Polymer Modification versus Styrene-Butadiene-Styrene (SBS) Polymer Modification

6 Statistical Analysis

Figures 5-8 show general trends, but an important component is missing. That component is the error from the three replicates for each test. Therefore, the Coefficient of Variation (COV) was found for each set of data. The COV was calculated by taking the standard deviation of each slope point, and dividing this standard deviation by the average of each slope point. The absolute value of these COV values was taken (as some of the slopes were negative), and the average of all eighteen COV values was found. Table 2 provides a summary of the COV values.

Table 2. Coefficient of Variation (COV) of Eight Mixtures

Set	Air Voids	Asphalt Cement			COV (%)
		Content	Type	Polymer Modification	
1	4%	Optimal	PG58-28	No	21.8
	7%				35.7
2	4%	Optimal	PG58-34	Yes	19.3
		Optimal + 0.5%			20.1
3	4%	Optimal	PG58-28	No	21.8
			PG64-28		19.9
4	4%	Optimal	PG64-28	No	19.9
				Yes	25.6

With the exception of the 7% air void mix in Set 1, the COV values fall within typical testing variation of asphalt concrete mixtures, which is 10-20%.

7 Conclusions

Cracking is a significant distress of asphalt pavements. Fracture tests are becoming a more common type of test to try and quantify cracking in the laboratory. To date, the majority of fracture test analysis has captured a single number, fracture energy. However, fracture energy is a single number. On occasion, two mixtures with different properties will have very similar fracture energies. Therefore, more information needs to be extracted from current fracture tests. This research focused on developing a new method of analysis of fracture data in order to better understand the cracking characteristics of asphalt concrete.

Four sets of mixtures were investigated. Each set had very similar fracture energy values, but had one key characteristic that varied between the two mixtures. For Set 1, the air voids were 4 and 7%. For Set 2, one asphalt cement content was 0.5% higher than the other. For Set 3, two different types of asphalt cement were used. Finally, for Set 4, one asphalt cement had polymer modification while the other was unmodified. By comparing the slopes of the Load/Displacement curves collected during the fracture test, the following conclusions were found for the four sets of mixtures studied:

- 7% air voids was less crack resistant than 4% air voids
- Adding 0.5% asphalt cement above the Superpave optimal asphalt content increased the cracking resistance of the mixture
- Changing the Superpave binder grade from a PG64-28 to a PG58-28 did not have a clear effect on the cracking resistance
- The addition of Styrene-Butadiene-Styrene appeared to increase the cracking resistance of the asphalt mixtures
- The Coefficient of Variation of the slopes fell within typical asphalt concrete mixture testing, with values around 20%

This method of analysis has several drawbacks. First, the analysis was only performed at one testing temperature, giving a small window of information into the cracking characteristics of each asphalt mixture. Second, the slope of the Load/Displacement curves does not have physical significance nor is it a fundamental engineering property. Third, only eight mixtures with three replicates each were analyzed in this data set. So while some general trends can be observed by analyzing data in this method, it is difficult to extract fundamental properties. These two concerns need to be addressed in order to create better methodologies for analyzing the cracking mechanisms of asphalt concrete mixtures.

References

[1] Miller, J.S., Bellinger, W.Y.: Distress identification Manual for Long-Term Pavement Performance Program, 4th Revised Edition, Federal Highway Administration, FHWA-RD-03-031 (2003)

[2] AASHTO T322, Standard Method of Test for Determining the Creep Compliance and Strength of HMA Using the Indirect Tensile Test Device, American Association of State Highway and Transportation Officials, AASHTO (2007)
[3] AASHTO T321, Standard Method of Test for Determining the Fatigue Life of Compacted Hot-Mix Asphalt (HMA) Subjected to Repeated Flexural Bending, American Association of State Highway and Transportation Officials, AASHTO (2007)
[4] Braham, A.F., Buttlar, W.G., Marasteanu, M.O.: Transportation Research Record: Journal of the Transportation Research Board (2001), 102–109 (2007)
[5] Wagoner, M.P., Buttlar, W.G., Paulino, G.H.: Journal of Testing and Evaluation 33(6), 452–460 (2005)
[6] Li, X., Marasteanu, M.: Cohesive Modeling of Fracture in Asphalt Mixtures at Low Temperatures. International Journal of Fracture 136, 285–308 (2005)

Cracking Behaviour of Bitumen Stabilised Materials (BSMs): Is There Such a Thing?

Kim Jenkins

Stellenbosch University, South Africa

Abstract. The behaviour of bitumen stabilised materials (BSMs) is uniquely different from all other materials used to construct road pavements. Unlike asphalt, where the bitumen as a continuum binds all the aggregate particles together, the bitumen in a BSM is dispersed selectively amongst only the finer particles, regardless of whether bitumen emulsion or foamed bitumen is used as the stabilising agent. When compacted, the isolated bitumen-rich fines are mechanically forced against their neighbouring aggregate particle, regardless of size, resulting in localised bonds which are non-continuous.

The purpose of this paper is to explore the question: Is the mode of failure of BSMs purely permanent deformation, similar to granular materials? Some engineers continue to argue that, similar to thick asphalt layers, BSMs fail in fatigue, supporting this stance by means of repeated-load tests carried out in a laboratory on beam specimens.

In this paper, the principles of fracture mechanics are employed to demonstrate that non-continuously bound materials experience different modes of failure to continuously bound materials. For a crack to propagate through a layer, the material must have sufficient internal cohesion to allow applied stresses to concentrate at the crack tip. In addition, the performance of several pavements, constructed at least five years ago, each with a base layer of high quality BSM material, is reviewed. Deflection measurements taken at regular intervals show none of the symptoms that would indicate deterioration due to fatigue initiation. Deflection measurements suggest that the pavement stiffness increases in the first year after construction.

1 Introduction

Bitumen Stabilised Materials or BSMs have emerged over the past 15 years as attractive base materials for pavement structures carrying medium to heavy traffic volumes. Often the use of BSM technology forms part of pavement rehabilitation, although it can also form part of new construction. BSMs incorporate either bitumen emulsion or foamed bitumen as binder, often supplemented with a small percentage of active filler, typically 1% cement or lime.

One of the challenges facing the pavement engineer, is that BSM components comprise all of the primary ingredients of pavement materials i.e. aggregates (virgin or recycled, granular, cemented or reclaimed asphalt), bitumen, active filler, water and air. This leads to visco-elasto-plastic behaviour of a BSM. How, then, does one develop appropriate response models and damage models for BSMs?

One of the distinguishing features of BSMs is the nature of dispersion of the bitumen. Asphalt consists of a continuum of bitumen binding aggregate together whilst BSMs comprise non-continuously dispersed bitumen within the mix, with selective dispersion amongst the fine aggregate. Typically, 1.8% to 2.5% bitumen is used to produce BSMs which is approximately half of the binder content of equivalent HMA. The binder in BSMs (both bitumen and the small amount of hydraulic binder) increases the shear strength predominantly through improved cohesion associated with typically a small reduction in friction angle of the parent material. These characteristics indicate the granular-type behaviour of BSMs.

2 BSM Characteristics and Response Models

Extensive laboratory investigations have been carried out on the dynamic response properties of BSMs. Dynamic triaxial testing illustrates the granular-type, stress-dependent behaviour of BSMs, see Figure 1. The resilient modulus of a BSM can double within a typical range of stresses experienced by such a layer.

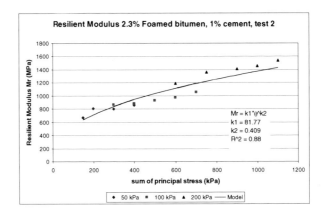

Fig. 1. Dynamic triaxial tests on BSM-foam at 25°C [1]

However, BSMs can also be expected to behave visco-elastically given their inclusion of bitumen. This has been verified by the master curves from dynamic flexural beam testing. Figure 2 verifies the dependency of BSMs on loading time and temperature. Data for two BSM-emulsions and one BSM-foam is shown in the figure. At the same time, the lower binder contents and non-continuous dispersion of the bitumen in the BSMs reflect in the flatter gradient of the master curves compared with the equivalent Half-warm (HW) and HMA mixes.

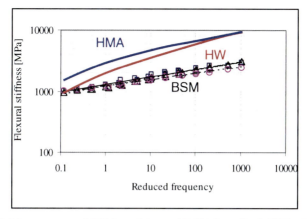

Fig. 2. Master curves of BSM-emulsion and BSM-foam Tref=20°C, after [2]

3 Damage Models for BSMs

In May 2009, South Africa launched the Second Edition of TG2 the Technical Guideline for the design and construction of BSMs, [3]. This guideline specifies permanent deformation as the design mechanism of failure. Does this mean that fatigue failure does not occur? This section explores the relevance of fatigue as a BSM damage mechanism.

3.1 Fatigue Testing of BSMs

Mathaniya *et al.* [2] carried out a range of four point beam tests on Bitumen Stabilised Materials with either emulsion or foamed bitumen as binders up to 2006. Firstly, it should be noted that these mixes included 3.6% residual binder contents, which is higher than the current practice norms for BSMs which applies typically 2 to 2.5% residual binder. The cement content used in the research i.e. 1%, complies with TG2 [3] recommendations.

The extended experimental design clearly showed that fatigue testing of BSMs is achievable, an example of which is shown in Figure 3. Mathaniya *et al.* also showed the weakness of strain-controlled testing, where BSM-foam mixes provided extended fatigue lives, primarily due to the lower initial flexural stiffness S_i than BSM-emulsion mixes, as seen in Table 1.

Table 1. Average initial stiffness (S_i) and failure stiffness (S_f) of beams at 5°C [2]

	Aggregate blend					
	25% RAP, 0% cem		25% RAP, 1% cem		75% RAP, 0% cem	
	S_i [MPa]	S_f [MPa]	S_i [MPa]	S_f [MPa]	S_i [MPa]	S_f [MPa]
A – emulsion	2432	1216	2746	1373	1612	806
B – emulsion	2521	1261	2119	1059	2119	1059
C – foamed bitumen	1590	795	1592	796	1045	522

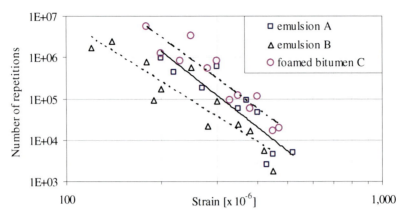

Fig. 3. 4 Point Beam Fatigue of BSMs: 25% RAP + 1% Cem, 5°C and 10 Hz [2]

3.2 Fracture Mechanics Considerations

The principles of fracture mechanics have been successfully applied in pavement engineering to the design of asphalt overlays where reflective cracking requires analysis. Collings and Jenkins [4] showed how fracture mechanics can provide insight into the behaviour of BSMs. Paris' Law shown in Equation 1, is used to describe crack growth in a material.

$$\frac{dc}{dN} = A \cdot K^n \qquad (1)$$

where, $\frac{dc}{dN}$ = increase in crack length per load cycle
K = stress intensity factor at the tip of the crack in bending or shear
A, n = material constants

Paris' Law is appropriate for asphalt if analysed as a homogeneous material incorporating bitumen that is distributed in a continuum. The asphalt layer can be treated, in effect, as a beam, since it is a continuously bound material. Besides other factors, the crack intensity factor is dependent on the ratio of crack length to beam thickness, c/d shown in Figure 4.

BSMs do not have a continuum of bitumen and less homogeneous than asphalt, especially when they include recycled material e.g. RAP. This is shown conceptually in Figure 5. Discrete distribution of bitumen splinters in BSM-foam makes classical fatigue and fracture mechanics inapplicable. If shear deformation between individual particles ruptures a "spot weld" of bitumen, there is no continuity of bound material that will allow a crack to develop, so c/d becomes meaningless. There is neither opportunity for a crack "head" nor stress intensity at

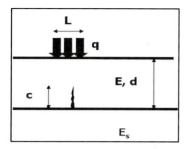 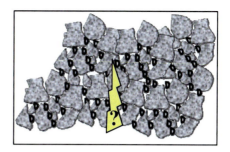

Fig. 4. Beam approach to crack-growth analysis [4]

Fig. 5. Non-continuous bound BSM-foam [4]

the tip to develop. A broken spot weld will result in particles re-orientating (micro-shearing), resulting in permanent deformation, as with granular material.

3.3 In-Service Behaviour of BSMs

The fact that fracture mechanics and classical fatigue do not apply does not imply that BSMs will not experience change in stiffness with time. The rupture of spot welds can influence the effective stiffness of a BSM layer. However, there are other factors such as curing that influence the in-service behaviour of BSMs. This section uses the example of a highway near Cape Town, South Africa, namely the N7, to explore full-scale BSM behaviour.

Part of the N7 highway was rehabilitated by recycling with BSM-foam (2.3% bitumen and 1% cement) on the Southbound Carriageway in 2002. The Northbound Carriageway was rehabilitated by recycling with BSM-emulsion in 2007 (2% residual bitumen and 1% cement). Both carriageways were recycled in situ and stabilised to a depth of 250mm. This allows for useful comparisons to be made between these two forms of BSMs.

Curing. Compaction of BSM layers is followed by a natural reduction in moisture content, known as curing. This phenomenon predominates in the first year of the BSM layer's life, followed by cyclic variations in moisture.

Physical moisture measurements of the BSM-emulsion base layer were made by Moloto who retrieved samples from the layer and tested them in the laboratory. This was supplemented by moisture button monitoring in the layer [5]. In addition, Portable Seismic Pavement Analyser (PSPA) measurements were taken on the BSM base over time, in order to evaluate the change in modulus of the base with time.

During the first seven months, the moisture content reduces asymptotically, as seen in Figure 6. The change in moisture content of the BSM-emulsion concurs with the trends found for BSM-foam curing rates, as shown by Malubila [6].

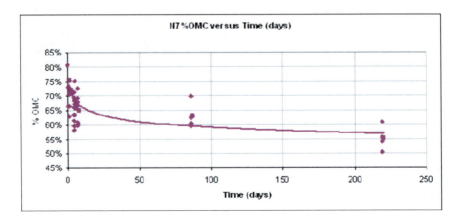

Fig. 6. Moisture content (oven dried) in BSM-emulsion base in-service [5]

The change in resilient modulus of the BSM layer was measured concurrently with the moisture sampling, by Moloto [5]. The PSPA measurements have the advantage of imparting pressure waves almost exclusively in the 250mm BSM layer, allowing for reliable analysis of relative changes in stiffness of that layer. Analysis of the PSPA results, as illustrated in Figure 7, shows the modulus to be inversely proportional to the moisture content of the BSM-emulsion layer. From these results, there are no signs of fatiguing nor stiffness reduction during the first seven months of this layer's life. Rather, the change in moisture content of the BSM with time, has a significant impact on the stiffness in the first year of a BSM's life.

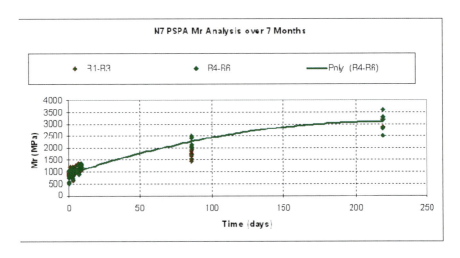

Fig. 7. Modulus of BSM-emulsion base measured with PSPA [5]

The findings of Moloto are verified by measurements on a major 6 lane Greek Highway between Athens to Corinth, which was rehabilitated in 2002/2003 using in place recycling with 2.3% foamed bitumen and 1% cement. The National Technical University of Athens NTUA carried out FWD measurements on the pavement initially as part of the rehabilitation investigation and then subsequently at 1 month, 6 months and then at yearly intervals until 4 years after construction [7]. The reduction in maximum deflection measured using the FWD is plotted in Figure 8 for the slow lane on both carriageways. The new layers in the pavement structure included only BSM-foam and HMA, so the stiffening of the pavement structure could only have emanated from the BSM layer. During the period of the deflection measurements, the pavement was exposed to some 60,000 vehicles per day (20% heavy vehicles with a legal axle load of 130 kN). The asymptotic reduction in maximum deflection with time shows that the BSM-foam layer gains the majority of its stiffness in the first 6 to 9 months after construction.

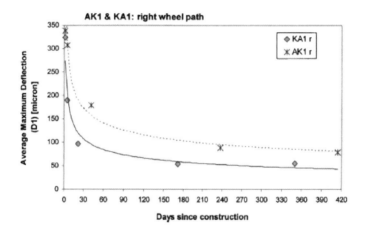

Fig. 8. Change in FWD Maximum Deflection with time [7]

BSM Stiffness more than 1 Year after Construction. Once the moisture content has stabilised in the BSM layer within the first year, the issue of the long term stiffness becomes important. Does stiffness reduction begin to manifest itself in the longer term in a BSM layer?

Network level FWD analysis was carried out on the N7 Highway every four to six years at 200 metre intervals. The back-calculated modulus of the base is provided in Figure 8. Up to and including 2000, the base layer comprised a high quality crushed stone layer that experienced seasonal variations in moisture content. From 2002 onwards, this section incorporated a BSM-foam layer as base. The back-calculated modulus of the 90th percentile maximum deflection and closest 6 bowls, is reflected in Figure 8.

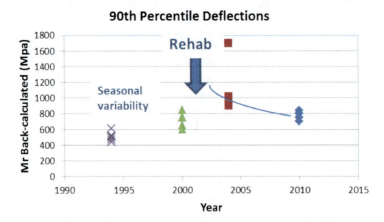

Fig. 8. Back Modulus of Base after Rehabilitation on N7

During the period of 2002 to 2010 the N7 Highway experienced close to its projected design traffic of 9 million 80 kN ESALs. The increase of the BSM-foam layer stiffness during the first year after construction, is not reflected in these results, which don't span sufficiently small intervals. Nevertheless, the longer term stiffness stabilises at approximately 775 MPa which reflects a reduction from the typical initial values of 1200 MPa that were obtained in the mix design [1]. There is evidence of rupturing of the foamed bitumen "spot welds" under the influence of traffic, therefore, but with a residual benefit of the bitumen stabilisation remaining evident.

The reliability of the FWD back-analysis was tested by correlating the resilient modulus with the inverse of the Base Layer Index BLI i.e. $\delta_{max} - \delta_{300}$. Figure 9 reflects a strong correlation, with one outlier, thereby verifying the derived moduli.

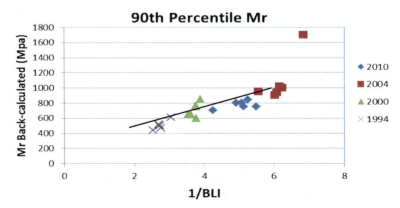

Fig. 9. Correlation between Mr back-calculated and FWD Base Layer Index

4 Conclusions

Bitumen Stabilised Materials BSMs can neither be modelled as purely asphaltic nor granular materials. These materials clearly need to be recognised by their own unique characteristics. Both laboratory and field analysis of these materials lead to the following conclusions:

- BSMs depict both stress-dependent characteristics of granular materials and temperature and loading-time dependency of bituminous materials. As such, BSMs comprise visco-elasto-plastic behaviour.
- Curing behaviour manifest in moisture reduction, dominates the gain in resilient modulus during the first year after construction.
- The non-continuous bound nature of BSMs precludes the applicability of classical fatigue behaviour under repeated loading.
- The long-term resilient modulus of BSM shows signs of reduction under traffic loading, although residual benefits of the binder remain. The stiffness reduction is probably a result of rupture of the adhesion of the bitumen to the aggregate.

References

[1] Jenkins, K.J., Robroch, S.: Laboratory Research of Foamed Bitumen and Emulsion Mixes (both with Cement) for Cold Recycling of N7 near Cape Town, Contract TR1 1/1. Institute for Transport Technology ITT Report 3/2002 for Jeffares and Green. Stellenbosch University, South Africa (2002)
[2] Mathaniya, E.T., Jenkins, K.J., Ebels, L.J.: Characterisation of Fatigue Performance of Selected Cold Bituminous Mixes. In: International Conference on Asphalt Pavements ICAP, Quebec, Canada (2006)
[3] Asphalt Academy, Bitumen Stabilised Materials, A Guideline for the Design and Construction of Bitumen Emulsion and Foamed Bitumen Stabilised Materials. TG2 Technical Guideline, Pretoria, South Africa (2009)
[4] Collings, D.C., Jenkins, K.J.: The Long-term Behaviour of Bitumen Stabilised Materials. In: Conference on Asphalt Pavements for Southern Africa CAPSA 2011, Drakensberg, South Africa (2011)
[5] Moloto, P.K.: Accelerated Curing Protocol for Bitumen Stabilised Materials. MScEng thesis. Stellenbosch University, South Africa (2010)
[6] Malubila, S.M.: Curing of Foamed Bitumen Mixes. MEng. thesis. Stellenbosch University (2005)
[7] Loizos, A., Papavasiliou, V.: Evaluation of Foamed Asphalt Cold In-Place Pavement Recycling using Non-destructive Techniques. In: International Conference on Advanced Characterisation of Pavement and Soil Engineering Materials ICACPSEM, Athens, Greece (2007)

Experimental and Theoretical Investigation of Three Dimensional Strain Occurring Near the Surface in Asphalt Concrete Layers

Damien Grellet[1], Guy Doré[1], Jean-Pierre Kerzreho[2], Jean-Michel Piau[2], Armelle Chabot[2], and Pierre Hornych[2]

[1] Department of Civil Engineering, Laval University, Québec (QC), Canada
[2] L'UNAM Université, IFSTTAR, CS4, F-44344 BOUGUENAIS Cedex, France

Abstract. Several pavement failures have been observed to be initiated at or near the surface of the hot-mix asphalt layers and some of them propagate downward through the surface layer (top-down cracking). These modes of failure are affected by heavy vehicular loading configuration, pavement structure and their interaction at the tire-pavement contact. This paper documents an experimental investigation of surface strain induced under the entire tire by using specific instruments based on fiber optic sensors. Two innovative retrofit techniques which allow measuring strains in the upper parts of the asphalt layer have been used on the IFSTTAR's test track facility. The association of these two techniques allows obtaining the strains, few centimeters below the surface, in three directions: longitudinal, transverse and vertical. Two pavement structures with two temperatures (moderate and hot) have been tested. Shape of the signal under the tire and magnitude of strain are compared with viscoelastic model pavement calculations.

1 Introduction

Fatigue cracking and rutting are common pavement failures resulting from traffic loading and climate environment. One type of cracking is initiated at the bottom of the asphalt layer and propagates towards the surface (bottom-up cracks). Another type is initiated at or near the surface of the hot mix asphalt layers and propagates downward through the bound layers (top-down cracking). All these cracks can also propagate among interfaces between layers. To better understand these modes of failure, the knowledge of strain distribution through the asphalt layer is necessary. This paper is part of a collaborative project between Laval University (Québec, Canada) and the IFSTTAR (Nantes, France) with the main objective to characterize the strain occuring through the asphalt layers under several loading conditions. Two different tires, with various inflation pressures, four applied loads and two asphalt temperatures have been tested. Fiber optic strain sensors were installed at diffcrent depth of the pavement structure. The objectif of this paper is to contribute to a better understanding of the near-surface strains induced by dual tires in the asphalt layer. More specifically, the effect of asphalt temperature on the transverse and vertical strains is analysed. An asphalt layer at moderate temperature around 18°C and at hot temperature around 40°C has been tested. The discussion is based on experimental results and theorical solutions computed by the *ViscoRoute2.0©* software using a viscoelastic model[1].The paper includes a description of the pavement structure, the material properties and the sensors. The experimental results are presented and analysed in comparison with computed strain signals from the upper part of the first layer.

2 Experiment Description

2.1 IFSTTAR's Facility, Pavement Structure

The tests were conducted at the IFSTTAR's accelerated pavement testing facility [2]. The outdoor test track is a large scale circular track and a loading system with a mean radius of 19 m, a width of 3 m and a total length of 120 m. The device includes a central motor unit and four arms. Each arm lies on a module wich can be equipped with various load configurations. Each module can move during revolution around a mean position to simulate the lateral wandering of the traffic.

Two pavement sections, representing 1/3 of the whole test track, were built in February 2011 and instrumented in May 2011 . The two sections present different structure in terms of asphalt concrete thickness. The first structure A (figure 1) is 22 m long and it consists of the following layers: 70 mm bituminous wearing course (layer N°1), a 60 mm binder course (layer N°2), a 300 mm granular subbase (layer N°3) and a sandy subgrade soil (layer N°4). The second structure B (figure 2) is 18 m long and it includes only one 70 mm bituminous wearing course resting on a base and a subgrade soil similar to the first structure.

2.2 Material Properties

Each layer of the structure is considered to be homogeneous and linear. The subbase is divided into three 100 mm-thick layers (layer N°3.1 to 3.3). The mechanical behaviour of the soil (layer N°4) and the unbound granular material are assumed to be elastic and the Poisson's ratio value is fixed to 0.35. With the help of elastic back calculations, their elastic modulus are supposed to be respectively $E_4 = 80$ MPa, $E_{3.3} = 160$ MPa, $E_{3.2} = 320$ Mpa and $E_{3.1} = 640$ Mpa. The bituminous wearing course is an asphalt concrete 0/10 with 5.48% of 35/50 bitumen and 7% voids. The binder course is bituminous mix 0/14 with 4.49% of 35/50 bitumen and 9.6% voids. The mechanical behaviour of asphalt materials is modelled using the Huet-Sayegh model [3] [4]. The complex modulus is represented by five viscoelastic coefficients: E_0 (static elastic modulus), E_∞ (instantaneous elastic modulus), k and h (exponents of the parabolic dampers), δ (positive dimensionless coefficient) and three thermal coefficients A_0, A_1, A_2. At frequency ω and temperature θ, the complex modulus is Eqn (1):

Experimental and Theoretical Investigation of Three Dimensional Strain

$$E^*(\omega,\theta) = E_0 + \frac{E_\infty - E_0}{1 + \delta(j\omega\tau(\theta))^{-k} + (j\omega\tau(\theta))^{-h}} \quad \text{and} \quad \tau(\theta) = \exp(A_0 + A_1\theta + A_2\theta^2) \quad (1)$$

According to French standards, complex modulus tests have been performed on the two bituminous materials. The software *Viscoanalyse* [5] was used to fit the laboratory data and calculate the eight parameters of the model. These values are shown in Table 1. Poisson ratio ν is assumed to be equal to 0.32 at moderate temperature and equal to 0.38 at hot temperature for every asphalt material.

Table 1. Huet-Sayegh parameters for bituminous materials

	E_0(MPa)	E_∞(MPa)	δ	k	h	A_0(t)	A_1(t °C^{-1})	A_3(t °C^{-1})
Layer 1	32	28 440	2.53	0.25	0.71	2.1688	-0.3671	0.00196
Layer 2	24	29 155	2.35	0.23	0.69	1.9706	-0.3670	0.00196

2.3 Sensors

Two innovative retrofit techniques which allow measuring strains in the upper and lower parts of the asphalt layer and also at the interface between two layers have been used on the two structures. These two technologies are: an asphalt concrete core specially trimmed for the installation of optic fiber gauges [6]; and a thin polymeric plate instrumented and fixed inside a saw cut in the asphalt layer [7]. Both systems use polymeric proof bodies selected to be mechanically (elastic modulus) and thermally (coeffficient of thermal expansion) compatible with asphalt concrete. The association of these two techniques allows obtaining the strains in three directions: longitudinal (ε_{xx}) transverse (ε_{yy}) and vertical (ε_{zz}).

Two optic fiber sensors are inserted into a polymeric proof body (gray area on figure 1 and figure 2) at orthogonal directions and fixed on the concrete core which allows measuring longitudinal and transverse strains. On the structure A, the sensors are first, positioned at the bottom of the first layer (figure 1-a) at Z=65 mm, then at the bottom of the second layer (figure 1-b) at Z=125 mm. On the structure B, the sensors are installed at the bottom of the layer (figure 2-b) at Z=65 mm and near the top (figure 2-a) at 10 mm.

The polymeric plates have been designed to be instrumented at various positions and installed in a saw cut perpendicular to the direction of travel. To evaluate the interface conditions between the asphalt layers of the structure A, ten horizontal sensors are placed on both sides of the interface. Five are at the bottom of the first plate (figure 1-c) at Z= 65 mm, and five at the top of the second plate (figure 1-d) at Z=75 mm. The first plate is also instrumented with five horizontal sensors at Z=15 mm and five vertical sensors at Z=20 mm. Finally, five horizontal sensors are installed at the bottom of the second layer at Z=125 mm. On the structure B, the plate is instrumented with twenty one sensors (figure 2-c), seven horizontal sensors at Z=10 mm, seven vertical sensors at Z=20 mm and seven horizontal at Z=65 mm. All the plates are symmetrical, the sensors are spaced six centimeters and the central sensor is positioned directly under the center line (Y= 0 mm). Thermocouples are located at 48,38,23,12,9,6,3,0 cm depth inside the pavement.

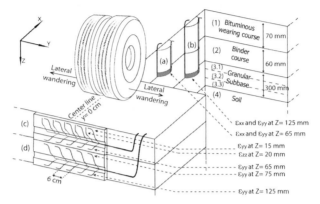

Fig. 1. Configuration and instrumentation of the structure A (two instrumented concrete cores (a and b) and two instrumented polymeric plates (c and d))

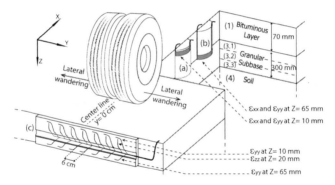

Fig. 2. Configuration and instrumentation of the structure B (two instrumented concrete cores (a and b) and one instrumented polymeric plate (c))

2.4 Experimental Protocol

The moving load is applied using a dual tire 12.00R20. The applied load (65 kN), the inflation pressure (850 kPa) and the revolution speed (6 rpm, which correspond to about 42km/h) were maintained during the whole experimental program. The data acquisition was performed at a 1 000 Hz sampling rate. The measurements were carried out for:

- eleven lateral tire positions (five on each side of the center line spaced by 10.5 cm) aim to determine the strain basin. As shown on figure 3, this protocol allows to obtain 55 measurement points for the structure A (5 sensors multiplied by 11 positions) and 77 points for structure B (7 sensors multiplied by 11 positions).

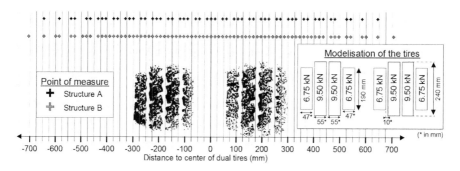

Fig. 3. Distribution of measurement points under the footprint of the tire

- two pavement temperatures. The tests took place on May 2011, allowing for moderate temperatures in the morning and hot in the afternoon. Table 2 summarizes the temperatures at different depths and their variation during tests.

Table 2. Temperature at different depths of the tested pavements

Temperature (°C)		surface	at 3 cm	at 6 cm	at 9 cm	at 12 cm
Structure A	Moderate	15.6° ± 2.6	17.1° ± 1.3	17.4° ± 0.7	18.6° ± 0.3	20.6° ± 0.2
(25/05/2011)	Hot	41.6° ± 0.9	42.4° ± 1.2	40.3° ± 1.0	37.8° ± 0.5	33.6° ± 0.3
Structure B	Moderate	16.1° ± 0.8	18.6° ± 0.2	19.3° ± 0.3	20.8° ± 0.3	22.4° ± 0.3
(24/05/2011)	Hot	38.6° ± 3.2	38.2° ± 1.3	36.6° ± 1.1	34.9° ± 0.7	31.6° ± 0.1

3 Results and Data Analysis

3.1 Three Dimensional Strain Occurring Near the Surface of Structure B

For each gauge, an elementary data acquisition consisted in recording the signal during three rotations of the carousel allowing a verification of the repeatability of the signal. The dual tire moves in the X-direction. Thus, the signal which depends on time can be converted into a signal depending on spatial position X. Figure 4 presents typical responses of strain sensors that have been recorded near the surface, on structure B, under the right tire (in the plane y=100mm). The measured longitudinal, transverse and vertical strain are given for two temperatures. In these figures, contraction is represented by negative values of strains.

The following general observations can be made:

- The shape of longitudinal strains is not perfectly symmetrical. There is two zones of extension strains, one before the passage of the tire over the gauge (X positive) and the other after the passage (X negative). The magnitude of extension strain after the passage is the largest. The two magnitudes depend on

the temperature. Due to the thermo-viscoelasticity of the material, the magnitude at higher temperatures is greater. The compressive strain zone is between both extension strain zones and it is associated with the passage of the load.
- The shape of transverse strains presents only a compressive zone. The magnitude of strain increases before the passage of the load to reach a peak under the tire and decreases slowly to zero.
- The vertical strains are extension at the front of the tire and decrease while the wheel is passing over. Despite the application of a vertical compressive stress caused by the wheel, the strain remains positive due to the Poisson's effect. After the load has passed, the strain reaches a second maximum and then decreases to zero. In contrast to the moderate temperature case, there is a contraction zone on the hot temperature curve, which occurs just in front of the tire. The signal reaches a negative peak, then decreases in two steps. The reduction is first rapid as the wheel leaves and then slower due to the relaxation of the material.

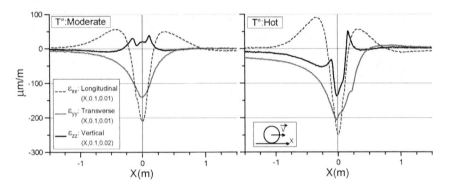

Fig. 4. Strains near the surface for two temperatures (plan y=100 mm)

3.2 Modeling of the Problem

A large number of transverse and vertical gauges were analysed and results denote a high sensitivity to the position under the passing load for both structure. Depending of the position under the tire, the signal presents different shapes and magnitudes. This variation can be induced by different stress states under the grooves and ribs of the tire. Under the tire, two characteristic curves have been identified and are presented in the next sections. In order to compare the measured strain signals, the software *ViscoRoute 2.0©* [1] has been used. In the software, the structure is represented by a multilayered half-space. This program integrates the viscoelastic behaviour of asphalt materials through the Huet-Sayegh model. The layers are assumed to be perfectly bonded (no slip). The influence of sliding interface condition on the response can be evaluated using the research version of *ViscoRoute* [8]. The load is assumed to move at a constant speed of 11.94 m.s^{-1}. The dual tire is discretized into eight rectangular shaped surfaces. Each rectangular load represents a tire tread as illustrated with the footprint on figure 3. Only a uniform vertical load is applied over the rectangular areas. To take account of the thermal gradient through the pavement, temperatures at 3 cm and 9 cm depth are used to model respectively the layer 1 and the layer 2.

3.3 Transverse Strain Occurring Near the Surface of Structure A

For structure A, two types of signal shapes have been identified (figure 5):

- The shape N°1 for $\varepsilon_{yy}(x,0.12,0.015)$ is the similar to the one previous illustrated on figure 4. The measured strain curves are showing time retardation and asymmetry that result from the viscoelastic properties of the material. The curves do not peak at X equals 0 but slightly after. The time delay is greater at higher pavement temperatures. Calculations did not result in the exact same shape as the measured strain curves and show more viscoelasticity and a wider zone of influence. The selected calculated curve was directly under a tire rib.
- The shape N°2 for $\varepsilon_{yy}(x,-0.09,0.015)$ presents two peaks located at the front and the rear of the tire. The passage of the load imposes an inversion of the compressive strain curve. The maximum decrease is observed slightly after X equals 0. This time retardation is particularly pronounced at higher temperatures.

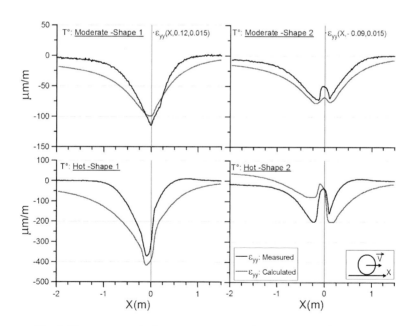

Fig. 5. Transverse strain for two temperatures (structure A) at Z= 15 mm

3.4 Vertical Strain Occurring Near the Surface of Structure A

For vertical strain, two shapes of the signal have been identified (figure 6):

- At moderate temperature, the curves present two peaks located before and after the passage of the wheel. Depending on the position under the tire, the vertical compressive stress can cause a significant decrease of the extension strain. The calculated strains have the same curve in the tensile zone preceding the passage of the tire. For the zone at the rear of the tire, the magnitude of calculated strains is lower. Nevertheless, the two curves are of similar order of magnitude.
- At hot temperature, the shape of the signal is the same as that described previously. Only the maximum magnitude in the compressive zone changes depending on the position under a groove (shape 1) or a rib (shape 2).The calculated strains in the tensile zone at the front of the tire are always higher than the measured ones.

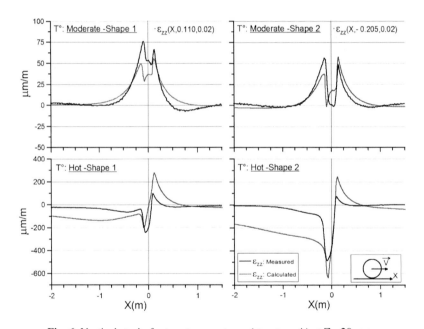

Fig. 6. Vertical strain for two temperatures (structure A) at Z= 20 mm

3.5 Strain Basin under the Tire

All signals from transverse and vertical gauges were analysed and the value of the characteristic point near X= 0 mm was selected. Knowing the Y position under the tire, all the value are placed on the same graphic to form the strain basin (figure 7) and compared with the calculated curve. It can be observed that:

- For transverse strain: Away from the tire edges, the strains are positive indicating an extension strain. The measured values are higher than the calculated ones. At moderate temperature, the strain under the tire remains negative and experimental values are grouped. At hot temperature, the measured values are more dispersed due to the two possible shapes of curves. This dispersion is explained by the influence of grooves and ribs of the tire.
- For vertical strain: away from the tire edges and between dual tires, the strains are positive. The maximums are measured under the outside tires edges and between the two tires. This extension strain is a consequence of the Poisson's ratio. The experimental values are lower than the calculated ones. Under the tire the measured and calculated values are of similar order of magnitude.

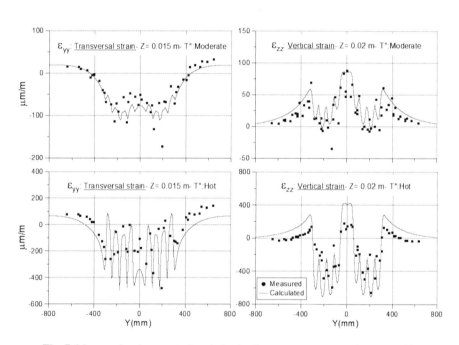

Fig. 7. Measured and computed strain basins for two temperatures (structure A)

3.6 Discussion

The structure behaviour modeling computations were made with a constant uniform temperature for each layer. As shown on Table 2, during measurements at moderate temperature, a negative temperature gradient was present in the layers 1 and 2. In contrast, the temperature gradient was positive at hot temperature. The stiffness differential in each asphalt layers has not been considered. Under each rib of a tire, three dimensional contact stresses are applied. In the modeling only a uniform vertical stress was considered. A more detailed load distribution with non-uniform transverse and longitudinal stresses could result in an even better fit.

The two asphalt layers have been considered fully bonded together. Recent research indicates that debonding between asphalt layers increases shear stresses near the surface[9] [10]. The interface conditions are not the same between moderate temperature and hot temperature. The effect of interface bonding condition could be integrated into the model using the information obtained with the gauges installed on each side of the interface. The measured data are the average strains over the lengths of the gauges (10mm).Thus the recorded signal is not necessarily equal to the calculated strain at the same position under the tire. The presence of an aggregate near the sensor can increase the local modulus and modify the local visco-elasticity properties.

4 Conclusions

The use of optic fibre sensors positioned at two different depths within the asphalt concrete layer allowed to adequately characterize the strains occurring within this layer. The analysis of strain curves clearly demonstrated the differences existing between the two temperatures and the necessity to use a viscoelastic model. A detailed analysis of the strain basins revealed critical zones under or outside the tire. The same analysis will be done for all experimental conditions in order to evaluate the impact of each criterion on the pavement damage.

References

[1] Chabot, A., Chupin, O., Deloffre, L., Duhamel, D.: ViscoRoute 2.0: a tool for the simulation of moving load effects on asphalt pavement RMPD Special Issue on Recent. Advances in Num. Simul. of Pavements 11(2), 227–250 (2010)

[2] Hornych, P., Kerzreho, J.P., Chabot, A., Bodin, D., Balay, J.M., Deloffre, L.: The LCPC's ALT facility contribution to pavement cracking knowledge. In: Proc. of the Sixth Internat. RILEM CP Conference on Pavement Cracking, Chicago, USA, pp. 13–23 (2008) ISBN 978-0-415-47575-4

[3] Huet, C.: Etude par une méthode d'impédance du comportement viscoélastique des matériaux hydrocarbonés. Ph.D dissertation. Université de Paris (1963)

[4] Sayegh, G.: Contribution à l'étude des propriétés viscoélastiques des bitumes purs et des bétons bitumineux. Ph.D dissertation. Faculté Sciences de Paris (1965)

[5] Chailleux, E., Ramond, G., Such, C., de la Roche, C.: A mathematical-based master-curve construction method applied to complex modulus of bituminous materials (EATA Special Issue). RMPDG 7, 75–92 (2006)

[6] Doré, G., Duplain, G., Pierre, P.: Monitoring mechanical response of in service pavements using retrofitted fiber optic sensors. In: Proc of the Intern. Conf. on the Advanced Charact. of Pavement and Soil Eng. Materials, Athens, Greece, pp. 883–891 (2007) ISBN 978-0-415-44882-6

[7] Grellet, D., Doré, G., Bilodeau, J.-P.: Effect of tire type on strains occurring in asphalt concrete layers. In: Proc. of the 11th Intern. Conf. on Asphalt Pavements, Nagoya, Japon (2010)

[8] Chupin, O., Chabot, A., Piau, J.-M., Duhamel, D.: Influence of sliding interfaces on the response of a layered viscoelastic medium under a moving load. International Journal of Solids and Structures 47, 3435–3446 (2010)
[9] Hammoum, F., Chabot, A., St-Laurent, D., Chollet, H., Vulturescu, B.: Effects of accelerating and decelerating tramway loads on bituminous pavement. Materials and Structures 43, 1257–1269 (2010)
[10] Wang, H., Al-Qadi, I.L.: Near-Surface Pavement Failure Under Multiaxial Stress State in Thick Asphalt Pavement. Journal of the Transportation Research Board 2154, 91–99 (2010)

Reasons of Premature Cracking Pavement Deterioration – A Case Study

Dariusz Sybilski[1], Wojciech Bańkowski[1], Jacek Sudyka[2], and Lech Krysiński[2]

[1] Road&Bridge Research Institute, Lublin University of Technology
[2] Road&Bridge Research Institute

Abstract. Cracking deterioration was observed on a motorway after only few years after construction. The motorway pavement was semi-rigid with a relatively thin thickness: 4 cm wearing layer, 11 cm asphalt base course, 20 cm lean concrete base course, 20 cm cement stabilized base course, 20 cm drainage layer, subgrade. lean concrete base course was cut in spacing 2,5 m to reduce the reflective cracking in asphalt layers. Testing program included: coring and materials composition testing, FWD testing to evaluate pavement layers stiffness modulus, radar testing to measure layer thickness, and to detect water presence in pavement layers, evaluation of cracks type and spacing, interlayer bond testing, low temperature resistance of asphalt wearing course. Testing results led to following conclusions: extremely low winter temperature in combination with relatively thin pavements thickness and semi-rigid pavement type was the reason of transverse cracking, longitudinal cracking observed on one of the sections were evaluated as top-down fatigue type cracking, low interlayer bond between asphalt base layers was one of the main reason of lower fatigue resistance of the pavement, lack of drainage in motorway median caused the presence of water in the pavement layers, and increased the danger of premature pavement deterioration, unpredicted increase in road traffic caused longitudinal top-down fatigue cracking.

1 Introduction

Three motorway sections of total length of 150 km were constructed in the period of 2002-2004 years. Section 1 was a semi-rigid pavement with use of old pavement constructed in late 1970-ties. Sections 2 and 3 were new constructions:

- asphalt wearing course 4 cm AC 16 PMB 45/80-55 (PG 64-22)
- asphalt base course 11 cm AC 25 35/50 (PG 64-16)
- cement bound base course 20 cm, R_{28} 3,5 – 7,0 MPa
- cement bound base course 20 cm, R_{28} 2,0 – 3,5 MPa.

The notches were transversely cut in the upper cement bound base course in both Sections with frequency 2-3 m and filled with bitumen to minimise the risk of reflective cracking in the semi rigid pavements (according to experience [1, 2]). At this frequency of notches, the width of crack opening is as small that neighbouring rigid plates co-work providing the load transfer. Hence, the pavement bearing capacity and fatigue distress is not reduced.

Pavement strengthening was foreseen for all sections after the traffic of 10 million 115kN axel loading. In reality, the traffic density increase was far faster and pavements have been strengthened with two asphalt layers in years 2007 (section 1), 2008 (section 2), 2009 (section 3).

2 Pavement Deterioration

Transverse cracking on section 2 and block (transverse and longitudinal) cracking on section 3 appeared in January 2006 r..Cracks appeared also on section 1 but in a very low number. Further cracking intensity was observed in 2008, especially longitudinal cracking on section 3. Cracks were not significantly dangerous for road users (do not create danger or driving comfort). Cracks were systematically and effectively sealed with polymer-bitumen. Sealing reduced also the risk of water penetration through cracking into the lower layers. However, the increasing cracking frequency leads to distress intensification, especially during the winter time due to water penetration and freezing as well as brine penetration (distress of hydraulically bound subbase),

3 Testing Programme

In June 2008 an extensive pavement testing programme have been performed on sections 2 and 3. Previous tests results performed in 2006 have been used in the final analysis. Testing programme included: FWD layers deflection and moduli, georadar tests to measure layers thickness and moisture, cracking inventory, cracking origin analysis, laboratory testing of pavement layers samples, properties of bituminous binders of section 3, low temperature resistance of wearing course asphalt mixture.

4 Testing Results and Analysis

4.1 Visual Evaluation

Section 3 pavement condition was significantly worse in comparison to section 2. Distress of section 2 was of low harmfulness – lower frequency of single transverse or longitudinal cracks. Distress on section 3 was of higher harmfulness – higher frequency of single transverse and longitudinal cracks, block cracks, and cavities of loss of aggregates and binder.

4.2 Pavement Structure Diagnosis Based on RADAR Measurements

Based on the inspection of pavement boreholes in Section 2, no significant deviations from the pavement design were identified. Pavement boreholes in Section 3 revealed three asphalt layers and two layers of hydraulically bound base course, while the design provided for two asphalt layers and two layers of hydraulically bound base course. The radar measurements were performed using the Ground Penetrating Radar (GPR). Pavement structure layer thickness tests were performed on the outer lanes, with the use of a 1 GHz 'air-coupled' antenna and a 400 MHz 'ground-coupled' antenna. Measurements were taken at 20 cm intervals. The example profile in Fig. 1 shows a record of electromagnetic waves from both antennas and the results of a quantitative assessment.

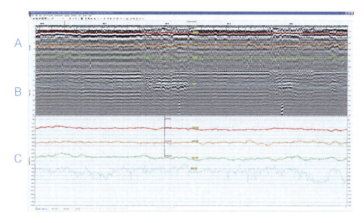

Fig. 1. Measurement data profile; A- record of electromagnetic waves from the 1 GHz antenna, B – record of electromagnetic waves from the 400 MHz antenna, C – calculated layer thicknesses incl. data from calibration boreholes

Measurement results were analysed with the use of RoadDoctor 2.0 [3] software. Layer thicknesses were calculated based on a dielectric constant determined during equipment calibration by measuring the amplitude of the signal reflected from the pavement surface and the amplitude of the signal reflected from a metal plate. In addition, pavement boreholes were used for absolute calibration of the results obtained.. The total thickness of the asphalt layer package, the hydraulically bound base course, the cement-treated subgrade of approx. 15-20 cm in thickness, and the drainage sand layer of a similar thickness was estimated (Fig. 2). In addition, at a depth of approx. 2.0-2.2 m, a reflected signal was recorded, which indicated the presence of a groundwater table at that depth.

Under the structural layers there is a layer of cement-treated stabilised subgrade and a sand drainage layer. The dielectric constants calculated (from 6 through 8) did not indicate any moisture accumulation in those layers.

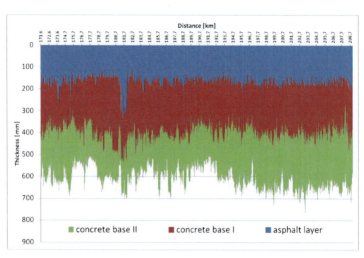

Fig. 2. Thickness of layers based on GPR measurement on the right carriageway, Section 2

4.3 Evaluation of the Degree of Pavement Moisture Accumulation from the Median Strip Based on Radar Measurements

Radar measurements were performed with the GPR System featuring a 1.0 antenna and a 2.2 GHz antenna. Evaluation of moisture content in the asphalt layers was performed on the inner lanes. The example profile in Fig. 3 shows a record of electromagnetic waves from both antennas and the results of the

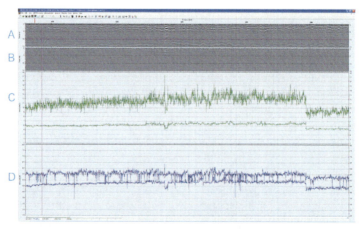

Fig. 3. Measurement data profile; A- record of electromagnetic waves from the 1 GHz antenna, B – record of electromagnetic waves from the 400 MHz antenna, C – calculated dielectric constants for layers to the depth of approx. 6-8 cm, D - calculated dielectric constants for layers to the depth of approx. 3-5 cm.

dielectric constant calculations. Dielectric constants were calculated based on the equipment calibration by measuring the amplitude of the signal reflected from the pavement surface and the amplitude of the signal reflected from a metal plate. Two dielectric constants were determined in each pass, i.e. for the layer at the depth of approx. 3-5 cm (wearing course) and for the layer at the depth of approx. 6-8 cm (asphalt base course).

In evaluating moisture content in the asphalt layers, the criteria were applied as specified in [4] (Table 1).

Table 1. Classification of moisture content in asphalt layers based on dielectric constant

Layer condition	Dielectric constant
dry layer	< 9
dry layer or slightly moist layer	9-12
moist layer	12-16
wet layer	> 16

It was determined that the dielectric constants obtained for the wearing course of Section 3 averaged at 7 while those of Section 2 averaged at 6. The wearing course in both sections was dry. On the other hand, the asphalt base course was found to be moist – the dielectric constant for Section 3 was 13 (max. 16) and that for Section 2 ranged between 10 and 14.

4.4 Layer Modulus Calculation Based on FWD Measurements of Pavement Deflections

The total equivalent modulus of elasticity for all the carriageway pavement layers was as follows:

- Section 3, right (south) carriageway   2,722 MPa
- Section 3, left (north) carriageway   3,016 MPa
- Section 2, right (south) carriageway   2,802 MPa
- Section 2, left (north) carriageway   2,899 MPa.

The pavement modulus of elasticity for motorway Sections 2 and 3 has a high value, twice that of the required minimum (1,100 MPa). This demonstrates a very good load-bearing capacity of the structure.

The moduli of elasticity (rigidity) were calculated for pavement structural layers using ELMOD 5.0. To determine the calculation model, information about the arrangement of the structural layers obtained through GPR measurements (example in Fig. 4) were used. Tables 2 and 3 show mean values and the calculated standard deviation of the layer modulus converted for the equivalent temperature of 10°C. The moduli for pavement structural layers and for subgrade soil are very high. The modulus for asphalt layers in the inner lanes is greater than that in the outer lanes, on the average by approx. 25 %. It is obviously related to higher fatigue damage due to larger heavy vehicles traffic on outer lanes.

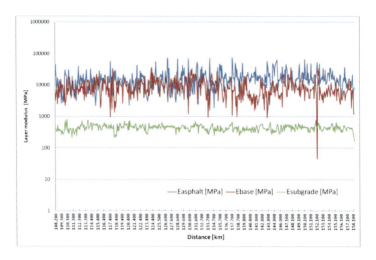

Fig. 4. Layer modulus, Section 3, right carriageway, right lane

Table 2. Mean modulus values and standard deviations – Section 2

	$E_{asphalt}$ [MPa]	E_{base} [MPa]	$E_{subgrade}$ [MPa]
Right carriageway			
Right lane	19,298±*13,841*	7,749±*5,313*	438±*133*
Left lane	22,516±*15,735*	7,020±*5,226*	421±*128*
Left carriageway			
Right lane	20,176±*14,030*	10,331±*6,866*	443±*133*
Left lane	19,539±*11,558*	10,329±*8,118*	453±*125*

Table 3. Mean modulus values and standard deviations – Section 3

	$E_{asphalt}$ [MPa]	E_{base} [MPa]	$E_{subgrade}$ [MPa]
Right carriageway			
Right lane	15,585±*11,304*	9,076±*5,853*	444±*100*
Left lane	20,107±*14,800*	9,752±*5,515*	463±*100*
Left carriageway			
Right lane	12,037±*7,943*	14,157±*7,156*	386±*80*
Left lane	15,290±*8,858*	17308±*9,475*	403±*87*

The moduli for the asphalt layers and base course featured high variability. The standard deviation often reached a value exceeding 50 % of the mean value. The subgrade soil moduli display definitely lower variability, with the standard deviation representing up to 30 % of the mean value. The greatest differences in layer moduli occur in Section 3. In Section 2, the modulus differences are much smaller.

5 Laboratory Tests of Bituminous Mixtures

5.1 Evaluation of Shear Bond Strength

Tests on shear bond strength between bituminous layers were performed with use of Leutner method. Results show very good bonding between bitumen wearing course and base course on Section 2 and Section 3 - shear bond strength 2,1 MPa or 2,6 MPa on average, respectively. In both cases the bond considerably exceeds recommended limits of 1,3 MPa. Lower bonding was found on Section 3 between two layers of bituminous base (1,3 MPa on average in cracking zone and 1,5 MPa outside this area) due to coarser grading in relation to the thickness of each bituminous layer and lower binder content.

5.2 Tests and Evaluation of Properties of the Binder on Section 3

Properties of polymer modified bitumen PMB 25/55-60 from the wearing course are shown in Table 4. Rheological properties evaluated by DSR didn't show any significant differences between bitumen samples, apart from specimen number 2, that had the highest shear modulus G^* (Fig. 5).

Fourier Transform Infrared (FTIR) Spectroscopy was performed in KTH, Stockholm, according to the method described in [5]. FTIR spectra are shown in Figures 6 and 7. Peaks at 966 and 699 cm^{-1} are characteristic of SBS polymer modified bitumen (Figure 6). To evaluate polymer content, IR absorbance at these specific wave numbers was calculated and compared with SBS calibration curves. The approximate polymer content was determined in the range from 3,5% to 5,1%.

The tested binders reveals properties showing their aging. The hardest binder (4) shows the lowest penetration, the highest softening point and the lowest elastic recovery. It indicates the lowest effectiveness of polymer modification.

Table 4. Properties of binders recovered from wearing course on Section 3

	Property	Specimen no.					
		right carriageway JPPA			left carriageway JLPA		
		1	2	3	4	5	6
Lab 1	Penetration at 25 °C, 0,1mm	23	25	22	17	21	21
	Softening point R&B, °C	68	64,2	69,0	72,6	71,6	69
	Fraass breaking point, °C	-8	-9	-14	-12	-15	-9
	Elastic recovery at 25 °C, %	67,5	74,5	63,5	52	66	61
Lab 2	Penetration at 25 °C, 0,1mm	22,0	16,5	15,8	14,4	20,1	22,2
	Softening point R&B, °C	67,8	66,8	71	76,2	72,6	69,0
	Fraass breaking point, °C	-10	-4	-10	-8	-10	-12
	Elastic recovery at 25 °C, %	69	65	62,5	-	60	65

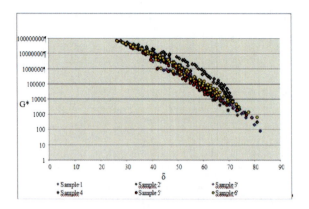

Fig. 5. Evaluation of modification degree of binders from wearing course

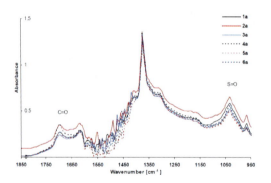

Fig. 6. FTIR spectra of binder samples (part 1)

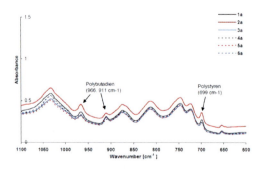

Fig. 7. FTIR spectra of binder samples (part 2)

5.3 Evaluation of Resistance to Low Temperature Cracking of Bituminous Mixture from Wearing Course

Low temperature cracking tests were performed with use of *Thermal Stress Restrained Specimen Test* TSRST method according to AASHTO TP10-93 on specimens cored from the wearing course in two locations – at the beginning and at the end of section (Table 5). Previous TSRST results for section 2 from 2006 showed 5°C lower cracking temperature. Low temperature properties of specimens from base and wearing course from Section 3 are worse than from Section 2.

Table 5. TSRST results, Section 3, 2008

Location	Average Force max [N]	Average Stress max [kPa]	Average Cracking Temperature [°C]
117+100	8598	3573	-20,0
145+900	7933	3233	-19,6

6 Conclusions

Transverse cracking on all three pavement sections are of thermal origin due to very low temperature, about -28°C, during winter 2006. It was confirmed by the analysis with use of MEPDG system. Cracks in semi-rigid pavement are of the dual origin – thermal in combination with reflective cracking effect (copy of dilatation notches) [6]. Relatively thin thickness of asphalt layers (15 cm) resulting from the stage construction of the pavement contributed to creation of thermal cracking. Low harmfulness of transverse cracks does not weaken the pavement bearing capacity, thanks to the pavement slabs load transfer.

Longitudinal cracks are limited to wearing course depth. Their origin is of top-down fatigue cracking. It may be presumed that the reason was the lower effectiveness of PMB binder.

Transverse and block cracking on sections 3 may have three origins. Firstly: significant growth of road traffic, exceeding 5-ve times prediction in the pavement design). Secondly: moisture in pavement asphalt subbase due to lack of drainage in motorway median (which is a known appearance eg. [7]). Thirdly: weakening of pavement bearing capacity due to dividing asphalt subbase layer into two sub-layers, which resulted in weaker compaction and interlayer bonding – with the final effect of lower pavement fatigue resistance.

After strengthening and drainage construction in motorway median, the pavement is in a good condition under a heavier traffic loading. It may be expecting that the semi rigid asphalt pavement may fulfil requirements of long life pavement, requiring only the wearing course renewal [8].

References

[1] Prevention of Reflective Cracking in Pavements. State-of-the-Art Report of RILEM Technical Committee 157 PRC. RILEM Report 18. E&FN Spon (1997)
[2] Conception et dimensionnement des strucutres de chaussée. Guide technique. Setra-LCPC (1994)
[3] Road Doctor User's Guide, Roadscanners Oy, Rovaniemi, Finland (2001)
[4] Saarenketo, T.: Electrical properties of road materials and subgrade soils and the use of ground penetrating radar in traffic infrastructure surveys. In: Faculty of Science, Department of Geosciences, University of Oulu (2006)
[5] Masson, J.-F., Pelletier, L., Collins, P.: Rapid FTIR method for quantification of styrene-butadiene type copolymers in bitumen. National Research Council Canada, NRCC-43151,
[6] Sybilski, D.: The dualism of bituminous road pavements cracking. In: 3rd RILEM Conference Reflective Cracking in Pavements, Maastricht (1996)
[7] Kandhal, P.S., Rickards, I.J.: Premature Failure of Asphalt Overlays from Stripping: Case Histories. In: Proc. AAPT (2001)
[8] Dumont, A.-G., Beuving, E., Christory, J.-P., Jasienski, A., Ortiz Garcia, J., Piau, J.-M., Sybilski, D.: Long Life Pavements and success stores. World Road Association (PIARC/AIPCR), Technical Committee C4.3 report (2007)

Effect of Thickness of a Sandwiched Layer of Bitumen between Two Aggregates on the Bond Strength: An Experimental Study

Subrata Mondal[1], Animesh Das[2,*], and Animangshu Ghatak[3]

[1] Former Master's student, Department of Civil Engineering, Indian Institute of Technology Kanpur,Kanpur, 208 016
sbdmondal@gmail.com
[2] Associate Professor, Department of Civil Engineering, Indian Institute of Technology Kanpur, Kanpur, 208 016
adas@iitk.ac.in
[3] Associate Professor, Department of Chemical Engineering, Indian Institute of Technology Kanpur, Kanpur, 208 016
aghatak@iitk.ac.in

Abstract. Understanding bond between bituminous binder and aggregate is an important consideration for bituminous mixes. The cohesion within the binder and the adhesion between the binder and the aggregate interface, affect the mix performance in terms of load transfer, propagation of crack, moisture sensitivity etc. In the present work, bond strength between aggregate and bituminous binder while varying bitumen film thickness has been studied. The test temperature and the displacement rate have been kept fixed as $23\pm1^{\circ}C$ and 1 mm per minute respectively. Polished surface of one aggregate is entirely covered with thin film of bitumen and then another aggregate is placed with its polished surface over the bitumen film. Vertical pulling load is then applied to this assembly. Load displacement data is recorded for various samples for different bitumen film thicknesses. The variation of peak stress, energy (toughness) are plotted with respect to the bitumen film thickness. The conditions for cohesion and adhesion failures are studied. It is envisaged that such studies would evolve as a useful input for the micro-mechanical modeling of bituminous mix.

1 Introduction

Bituminous mix is composed of aggregates, bitumen and air-voids. A good bonding between the aggregates and bitumen is important for satisfactory performance of the bituminous mix in terms of proper load transfer and resistance to fatigue, ravelling, moisture sensitivity etc.

---
* Corresponding author.

A large number of studies have been conducted to understand the bond between aggregate and bitumen. Contact angle measurement [1-4], inverse gas chromatography [5], micro-calorimetry [6, 7] methods have been used to estimate surface energy of aggregates-bitumen bonding. Some research studies report tensile strength testing between bitumen and aggregate [8, 9] or bitumen and other materials [9-13].

Marek and Herrin [9] conducted tensile strength test on bitumen film and found that with the increase in bitumen film thickness, tensile strength first increases to a peak value and then decreases and almost becomes constant. In their study, the peak strength was observed at bitumen film thickness of about 20 µm [9]. Frolov et al. [8] found that for higher values of film thickness, cohesive strength is independent of film thickness. Canestrari et al. [10] performed a number of tests with Pneumatic Adhesion Tensile Testing Instrument (PATTI) test set-up on different types of asphalt binders and aggregates. They found that samples under dry condition primarily showed cohesive failure and water conditioned samples showed adhesive or cohesive failure depending on the affinity between aggregate-bitumen system [10]. Poulikakos and Partl [13] conducted tensile strength testing on bituminous film placed between steel or aggregates. In their study, cohesive failure was observed at 23°C and primarily adhesive-failure was observed at -10°C [13].

As can be seen from the above discussions that limited literature is available on study of tensile strength between aggregate and bitumen. This has motivated the present researchers to initiate a further study in this direction. Thus, the scope of the present study [14] is identified as:

- Study of bond strength between aggregate and bitumen using tensile strength testing.
- Study on effect of bitumen film thickness on aggregate-bitumen-aggregate bond.

2 Experimental Study

Locally available sand-stone aggregates and bitumen of grade VG30 [15] (penetration value obtained between 60 and 70) are used in the present study. Irregular shaped aggregates are cut into small cubes (approximately 25 mm each side) by using a diamond cutter. The surface of the aggregate is polished thoroughly by using a polishing equipment with abrasive silicon carbide powder (of size 180 mesh) and water. Followed by polishing, the samples are rinsed with water to remove any abrasive powder stuck onto the polished surface. Roughness of polished surface of five representative aggregate samples are measured at the Manufacturing Sciences Laboratory, IIT Kanpur. The arithmetic mean parameter (R_a) is obtained as 6.78 µm.

In order to form a thin sandwiched layer of bitumen, two aggregate samples are heated in an oven to about 110°C and the polished surface of one of them is dipped inside a pool of hot and molten bitumen maintained at a temperature of about 150°C. The aggregates are brought closer (with the bitumen dipped surface facing the polished surface of the other aggregate) and they are mildly pressed. Excess bitumen is removed using a blade. Figure-1 shows a side view of a typical aggregate-bitumen-aggregate sample.

Fig. 1. Sandwitched layer of bitumen between two polished aggregate

The sample is brought down to ambient temperature. The sample is kept immersed inside a water bath at 23±1°C. Care is taken so that water does not come in contact with bitumen. This is done by sealing the sample with water-tight plastic bags. The thickness of bitumen layer is measured using a slide caliper, considering the difference of total height of the two aggregates with and without bitumen film in-between. The measurement is repeated number of times and the average value is taken.

A schematic diagram of a typical sample holder is shown in Figure 2. It comprises of four screws placed in all four sides for holding the aggregate specimen properly secured. Two such sample holders are used for holding the two aggregate pieces. Suitable adjustments are done using the screw system, so that the polished faces of the aggregates (facing each other) remains horizontal. The sample holders are fixed to the jaw of an instron machine (in the ACMS Laboratory, IIT Kanpur) used for carrying out the tensile testing. The experimental set-up is shown in Figure 3(a). The load is applied vertically. In the present study, test results of about 60 samples are reported, while plotting three data points have been removed because of being outliers. the average bitumen film thickness varied between 0.011 mm to 0.64 mm.

During the test, tensile load is applied at constant loading rate as 1 mm per minute, and load versus displacement data is recorded for each experiment. All tests are conducted at a constant temperature of 23±1°C. Figure 3(b) presents a close-up view of the sample being tested.

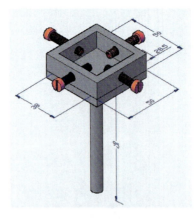

Fig. 2. Schematic diagram of a sample holder (dimensions in mm)

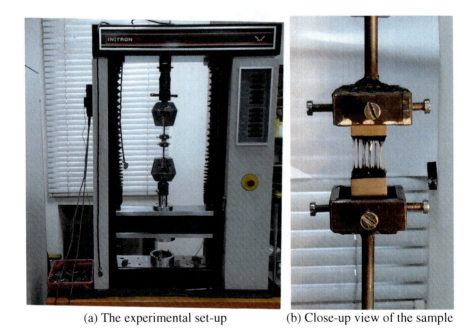

(a) The experimental set-up (b) Close-up view of the sample

Fig. 3. Test being conducted on aggregate-bitumen-aggregate sample

3 Results and Discussions

Failure of sample occurs due to development of crack at the weakest point of the aggregate-bitumen-aggregate junction. The crack may develop either at the aggregate bitumen interface (i.e. adhesive failure) or within the bitumen (i.e. cohesive failure) or combination of both. As the vertical deformation increases the size of crack also increases. These results in separation of two aggregates from each other and the material fail to carry any further load. Figures 4(a), (b) and (c) show aggregate surfaces after adhesive, cohesive and combined failures respectively.

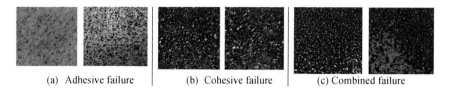

(a) Adhesive failure   (b) Cohesive failure   (c) Combined failure

Fig. 4. Types of failure observed after testing

Figures 5(a) and (b) show typical variation between load and deformation for adhesive and cohesive failures respectively. From these figures and also from the all other experimental results, it can be seen that failure is sudden for adhesive failure and gradual for cohesive failure.

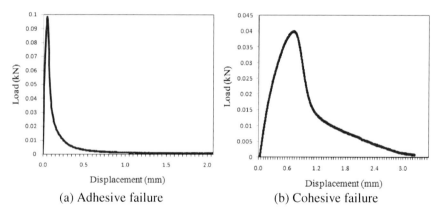

(a) Adhesive failure   (b) Cohesive failure

Fig. 5. Typical load deformation curve for adhesive and cohesive failure

As the displacement rate, test temperature, type of bitumen and aggregate are kept unchanged in the present study, it can be postulated that in the present case, it is only the bitumen film thickness which governs the adhesive or cohesive failure. In the present study, it is observed that samples with bitumen film thickness less than 0.15 mm have primarily undergone cohesive failure, samples with bitumen film thickness more than 0.2 mm have primarily undergone adhesive failure, and samples with bitumen film thickness in between 0.15 mm to 0.2 mm have primarily undergone combined adhesive-cohesive mode of failure. This aspect is schematically presented in Figure 6.

At lower bitumen film thickness cavities may form due to development of negative pressure (opposite to adhesive force). Bitumen between cavities behaves like individual column, thus reduce the film cross sectional area. But, at interface the film area is constant and equal to aggregate surface area. Reduction in film area in between two aggregates may result in stress concentration and the sample may fail by detachment of bitumen film instead of interfacial failure. This may be postulated as a possible explanation of why cohesive failure is prominent at lower film thickness [16]. As bitumen film thickness increases the possibility of cavity formation decreases and therefore chances of adhesive failure increases.

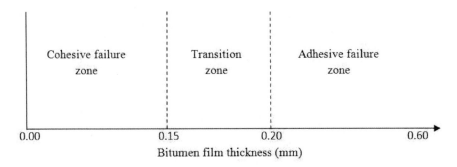

Fig. 6. Dependence on cohesive or adhesive failure on bitumen film thickness

The samples which showed more than 80% of cohesive failure and the bitumen film thickness is less than equal to 0.15 mm are studied separately. The variation of peak stress versus thickness and variation of toughness versus thickness are plotted in Figures 7 and 8 respectively. Data fitting on Figure 7 suggests that peak stress is approximately inversely proportional to the 0.29 power of bitumen film thickness. Analytical derivation shows for rigid substrate and compressible, elastic adhesive material the stress is inversely proportional to square root of film thickness [17].

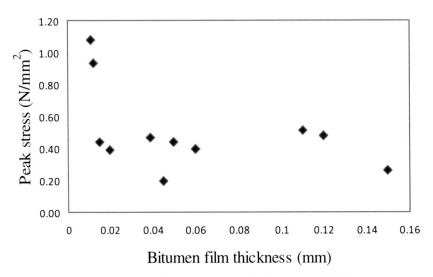

Fig. 7. Variation of peak stress with thickness for cohesive failure

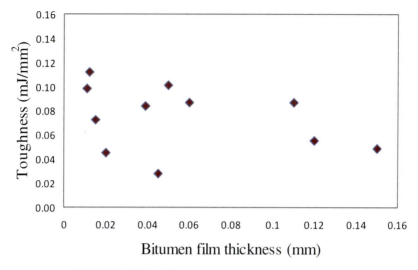

Fig. 8. Variation of toughness with thickness for cohesive failure

The samples which showed more than 80% of adhesive failure and the bitumen film thickness is more than equal to 0.20 mm are studied separately. The variation of peak stress versus thickness and variation of toughness versus thickness are plotted in Figures 9 and 10 respectively. Data fitting on Figure 9 suggests that peak stress is approximately inversely proportional to the 0.57 power of bitumen film thickness.

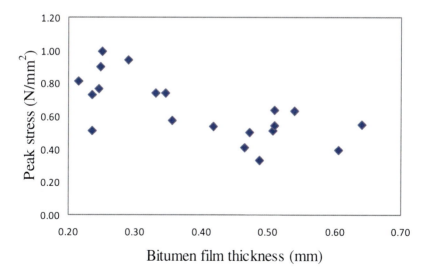

Fig. 9. Variation of peak stress with thickness for adhesive failure

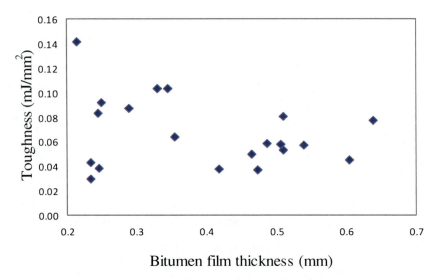

Fig. 10. Variation of toughness with thickness for adhesive failure

4 Conclusion

In the present study experiments on tensile strength has been performed (with deformation rate as 1 mm per minute and test temperature as 23±1°C) to study the failure in aggregate-bitumen-aggregate bond. Sand-stone aggregates and VG 30 grade of bitumen has been used. In the present study, it is observed that samples with bitumen film thickness less than 0.15 mm have primarily undergone cohesive failure, and samples with bitumen film thickness more than 0.2 mm have primarily undergone adhesive failure. For both cohesive and adhesive failures, the peak stress at failure is observed to be inversely proportional to the bitumen film thickness, however for adhesive failure the dependency is found to be stronger.

References

[1] Bhasin, A., Hefer, A.W., Little, D.N.: Bitumen surface energy characterization using a contact angle approach. Journal of Materials in Civil Engineering 18, 759–767 (2006)
[2] Cheng, D., Little, D.N., Holste, J.C.: Use of surface free energy of asphalt-aggregate systems to predict moisture damage potential. In: Proceedings of the Association of Asphalt Paving Technologies, vol. 71, pp. 59–84 (2002)
[3] Peltonen, P.V.: Road aggregate choice based on silicate quality and bitumen adhesion. Journal of Transportation Engineering 118(1), 50–61 (1992)
[4] Xiao, Q.Y., Wei, L.Y.: A precise evaluation method for adhesion of asphalt aggregate. International Journal of Pavement Research and Technology 2(6), 270–274 (2009)

5. Hefer, A.W., Little, D.N., Herbert, B.E.: Bitumen surface energy characterization by inverse gas chromatography. Journal of ASTM International (2005)
6. Bhasin, A., Little, D.N.: Application of micro calorimeter to characterize adhesion between asphalt binders and aggregates. Journal of Materials in Civil Engineering 21(6), 235–243 (2009)
7. Powell, M.W., Nowell, D.V., Evans, M.B.: Determination of adhesion in bitumen-mineral systems by heat-of-immersion Calorimetry II. Correlation of chemical properties with adhesion. Journal of Thermal Analysis 40, 121–131 (1993)
8. Frolov, A.F., Vasiéva, V.V., Frolova, E.A., Ovchinnikova, V.N.: Strength and structure of asphalt films. Chemical Technology Fuel Oil 19, 415–419 (1983)
9. Marek, R.C., Herrin, M.: Tensile behaviour and failure characteristics of asphalt cement in thin films. Proceedings of Association of Asphalt Paving Technologists 37, 387–421 (1967)
10. Canestrari, F., Cardone, F., Graziani, A., Santagata, F.A., Bahia, H.U.: Adhesive and cohesive properties of asphalt-aggregate systems subjected to moisture damage. Road Materials and Pavement Design, 11–32 (2010)
11. Cebon, D., Harvey, J.A.F.: Fracture tests on bitumen film. Journal of Materials in Civil Engineering 17(1), 99–106 (2005)
12. Masad, E., Howson, J., Bhasin, A., Caro, S., Little, D.N.: Relationship of ideal work of fracture to practical work of fracture: background and experimental results. Journal of the Association of Asphalt Paving Technologists 79, 81–218 (2010)
13. Poulikakos, L.D., Partl, M.N.: Micro scale tensile behaviour of thin bitumen films. Experimental Mechanics 51(7), 1171–1183 (2010)
14. Mondal, S.: Effect of thickness of a sandwiched layer of bitumen between two aggregates on the bond strength: an experimental study, Master's thesis, Department of Civil Engineering (2011)
15. IS 73, Paving bitumen specification, Bureau of Indian Standards, 3rd Revision, New Delhi (2006)
16. Creton, C., Fabre, P.: The mechanics of adhesion, 1st edn. Elsevier, Amsterdam (2002)
17. Yang, F., Li, J.C.M.: Adhesion of a rigid punch to an incompressible elastic film. Langmuir 17, 6524–6529 (2001)

Hypothesis of Existence Semicircular Shaped Cracks on Asphalt Pavements

Dejan Hribar

Head of the Traffic Routes and Infrastructure, Building and Civil Engineering Institute ZRMK d.o.o., Dimičeva 12, SI -1000 Ljubljana, Slovenia
dejan.hribar@gi-zrmk.si

Abstract. There are many different causes of cracks on the surface of asphalt pavements. The primary cause of cracking in asphalt layer is increasing tension stresses and related strains to the point when the tensile strength of the material is exceeded.Semi-circular shaped cracks occur at the edges of the excavations and pavements where there is poor local sub-layers, thin layer of asphalt and excessive traffic loads. Semi-circular shaped cracks are very common. However, they are often confused with the netlike cracks. The typical form occurs when the crack progresses to a certain point of contact to the outside edge of the pavement, whereas in the opposite direction (inside) of the track it provides adequate resistance against the occurrence of crack. Cracks spread in the direction of poor sub-base and are formed gradually one next to the other or independently. This paper presents semi-circular shaped cracks with a single, double or several free edge. Semi-circular shaped cracks propagated from top to bottom at the cross section.

The analysis of numerical model shows that the line deflection and maximum tensile stresses on the toper surface are in a semi-circular, which indicates the formation of semi-circular shaped cracks. To determine the location of cracks it is necessary to look at the maximum tension stresses σij on the top surface of the model. The location of crack is somewhere in the area of maximum tension stresses and occur when the tensile strength of the material is exceeded.

1 Introduction

This paper shows what it is semi-circular shaped cracks. Why and how semi-circular shaped cracks occur. Why they have such characteristic shape. It also shows how such cracks are spread on the surface pavement and the cross section. Schematic is presented the cracks with a single (one), double and several free edges. At the end of this paper is presented the numerical model that illustrate this problem and is treated by the finite elements method (FEM).

Pavement cracking can be caused by several factors, such as: structural changes of the bituminous binder in bituminous mixtures because of the aging, large temperature changes during use, excessive traffic loads and/or deficiencies in the construction [1]. Cracks in asphalt pavements can take many forms. The most common types of cracking are [2]: fatigue cracking, longitudinal cracking, transverse cracking, block cracking, slippage cracking, reflective cracking and edge cracking. On the pavement you can find cracks in semi-circular shape (Fig .1). In the

American literature we find the concept of edge cracks which formed a kind of semi-circular shaped cracks. Edge cracks typically start as crescent shapes at the edge of the pavement. They will expand from the edge until they begin to resemble alligator cracking. This type of cracking is the result of lack of support of the shoulder due to weak material or excess moisture. They may occur in a curbed section when subsurface water causes a weakness in the pavement [3]. Longitudinal cracks with in one to two feet of the outer edge of a pavement are referred to as edge cracks [6], as shown in Fig. 2 left. Longitudinal edge cracks on the pavements occur if there is low bearing capacity on lower layers through the entire edge. These cracks are not considered as semi-circular shaped cracks. In addition, the semi-circular shape crack occur at the shaft (Fig. 2 right), only to have this as a full circle, but in our view have the same cause of the accident.

Fig. 1. Semi-circular shaped cracks at the edges of the pavements (left) and excavations (right)

Fig. 2. Longitudinal edge cracks at the edge of the pavements (left) and circular shaped cracks at the shafts (right)

2 Semi-Circular Shaped Cracks (SCSC)

2.1 Characteristics of the SCSC

Semi circular shaped cracks have certain characteristics that distinguish them from other cracks. We distinguish them by:
- shape and propagate of semi-circular shaped cracks,
- propagation through the cross section and
- cause of formation.

2.1.1 Shape and Propagation of Semi-Circular Shaped Cracks

When the vehicle (load) carries a weak location in the pavement occur the crack that propagation into the interior (Fig. 3). A typical semi-circular crack on the pavement surface occurs when the crack progresses and stops at a particular point (crack a.) or progression route, making the characteristic circular shape, to the outer edge of the pavement (crack b.), because in the opposite direction of the pavement is more resistant to cracking. It does not run parallel to the pavement edge because at some point the support gets stronger, and so the crack turns outward [4].

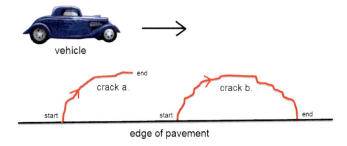

Fig. 3. Schematic presentation of semi-circular shaped cracks propagated in the direction of low bearing capacity of lower layer with single free edge crack a. that extends into the interior and crack b. witch complete in the edge of the pavement.

Fig. 4 shows how SCSC propagates on surface of the pavement. First is formed "1." crack and then next "2." crack occurs at the first and touches it at a different point or they can appear completely independently. So, SCSC propagated next to each other in the direction of low bearing capacity of lower layers (sub-base or base). Fig. 4 also presents semi-circular shaped cracks with a single, double or several free edges. Semi-circular shaped cracks with a single (one) free edge are "1." crack and the beginning and the end of the crack is at the same height. Crack with two free edges (e.g. i. crack) has the beginning and the end of the cracks at different heights and one of free edge is already formed crack, in this case "2." crack. Crack with several free edges is crack "n." crack.

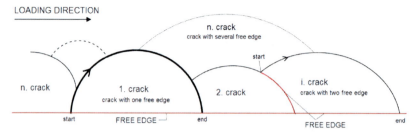

Fig. 4. Schematic presentation of semi-circular shaped cracks propagated in the direction of low bearing capacity of sub-base or base layer with single (1. crack) and two free edge (i. crack)

2.1.2 Propagation through the cross section

On Slovenian local road, the asphalt core was drilled on semi- circular shaped crack (Fig. 5 left). Fig. 5 right shows cross sections of this drilled core and we see that the crack propagation from top to bottom. So, at the top surface we have tensile tension. This is not typical of fatigue cracks. Also, Shuler (2005) in final report concludes that edge crack propagation from top to bottom. This means that upward movement from below and/or bending is causing excessive tensile strain at the surface of the pavement [5].

Fig. 5. The location of asphalt core taken on semi-circular shaped cracks (left) and presentation of semi-circular shaped cracks propagated top to bottom through the cross section (right)

2.1.3 Cause of formation

Semi-circular shaped cracks occur at the edge of the pavement - edge cracks (near unpaved shoulder) or at the edge of the excavations on asphalt pavements. Fig. 6 shows schematically the formation of these tensile cracks in the area near the maximum tensile stress of asphalt layers. The creation of those cracks is mostly attributable to:

- lower bearing capacity of lower layers (increased flexibility); the material is very dirty (contains clay) and can keep water in it (higher moisture) or because

of weaker drainage capability at the edge of the pavement or the local lower density - called "nest" (Fig. 7),
- thin asphalt layer (the pavement is too thin for the traffic loads),
- narrow pavement (heavy vehicles drive closer to the edge of the pavement),
- heavy traffic load during spring thaw (very weak and unstable shoulder and base),
- structural (hardening of bitumen) and thermal changes (cryogenic tension) in the asphalt layer.

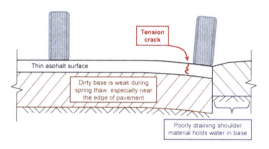

Fig. 6. Schematic view of cracking at the edge of pavement [4]

Fig. 7. Excavation on existing pavement – formation »nest«

3 Numerical Model of Semi-Circular Shaped Cracks

Our numerical model represents formation of semi-circular cracks. It is prepared and processed in programming environment SAP2000 with finite element method (FEM), where we analyze the size and shape of stress and strain.

With this numerical model we want to answer why semi-circular shape occur in relation to the stresses and strains.

The assumptions for our model are:

- shell-thin layer (asphalt layer thickness = 7 cm),
- material properties of asphalt (elastic modulus E = 4000 MPa, Poisson ratio $v = 0.35$),

- joint springs are flexible connections to ground (linear elastic) as sub-layer with modulus of soil reaction k1 = 150 MPa (good compacted)
- point force F = 50 kN as wheel load at the free edge and under is area of weak sub-layer with spring modulus of soil reaction k2 = 100, 130, 140 MPa.

With this assumption we create the worst possible conditions for occur semi-circular shaped crack. The numerical model is presented in Fig. 8.

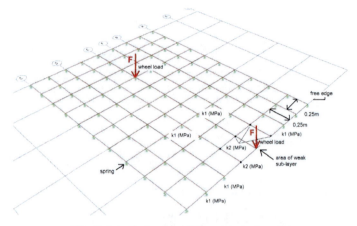

Fig. 8. Schematic view of the numerical model

The variable in this model was a modulus of soil reaction k2 = 100, 130, 140 (MPa), because area of weak sub-layer it is not always the same.

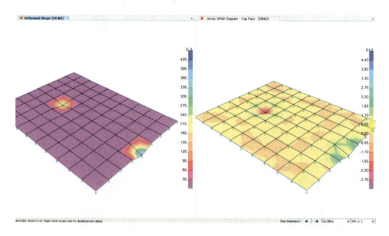

Fig. 9. The results of deflection (left) and tension stress on the top surface (right); k1= 150 MPa and k2 = 100 MPa at force F = 50kN

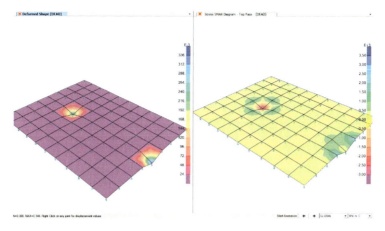

Fig. 10. The results of deflection (left) and tension stress on the top surface (right); k1= 150 MPa and k2 = 130 MPa at force F = 50 kN

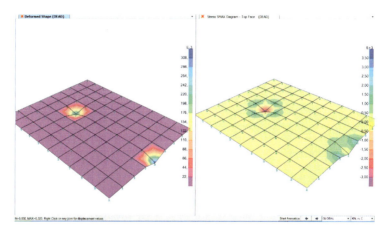

Fig. 11. The results of deflection (left) and tension stress on the top surface (right); k1= 150 MPa and k2 = 140 MPa at force F = 50 kN

On Fig. 9-11 is shown the line of deflection and maximum tensile stresses. Semi-circular lines are clearly visible at the free edge, which indicates the creation of semi-circular shaped cracks. It is only questions where is the location of crack and when occur. To determine the location of cracks it is necessary to look at the maximum tension stresses σij on the top surface of the model. The location of crack is somewhere in the area of maximum tension stresses and occur when the tensile strength of the material is exceeded. Also, Fig. 9-11 clearly shows the area of maximum tension stresses (light to dark green) extending from the extreme edges towards the centre of the semi-circular form. On the bottom surface is compressive stresses, conclude that the crack propagated from the top to bottom.

Fig. 9 and 10 shows that with increasing k2 decreases deflection and tensile stresses and vice versa. If we compare Fig. 9 (k2 = 100 MPa) and Fig. 10 (k2 = 140 MPa) we see that the deflection is lower at higher modulus k2 and maximum tension stresses is for almost half size higher at lower k2.

4 Conclusion

This paper shows what it is semi-circular shaped cracks. Why and how semi-circular shaped cracks occur. Why they have such characteristic shape. It also shows how such cracks are spread on the surface pavement and the cross section. Schematic is presented the cracks with a single (one), double and several free edges. At the end of this paper is presented the numerical model that illustrate this problem and is treated by the finite elements method (FEM).

Semi circular shaped cracks have certain characteristics that distinguish them from other cracks. This paper shows shape and propagate of semi-circular shaped cracks, propagation through the cross section and cause of formation.

Semi-circular shaped cracks are very common, but often confused with the netlike cracks. The typical form occurs when the crack progresses to a certain point of contact to the outside edge of the pavements, whereas in the opposite direction (inside) of the track it provides adequate resistance against the occurrence of cracks. The creation of those cracks is mostly attributable to: lower bearing capacity of sub-layers, thin asphalt layer, narrow pavement, heavy traffic load during spring thaw, structural and thermal changes in the asphalt layer. On surface of the pavement they propagated in the direction of low bearing capacity sub-layers layer, and are formed gradually one next to the other or independently. SCSC propagated from top to bottom at the cross section. This means that upward movement from below and/or bending is causing excessive tensile strain at the surface of the pavement.

The analysis of numerical model shows that the line deflection and maximum tensile stresses on the toper surface are in a semi-circular, which indicates the formation of semi-circular shaped cracks. To determine the location of cracks it is necessary to look at the maximum tension stresses σ_{ij} on the top surface of the model. The location of crack is somewhere in the area of maximum tension stresses and occur when the tensile strength of the material is exceeded.

References

[1] Žmavc, J.: Vzdrževanje cest. In: UL FGG in DRC – Družba za Raziskave v Cestni in Prometni Stroki Slovenije, Ljubljana, vol. 3, pp. 32–71 (2010)
[2] Read, J., Whiteoak, D.: The Shell Bitumen Handbook, 5th edn., ch. 10, pp. 195–209 (2003)
[3] Orr, D.P.: Pavement Maintenance 4, 17–27 (2006)

4. Lynne, H.I.: Those Cracks on the Edge of the Road...What Causes Them? Corenll Local Roads Program, Centerlines, 2–4 (2008)
5. Shuler, S.: Edge Cracking in Hot Mix Asphalt Pavement - Final report, Colorado Asphalt Pavement Association Englewood, Colorado, Summary of Findings, pp. 37–40 (2005)
6. Grass, T.,, P.: The Asphalt Handbook, 7th edn., vol. MS-4, ch. 11, pp. 543–545 (2007)

Quantifying the Relationship between Mechanisms of Failure and the Deterioration of CRCP under APT: Cointegration of Non-stationary Time Series

Ebenhaezer Roux de Vos

BKS (Pty) Ltd, Hatfield Gardens, 333 Grosvenor Street, Pretoria, South Africa

Abstract. A statistical property of non-stationary time series is applied to quantify the long-run equilibrium relationship between mechanisms of pavement failure and deterioration. The concept is applied to measured pavement responses from an Accelerated Pavement Test (APT) conducted on the thin (185 mm) Continuously Reinforced Concrete Pavement (CRCP) of the Ben Schoeman freeway in South Africa. The statistical property is that of cointegration, which allows testing for the existence of equilibrium relationships among trending variables within fullydynamic specification frameworks. The predominant failure mechanisms leading to punchout are known to be loss of load transfer capability at transverse cracks and the deterioration of substructure support. The long-term relationship between proxy variables for the mechanisms leading to punchout and pavement deterioration is determined. The average long-term relationship between a change in relative movement and a change in surface deflection at transverse cracks is found to be 17.5. This means that on average a 1.0 % increase in relative movement will translate into a 17.5 % increase in surface deflection and *vice versa*. The empirical quantification of the long-term equilibrium relationships between pavement mechanisms of failure and deterioration offers the potential to improve the reliability of design systems and may also be used in pavement management.

1 Introduction

Concrete pavements are complex and dynamic systems. The concrete pavement system as a whole exhibits different characteristics than the individual components it is comprised of. The interrelations and dependencies between the various components of the concrete pavement system are difficult to specify and quantify.

In the case of a thin (150 to 185 mm) CRCP, punchout is caused by the loss of load transfer capability at transverse cracks; and the deterioration of the support structure [1-3]. The interdependence of these mechanisms and the relative contribution to deterioration of pavement performance are largely unknown.

This gap exists as one is unable to solve such a complex system in discrete time. It is fundamentally a system of simultaneous equations that is under-identified and of which the structural coefficients therefore cannot be recovered.

Researchers commonly focus on the characterisation of a specific mechanism in terms of material and/ or environmental parameters and conditions [4-6]. Whereas such studies have been achieved successfully, it is a challenge to capture and encompass the interdependence between failure mechanisms and their specific contributions to pavement deterioration.

The quantification of such interrelationships has the potential of improving reliability and confidence in pavement design. It will furthermore reduce the variance between the individual functions used to characterise the failure mechanism and the pavement-specific performance function as used in a risk-based design method [7].

The interrelationship between the different mechanisms may be quantified using the property of cointegration. The objective of this paper is to illustrate this concept by applying its principles to CRCP responses measured under APT.

2 Cointegration

Cointegration is a statistical property of non-stationary time series variables, and allows one to view the problem in continuous instead of discrete time. This enables the quantification of the stated interrelationships in a dynamic environment.

2.1 Background

Cointegration developed within the discipline of economics, where a substantial part of theory is concerned with long-run equilibrium relationships, generated by market forces and behavioural rules. As a result most empirical econometric studies involving time series' may be interpreted as attempts to evaluate such relationships in a dynamic framework [8].

Granger [9] points out that a vector of variables, of which all achieve stationarity after differencing, may have linear combinations that are stationary in levels. Engle & Granger [10] formalise the idea of integrated variables sharing an equilibrium relation that turns out to either be stationary or to have a lower degree of integration than the original series. This property is denoted as cointegration, signifying co-movements among trending variables that may be exploited to test for the existence of equilibrium relationships within a fully dynamic specification framework.

2.2 Integrated and Cointegrated Processes

If a process x_t is stationary, the process

$$y_t = x_t + y_{t-1} = \sum_{s=0}^{\infty} x_{t-s} \tag{1}$$

is called integrated of order one, $I(1)$. y_t has the property by construction that its first difference is x_t and is therefore stationary ($I(0)$).

$$\Delta y_t = y_t - y_{t-1} = x_t \qquad (2)$$

A white noise series and a stable first-order autoregressive $AR(1)$ process are well-known examples of $I(0)$ series; a random walk process is an example of an $I(1)$ series, which when accumulating a random walk leads to an $I(2)$ series.

A (n x 1) vector time series $Y_t = [y_{1t}, ..., y_{nt}]'$ is said to be cointegrated if each of the series' is individually I(1); while some linear combination of the series $a'Y_t$ is stationary, $I(0)$, for some non-zero (n x 1) vector a. Generally speaking; if two or more integrated variables have a common stochastic trend, so that a linear combination has no stochastic trend, then they cointegrate.

2.3 Application to the Domain of Pavement Engineering

For the case of thin CRCP; given that one has three variables, maximum deflection (jd), permanent deformation (pd) and relative movement (rm), that one expects to be cointegrated, the expected cointegration relation could be algebraically stated as Eqn. (3).

$$\alpha jd_t + \beta pd_t + \gamma rm_t + C = z_t \qquad (3)$$

Where z_t is the equilibrium error, the distance that the system is away from equilibrium at any point in time. In the equation a variable pd_t typically refers to its entire sequence $\{pd_t : t = 0, ..., \infty\}$.

In Eqn. (3) the coefficients α, β, γ and C constitute the cointegration vector. If one finds such a vector of coefficients where the resulting linear combination is stationary for the duration of the period under investigation, or zero if a constant is included, the time series variables are regarded as cointegrated; and may have a specific meaning.

3 Accelerated Pavement Testing and Instrumentation

Using the Heavy Vehicle Simulator (HVS), the APT was conducted by the Council of Scientific and Industrial Research (CSIR). The test was on the 1.5 m wide CRC shoulder of the in-service Ben Schoeman freeway, with the centre of the dual wheel at a distance of 420 mm from the CRC edge. The pavement structure of the section comprises of 185 mm CRCP; 30 mm open-graded asphalt overlay; 30 mm continuously-graded asphalt surfacing; 200 mm crushed stone base; 150 mm stabilised gravel subbase and two 150 mm layers of selected subgrade. The test section had six transverse cracks of which five were instrumented.

A total of 2750200 half–axle loads were applied in canalised, bi-directional trafficking mode over a period of five months ending in May 2010. The applied loads comprised load levels of 40, 60, 80 and 100 kN respectively. Loading was applied with a dual-wheel configuration of two 12R22.5 tyres at pressures of 800 kPa. The section was artificially and continuously watered for a two-day period in order to simulate rainfall; followed by six dry days.

The maximum surface deflection and permanent surface deformation were measured by means of Joint Deflection Measuring Devices (JDMDs). The relative movement across each transverse crack, with the entire loading wheel on one side of the crack, was calculated from the surface deflection data. Thermocouples were installed to measure environmental and pavement temperatures. Environmental conditions and pavement responses were measured at 30-minute intervals.

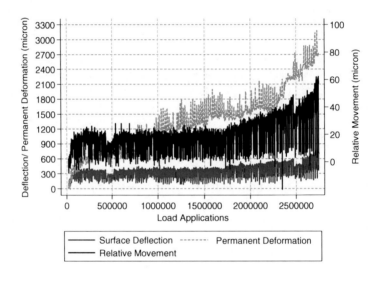

Fig. 1. Pavement responses measured under accelerated loading

4 Data Description and Model Development

It is postulated that a statistically significant, stable and long-term linear relation exists between deflection, permanent deformation and relative movement.

This is presented in Eqn. (4), with ε representing the stochastic error term.

$$jd = c + \alpha pd + \beta rm + \varepsilon \qquad (4)$$

APT data, containing 6953 sets of measurement, were used to estimate the parameters of Eqn. (4). Assumptions were made with regard to the proxies for the specific mechanisms of pavement failure and deterioration.

Maximum elastic surface deflection is the optimal proxy of pavement response and therefore performance under wheel loading. The response is a function of the structural condition of the pavement as a whole, and entails both the load transfer efficiency of the transverse cracks and the condition of the substructure.

Relative movement serves as the proxy for the load transfer capability of the transverse cracks in the pavement. The measured relative movement incorporates the effects of loss of load transfer, due to dowel action and aggregate interlock, and concrete shrinkage and curling due to environmental conditions. The condition of the substructure also influences the relative crack movement.

Due to the lack of alternatives, permanent surface deformation serves as a proxy for the deterioration of the CRC support. As the concrete layer cannot be compressed itself, permanent deformation reflects changes in the substructure. Permanent deformation is not, however, the optimal proxy variable, as it is unable to discern between sub-layer compaction, possible void formation and loss of material.

4.1 Non-stationarity of Individual Endogenous Variables

The non-stationarity of the individual variables is established through the Augmented Dickey-Fuller (ADF) test on the first difference of the variable in question. If the first difference is stationary, I (0), the variable in level is non-stationary and integrated of order one, I (1). Asymptotic critical values for unit root tests are used for tests of non-stationarity in the first difference [11]. Test statistics and the critical values are summarised in Table 1.

4.2 Basic Regression

Deflection is regressed on permanent deformation, relative movement and an intercept, robust for heterogeneity. The fitted values based on the estimated parameters are then predicted and saved. The residuals, the difference between estimated and actual values, are determined at each point in time.

Table 1. Test statistics and critical values for unit root tests

Crack nr	jd*	pd*	rm*	residual**
1	-114.526	-34.649	-120.433	-23.05
2	-110.353	-33.045	-118.754	-21.13
3	-110.082	-31.656	-121.98	-26.08
4	-112.42	-28.383	-122.564	-23.19
5	-113.136	-29.747	-120.986	-25.26

* Critical values at 10 %, 5 % and 1 % are -2.57, -2.86 and -3.43 [12]

** Critical values at 10 %, 5 % and 1 % are -3.45, -3.74 and -4.29 [12]

The stationarity of the residuals are tested by regression of the first difference on the lagged values. This regression is equivalent to the ADF-test, but should be conducted as estimated values are used in lieu of actual values [13].

The resulting test statistic, Table 1, is subsequently evaluated against asymptotic critical values [13]. If the residuals are found to be stationary, the first of the two conditions for the existence of a cointegrating relationship is met.

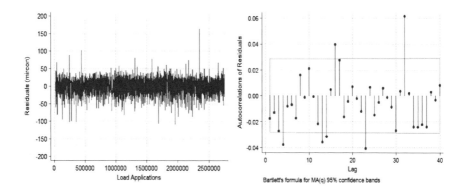

Fig. 2. Residuals (a) and autocorrelation of residuals (b) from the Autoregressive Distributed Lag (ADL) regression, Crack 3

5 General Autoregressive Distributed Lag Regression

A general Autoregressive Distributed Lag (ADL) regression, Eqn. (5), is conducted where lags of all three endogenous variables are included in the basic regression.

$$A(L)jd = B_1(L)pd + B_2(L)rm + c \tag{5}$$

Where $A(L), B_1(L)$ and $B_2(L)$ are lag operators. The ADL-regression is conducted to dispense with the autocorrelation in the residuals, Figure 2. Autocorrelation in the residuals causes underestimation of the standard errors that might lead to erroneous inference.

An F-test is conducted to establish if the sum of the coefficients of each endogenous variable and its lags are statistically significant. The sums of the estimated coefficients of $A(L), B_1(L)$ and $B_2(L)$ are $\hat{\alpha}$, $\hat{\beta}$ and $\hat{\gamma}$ respectively.

If the sum of the coefficients is significantly different from zero, a definitive cointegration relation exists. As can be seen from Table (2), all coefficients are significant at one percent (1 %), except the permanent deformation coefficient of Crack 4, which is at five percent (5 %). The permanent deformation coefficients of Cracks 2 and 5 are insignificant.

Table 2. Sums of ADL Coefficients and p-values of the Corresponding F-tests

Crack nr	Full specification						Truncated specification			
	$\hat{\beta}$	p-value	$\hat{\gamma}$	p-value	$\hat{\alpha}$	p-value	$\hat{\beta}$	p-value	$\hat{\gamma}$	p-value
1	-0.00399	0	0.876	0	-0.0429	0	1.572	0	-0.0928	0
2	2.14E-05	0.982	1.441	0	-0.595	0	2.366	0	-0.0906	0
3	-0.00512	0.007	0.985	0	-0.48	0	0.989	0	-0.0998	0
4	0.00178	0.0215	0.969	0	-0.056	0	1.564	0	-0.0781	0
5	-5.1E-05	0.936	1.018	0	-0.067	0	1.243	0	-0.0823	0

5.1 Implied Long-Run Relationship

The implied long-run equilibrium relationships, Table 3, in the form of Eqn. (8) are determined from the sums of estimated coefficients as in Table 2.

$$jd = \beta^* pd + \gamma^* rm + C^* \qquad (8)$$

In Eqn. (8), $\beta^* = -\frac{\hat{\beta}}{\hat{\alpha}}$, $\gamma^* = -\frac{\hat{\gamma}}{\hat{\alpha}}$ and $C^* = -\frac{\hat{C}}{\hat{\alpha}}$.

Table 3. Coefficients of the Long-Run Relationships

Crack nr	Full specification			Truncated specification	
	β^*	γ^*	C^*	γ^*	C^*
1	-0.0914	20.373	0	16.93	0
2	0	28.31	172.063	26.114	189.29
3	-0.067	12.943	190.802	9.909	159.218
4	0.031786	17.304	62.482	20.025	135.595
5	0	15.194	128.597	14.678	200.378
Average	-0.0254	18.825	110.788	17.533	136.897
Standard Deviation	0.0516	5.967	79.186	6.051	80.647

5.2 Comparison with Findings from Diagnostic Investigation

Analysis of the full specification yielded results that were contrary to expectation. Cointegration between the permanent deformation and the other two endogenous variables broke down in two of the five instances. Two of the three significant, permanent deformation coefficients were furthermore negative in sign. The general expectation was that the permanent deformation coefficients would be significant and positive.

Diagnostic coring confirmed the expectation. It indicated the stripping and loss of material of the open-graded asphalt layer below the CRC. It is further assumed that the loss of the fine material, manifested through permanent deformation, will lead to increased surface deflections; a positive β^*-coefficient.

Possible causes for the unexpected coefficient sign were identified, investigated and delineated to the following:

- Permanent surface deformation is a poor proxy variable for the deterioration of the substructure and the bond between CRC and the underlying asphalt. Measurements at the surface gave no indication regarding the stripping of asphalt layers and loss of fine material.
- The permanent deformation data was of poor quality. Instrument drift might have influenced measured levels. The standard deviation in the first difference of the permanent deformation series was 30 micron.

5.3 Truncated Model

As result of the inconsistency between the expected and estimated coefficients, the model was adapted by omitting permanent deformation. The resulting truncated specification is presented in Eqn. (9) below. The same statistical procedure was followed to determine the implied long-run relationship, with results in Tables 2 and 3.

$$jd = c + \beta rm + \varepsilon \qquad (9)$$

6 Interpretation of Results

The average long-term relationship between a change in relative movement and a change in maximum surface deflection is 17.5, Table 3. A one percent (1 %) change in relative movement at a transverse crack is accompanied by a 17.5 % change in maximum surface deflection. This relationship quantifies the impact of a loss of load transfer capability, albeit small, on surface deflection and ultimately pavement performance.

The estimated coefficients should be viewed as elasticities in an economic sense, and may be presented as the partial derivatives of the general specification. The elasticities are the absolute effects, and the effect of a change in relative movement through the channel of permanent deformation on deflection, should not be considered. Indirect effects are accounted for through solving of the simultaneous system as a whole, even although each of the endogenous variables is implicitly a function of the other two.

The analysis methodology is advantageous as it quantifies the empirical relationship between endogenous variables that act as proxies for CRCP performance and failure mechanisms.

7 Conclusions

The researcher describes the development of a model to capture the long-run equilibrium relationship between the two predominant failure mechanisms and pavement performance deterioration of a thin CRCP under APT.

The average long-term relation between a change in relative movement and a change in maximum surface deflection was found to be 17.5 for the pavement section and structure tested. The relationship was found to be statistically significant at a level of one percent for each of the five transverse cracks analysed.

The discrepancy between the results of the cointegration analysis, for the full specification of both failure mechanisms and pavement performance, and the findings of the investigation indicate one of two things:

1. Deterioration of the substructure support is at times negligible compared to the deterioration of load transfer capacity of the transverse crack, although statistically significant in the cointegration model.
2. Permanent surface deformation is not the optimal proxy variable for the deterioration of the substructure. Although statistically significant in some cases, the negative sign of the coefficient did not correspond to the stripping of the open-graded asphalt, as evident from the diagnostic investigation.

8 Recommendations

Based on the analysis methodology presented, it is recommended that the utilisation and application of the cointegrating relationships be investigated further through:

1. Application of the methodology to historic APT-datasets to evaluate and characterise the distributions of cointegrating relationships for specific pavement types.
2. Evaluation of alternative proxy variables for substructure support; particularly Multi-Depth Deflectometer data from similar APT tests.

References

[1] Selezneva, O., Rao, C., Darter, M.I., Zollinger, D., Khazanovich, L.: Transp. Res. Rec. (1896), 46–56 (2004)
[2] Strauss, P.J., Lourens, J.P.: In: Proceedings of the 8th International Symposium on Concrete Roads, Lisbon (1998)

[3] Steyn, W.J., Strauss, P.J., Perrie, B.D., Du Plessis, L.J.: In: Proceedings of the 8th International Conference on Concrete Pavements, Colorado Springs (2005)
[4] Kohler, E., Roesler, J.: Transp. Res. Rec. (1974), 89–96 (2010)
[5] Strauss, P.J., Perrie, B.D., Du Plessis, J.L., Rossmann, D.: In: Proceedings of the 8th International Conference on Concrete Pavements, Colorado Springs (2005)
[6] Jung, Y., Zollinger, D., Wimsatt, A.: In: Proceedings of the 89th Transportation Research Board Annual Meeting, Washington, D.C. (2010)
[7] Strauss, P.J., Slavik, M., Perrie, B.D.: In: Proceedings of the 7th International Conference on Concrete Pavements, Florida (2001)
[8] Dolado, J.J., Gonzalo, J., Marmol, F.: In: Baltagi, B.H. (ed.) A Companion to Theoretical Econometrics, pp. 283–290. Blackwell, Oxford (2000)
[9] Granger, C.W.J.: Journ. of Econom. 23, 121–130 (1981)
[10] Engle, R.F., Granger, C.W.J.: Econom. 55, 251–276 (1987)
[11] MacKinnon, J.G.: In: Engle, R.F., Granger, C.W.J. (eds.) Long-run economic relationships: Readings in Cointegration. Oxford University Press, Oxford (1991)
[12] Davidson, R., MacKinnon, J.G.: Estimation and Inference in Econometrics. Oxford University Press, Oxford (1993)
[13] Johnston, J., DiNardo, J.: Econometric Methods. McGraw-Hill, Singapore (1997)

Influence of Horizontal Traction on Top-Down Cracking in Asphalt Pavements

C.S. Gideon and J. Murali Krishnan[*]

Department of Civil Engineering, Indian Institute of Technology Madras,
Chennai 600036, India
jmk@iitm.ac.in

Abstract. Traditionally, in pavement analysis and design, stress-strain analysis is carried out by assuming wheel loads as uniform vertical traction acting on a circular area. However the contact forces at the tire-pavement interface are not purely vertical. Considerable amount of horizontal traction is developed when a tire moves on the surface of the pavement. It is expected that the state of stress and strain in a pavement in such a case might be completely different. In this study, three dimensional finite element stress-strain analysis of a multi-layered pavement system subjected to both vertical and horizontal tractions was carried out. A linear viscoelastic model was used to characterize asphalt layers and a Drucker-Prager model with a linear hardening rule was used to characterize the elastic-plastic response of the granular material of the base layer. It was observed that due to the application of horizontal traction considerable horizontal tensile strains are produced at the surface of the asphalt layer. These horizontal strains at the surface are tensile to the rear of the loaded area and compressive to the front of the loaded area. Also, the maximum horizontal tensile strain occurs at the surface and not at the bottom of the asphalt layer. In order to investigate the influence of the increased horizontal tensile strains at the surface on top-down cracking, the strains obtained from the finite element analysis were used in typical top-down crack prediction models and the number of repetitions to failure was determined. It was found that the addition of horizontal traction significantly reduces the allowable number of repetitions.

1 Introduction

Fatigue cracking in asphalt pavements is normally related to the magnitude of the critical tensile strain in the pavement structure due to load application. The magnitude of the critical tensile strain and the location where it occurs depends on the load configuration, the structural configuration of the pavement system and the nature of mechanical response of the materials of the layers. These critical tensile strains dictate the nature in which the cracks initiate and propagate. It is normally assumed that the critical tensile strains occur at the bottom of the asphalt layer and

---

[*] Corresponding author.

hence cracks initiate at the bottom of the asphalt layer and propagate to the top. This type of cracking is called as bottom-up cracking. However several studies have clearly shown that cracks can initiate at the top and propagate to the bottom [1-2]. These are increasingly observed in flexible pavements and are reported to be the most prominent pavement distress in many parts of the world. This type of cracking known as top-down cracking occurs because the critical strains occur at the top of the asphalt layer. Top-down cracking has been attributed to a number of factors and the most important of them are tensile strains at the surface of the asphalt layer, asphalt aging, spatial temperature gradients, asphalt mix properties and segregation [1].

One important aspect related to the determination of tensile strain at the surface is the role of the stress analysis procedure and the appropriate constitutive model used. The multi-layered elastic theory based stress-strain analysis with only the normal component of traction predicts the tensile strains at the bottom of the asphalt layer. Therefore top-down cracking cannot be predicted using this approach of stress-strain analysis. For this purpose, a rigorous stress-strain analysis that considers appropriate material properties and realistic loading conditions is needed. It should be pointed out here that tensile strains at the surface of the asphalt layer can be observed when the viscoelastic nature of the asphalt layers [2] and horizontal traction [3] is considered.

In the following sections, we present a brief overview of the existing literature related to the influence of horizontal traction on top-down cracking. We then detail the computational model, material properties and the load cases considered. The subsequent section is devoted to a discussion on the results of the finite element analysis. In the final section, the strain measures determined by the finite element analysis are used in appropriate pavement distress models and the influence of horizontal traction on the allowable number of repetitions to failure is analyzed.

2 Literature Survey

A vast body of literature on the issues related to fatigue cracking and the possible causes of fatigue cracking exists. The influence of horizontal traction on the development of tensile strain and top-down fatigue cracking has been investigated by several workers [3-5]. Collop and Cebon [6] used the linear elastic fracture mechanics framework and the finite element analysis for stress-strain analysis and reported that horizontal traction can cause short surface cracks in the asphalt layer. Groenendijk [7] investigated the influence of non-uniform contact stresses that included vertical, longitudinal and transverse stresses as given by De Beer [8] on the stress distribution at the surface and concluded that horizontal traction can cause critical tensile stresses at the surface. Several other researchers have reported that top-down cracking is initiated due to the surface tensile stresses that are induced because of horizontal traction [9-12].

The magnitude and orientation of the induced horizontal traction depends on the nature of vehicle movement, the frictional characteristics of the pavement and tire, the mechanical response of the pavement structure and the structural properties of the tire. Extensive work in the area of quantifying tire-pavement contact stresses is due to De Beer and coworkers [8] who have used Stress-in-Motion technology to capture the spatial and temporal variation of stresses, both vertical and horizontal. It has been shown that considerable horizontal stresses develop both in the longitudinal (in the direction of movement of tire) and transverse (perpendicular to the direction of movement of tire) directions. Experiments conducted by De Beer et al. [13] suggest that the horizontal traction can be 20% of the vertical traction. Taramoeroa and de Pont [14] indicated that the magnitude of the horizontal loads for a non-steering wheel in a cornering maneuver can be as high as 100% of the vertical loads, the percentage of the horizontal loads transferred as horizontal traction dictated by the coefficient of friction. Perret [15] carried out stress-strain analysis using the finite-element method and reported that horizontal traction applied on the pavement surface has a non-negligible effect on the near surface stress regime. Novak et al. [16] used the tire-pavement contact stress distribution data from Pottinger [17] to evaluate the effect of horizontal load and concluded that horizontal traction causes high transverse shear stresses in the top asphalt layer. All the above investigations assumed that all the pavement layers were linearized elastic in nature. In this study, a linearized viscoelastic model was used to model the mechanical response of the asphalt layers while a Drucker-Prager model with a linearly hardening rule was used to model the elastic-plastic mechanical response of the base layer.

3 Finite Element Model

A typical pavement cross-section corresponding to a design traffic of 150 million single axles and a California bearing ratio of 10, suggested by the Indian Roads Congress design guideline IRC-37:2001 [18] was selected. The selected cross-section consisted of two asphalt layers and three granular layers and the geometry of the pavement model is shown in Figure 1. The top two asphalt layers are Bituminous Concrete (BC) and Dense Bituminous Macadam (DBM). The dimensions of the finite element model are $5 \times 5 \times 1.25$ m. An 8-node linear brick element (C3D8R) with reduced integration and hourglass control was used to mesh the pavement model and the final pavement model consisted of 57072 elements. The finite element package, ABAQUS [19] was used to model the pavement system and to carry out the finite element analysis. It was assumed that there is no slip between the layers. The materials properties are chosen from the literature and are shown in Table 1, 2 and 3. A Poisson's ratio of 0.35 was assumed for the materials of all the layers. The viscoelastic material properties of the asphalt layer correspond to a temperature of 40 °C. The materials of the sub-base and the sub-grade were assumed to exhibit linearized elastic behavior.

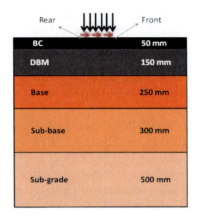

Table 1. Elastic Material properties [20]

Material	(MPa)
Bituminous Concrete(BC)	4479
Dense Bituminous Macadam(DBM)	3956
Base	200
Sub-base	150
Prepared sub-grade	60

Fig. 1. Model geometry (Not to scale)

Table 2. Viscoelastic material properties in terms of Prony series parameters [20]

Asphalt Layer	g_1	g_2	τ_1	τ_2
BC	0.674	0.326	0.116	2.811
DBM	0.232	0.768	0.096	1.355

Table 3. Drucker-Prager model parameters for the base layer [21]

Shear Criterion	Linear
Hardening Behaviour	Compression
Angle of Friction	40 °
Dilation Angle	40 °
Flow Potential eccentricity	0.1
Flow Stress Ratio	1
Yield Stress (kPa)	610
Absolute Plastic Strain at yield	0

4 Load Cases

The stress-strain analysis was carried out for different ratios of horizontal to vertical traction. A constant pressure of 650 kPa was applied on a circular area in the vertical direction and the horizontal traction was varied from 0 to 100% of the vertical traction in steps of 25% (Table 4). All the tractions were applied uniformly over a circular area of radius 16 cm. The horizontal traction was assumed to act in the direction of movement of the tire. The magnitude of the vertical traction assumed here corresponds to the contact pressure transferred for a standard axle load in pavement design.

Table 4. Load cases

Load Case	Vertical traction (MPa)	Horizontal traction (MPa)
L-1	0.6500	0.0000
L-2	0.6500	0.1625
L-3	0.6500	0.3250
L-4	0.6500	0.4875
L-5	0.6500	0.6500

It is well known that during the movement of the tire on the pavement, the pavement is subjected to horizontal traction (longitudinal and transverse) over and above the vertical traction. Experimental investigation conducted several decades back [22] have clearly shown that one can simulate the response of pavement due to vertical traction alone by assuming a haversine load form acting in a specific zone of influence. What is not known, however, is the extent and the manner of variation of horizontal traction at a specific point in the pavement structure during the movement of the tire. In the investigation reported here, we assumed that the horizontal and vertical traction vary in the same manner for simulation purposes and a loading time of 5 s and a rest period of 5 s was given for both the tractions. The load was applied for ten cycles. However, as can be seen later, the state of stress is tensile at the rear of the tire and compressive at the front due to horizontal traction and as the tire moves on the surface of the pavement, the tensile and compressive stresses is expected to vary in a cyclic manner.

5 Results and Discussion

It was found that of all the strains the horizontal tensile strain in the direction of application of the horizontal traction, ε_{xx}, is the most influenced by the application of horizontal traction. Therefore, the variation of ε_{xx} was studied in detail. The results of some of the simulations are shown in Figures 2 to 5. It is to be noted that in all these figures, tensile stresses and strains are taken as positive.

Figure 2 shows the variation of ε_{xx} with depth after 5 s of loading to the front and rear of the loading area. Considerable horizontal tensile strains were produced at the surface of the asphalt layer when horizontal traction was considered. These horizontal strains at the surface are tensile to the rear of the loaded area and compressive to the front of the loaded area. It is interesting to note that for L-1, at t= 5 s, the maximum tensile strain occurs at the bottom of the DBM layer whereas for L-2 to L-5, the maximum tensile strains occur at the top of the BC layer. It can be observed from Figure 2 that the influence of application of horizontal traction is restricted to the asphalt layers and is negligible in the granular layers.

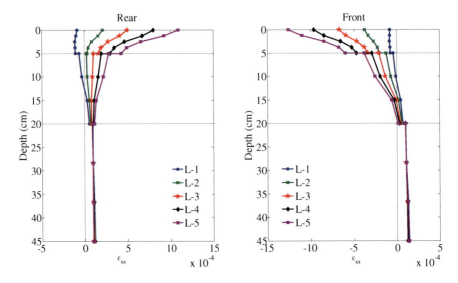

Fig. 2. Variation of ε_{xx} for different ratios of horizontal to vertical traction at t=5 s

Figure 3 and 4 shows the variation of ε_{xx} at different instants of time for L-1 and L-5. It can be seen from Figure 3 that for L-1, during the first cycle of loading, the maximum tensile strains occur at the bottom of the DBM layer. With the application of subsequent cycles of loading, it was observed that the maximum tensile strains occur at the top of the BC layer. Tensile strains were produced at the top of the BC layer even for L-1, the load case where no horizontal traction is applied. This correlates well with the observations of Kim et al. [2] who have showed that when the viscoelastic nature of the asphalt layers is accounted for, tensile strains at the top of the asphalt layer can be observed with the increase in the number of load applications. It should be pointed out that when the multi-layered elastic theory is used, tensile strains are predicted at the bottom of the asphalt layer and only when the viscoelastic nature of the asphalt layer is considered, for all the load cases from L-1 to L-5, tensile strains can be predicted at the top of the asphalt layer. For L-5, it can be observed that the horizontal strains to the rear of the load are tensile right from the first load cycle. The strain at the top of the BC layer increases at a much faster rate, thereby leading to considerably high tensile strains at the end of ten cycles.

In highways, the speed of the vehicle and the inter-arrival time of vehicles are all random in nature. Due to the repetitive nature of traffic loading, the materials in the layers are subjected to cycles of loading/unloading. In the rest period between load applications, stress relaxation and strain recovery takes place. The manner in which the pavement recovers all the strain or relaxes all the stresses depends on the rate at which these processes occur as well as on the inter-arrival time of the next vehicle. If the subsequent vehicle arrives before all the strain recovery takes place, accumulation of strain takes place. Hence, how the strain recovery takes place during rest period gives a fair idea about the likely strain accumulation.

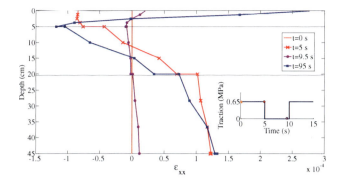

Fig. 3. ε_{xx} at different instants of time to the rear of the load for L-1

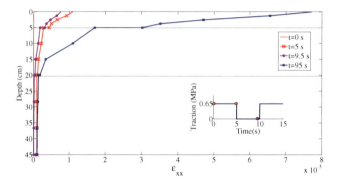

Fig. 4. ε_{xx} at different instants of time to the rear of the load for L-5

In the investigation, the strain variation with time was studied at four locations in the BC layer and is shown in Figure 5. It was observed that the strain accumulation at the top is much higher than the strain accumulation at the bottom of the BC layer. There is a tensile strain accumulation to the rear of the load and compressive strain accumulation to the front of the load. The application of horizontal traction significantly increased the rate of strain accumulation. The accumulation of the tensile strains at the top of the asphalt layer can eventually lead to initiation of top-down cracks.

6 Pavement Distress Model

In the mechanistic-empirical pavement design procedure, the mechanistically determined pavement response is used in pavement distress models to predict the performance of the pavement. The pavement performance predicted by these pavement distress models depends on the strain measures used and if the strain measures are not realistic enough the distress prediction may be too conservative or otherwise. The strain measures that are determined using the stress-strain analysis depend to a great extent on the material properties and the loading condition assumed. When the asphalt layers are assumed to exhibit viscoelastic response, the state of stresses and strains is markedly different from the case where the asphalt layers are assumed to exhibit elastic response. Similarly, when horizontal traction is considered in addition to vertical traction, the state of stresses and strains and the manner in which the strains are accumulated and recovered are completely different from the case where only vertical traction is assumed. It will be interesting to observe the pavement performance prediction by the currently used pavement distress models by using the increased strains obtained by considering horizontal traction. The strains obtained from the finite element analysis were used in typical top-down crack prediction models and the influence on the number of repetitions to failure was analyzed.

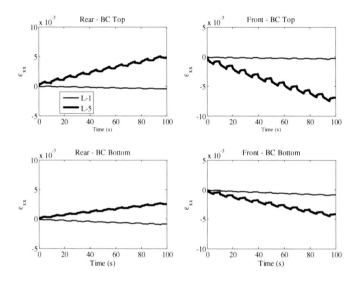

Fig. 5. Strain evolution with time

The pavement distress model to predict fatigue cracking suggested by the M-E PDG guideline [23] is of the following form:

$$N_f = Ck_1 \left(\frac{1}{\varepsilon_t}\right)^{k_2} \left(\frac{1}{E}\right)^{k_3}, \tag{1}$$

where, N_f = number of repetitions to fatigue cracking, ε_t = tensile strain at the critical location, E = Young's modulus of the material, k_1, k_2, k_3 = laboratory regression coefficients and C = laboratory to field adjustment factor. It is interesting to note that instead of using the tensile strains at the bottom of the asphalt layer, the tensile strains at the critical locations are used in M-E PDG [23]. These critical locations generally occur at the surface of the asphalt layer resulting in top-down cracking or at the bottom of the asphalt layer which can be related to bottom-up cracking. The model suggested by Asphalt Institute [24] is of the following form:

$$N_f = 0.00432C \left(\frac{1}{\varepsilon_t}\right)^{3.291} \left(\frac{1}{E}\right)^{0.854}, \qquad (2)$$

$$C = 10^M, \qquad (3)$$

$$M = 4.84 \left(\frac{V_b}{V_a + V_b} - 0.69\right), \qquad (4)$$

where V_b = effective binder content (%) and V_a = air voids (%). Assuming an effective bitumen content of 5% and air voids of 3%, one can rewrite eqn. 2 as,

$$N_f = 0.002093234 \left(\frac{1}{\varepsilon_t}\right)^{3.291} \left(\frac{1}{E}\right)^{0.854}. \qquad (5)$$

Using the Asphalt Institute model (Eqn. 5) and the maximum tensile strains at the top and bottom of the BC layer at time t=5 s, the allowable number of repetitions N_f are calculated and shown in Figure 6. The authors recognize the fact that the pavement distress model used here is calibrated for a particular temperature and not for the temperature of the asphalt layer assumed in this study (40 °C). Also, the pavement distress models are calibrated by assuming only vertical traction. Therefore the results shown in Figure 6 should be read in a qualitative manner. The difference between the numbers of repetitions predicted based on the strains at the top and bottom of the BC layer in some cases is of the order of 10^3. It can be seen that there is considerable decrease in the allowable number of repetitions when horizontal traction is considered. For L-1, the allowable number of repetitions based on the strains at the bottom of the BC layer is much lower when compared to the number of repetitions based on the strains at the top of the BC layer and hence governs the design. However, for L-2 to L-5, the allowable number of repetitions based on the strains at the top of the BC layer governs the

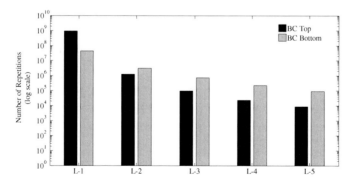

Fig. 6. Allowable number of repetitions as per the Asphalt Institute model

design. This finding highlights the importance of considering horizontal traction and the importance of considering the critical location of the strains while predicting pavement performance. The critical location for L-1 is at the bottom of the BC layer whereas for L-2 to L-5 is at the top of the BC layer.

7 Summary and Conclusions

When a vehicle moves on the surface of a pavement considerable amount of horizontal traction is induced on the pavement by the tire. This horizontal traction is not accounted for by current pavement design methods that are based on the multi-layered elastic theory framework. The multi-layered elastic theory predicts the critical tensile strains at the bottom of the asphalt layer when vertical traction is applied. When the viscoelastic nature of the asphalt layer is considered, tensile strains are produced at the surface of the asphalt layer, albeit, the strains at the top of the asphalt layer become tensile only after a number of load repetitions. In addition to considering the viscoelastic behavior of the asphalt layers, when both vertical and horizontal traction are applied, tensile strains can be seen at the surface of the asphalt layer right from the first cycle of load application. Moreover, the critical tensile strains occur at the top of the asphalt layer and not at the bottom of the asphalt layer. Distresses such as top-down cracking are due to crack initiation at the top of the asphalt layer and can be attributed to the high tensile strains at the top of the asphalt layer. Only when proper constitutive models that capture actual material response as well as realistic load conditions are used, can tensile strains be predicted at the top of the asphalt layer. Besides, the state of stresses and strains and the manner in which the strains accumulate are entirely different when horizontal traction is considered. Pavement distress predictions can be misleading if horizontal traction is not considered in the stress-strain analysis.

References

[1] Pellinen, T., Rowe, G., Biswas, K.: Evaluation of surface (top down) longitudinal wheel path cracking. Tech. Rep. 93. Joint Transportation Research Program. Purdue University, West Lafayette (2004)
[2] Kim, J., Roque, R., Byron, T.: Viscoelastic analysis of flexible pavements and its effects on top-down cracking. Journal of Materials in Civil Engineering 21(7), 324–332 (2009)
[3] Jacobs, M.M.J., Moraal, J.: The influence of tyre characteristics on the normal stresses in asphalt concrete pavements. In: Third International Symposium on Heavy Vehicle Weights and Dimensions, pp. 218–224. Thomas Telford Limited, Cambridge (1992)
[4] Hugo, F., Kennedy, T.: Surface cracking of asphalt mixtures in southern Africa. Proceedings of the Association of Asphalt Paving Technologists 54, 454–496 (1985)
[5] Matsuno, S., Nishizawa, T.: Mechanism of longitudinal surface cracking in asphalt pavement. In: Proceedings of the 7th International Conference on Asphalt Pavements (Nottingham), pp. 277–291 (1992)
[6] Collop, A., Cebon, D.: A theoretical analysis of fatigue cracking in flexible pavements. Journal of Mechanical Engineering Science 209, 345–361 (1995)

7. Groenendijk, J.: Accelerated testing and Surface Cracking of Asphaltic Concrete Pavements. Ph.D. thesis. Department of Civil Engineering, Delft University of Technology, The Netherlands (1998)
8. De Beer, M.: Measurement of tyre/pavement interface stresses under moving wheel loads. Heavy Vehicle Systems, Special Series, International Journal of Vehicle Design 3(1-4), 97–115 (1996)
9. Molenaar, A.: Fatigue and reflective cracking due to traffic (with discussion). Proceedings of the Association of Asphalt Paving Technologists 53, 440–474 (1984)
10. Gerritsen, A., van Gurp, C., van der Heide, J., Molenaar, A., Pronk, A.: Prediction and prevention of surface cracking in asphaltic pavements. In: Proceedings of the 6th International Conference on Asphalt Pavements. Ann Arbor, pp. 378–391 (1987)
11. Kunst, P.: Surface cracking in asphalt layers. CROW Record, 4th edn., The Netherlands (1990)
12. Myers, L.A., Roque, R., Ruth, B.E.: Mechanisms of surface-initiated longitudinal wheel path cracks in high-type bituminous pavements. Journal of the Association of Asphalt Paving Technologists 67, 401–432 (1998)
13. De Beer, M., Fisher, C., Kannemeyer, L.: Towards the application of Stress-In-Motion (SIM) results in pavement design and infrastructure protection. In: Eight (8th) International Symposium on Heavy Vehicles, Weights and Dimensions. Loads, Roads and the Information Highway, Gauteng, South Africa (2004)
14. Taramoeroa, N., de Pont, J.: Characterising pavement surface damage caused by tyre scuffing forces. Tech. Rep. 374. Land Transport New Zealand, Wellington, New Zealand (2008)
15. Perret, J.: The effect of loading conditions on pavement responses calculated using a linear-elastic model. In: 3rd International Symposium on 3D Finite Element for Pavement Analysis, Design and Research, Amsterdam, April 2-5 (2002)
16. Novak, M., Birgisson, B., Roque, R.: Three-dimensional finite element analysis of measured tire contact stresses and their effects on instability rutting of asphalt mixture pavements. Journal of the Transportation Research Board 1853, 150–156 (2003)
17. Pottinger, M.: The three-dimensional contact stress field of solid and pneumatic tires. Tire Science and Technology 20(1), 3–32 (1992)
18. IRC-37, Guidelines for the design of flexible pavements(second revision). Indian Roads Congress, New Delhi (2001)
19. Hibbitt, Karlsson, Sorensen, P.: ABAQUS/CAE user's manual version 6.6 (2003)
20. Alagappan, P.: Stress strain analysis of flexible pavements. Master of Technology thesis. Department of Civil Engineering. Indian Institute of Technology Madras, Chennai, India (May 2010)
21. Zaghloul, S.M., White, T.D.: Use of a three dimensional dynamic finite element program for analysis of flexible pavement. Transportation Research Record 1388, 60–69 (1993)
22. Huang, Y.H.: Pavement Analysis and Design. Pearson Education India (2008)
23. NCHRP. (Mechanistic-Empirical) Pavement Design Guide, Tech. Rep. 1-37A. Federal Highway Administration, Champaign, Illinois, USA (2004)
24. Asphalt Institute, Asphalt pavements for highways and streets, Manual series no.1(MS-1) (1981)

Predicting the Performance of the Induction Healing Porous Asphalt Test Section

Quantao Liu[1], Erik Schlangen[1], Martin F.C. van de Ven[2], Gerbert van Bochove[3], and Jo van Montfort[4]

[1] DelftUniversity of Technology, Faculty of Civil Engineering and Geosciences,
 Micromechanics Laboratory, Stevinweg 1, 2628 CN Delft, The Netherlands
[2] Delft University of Technology, Faculty of Civil Engineering and Geosciences,
 Road and Railway Engineering, Stevinweg 1, 2628 CN Delft, The Netherlands
[3] Heijmans-Breijn, Rosmalen, Netherlands
[4] SGS-Intron, Sittard, Netherlands

Abstract. The induction healing concept of porous asphalt was developed at Delft University of Technology and was proven very successful in the laboratory. A porous asphalt test section with this self healing concept was also paved on Dutch highway A58. This special porous asphalt contained 4% steel wool (by volume of bitumen). A number of cores were drilled from this test section to predict its performance. Beams were also prepared with the same materials as used in the test section. Experiments were done on these specimens to study the mechanical, heating and healing properties. It is found in Cantabro test that the particle loss resistance of porous asphalt concrete is improved by addition of steel wool. The improvement in particle loss resistance will delay ravelling on the pavement. It is also proven that the cores containing steel wool can be heated quickly with induction energy. Finally, it is found that the fatigue life of the beamsis extended greatly by applying induction heating during the rest period. The damage (cracking) in the porous asphalt beams ishealed by induction healing. Based on these findings, it is concluded that the life time of the test section will be extended by the reinforcement of steel wool and induction heating.

Keywords: porous asphalt, test section, steel wool, induction heating, self healing.

1 Introduction

Porous asphalt concrete is used very commonly as a surface material on Dutch motorways. At present, about 90% of the Dutch motorways are surfaced with porous asphalt wearing course. Noise reduction, driving comfort and driving safety are the main reasons for this extensive application of porous asphalt wearing course. However, the durability of porous asphalt wearing course has been a matter of concern, because the attractive features of porous asphalt do not last long due to clogging, stripping, and accelerated aging. Ravelling, which is the loss of aggregates from the road surface, is the main defect on porous asphalt surface wearing course [1, 2]. When ravelling occurs, the acoustical benefits and skid resistance of porous asphalt are diminished. Moreover, Ravelling requires early maintenance. It is report that ravelling, in about 76% of the cases, is the cause for maintenance or renewing of the top layer [1]. Duo to ravelling, the service life of porous asphalt is much shorter than that of dense graded asphalt road. To extend the lifetime of porous asphalt, ravelling should be prevented.

Asphalt concrete can repair the damage autonomously. Asphalt concrete has a potential to restore its stiffness and strength, when subjected to rest periods. This self healing capability of asphalt concrete has been shown both with laboratory tests and in the field since 1960s[2-6]. As a consequence of healing, asphalt concrete can restore its stiffness and strength, and extend its fatigue life when subjected rest periods. Besides, self healing of asphalt concrete is a temperature dependent phenomenon. The temperature sensitivity of the self healing rate of asphalt concrete is highly non-linear[7]. An increase in the temperature increases the healing rate and shortens the time needed to full healing. Grant implied that, healing is immediate above a certain temperature [8]. So, it is logical to enhance the self healing rate of asphalt concrete by increasing the temperature. As a result of healing, cracks can be closed and the durability of the pavement will be improved.

To enhance the self healing rate of porous asphalt concrete, an induction healing approach was developed in Microlab, Delft University of Technology. The idea is to make porous asphalt mixture electrically conductive by addition of steel wool. When micro cracks occur in the mortar of porous asphalt pavement, induction heating will be used to heat the pavement to activate the self healing capacity of asphalt concrete. The cracks will be closed because of the high temperature self healing of bitumen and the lifetime of the pavement will be extended. This approach, using induction heating to activate and enhance the self healing rate of porous asphalt concrete, is named induction healing.

In the previous laboratory experiments, it has been proven that steel wool can reinforce porous asphalt concrete by increasing its particle loss resistance, indirect tensile strength, and fatigue resistance [9, 10]. The reinforcement of steel wool will delay ravelling and cracking on the pavement. It was also found that the self healing potential of steel wool reinforced porous asphalt was significantly improved by induction heating [11, 12]. The enhanced healing potential will also prevent ravelling. Considering the reinforcement of steel wool and the enhanced induction healing rate, it can be expected that ravelling on the pavement will be delayed via induction heating.

To apply this induction healing approach, an induction healing porous asphalt test section was paved on Dutch motorway A58 near Vlissingen in December 2010. This special porous asphalt mixture contained standard porous asphalt PA 0/16 and 4% steel wool (by volume of bitumen). The test section survived the cold winter of 2010-2011 perfectly and no damage or even small cracks could be observed by the inspection of experts.

To predict the performance of this test section, a large number of cores (ø100 mm × 50 mm) were drilled from this test section. For comparison purpose, reference cores without steel wool were also drilled. Besides, beams (50 × 50 × 400 mm^3) were prepared by the contractor Breijn-Heijmans with the same materials as used in the test section. Experiments were done on these samples to study their mechanical, heating and healing properties.

2 Experiments

2.1 Particle Loss Test

For durability purpose, porous asphalt wearing course should have good particle loss (ravelling) resistance in itself. The particle loss resistances of the steel wool reinforced cores and reference cores were compared with Cantabro test. The test was done at 21.5 °C in a Los Angeles abrasion machine without steel balls according to the European Norm EN 12697- 17. Each specimen was initially weighed (W_1) and placed separately into a Los Angeles drum. Thereafter, each specimen was weighed again after 300 revolutions of the drum (W_2) in order to determine the weight loss during testing. To check reproducibility of the experiment, the tests were repeated five times for both mixtures. The test results were expressed as a percentage of weight loss in relation to the initial weight (W_2/W_1) This weight loss is an indication of the cohesive properties of the mix. Lower weight loss means better cohesion and better ravelling resistance.

2.2 Induction Heating Speed Measurement

To make induction healing work, the test section should be induction heated at an acceptable heating speed. To check if the cores from the test section can be heated with induction energy, the induction heating experiments were performed using a RF-generator big 50/100 (Hüttinger Electronic, German) at a frequency of 70 kHz. The distance between the sample and the coil of the induction generator was 10 mm. The cores were induction heated for 3 min. A full colour infrared camera was used to monitor the temperature variations of the sample during heating.

Fig. 1. The induction generator used in this research

2.3 Induction Healing Effect Detection

As there is no permanent deformation during four point bending fatigue test, beams instead of field cores were used to study the induction healing effect of porous asphalt concrete. Firstly, porous asphalt concrete beams were damaged with a strain amplitude of 300 microstain (8Hz) in four point bending test. The fatigue criterion was half reduction in the stiffness. Then the beams were induction heated to 70 °C/85 °C/100 °C and rested at 20 °C for 18 hours or directly rested at 20 °C/ 5 °C for 18 hours. Finally the extra fatigue lives of the healed beams were measured again with the same microstrain. After the test, the fatigue life extension ratio (the extra fatigue life divided by the original fatigue life) was used as a healing indication. This experiment was also conducted on aged beams (10 days ageing at 85 °C in the oven). This ageing method was equivalent to 5 years field ageing [13].

The possibility of multiple times induction heating was also examined to show that induction heating could be repeated when cracks return. A strain amplitude of 300 microstrain, at a frequency of 8 Hz, was applied on the beams for 50,000 cycles. Then, the samples were induction heated to 85°C and rested for 18 h or directly rested for 18 h for the first time. After that, another 50,000 cycles fatigue loading was applied on the beams, followed by the second time heating/resting or resting alone process. The damaging, heating/resting or resting alone and re-damaging process was repeated a few times. Finally, the beams werefatigued until the stiffness reduced to half of its initial value.

3 Results and Discussion

3.1 Particle Loss Resistance of the Cores

The particle loss values of the reference cores and steel wool reinforced cores are presented in Figure 2. The particle loss of reference cores scatters more than that of steel wool reinforced cores. The average particle loss value of five measurements is used as an indication for ravelling resistance. The average particle loss of steel wool reinforced cores is 23.6%, much lower than that of references cores 34.7%. So, steel wool improves the particle loss value of porous asphalt concrete. As a result, ravelling on the test section will be delayed.

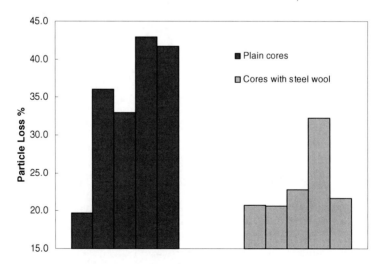

Fig. 2. Particle loss values of the reference and steel wool reinforced cores

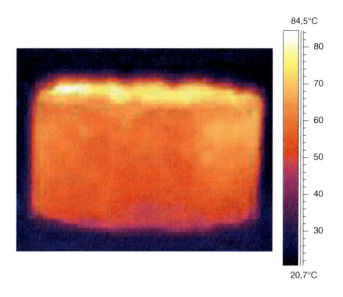

Fig. 3. Infrared image of the heated field core

3.2 Induction Heating Speed of Field Cores from the Test Section

To have a better understanding of induction heating, an infrared image of the heated core is shown in Figure 3. As ravelling is a surface defect on porous asphalt wearing course, the temperature at the surface of the pavement is important for ravelling. So, after the experiments, the mean heating speeds at the surface of the samples were calculated (the temperature increase divided by the heating time). Given the heating speed, the heating time needed to get a certain temperature can be calculated. The heating speeds of three steel wool reinforced cores were measured and the results are summarized in Table 1. The average value of the three measurements is 0.3442 °C/s. With the induction heating speed, it takes 190 s to heat the sample from 20°C to 85 °C with the temporary heating machine. The heating speed can be improved more by optimizing the induction coil, which is currently being investigated by the research partner SGS-Intron. In the latest study, the field cores can be heated at a great speed of 2.5°C /s with another heating device in the induction generator manufacture company in Germany.

Table 1. Induction heating speed of the steel wool reinforced cores

Heating speed 1	Heating speed 2	Heating speed 3	average heating speed
0.3144 °C/s	0.3383 °C/s	0.38 °C/s	0.3442 °C/s

3.3 Induction Healing Effect of the Beams

Figure 4 shows how fatigue life extension changes as a function of resting/heating temperature. As shown in Figure 4, the fatigue life extension ratio of the beams (healing) is very temperature dependent. The fatigue life extension ratio is quite low at a low resting temperature of 5 °C and increases with increase of the resting temperature. When the heating temperature increases from 70 °C to 85 °C, the healing is improved strongly. The increment of fatigue life extension ratio is a proof of induction healing. After that, further increase of the heating temperature to 100 °C results in a decrease of the fatigue life extension ratio. The reason for this decrease can be attributed to the geometry damage and the binder drainage problem caused by overheating. In this case, a swelling problem appears in the mortar of the sample, because the mortar cannot bear the excess expansion caused by temperature increase. Besides, the binder in the sample tends to drain down at such a high temperature, reducing the healing rate. Based on the results, it is concluded that 85 °C is the optimal heating temperature for best healing effect.

However, these healing effects are not complete. The first reason for this is that the temperature increment after induction heating is very limited, because the temperature decreases to the resting temperature of 20 °C very quickly in 3 hours. The temperature difference is only in the first 3 hours, so the healing increment is limited. Another reason for the limited increment of healing can be attributed to the temperature gradient in the sample. The sample is fully damaged over its height, but induction heating tends to only heal the damage in the top part of the beam, where the temperature is much higher than in the lower part after induction heating. The tendency of healing the surface damage is just what we need to prevent surface ravelling without damaging the stone structure in the lower part.

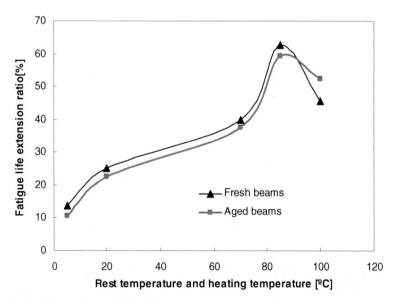

Fig. 4. Fatigue life extension of the beams against rest and heating temperature

It also can be seen from Figure 4 how ageing influences the fatigue life extension ratio of the beams. Ageing slightly decreases the fatigue life extension ratio by 3%. It means that ageing don't decrease the healing rate of porous asphalt concrete too much. As for the severe ravelling of aged porous asphalt wearing course, the cumulated damage during the service (ageing) process caused by traffic and environmental loading plays an important role. With a heating temperature of 100 °C, the healing rate of the ageing beam is higher than that of fresh beam. The reason for this can be attributed to ageing hardening behaviour of the binder. With ageing, the binder becomes harder and its softening point increases, reducing the binder drainage at 100 °C. As a result of less binder drainage, the healing of aged beam increases a bit.

The original fatigue curve, the natural healing (with rest periods) modified curve and the multiple times induction heating modified fatigue curve of the samples are compared in Figure 5 to show the fatigue life extension caused by multiple times induction heating. The original fatigue life of the sample is 95,700 cycles. With rest periods alone, the sample was fatigued after resting two times. The fatigue life of the sample with natural healing is 149,860 cycles, showing a fatigue life extension ration of 56.7%. With 4 times damage loading of 50,000 cycles followed by 4 times induction heating and resting, the modified fatigue life of the sample is 277,720, which is much longer than the original fatigue life and natural healing modified fatigue life. The fatigue life extension ratio in this case is 190%. It is definite that multiple times heating can greatly extend the fatigue life of porous asphalt concrete. It can be expected that the test section with the same materials as the beams will show a better healing effect with induction heating.

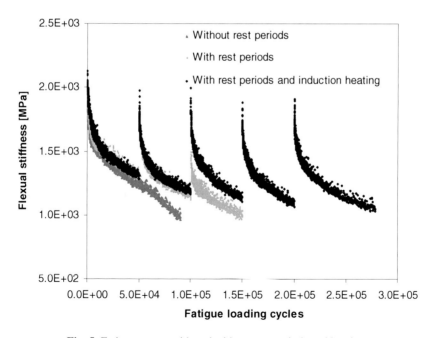

Fig. 5. Fatigue curves with and without rest periods and heating

4 Conclusions

Based on the results of the experiments, the following conclusions are drawn: 1) The Particle loss (ravelling) resistance of porous asphalt concrete is improved by addition of steel wool, which will delay ravelling on the test section. 2) Porous asphalt cores from the test section can be heated at a speed of 0.3442°C/s with the induction generator used at present. 3) The healing effect of porous asphalt beams is greatly enhanced by induction heating, so induction heating can be used to prevent ravelling on the test section. 4) The optimal heating temperature is 85 °C for best healing effect. 5) Ageing slightly decreases healing of porous asphalt concrete. 6)Multiple times heating can greatly extent the fatigue life of porous asphalt concrete. 7) It can be expected that the durability of the test section will be improved by the reinforcement of steel wool and induction heating.

Acknowledgement. The corresponding author would like to acknowledge the scholarship from China Scholarship Council. The work on the test section is financed by Rijkswaterstaat, Ministry of Transport, the Netherlands.

References

[1] Padmos, C.: Over ten years experience with porous road surfaces. In: ISAP Ninth International Conference on Asphalt Pavements, Copenhagen, Denmark (2002)
[2] Voskuilen, J.L.M., Huurman, M.: Conversations, Centre for transport and Navigation of the Dutch Ministry of Transport. Public Wprks and Water Management, Delft, The Netherlands (2009)
[3] Bazin, P., Saunier, J.: Deformability, fatigue, and healing properties of asphalt mixes. In: Proceedings of the Second International Conference on the Structural Design of Asphalt Pavements, pp. 553–569. Ann Arbor, Michigan (1967)
[4] Van Dijk, W., Moreaud, H., Quedeville, A., Uge, P.: The fatigue of bitumen and bituminous mixes. In: Proceedings of the 3rd International Conference on the Structure Design of Pavement, London, pp. 354–66 (1972)
[5] Raithby, K.D., Sterling, A.B.: The effect of rest periods on the fracture performance of a hot rolled asphalt under reversed axial loading. In: Proceeding of AAPT, vol. 39, pp. 134–147 (1970)
[6] Francken, L.: Fatigue performance of a bituminous road mix under realistic bestconditions. Transport Res. Rec. 712, 30–34 (1979)
[7] Kim, B., Roque, R.: Evaluation of Healing Property of Asphalt Mixture. In: Proceeding ofthe Annual Transportation Research Board Meeting CD-ROM (2006)
[8] Grant, T.P.: Determination of Asphalt Mixture Healing Rate Using the Superpave Indirect Tensile Test. Master thesis. University of Florida, America (2001)
[9] Liu, Q., Schlangen, E., van de Ven, M., García, A.: Induction heating of electrically conductive porous asphalt concrete. Constr. Build. Mate. 24(7), 1207–1213 (2010)
[10] Liu, Q., Schlangen, E., van de Ven, M., Poot, M.: Optimization of steel fiber used for induction heating in porous asphalt concrete. In: Mao, B., Tian, Z., Huang, H., Gao, Z. (eds.) Traffic and Transportation Studies, pp. 1320–1330. American Society of Civil Engineers, USA (2010)

11. Liu, Q., García, A., Schlangen, E., van de Ven, M.: Induction healing of asphalt mastic and porous asphalt concrete. Constr Build Mater 25(9), 3746–3752 (2011)
12. Liu, Q., Schlangen, E., van de Ven, M., van Bochove, G., van Montfort, J.: Evaluation of the induction healing effect of porous asphalt concrete through four point bending fatigue test. Constr. Build. Mate. 29(4), 403–409 (2010)
13. Liu, Q., Schlangen, E., van de Ven, M., Poot, M.: Performance prediction of the self healing porous asphalt test section. Report. Delft University of Technology, The Netherlands (2012)

Determining the Healing Potential of Asphalt Concrete Mixtures--A Pragmatic Approach

S. Erkens[1], D. van Vliet[2], A. van Dommelen[1], and G.A. Leegwater[2]

[1] Rijkswaterstaat, Dutch Ministery of Infrastructure and Environment
[2] TNO Built Environment

Abstract. Most design methods for pavements use a factor explaining the difference between pavement life predictions from design models and performance in the road [1]. Part of this correction factor is healing, the natural capacity of asphalt concrete to recover in rest periods, which generally are not present in laboratory fatigue tests in order to limit the test time. In the design method used for Dutch highways [2] a shift factor of 4 is traditionally used. This factor is based on in-practice behaviour of pavements with mixes using straight run bitumen of limited softness (mostly 40/60 and 70/100 pen bitumen). Currently many developments regarding polymer and chemical modification and the use of hard binders raise questions about the value of the shift factor for these mixtures as well as questions about how much of this factor is actually related to the healing potential of the mixes.

In the project described in this paper it is investigated whether the four point bending test used to assess the fatigue resistance in the European standard can be used in a discontinuous way to determine the healing capacity of a mix. In order to do this, a set of continuous and discontinuous tests was performed on two mixes that deviate only in the softness of their binder, 70/100 or 10/20. Although the approach appears promising in its simplicity and the consistency of the results, finding the correct interpretation remains difficult and may require linking the pragmatic with more fundamental understanding of the mechanism behind healing.

1 Introduction

The self healing capacity of bitumen and bituminous mixtures is a well known phenomenon, which is an integral part of most design methods [1]. In the design method used by Rijkswaterstaat in de the Netherlands [2], healing is part of a factor that describes the differences between laboratory results and practice. Originally, in the early eighties, this factor was derived by calibrating the design models to performance in reality, resulting in a factor 4. For a standard structure, this results in a reduction of pavement thickness of about 6,5 cm, which means that the pavement of a standard highway with two lanes and an emergency lane is over 100.000 euro's per kilometer cheaper than if the healing was not taken into account (healing factor equals 1).

The standard mixtures used at the time were all designed using straight run penetration binders with penetrations between 40 and 100 x0,1mm. Over the years, other binders came into use, as well as the use of reclaimed asphalt. The variety in mixtures makes a laboratory-practice calibration impractical and calls for a sound laboratory testing method to take into account the healing capacity of mixes in order to obtain a technically sound, but cost-effective pavement.

For this reason, a project was initiated to assess if the current European standard test for fatigue that is used in the Netherlands (four point bending test, EN 12697-24 – Annex D), could also be used to determine the healing potential of a mixture. The aim was to develop a pragmatic test that could be part of the updated CEN standards, allowing the healing capacity to be assessed in any road engineering laboratory with standard four point bending equipment.

It was hoped that by determining both fatigue and healing under similar conditions, the problem of assessing two parameters in design under different conditions could be circumvented. Because of the variety in results from healing research [3,4], and the fact that the shift factor between laboratory and practice contains more than just healing it is difficult to define which results to expect from the test. However, there is wide spread agreement that softer binders have a higher healing capacity than harder ones. Therefore, the tests were carried out on two mixes that differed only in the type of binder used: one 70/100 penetration and one 10/20 penetration binder and the test would be considered potentially useful if the 70/100 mixture exhibited significantly more healing than the 10/20 mixture.

1.1 Initial Approach

It may seem logical to compare specimens tested with and without rest periods to determine the healing capacity of an asphalt mixture. However, the large natural variation of fatigue life between specimens would require a large number of tests for a reliable result. Therefore, in this project it is attempted to determine healing potential using a single specimen.

For the analysis, the Partial Healing model [5] was chosen initially. This model was developed to describe the behaviour of asphalt concrete in four point bending tests and describes the evolution of the stiffness in this test. As the name states, the model contains a certain healing component. Originally Pronk expected this model to predict healing based on a continuous test only. The model has the advantage that, unlike analyses methods based on dissipated energy [6] you do not need to measure the full response. The maximum load, and deflection per cycle are sufficient. That information, along with the phase angle, is standard output from most 4PB set-ups and as such it is something that can be used in a pragmatic approach.

The idea to use the PH-model to predict healing was based on a preliminary discontinuous four point bending test. Figure 1 shows the data from that test that was done at the Delft University of Technology. As can be seen from the measured data, the behaviour before the rest period consists of a first phase with an initially fast stiffness decrease which gradually slows down to a more or less constant decrease (the linear part) in a second phase. This will eventually be followed by a third phase of progressive stiffness decrease, but we try to avoid this in the healing tests described here as we presume that healing predominantly occurs before this progressive deterioration sets in. After the rest period we see a marked increase in stiffness (recovery), followed by a decrease similar to that observed during the first loading period. The PH model can describe this effect, based on the data from the first loading part only. However, the model predicts that the stiffness after the rest period decreases in such a way that the second phase coincides with the linear part from the first loading step (the red line). As we can see from the data, the actual response does again reach a stage of linear decrease, but it doesnot fall on the red dotted line.

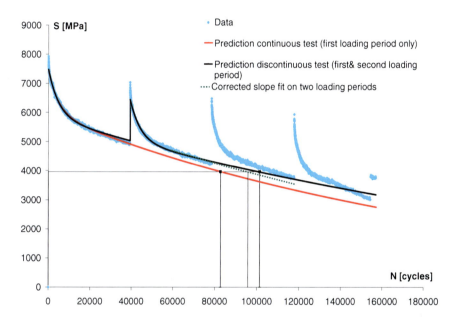

Fig. 1. Illustration of the determination of a healing factor on a single specimen

So, the PH model describes some stiffness recovery, but not all of it. The actual data stabilize at a higher stiffness level than the model predicts. This generated the idea that the PH model describes a visco-elastic phenomenon related to heating up and cooling down of the binder during the test and that the part of stiffness recovery that is not predicted, is the actual healing. This fits in with work done in France using 2PB tests, where it was also found that part of the recovery was a temporary stiffness recovery [7].

The "actual healing" we are looking for is the increase in fatigue life as the result of the rest period. Since a stiffness reduction of 50% is the end-of-life criterion in this test, this means that the number of load repetitions at which tests with and without rest reach 50% of the initial stiffness should be compared. However, running a test until this stiffness reduction is reached, leaving the specimen to recover and then repeat the process is quite time consuming. For that reason, it was decided to run the test for a fixed number of cycles both before and after the rest period and to use the the PH model to extrapolate the response. In this approach the model is , first fitted on the first loading period only (red line). In th second step it is fitted on both loading periods including the rest period (black line). From the comparison between these two fits at 50% of the initial stiffness a healing factor can be determined. When this is applied to Figure 1 we find: $N_1=82781$ and $N_2=101397$, which gives a healing of $H=N_2/N_1=1,2$. However, if we look at the fits and the data points, it can be seen that the healing value will be over-estimated because the second fit underestimates the slope of the linear response. This is due to the fact that the model expects only a temporary stiffness increase due to rest. Eventually, the stiffness is assumed to return to the original response, falling on the same straight line. Because of this, it fits a single linear line to describe the linear phase in both loading periods. Because in reality only part of the stiffness increase is lost again, fitting the same line to both e linear branches results in a line that is tilted upward to when compared to the actual data. The green dotted line shows the effect of using the same slope, while allowing for the increased stiffness level. This yields an N_2 of 93440 cycles and a healing of 1,1. This bias in describing the data, together with some other complications in applying the model fit lead tot the decision to adopt a more pragmatic approach. This approach where the healing is determined directly from the test data, without the use of models or data fitting, which is described in this paper.

2 Test Program

As mentioned in the introduction, an objective reference frame for the healing test results does not really exist. The current factor 4 in the Dutch design method is based on a correlation with practice that entails not only healing, but multiple factors. For this reason, the mixes used varied only in the type of binder used, with the expectation that the mixture with the softer binder would exhibit considerably better healing characteristics that the one with a hard binder.

2.1 *Mixtures Used*

The mixture used was an AC 16 base mixture (EN 13108-1), usually this mixture has less binder (4,3% by mass) when made with the softer bitumen and more (5,6%) when made with the hard one, but to ensure that in the tests the differences could be attributed solely to the type of binder, in this case an intermediate binder content of 4.5% was used. The mix information is given in Table 1. As described above, the intention was to use straight run bitumen in both mixtures. However, during specimen production there was a disruption at the Q8 plant which meant the harder binder was not available. It was replaced by a non-straight run binder from another producer. Unfortunately, this was not discovered until after the tests, when additional research into the binders was being prepared. Obviously, this has a significant impact on the project, to the extend where the assumption mentioned at the end of the introduction, that a 70/100 binder exhibits more healing capacity than a 10/20 mixture, may not necessarily hold true for these specific binders. This should be kept in mind when evaluating the healingfactors found.

Table 1. Sieve curve and mix composition of the mixes used

Sieve curve, % m/m through sieve		Mix composition	70/100 mix	10/20 mix
sieve	% m/m	Material type	% m/m	% m/m
C22,5	100.0	Bitumen 70/100	4.50	-
C16	97.6	Bitumen 10/20	-	4.50
C8	69.0	Wigras 40K filler	5.41	5.41
C2	43.0	Production dust	0.96	0.96
C125 μm	8.0	Putman natural sand	34.47	34.47
C63 μm	6.0	Scottish granite 2/8	23.36	23.36
		Scottish granite 8/16	31.30	31.30

2.2 Test Conditions

The test conditions in the discontinuous test are presented in Table 2. Temperature and frequency are equal to the standard conditions of the EN 12697-24 Annex D. The strain level used in these tests is for both mixes their ε_6 value (i.e. the strain level at which the mixture can withstand 1×10^6 load repetitions. This strain level was determined in continuous tests according to EN 12697-24 Annex D. The testing program used to determine ε_6 is shown in Table 3. The number of cycles in the loading period is determined on the basis of the continuous tests as well. It is chosen in such a way that is falls well within the linear branch, but remains sufficiently far away from the point of progressive damage. A rest period of 24 hours was selected as a workable rest period that, according to other researchers, sufficed to show healing [5, 8].

Table 2. Test conditions discontinuous tests

Pen mixture	10/20	70/100
Loading period	400,000 cycles	400,000 cycles
Rest period	24 h	24 h
Strain level	121 με	102 με
Temperature	20°C	20°C
Frequency	30 Hz	30 Hz

Table 3. Test conditions used for the determination of ε_6

Pen	# tests	strain level
70/100	3	100
70/100	3	150
10/20	1	105
10/20	2	15
10/20	3	140
10/20	2	150

3 Test Results

3.1 Continous Tests

The results of the continuous test are shown in Figure 2. From the figure it can be seen that one of the results for the 10/20 mixture deviates from the others to such an extent that it is considered to be an outlier. In the graph the point is shown with a circle around it. This point is not taken into account in the regression line fitted through the data. The ε_6 values for the two mixtures are 102 µε for the 70/100 mixture and 121 µε for the 10/20 mixture. The results are reported in detail in [9].

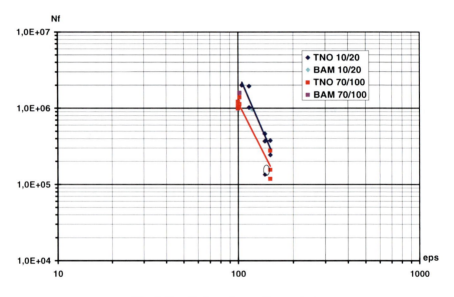

Fig. 2. Results from the continuous tests

3.2 Discontinuous Tests

In most cases the mixtures showed the expected response (Figure 3). The 70/100 mixture, however, occasionally exhibited tertiary response within the 400.000 cycles of the first loading period. Since this occurred at a strain level where this mixture, on average, can withstand 1×10^6 load repetitions, this clearly illustrates the large variability in four point bending test results. The complete series of test results is shown in [9].

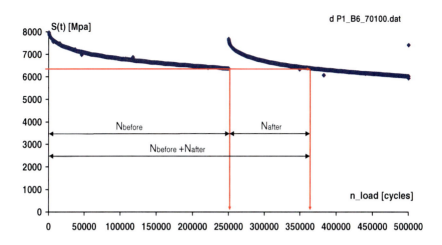

Fig. 3. Example of a typical discontinuous test (P1-B6-70100, 102 με)

4 Analysis

The approach that was used to determine the healing directly from the test data used the number of load repetitions needed for the stiffness in the second loading period to fall back to the stiffness at the end of the first loading period (Figure 3) as the increase in life time: H= (N_{before}+N_{after})/ N_{before}. This approach, that has also been used by other researchers [10], has the advantage that it allows for determining the healing capacity on a single specimen.

If the healing factor obtained using this approach is plotted as a function of the relative stiffness at the end of the first loading period, the graph in Figure 4 is obtained. The relative stiffness at any point can be seen as an indication of the amount of damage. That the healing capacity (in a fixed rest period) decreases when the damage is more extensive seems logical. The labels in this graph give the number of load repetitions until the rest period. Besides the original data points obtained with a loading period of 400.000 cycles there are additional points found using a different number of cycles or a given decrease in stiffness. These different conditions were used to get data throughout the healing-relative stiffness relation.

The consistency of the results in Figure 4 is striking considering the fact that the results are obtained in two different laboratories, at various damage levels (relative stiffnesses) and especially that in several tests the third phase behaviour had already set in at the start of the rest period. This consistency is attributed to the fact that the healing factor is determined upon a single specimen, excluding the effect of the huge scatter in fatigue life.

However this graph also raises the question at which relative stiffness S_{value}= S_{rest}/S_{ini} level the healing factor should be determined, and if this value should be different for different materials. Based on the approach described in this paper, three potential methods of comparison were considered.

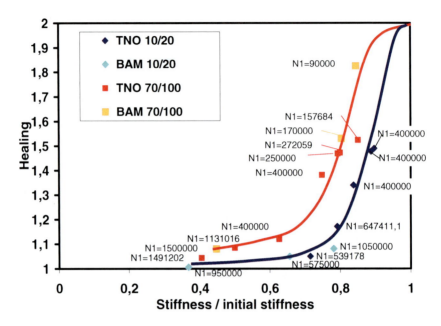

Fig. 4. Healing factors and their conceptual relation, using the pragmatic approach

First of all, compare the healing capacity at the end-of-life criterion in the fatigue test (50% S_{ini}), since that test is the basis for the approach described in this paper and in pavement design the healing capacity is used in combination with the fatigue life.

However, this doesnot take into account the healing capacity at both intermediate stiffness levels, or at levels below 50%. The fact that 50% stiffness reduction proved to be a useful end-of-life criterion in fatigue testing does not automatically mean it is also appropriate for the assessment of healing capacity. Integrating the healing factor over the relative stiffness range between zero and 1 makes for a more general assessment. However, this second approach does not account for the different rates of stiffness decrease in the mixtures. As can be seen from the graphs in Figure 9 and Figure 10, the stiffness of the 10/20 mixture drops less at the onset of the test en decreases more gradually in the linear part. It is only near the end of the curve that the stiffness starts to decrease rapidly (tertiary response). When we only look at the healing capacity as a function of the relative stiffness, we may be underestimating mixtures that retain their stiffness longer: a mixture that has 50% less healing capacity at a stiffness level of 80%, but that takes three times more load repetitions to reduce the stiffness to 80% can still be the better performing mix. To allow for the influence of the stiffness decrease, a healing factor that is weighted for the shape of the stiffness decrease was calculated as well. First, the average normalized stiffness decrease was determined for both mixtures (graphs in Figure 6 and Figure 7). The graph was divided in ten sections and for the average N/N_f ratio in each section (the red vertical lines and left column in the tables) the corresponding relative stiffness (S/S_{ini}) was determined (blue markers and centre column). The healing capacity at those relative stiff nesses was determined from Figure 4 (right hand column in the tables The overall healing is determined as the average value. This way, the healing is related to the life time through the rate of stiffness reduction. The values found using these three approaches are shown in Table 4.

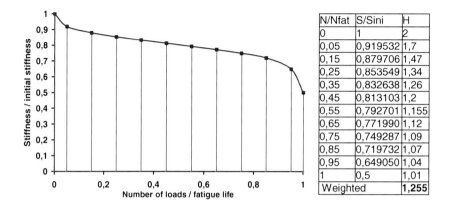

N/Nfat	S/Sini	H
0	1	2
0,05	0,919532	1,7
0,15	0,879706	1,47
0,25	0,853549	1,34
0,35	0,832638	1,26
0,45	0,813103	1,2
0,55	0,792701	1,155
0,65	0,771990	1,12
0,75	0,749287	1,09
0,85	0,719732	1,07
0,95	0,649050	1,04
1	0,5	1,01
Weighted		1,255

Fig. 5. Weighting of healing factor over stiffness decrease for 10/20 mix

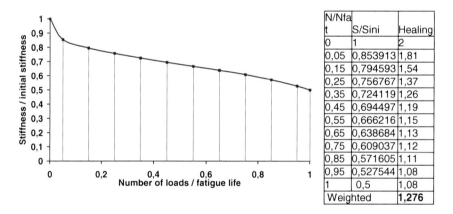

N/Nfat	S/Sini	Healing
0	1	2
0,05	0,853913	1,81
0,15	0,794593	1,54
0,25	0,756767	1,37
0,35	0,724119	1,26
0,45	0,694497	1,19
0,55	0,666216	1,15
0,65	0,638684	1,13
0,75	0,609037	1,12
0,85	0,571605	1,11
0,95	0,527544	1,08
1	0,5	1,08
Weighted		1,276

Fig. 6. Weighting of healing factor over stiffness decrease for 70/100 mix

Table 4. Values for healing using the three approaches described

	10/20	70/100	70/100:10/20
healing @S/Sini=50%	1,01	1,08	1,07
healing integrated over S/Sini	0,13	0,25	1,86
average healing over life time weighted by stiffness	1,24	1,28	1,03

5 Conclusions and Further Research

The project described in this paper aimed at a pragmatic evaluation method of the healing capacity, not on understanding of the mechanism of healing itself as other researchers aim to do [6,11,12]. This approach is driven by the fact that healing is an important factor in pavement performance, yet it can not be effectively assessed at the moment. The pragmatic approach described seems promising in its simplicity and consistency and proved a wealth of information. However, it seems difficult to interpret that information without more fundamental understanding of the mechanism. Of the tree approaches used to determine a healing factor based, two show hardly any difference between the two mixtures, while shows considerable difference. It is tempting to decide that in that case the approach that does distinguish between the mixtures must be the best one, because that is what we expected to find.

Especially since, due to an error in specimen production the differences between the mixtures were not as explicit as was intended. As a result, at the moment we cannot be sure which assessment of the healing capacity is the best one, nor whether the binders used should or should not be expected to perform differently in healing. Currently the binders that were used are being tested in more detail. Futhermore it is recommended to try and relate these pragmatic tests to more fundamental studies into healing in order to better understand the response. Getting results from specimens with the originally intended 10/20 binder and analyse those along with the data presented in this paper should also provide a clear indication of the usefulness of the approach. Those specimens should exhibit less healing than both mixtures described in this paper.

References

[1] Rijkswaterstaat, Ontwerp Specificaties Asfaltverhardingen, Rijkswaterstaat, Nederland (2011) (in Dutch)
[2] COST333, European Communities, Transport Research, Cost 333: Develoment of New Bituminous Pavement Design Method, ISBN 92-828-6796-X
[3] Westera, G.E.: Onderzoek naar het Healingsproces van asfalt beton., TWAO-F rapport (1989) (in Dutch)
[4] Westera, G.E., Bouman, S.R.: Studie over healing ten behoeve van het project ASFLT/TWAO; nadere beschouwing van het fenomeen healing bij asfalt KOAC-WMD, rapport 94.0002 (1994) (in Dutch)
[5] Pronk, A.C., Cocurullo, A.: Investigation of the PH model as a prediction tool in Fatigue bending tests with Rest Periods. In: Al-Qadi, Scarpas (eds.) Advanced Testing and Characterisation of Bituminous Materials, Loizos, Partl, Taylor & Francis Group, London (2009) ISBN 978-0-415-55854-9
[6] Shen, S., Carpenter, S.H.: Dissipated Energy Concepts for HMA Performance: Fatigue and Healing. Dept. of Civil and Environmental Engineering, Univ. of Illinois at Urbana-Champaign, Advanced Transportation Research and Engineering Laboratory, ATREL (2007)

[7] Breysse, D., de la Roche, C., Domee, V., Chauvin, J.J.: Influence of Rest tim on Recovery and Damage during Fatigue Tests on Bituminous Composites. In: Proceedings 6th RILEM Symposium PTEBM 2003, Zurich (2003)
[8] Bazin, P., Saunnier, J.B.: deformability, Fatigue and Healing Properties of Apshalt Mixes. In: Second Int. Conf. On the Structural Design of Asphalt Pavements, Ann Arbor, Michigan, pp. 553–569.
[9] Dommelen, A.E., van Erkens, S.M.J.G., van Vliet, D., Leegwater, G.: Healing van asfalt mengsels, onderzoek naar een pragmatische proefmethode. InfraQuest report IQ-W-2011-1 (2011) (in Dutch, concept)
[10] Daniel, J.S.: Rate Dependent Stiffnesses of Asphalt Concrete used for Field to Laboratory prediction and Fatiue and Healing Evaluation, MSc-thesis. North Carolina State University, Raleigh, NC (1996)
[11] Little, D.N., Lytton, R.L., et al.: An Analysis of the mechanism of Micro damage healing based on the of micro mechanics first principles of fracture and healing. AAPT 68, 501–542 (1999)
[12] Kringos, N., Pauli, T., Scarpas, A., Robertson, R.: A Thermodynamical Approach to Healing in Bitumen. In: 7th RILEM Conference on Advanced Testing and Characterization of Bituminous Materials, Rhodes, Greece (May 2009)

Asphalt Durability and Self-healing Modelling with Discrete Particles Approach

V. Magnanimo[*], H.L. ter Huerne, and S. Luding

Tire-Road Consortium, CTW, University of Twente, Netherlands
v.magnanimo@utwente.nl

Abstract. Asphalt is an important road paving material, where besides an acceptable price, durability, surface conditions (like roughening and evenness), age-, weather- and traffic-induced failures and degradation are relevant aspects. In the professional road engineering branch empirical models are used to describe the mechanical behaviour of the material and to address large-scale problems for road distress phenomena like rutting, ravelling, cracking and roughness. The mesoscopic granular nature of asphalt and the mechanics of the bitumen between the particles are only partly involved in this kind of approach. The discrete particle method is a modern tool that allows for arbitrary (self-)organization of the asphalt meso-structure and for rearrangements due to compaction/cyclic loading. This is of utmost importance for asphalt during the construction phase and the usage period, in forecasting the relevant distress phenomena and understand their origin on the grain-, contact-, or molecular scales. Contact models that involve visco-elasticity, plasticity, friction and roughness are state-of-the art in fields like particle technology and can now be modified for asphalt and validated experimentally on small samples. The ultimate goal is then to derive micro- and meso-based constitutive models that can be applied to modellingbehaviour of asphalt pavements on the larger scales. Using the new contact models, damage and crack formation in asphalt and their propagation can be modeled. Furthermore, the possibility to trigger self-healing in the material can be investigated from a micromechanical point of view.

1 Introduction

Asphalt mixtures are composite materials that consist of solid particles, viscous binder/fluid (bitumen) and pores filled with air. The fluid in the mixture (that can be hot or cold, i.e. less or more viscous) lubricates the contact surfaces between the particles and can make movement of the particles easier [1]. The multiphase material has properties that depend on those of the original components. The physical properties of the skeleton (e.g. shape, surface texture, size distribution, moduli), but also the properties of the binder (e.g. grade, relaxation characteristics, cohesion) and

---

[*] Corresponding author.

binder–aggregate interactions (e.g. adhesion, absorption, physiochemical interactions) characterize the material behavior[2].When looking at asphalt, it makes sense to distinguish between three different length scales, i.e. micro-, meso- and macro-scale. The interaction between the mortar (composition of bitumen and the smallest particles) and a single large stone is defined as the micro-scale. The interaction of multiple stones of various sizes and the mortar is defined as the meso-scale. On the macro-scale, the behavior of the whole road is accounted for. Kinematics at different scales apparently governs the behavior of the material: to gain thorough knowledge of asphalt pavement behavior, one has to focus on the three length scales. As common practice in the professional asphalt branch, fundamental constitutive models able to describe the micro- and meso-mechanical behavior are hardly used. Large-scale problems are addressed by using empirical models [1] for road distress phenomena like rutting, raveling or cracking. Our ambitious goal is to bridge the gap between discrete and continuous, macro-concepts. The material behavior at the grain-scale can be combined with the granular structure in order to identify the contact law for the asphalt components and relate such kinematics with the macroscopic response at the larger scale of the road. The major goal of the project, wherever we publish here in this paper, is finding a micro-based model with enough predictive quality on the macro-level in order to do forecasting of self-healing and durability capabilities.

On the particle-scale, the interaction of the mortar with the grains and between the grains can be efficiently investigated using a Discrete Element Method (DEM) [3]. Discrete element methods simulate particulate systems by modelling the translational and rotational degrees of freedom of each particle using Newton's laws, and the forces are calculated associating proper contact models with each particle contact.

In the last twenty years, attempts for a micromechanical modelling of asphalt have been done by other researchers: a contact law to reproduce particles connected by a binder and describe the elastic behavior of an assembly of bonded particles was proposed in Ref. [4]; a micromechanical description of rutting with intergranular and aggregate-binder interactions was given in Ref. [5]; 2D modelling based on image processing are described in [2].DEM studies on cemented particulate materials include the work by Rothenburg et al. [5], Chang and Meegoda[6], Sadd and Dai [7], Buttlar and You [8] and Ullidtz[9]. Nevertheless, a well establishedmultiscale description for the constitutive behavior of particle-bitumen systems is still missing. Particularly, very limited work has been done on 3D micromechanical modelling of the mixture, with proper visco-elasto-plastic, temperature-dependent contact/interaction models. Moreover, to our knowledge, no systematic numerical study on the microscopic processes that govern cracks (and eventually self-healing) in asphalt has been done. The novelty of our project is the application of a contact model where visco-elasto-plastic, temperature-dependent properties are taken into account to reproduce fracture and subsequent healing of asphalt at particle-level. The phenomena will be not "imposed" at macroscopic level, but fracture and healing will be modeled in large-scale processes only by mimicking proper micromechanical properties. The detection and analysis of damage as well as microscopic self-healing mechanisms have been achieved by particle based simulations already[10][11]. So that these results now can be applied to asphalt and translated to a practicable continuum model with predictive quality[12].

In this preliminary study an approach to modelling of cohesive, sintering and self-healing particulate materials are described [13]. The mechanical evolution in time of all the particles in the material is simulated by a DEM sintering contact model[10]. This contact model mimics the physical behavior of the interaction of the particles accounting for dissipative, elasto-plastic, adhesive, and frictional effects and the pressure dependency[10]. An additional temperature-dependence, as possibly resulting from diffusion of atoms and the resulting sintering effects, can be added on top of the elasto-plastic contact model [11]. In this very first work, only the approach that we want to follow is going to be described. In particular, we want to show how powerful the methodology is in reproducing the general behavior of asphalt, already in his basic formulation (visco-elasto-plastic contact model applied to an aggregate of dry spherical particles with no interstitial fluid). The major strength of this approach involves the easy re-structuring of the grains/particles relative to each other – as relevant during fracture, damage, shear-localization or creep, but often neglected in other approaches.

Initially, some particle-samples are prepared and (isotropically) compacted in order to resemble asphalt mixtures. The spherical particles deform plastically at contact and stick to each other forming a solid sample with (after releasing pressure) zero confining stress – on which uni-axial tension or compression is further applied. The numerical results are compared with laboratory uni-axial tests from [14] on samples of dense asphalt concrete. The comparison is purely qualitative, since in this first work no attempt to calibrate the DEM model with real experiments on asphalt has been made. The paper is supposed to be a reference report to describe the details of the methodology used to describe asphalt solid samples. Interestingly, the simple cohesive model is able to properly reproduce the shape (peak and subsequent softening) in the experimental data, using poly-disperse spherical particles without any background matrix. The behavior of the material can be easily tuned, changing the ratio between compressive and cohesive properties at contact level.In the final part of the work, self-healing of damaged samples is modeled through re-sintering, a process that globally increases further the contact adhesion between particles. The details of microscopic (physical and chemical) mechanisms are not taken in account here, but just the relation between the sudden change of strength at contact level and the increase in the global resistance of the sample[11][15][16]. If the re-sintering is applied after the sample has already experienced some damage under mechanical loading, the question is to which extent the damage can be healed.

2 Simulation Method

The Discrete Element Method (DEM) [3] for particle systems can be used to illustrate how the macroscopic response of a solid-like, sintered sample, resembling an asphalt mixture, depends on various micro- and meso-scopic properties such as the particle-particle contact network, the particle size, and the contact adhesion between particles (simulating the interaction of the bitumen with the particles), their contact friction and stiffness. In the present work, after solving the equations of motion at the particle level, the coupling between micro- and macro scale properties is performed at local level. It means that the response of the particle system (expressed in terms of macroscopic stress and strain) is obtained by averaging [17] the local quantities at the particle contact level (interparticle contact forces and displacements) over the assembly. The effective response then depends directly on the chosen particle contact model [10][18][19]. Even though recent studies have demonstrated that the accurate simulation of systems composed of non-spherical particles is possible[20][21], for simplicity we restrict ourselves here to spherical particles.

2.1 Sintering Contact Model

In the following, particulate material samples are (i) *prepared*, (ii) *deformed* and *damaged*, and then (iii) *self-healed* and (iv) *deformed* again. The non-linear model by Luding *et al.*[10][22][23] is used – see these references for more details, to describe the adhesive particle-particle interaction in the asphalt mixture. In Fig. 1, the normal contact force f (that is directed parallel to the line connecting the centers of two contacting particles) is plotted against the contact overlap (resembling the deformation between particles at the contact), $\delta > 0$. If $\delta < 0$, there is no contact between particles, and thus $f = 0$. This sign convention relates positive (*negative*) values of the contact displacement δ to overlap/deformation (*separation*), while positive (*negative*) values of the contact force f relate to compression/repulsion (*tension/attraction*).

A contact begins at $\delta = 0$ and, during initial compressive loading, the contact force increases with the overlap as $f=k_1\delta$, with k_1 the elasto-plastic contact stiffness. When the external compressive forces are compensated by the contact repulsive force at the maximum contact overlap, δ_{max}, for unloading, the contact stiffness increases to a value k_2, so that the elastic unloading force is $f = k_2(\delta-\delta_f)$. Elastic unloading to zero contact force leads to the (plastic) contact overlap $\delta_f = (k_2-k_1)\delta_{max}/k_2$. If the overlap is further decreased, the contact force gets tensile, with maximum tensile contact force $f_{t,max} = -k_t\delta_{t,max}$, realized at contact displacement $\delta_{t,max}$.

For the sake of brevity, the tensile softening parameter k_t hereafter is referred to as the "contact adhesion". Note that, mostly for practical reasons [10], for contact deformations above δ_{max}, the force follows the limit branch $f = k_2(\delta-\delta_f)$, since further loading is unrealistic anyway and would lead to much stiffer behavior if properly modeled. The extreme loading and unloading limit branches are reflected by the outer triangle in Fig. 1. Starting from the realized maximal overlap,

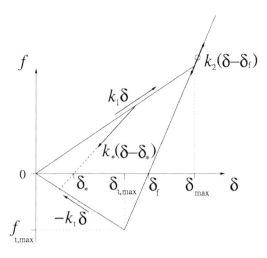

Fig. 1. Particle contact model plotted as force-displacement relation [11,22]

$\delta^*_{max} < \delta_{max}$, unloading occurs within the outer triangle, as characterized by a branch with stiffness $k_* = k_1 + (k_2 - k_1)\,\delta^*_{max}/\delta_{max}$, and (elastic, reversible) force, $f^* = k_*(\delta - \delta^*)$.

The intermediate stiffness k_* follows from a linear interpolation between k_1 and k_2 – which is our (arbitrary) choice due to the lack of experimental data on this (probably) non-linear behavior. In summary, the model has three ("stiffness") k-parameters that describe the three relevant physical effects at the contact: (1) elasticity, (2) plastic deformations, and (3) contact-adhesion. Furthermore, the model involves (4) a non-linear contact stiffness via the choice of k_*. This piece-wise linear model is a compromise between simplicity and the need to model physical effects. Except for some early theoretical studies, see [19] and the many works that are based on it, there is no experimental/numerical literature available to our knowledge that provides enough detailed information on the force-displacement relations, involving all four physical contact properties above and their nonlinear, history-dependent behavior. If this information becomes available, the present model can be extended and generalized.

The *tangential contact force* acts parallel to the particle contact plane and is related to the tangential contact displacement through a linear elastic contact law, with the tangential stiffness k_s. The tangential contact displacement depends on both the translations and rotations of the contacting particles. Coulomb friction determines the maximum value of the tangential contact force: During sliding the ratio between the tangential contact force and the normal contact force is assumed to be limited and equal to a (constant) dynamic friction coefficient μ_d; during sticking, the tangential force is limited by the product of normal force and static coefficient of friction μ_s.

Table 1. Material parameters from Ref. [10], rescaled in the right column

Property	Symbol	Values [14]	SI units
Time Unit	t_u	1	1 ms
Length Unit	x_u	1	1 m
Mass Unit	m_u	1	1 kg
Average radius	$\overline{R}$	$0.5 \cdot 10^{-3}$	0.5 mm
Material density	ρ	2000	2000 kg/m³
Elastic stiffness	$k = k_2$	$5 \cdot 10^4$	$5 \cdot 10^{10}$ kg/s²
Plastic stiffness	k_1/k_2	0.5	
Adhesion "stiffness"	k_c/k_2	[0, ..., 20]	
Friction stiffness	k_s/k_2	0.2	
Plasticity range	ϕ_f	0.05	
Coulomb friction	$\mu = \mu_d = \mu_s$	1	
Normal viscosity	$\gamma = \gamma_n$	$5 \cdot 10^3$	$5 \cdot 10^6$ kg/s
Tangential viscosity	γ_t/γ	0.2	
Background viscosity	γ_b/γ	4	
Background torque	γ_{br}/γ	1	

3 Results

The simulations reported here consist of four subsequent stages, namely (i) a sample preparation (isotropic compression), (ii) a uni-axial (tensile/compressive) loading, (iii) a self-healing, and (iv) the continuation of the uni-axial loading.

Six plane, perpendicular outer walls form a cuboidal volume, with side lengths of $L = 11.5$ mm. The samples are composed of about 10^3 poly-disperse spherical particles (see Fig. 2 as an example), with particle radii drawn from a Gaussian distribution around mean $\overline{R} = 0.5$ mm[10][24]. The particle density used in the simulations is $\rho=2000$ kg/m³, the maximum elastic contact stiffness is $k_2=5 \cdot 10^{10}$ N/m. The initial elasto-plastic stiffness (normalized by k_2) is $k_1/k_2=1/2$, and the

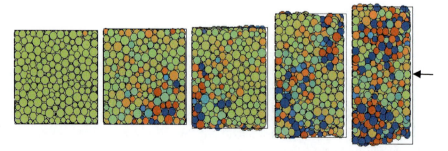

Fig. 2. Snapshots from a compression test with $k_1/k_2=0.5$. The circles are the particles with the greyscale coding the dimensionless average stress.

contact adhesion k_1/k_2 is varied. The other stiffness parameters, friction coefficients, and viscous damping parameters are summarized in Table I, most of them being dimensionless, like all quantities discussed in the rest of the paper. As final remark, we note that the choice of parameters is empirical – most of them kept fixed here, only adhesion is varied systematically.

3.1 Isotropic Loading

In this section the sample preparation by *pressure sintering*[22] is reported. During sintering, the particles deform plastically at contact and stick to each other due to strong, non-linearly increased van der Waals forces. At the same time, the sample shrinks, i.e. becomes denser. Such pressure-sintering results in a solid sample with bonded particles, similar to asphalt mixtures. Moreover, the sintering model can be temperature-dependent, resembling the effect of the temperature in the asphalt preparation with bitumen [11].

The process is characterized by two stages: the first stage reflects the application of a hydrostatic (or isotropic) pressure, $\sigma_s/\sigma_0 = 4.10^{-2}$, to a loose assembly of particles, with the reference stress $\sigma_0 = k_1/(2\overline{R})$. This desired isotropic stress is slowly applied to the six outer walls. In this stage both friction and adhesion between the particles are switched off. The hydrostatic loading process is considered to be finished when the kinetic energy of the sample is negligible compared to the potential energy. For our sample, the solid volume fraction (volume of the solid particles over total volume) at the end of the hydrostatic loading process is $v = 0.676$, which relates to a porosity (volume of fluid and air phases with respect to total volume, in the specific case of asphalt) of $1-v = 0.324$.

The second stage of the preparation (pressure sintering) process is reflected by a *stress relaxation* phase, where the external hydrostatic pressure is strongly reduced, while the adhesion between particles is now made different then zero. Due to the presence of particle contact adhesion, the lateral stability of the specimen remains preserved when the hydrostatic pressure is released, i.e., a coherent and stable particulate structure is obtained that can be subsequently used in the analysis of damage and healing under uni-axial loading conditions. The solid volume fraction of the sample after stress relaxation is decreased to $v = 0.63$. By means of the two subsequent steps, first the density of the granular/asphalt sample is carefully tuned (compression with zero friction and adhesion) and only later the stability of the aggregate is totally entrusted to the contact adhesion.

3.2 Response under Uni-axial Compression and Tension

Both during the uni-axial compression and tension test, one of the two outer walls, with its normal parallel to the axial (loading) direction, is slowly moved towards (away from) the opposite wall (see Fig. 2). The change of the wall displacement in time is prescribed by a cosine function with rather large period in order to limit inertia effects.

The response of the sample under uni-axial compression and uni-axial tension is shown in Fig. 3. The normal axial stress σ, normalized by the reference stress σ_0, is plotted as a function of the normal axial strain ε, where positive stress (*strain*) values relate to compression (*contraction*). The stress-strain curves are depicted for different values of k_t (normalized by k_2), which quantifies the adhesion at the particle contacts, see Fig. 1. A first interesting conclusion is detected by this numerical experiment: a larger particle contact adhesion increases the effective strength of the sample, both under uni-axial tension and compression. Furthermore, the overall strain at which the effective stress reaches its maximum increases with increasing contact adhesion, k_t.

Note that the maximum stress under uni-axial compression is order of five times larger than under uni-axial tension. A relatively high compressive strength in relation to the tensile strength is typical of various sintered materials, such as ceramics [25], and appears even in asphalt mixtures[14][26]. As further result, from Fig. 3, the softening branch under uni-axial tension is somewhat steeper than under uni-axial compression. Here we use a rate that is close to the quasi-static regime, as studied in more detail in [24]. The initial loading branch is linear up to large stress and the initial (elastic) axial stiffnessesσ_t in tension and compression are determined by the sample preparation procedure, and are approximately equal for all the cases considered here, i.e., σ_t/σ_0 =1.04. The tensile responses are all characterized by local failure at the center of the sample.

3.3 Comparison between Numerical Simulations and Physical Experiments

We refer to experiments in [14] to qualitatively compare our numerical investigation with laboratory tests on asphalt mixtures. The authors carry on uni-axial unconfined compression tests on compacted samples of dense asphalt concrete (DAC 0/5). The tests are performed in the displacement control mode, at constant axial deformation rates (refer to [14] for details). Mixtures such as DAC are continuously graded (as particles in our numerical sample) and derive their stability from the particles arrangement and the cohesion provided by the bitumen.

We report in Fig. 4 the stress-strain behavior for the aggregate matrix compressed at different strain rates. The comparison of Figs. 3 and 4 (axial stress versus axial strain on the right side of Fig. 4) shows that the elasto-plastic adhesive contact model is able to capture qualitatively the basic features in the behavior of the asphalt mixture. In fact both the peak and the softening behavior reported in the experimental data are perfectly reproduced in the numerical simulation. No direct comparison is made, as the dimensionless stress in Fig. 3 can notquantavely be compared with the stress in Fig. 4. Nevertheless, the analysis of the numerical simulation show that the stress level strongly depends on the particle adhesion k_t/k_2 and this quantity can be easily tuned in order to model different asphalt mixtures.

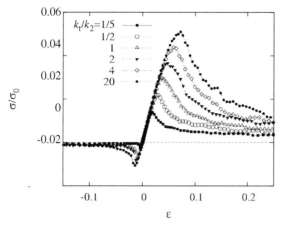

Fig. 3. Dimensionless axial stress versus axial strain during uni-axial compression, positive, and tension, negative, (after [13])

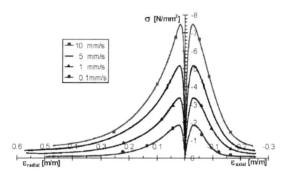

Fig. 4. Axial stress σ versus axial strain ε_{axial} and radial strain ε_{radial} for uni-axial compression tests on DAC samples at different rates and T=30°C (after [14])

It is worthwhile to point out here that, in this preliminary work no attempt to calibrate the DEM model with real experiments on asphalt has been made yet. The paper is aimed to show the details of the methodology used to model asphalt solid samples. Results are encouraging, since the contact-model is able to mimic the general stress-strain trend, already in his basic formulation (visco-elasto-plastic contact model applied to an aggregate of spherical particles with no interstitial fluid). A detailed comparison between numerical simulations and proper quasistaticuni-axial compression experiments is now in progress.

3.4 Self-healing under Uni-axial Compression and Tension

In this section we show a possible modelling of induced self-healing mechanisms in asphalt [12]. During uni-axial compression, the sample is stopped at various strains, see Fig. 5, and the self-healing is achieved by an instantaneous increase of the particle contact adhesion k_t, which is assumed to be the net-effect of a re-sintering cycle, like warming up an asphalt mixture. The details of the physio-chemical interactions driving the healing process are not highlighted here, but we want to focus on the effect that re-sintering/healing of the contacts can have on the macro stiffness of the aggregate. Technically, on the contact level, an increase of the contact adhesion k_t corresponds to an increase of the maximum tensile strength $f_{t,max}$, and a decrease of the corresponding displacement $\delta_{t,max}$, i.e. rupture occurs at large tensile strain. After the application of the re-sintering cycle (self-healing), uni-axial loading is resumed, where the effect of self-healing on the effective stress-strain response of the sample becomes apparent through a comparison of its response with that of both, the unhealed reference sample and a pre-emptively healed sample, which has the stronger contact adhesion from the beginning on.

Under uni-axial compression, see Fig. 5(a), or uni-axial tension, see Fig. 5(b), the self-healing of the initial sample with $k_t/k_2 = 1/5$ is activated by instantaneously increasing the contact adhesion to $k_t/k_2 = 1$ or 20, uniformly at all particle contacts. Fig. 5 shows the response curves after the initiation of self-healing (dashed lines, labeled with the abbreviation 'SH'), together with the stress-strain responses of the relatively weak ($k_t/k_2 = 1/5$, solid squares), strong ($k_t/k_2 = 1$, triangles) and very strong ($k_t/k_2 = 20$, solid circles) samples. The maximum compressive strength reached after self-healing is larger for healing at smaller deformation – and thus smaller damage. A further conclusion can be drawn here: astonishingly, for all self-healing cases considered, the response eventually converges to the response of the "strong" sample with $k_t/k_2 = 1$ or 20. The strong sample stress-response thus can be interpreted as the response of a pre-emptively "self-healed" sample, where the increase in contact adhesion is initiated at the onset of mechanical loading already. The response of the sample with $k_t/k_2 = 1$ or 20 acts as envelope for the responses of the self-healed samples with k_t/k_2 increased from 1/5 to 1 or 20, respectively.

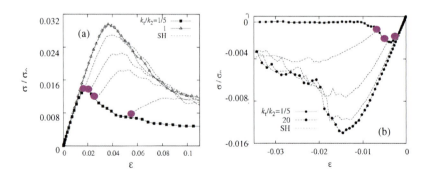

Fig. 5. Axial stress versus axial strain during (a) uni-axial compression and (b) uni-axial tension, for particle contact adhesions $k_t/k_2=1/5$ (solid squares). The self-healing stress responses are given (from the different strain points) by the dashed lines. The outer envelope corresponds to self-healing at zero strain (after [13]).

4 Conclusions

In this work, (1) isotropic preparation, (2) uni-axial deformation, and (3) self-healing processes in damaged adhesive granular assemblies have been studied using DEM simulations with the final goal of application to asphalt mixtures.

After isotropic compression, both uni-axial compression and tension was applied to the sample for determining the different (simulated) contact adhesion strengths. It appears from our analysis that, while the initial stiffness (slope of the stress-strain curves) is not much affected, the peak strength of the material is: the stronger the contact adhesion, the larger the peak strength. For compression, the strength appears about a factor of five larger than for tension and the softening branch is rather smooth for compression, while the tensile regime shows a sharper drop, resembling more brittle-like behavior. Moreover, at different strains, the uni-axial deformation is stopped and global self-healing is applied. The effect of self-healing is mimicked by a (global) sintering process, as modeled by increasing the particle contact adhesion from relatively "weak" to rather "strong". The stress-strain response obtained from such self-healed samples eventually converges to the envelope curve that represents the damage response of a sample that has the "strong" contact adhesion since the onset of loading. Another result is that the maximum sample strength reached after self-healing very much depends on the deformation level at which self-healing is activated.

This preliminary work shows how powerful DEM simulations can be in describing the constitutive behavior of asphalt mixtures. In fact, discrete simulations give insights on the material microstructure and link observable (macroscopic) phenomena with the kinematics of the interacting components. Very useful information can be extracted from such microscopic analysis, e.g. on how and when act to induce self-healing in a damaged material.

References

[1] ter Huerne, H.L., van Maarseveen, M.F.A.M., Molenaar, A.A.A., van de Ven, M.F.C.: Int. J. Pav. Eng. 9(3), 153 (2008)
[2] You, Z., Dai, Q.: Can. J. Civ. Eng. 34, 239 (2007)
[3] Cundall, P.A., Strack, O.D.L.: Geotechnique 29, 47 (1979)
[4] Zhong, X., Chang, C.: ASCE J. Eng. Mech. 125(11), 1280 (1999)
[5] Rothenburg, L., Bogobowicz, A., Haas, R.: In: Proceedings of the 7th International Conference on Asphalt Pavements, vol. I, pp. 230–245 (1992)
[6] Chang, G.K., Meegoda, N.J.: In: Proceedings of the 2nd International Conference on Discrete Element Method, pp. 437–448. MIT (1993)
[7] Sadd, M.H., Dai, Q.: Mech. Materials 37, 641 (2005)
[8] Buttlar, W.G., You, Z.: TRR 1757, 111 (2001)
[9] Ullidtz, P.: In: Proceedings of the 80th TRB Meeting, Washington, DC (2001)
[10] Luding, S.: Granul. Matter 10, 235 (2008)
[11] Luding, S., Manetsberger, K., Muellers, J.: J. Mech. Phys. Solids 53(2), 455 (2005)
[12] Van der Zwaag, S.: Self Healing Materials. In: Van der Zwaag (ed.) An Alternative Approach to 20 centuries of Materials Science, pp. 1–18. Springer, Dordrecht (2007)

[13] Luding, S.: Comp. Methods in Mater. Science 11(1), 53 (2011)
[14] Erkens, S.M.J.G.: Asphalt Concrete Response (ACRe) – determination, modelling and prediction, Ph.D. thesis – Delft University of Technology (2002)
[15] Takeuchi, T., Kondoh, I., Tamari, N., Balakrishnan, N., Nomura, K., Kageyama, H., Takeda, Y.: J. Electrochem. Soc. 149, A455 (2002)
[16] Shoales, G., German, R.M.: Metall. Mat. Trans. A 29a, 1257 (1998)
[17] Luding, S.: Int. J. Solids Struct. 41, 5821 (2004)
[18] Thornton, C., Antony, S.J.: Powder Technol. 109, 179 (2000)
[19] Tomas, J.: Particul. Sci. Technol. 19, 95 (2001)
[20] Alonso-Marroquin, F., Luding, S., Herrmann, H.J., Vardoulakis, I.: Phys. Rev. E 71, 051304 (2005)
[21] D'Addetta, G., Kun, F., Ramm, E.: Granul. Matter 4(2), 77 (2002)
[22] Luding, S., Suiker, A.: Philos. Mag. 88(28-29), 3445 (2008)
[23] Luding, S., Bauer, E.: Geomechanics and Geotechnics: From Micro to Macro. In: Jiang, M., Fang, L., Bolton, M. (eds.) IS-Shanghai Conference Proceedings, pp. 495–499. CRC Press/Balkema, NL (2010)
[24] Herbst, O., Luding, S.: Int. J. of Fracture 154, 87 (2008)
[25] Lee, W.E., Rainforth, W.M.: Ceramic Microstructures. In: Property Control by Processing, pp. 46–47. Chapman and Hall, London (1995)
[26] Muraya, P.M.: Permanent deformation of asphalt mixes, Ph.D. thesis – Delft University of Technology (2007)

Quantifying Healing Based on Viscoelastic Continuum Damage Theory in Fine Aggregate Asphalt Specimen

Sundeep Palvadi[1], Amit Bhasin[1], Arash Motamed[1], and Dallas N. Little[2]

[1] Department of Civil, Architectural and Environmental Engineering
University of Texas
Austin, TX 78712, (512) 471-3667
a-bhasin@mail.utexas.edu
[2] Zachry Department of Civil Engineering,
Texas A&M University
College Station, TX 77843, (979) 845-9847

Abstract. The ability of an asphalt mix to heal is an important property that influences the overall fatigue performance of the mix in the field. In this study, an experimental and analytical method based on viscoelastic continuum damage theory was developed to characterize the healing in an asphalt composite (fine aggregate matrix) as a function of the level of damage prior to the rest period and the duration of the rest period. Four different types of fine aggregate matrix (FAM) were tested to quantify overall healing at isothermal conditions. Two different verification tests were conducted to demonstrate that the percentage healing measured using the proposed method are independent of the sequence of loading or rest period. Results from the tests support the hypothesis that the healing characteristics determined using the proposed test method can be treated as a characteristic material property.

1 Introduction

Fatigue damage in asphalt mixtures is defined as the growth or accumulation of cracks under the action of repetitive loading. Extensive research during the past two decades has demonstrated the significance of self-healing while characterizing the fatigue cracking resistance of asphalt pavements [1-3]. Kim et al. [1] defined self-healing or micro-damage healing phenomena as partial or complete reversal of micro-damage or micro-crack growth induced due to fatigue loads. Micro damage healing in an asphalt mix is a function of the constituent binder's chemical and physical make up, level of damage prior to the rest period, duration of the rest period, mixture properties such as gradation, binder content, air voids and other external factors such as temperature.

Healing is most commonly quantified in terms of the percentage gain in the number of load cycles to fatigue failure or gain in the modulus as a function of rest period [4-7]. This approach is useful and provides a direct method to compare the relative ability of different asphalt mixes to heal. However, a limitation to this approach is that the percentage healing reported is a function of the rate of loading, type of loading (monotonic/cyclic) and mode of loading (controlled strain/controlled stress), rendering it useful only in the context of the specific test conditions. One approach to overcome this limitation is to use the work potential or viscoelastic continuum damage theory (VECD). VECD theory relates the reduction in stiffness under fatigue loads to an internal state variable, S, that represents the overall damage within the specimen. This internal state variable is related to the loading and stiffness by a damage evolution law. The closed form relation between pseudo stiffness (C) and a damage parameter (S) was demonstrated to be independent of loading characteristics and unique for a particular mix [8-10]. In this study, a method was developed to use this relationship to quantify healing as a function of rest periods and damage levels prior to the rest periods.

This paper presents the findings from a study conducted to (i) evaluate the change in mechanical properties of fine aggregate matrix (FAM) specimens during a rest period and (ii) employ the VECD theory to quantify the percentage of healing as a function of the damage level immediately before the rest period as well as the duration of the rest period.

2 Preliminary Investigation on Healing in FAM Specimens

2.1 Materials

Four different FAM mixes manufactured using one type of aggregate passing ASTM sieve no 16, two different types of asphalt binders (PG 67-22 and 64-16) and two different binder contents (10% and 12%) were used in this study. All tests were conducted on cylindrical FAM specimens that were 20 mm in diameter and 50 mm in height. These specimens were cored from one large 100-mm diameter and 75-mm high specimen compacted using the Superpave gyratory compactor. The cylindrical FAM test specimens were glued to end plates and used with the DSR for testing. All tests in this study were conducted at a temperature of $25°C$.

2.2 Test Method and Preliminary Results

The self-healing characteristics of the FAM specimens were quantified based on the following method and metrics. A time sweep test following a sinusoidal waveform in shear was conducted with a frequency of 5Hz and constant stress amplitude of 210kPa. The test was continued until the measured complex shear modulus was 50% of the linear viscoelastic complex shear modulus measured at low stress amplitudes. At this stage, the test was stopped for a duration of 30 minutes. During this rest period, the linear viscoelastic complex shear modulus was measured by applying a shear stress following a sinusoidal wave form at a frequency of 5Hz and a stress amplitude of 10kPa for 50 cycles. The linear viscoelastic shear modulus was measured at 0.5, 1, 3, 5, 10 and 30 minutes after the start of the rest period.

After the completion of the 30-minute rest period, the complex shear modulus of the specimen had significantly increased. At this time, the application of the high amplitude shear stress following a sinusoidal wave form with a frequency of 5Hz and stress amplitude of 210kPa was resumed. The high stress amplitude fatigue test was continued until the specimen again reached 50% of its linear viscoelastic shear stress amplitude. A second rest period of 30 minutes was introduced and the change in complex shear modulus was recorded as before. The process was repeated for a total of four rest periods with the only exception that the fourth rest period was for 60 minutes instead of 30 minutes. Figure 1 illustrates the typical increase in complex modulus as a function of time during each of the four rest periods.

Examination of results from these preliminary tests revealed two important findings. First, the increase in complex modulus during the rest period (introduced at the same level of damage) did not change significantly from the first rest period to the fourth. Second, the damage evolution curves (complex modulus versus number of cycles) following the 30-minute rest periods were similar to each other but different from the original intact specimen. In other words, although the complex modulus of the specimen increases during the rest period, the improvement in fatigue cracking resistance was not commensurate with the gain in stiffness. Based on these findings, further studies were conducted to quantify healing in terms of the state of damage in the specimen instead of the specimen stiffness.

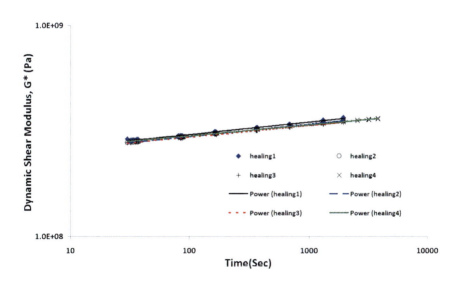

Fig. 1. Typical healing vs. time curves from four rest periods

3 Analytical and Experimental Method to Quantify Healing Based on State of Damage

3.1 Definitions and Hypothesis

The work potential theory developed by Schapery [11], also referred to as the viscoelastic continuum damage theory (VECD) [8-10], was used to characterize fatigue cracking in asphalt concrete mixtures. According to the work potential theory, damage is defined as a process that results in an increment in the value of the internal state variable S. Studies [8-10] demonstrate that the relationship between the "damage parameter" S based on power law of crack growth, and the pseudo stiffness C is unique for a given mix design and temperature. Details pertaining to the determination of the $C(S)$ function from fatigue tests can be found in other literature [8-10, 12]. Healing during a rest period, which is in essence the reversal of micro-damage, produces a net reduction in the value of the state variable S. Therefore, in this study healing was quantified in terms of the relative percentage reduction in the damage parameter S as:

$$\% \text{ Healing}(C, t) \equiv \frac{(S_f - S_i)}{(S_i)} * 100 \quad (1)$$

Where, C represents the pseudo stiffness immediately before the rest period, t denotes the duration of the rest period, and $S_i(C)$ and $S_f(C, t)$ correspond to the internal state variable representing the state of material immediately before and after the introduction of a rest period of time t, respectively.

3.2 Materials and Test Procedure

This study was conducted using four different FAM specimens as described in the previous section. A test protocol was developed to determine $S_i(C)$ and $S_f(C, t)$ for FAM specimens. Each FAM specimen was first subjected to a creep load of 1 kPa for one minute followed by a recovery period of 15 minutes. The creep-recovery was followed by application of cyclic loads following a sinusoidal waveform with stress amplitude of 1kPa at 10Hz. These tests were conducted to obtain the linear viscoelastic properties of the FAM specimen. The specimen was then subjected to cyclic torsion following a sinusoidal waveform with a stress amplitude of 210 kPa to induce fatigue damage. In order to obtain the healing characteristics, a rest period was introduced when the specimen reached a predefined fraction of its initial stiffness. Following the rest period, the specimen was subjected to cyclic torsion until it reached the next predefined fraction of its initial stiffness and the process was repeated. The predefined levels of stiffness used in this study were $0.8C_1$, $0.7C_1$, and $0.6C_1$ where, C_1 is the initial undamaged pseudo stiffness of the specimen. This procedure was applied to four different specimens; each specimen being subjected to a different duration of the rest period. The rest periods applied in this study were 5, 10, 20 and 40 minutes in duration. Figure 2 illustrates a schematic of the load sequence for a specimen with a 5-minute rest period that was introduced at different fractions of its initial stiffness. At least two replicate sets consisting of four specimens each were tested following the above procedure.

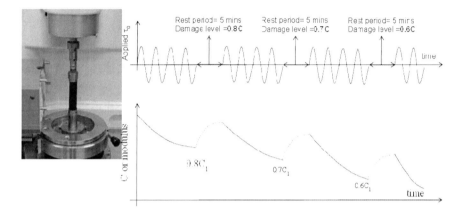

Fig. 2. Test protocol used to characterize healing as a function of pseudo stiffness

3.3 Analysis

A simplified version of the VECD analysis was used to determine the pseudo stiffness C of the specimen C_N at the Nth cycle [13]:

$$C = \frac{G^*}{G^*_{lve}} \quad (3)$$

Where, G^* is the measured complex modulus at N^{th} cycle and G^*_{lve} is the linear viscoelastic G^* measured using the low stress amplitude. The reduction in pseudo stiffness with increasing number of load cycles was then used to determine the increase in S, the internal state variable for damage, following the procedure described by Lee and Kim [8-10]:

$$S \cong \sum_{i=1}^{n} [0.5 * (\epsilon^R{}_i)^2 * (C_{i-1} - C_i)]^{\frac{\alpha}{1+\alpha}} * (t_i - t_{i-1})^{\frac{1}{1+\alpha}} \quad (4)$$

Where, α is related to the material's linear viscoelastic creep exponent m (determined from creep-recovery) and is suggested to take the value of $\left(1 + \frac{1}{m}\right)$. Thus equation (4) was used to determine the characteristic damage evolution curve or $C(S)$ function for each FAM mixture.

Based on equation (1), the two parameters that are required to characterize healing are $S_i(C)$ and $S_f(C,t)$. $S_i(C)$ is the value of S calculated just prior to the provision of a rest period and can be readily obtained from the $C(S)$ function.

However $S_f(C,t)$ is not the value of S corresponding to the pseudo-stiffness measured immediately after the rest period for the following reason.

The results from this study indicate that the evolution of damage in the specimen immediately following the rest period was very different as compared to the original specimen (Curve 2, Region-I in Figure 3). However, as the loading is continued the damage evolution in the specimen following the rest period eventually becomes similar to the damage evolution in the intact specimen (Region-II in Figure 3). This is typically after the pseudo stiffness of the post-healed specimen reduces to less than the pseudo stiffness prior to the rest period. Therefore the stiffness immediately following the rest period (starting point of Curve 2) is not a good representation of the level of damage reversal in the matrix and is not recommended to determine $S_f(C,t)$. Consequently, a direct measurement of the pseudo stiffness, C_f, and corresponding $S_f(C,t)$ will over predict healing. In order to avoid this over prediction, the pseudo stiffness used to determine $S_f(C,t)$ from the $C(S)$ function was not the measured pseudo stiffness immediately after healing C_f but a *reduced pseudo stiffness, C_f'*. This reduced pseudo stiffness represents the equivalent effect of both partially healed and fully healed interfaces within the matrix after the rest period. The procedure used to obtain the reduced pseudo stiffness was as follows.

Curves 1 and 2 in the schematic figure 3 (denoted as C^3 and C^2) represent measured damage evolution in an intact specimen and damage evolution in the partially healed specimen, respectively. Curve 3 (denoted as C^3) was developed using data from Region II such that it has the same functional form as Curve 1 (which is known apriori) and extrapolated backwards to predict the reduced pseudo stiffness C_f' and determine $S_f(c,t)$ using the $C(S)$ relationship. Finally, $\%Healing\ (C,t)$ was computed using equation (2).

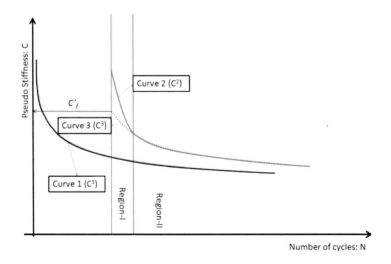

Fig. 3. Actual and modified C vs. N curve represented by Curves 2 and 3

4 Results

Using the experimental and analytical procedure listed in previous section, the percentage healing was calculated for the four different FAM mixes. The percentage healing is determined as a function of the pseudo stiffness at which the rest period was introduced as well as the duration of the rest period. A function of the form presented in equation 5, was used to describe the percentage healing as function of rest period for each damage level at which the rest period was introduced.

$$\% \, Healed = m_1 * (1 - \exp(-m_2 * t)) \tag{5}$$

Figures 4 and 5 show the typical final results for the percentage healing as a function of the rest period and pseudo stiffness (or level of damage) immediately preceding the rest period. The results from both the replicates are shown as points and the results from the model as the solid line. These figures also illustrate verification points. These verification points correspond to healing measured on specimens subjected to a randomized sequence of rest periods or a mode of loading different from the one used to obtain the characteristics healing functions. The verification points demonstrate that the characteristic healing function is unique and not dependent on the mode of loading used to induce fatigue damage to the specimen or sequence and duration of rest periods.

The results clearly indicate that, as expected, longer rest periods introduced at a similar level of damage (or pseudo stiffness) translate into a higher healing, a trend which can also be seen in binder healing test results reported earlier by Bhasin et al. [14]. Also, higher percentage of healing is achieved when the rest periods are introduced at lower levels of damage (or higher relative pseudo stiffness) in the specimen.

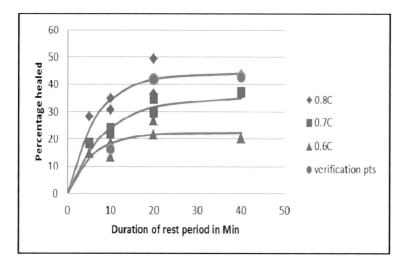

Fig. 4. Percentage healing in FAM mixes containing 10% PG 67-22 binder

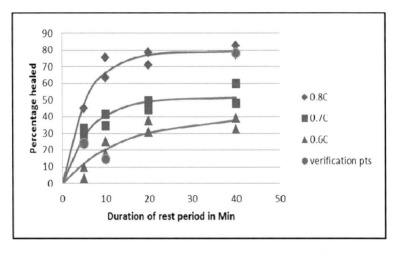

Fig. 5. Percentage healing in FAM mixes containing 12% PG 64-16 binder

5 Conclusions

This paper presented an experimental and analytical method to quantify healing in FAM mixes as a function of the duration of rest period and the level of damage preceding the rest period. The analytical method builds on the work potential theory and the viscoelastic continuum damage approach that was developed by Kim and co-workers [8-10] to characterize fatigue cracking and healing in asphalt mixtures. The method was applied to four different FAM mixes. Verification tests were conducted on two FAM mixes and the results demonstrated that healing characteristics obtained using the proposed method were independent of the test method or mode of loading used to induce damage in the specimen.

Acknowledgements. The authors would like to acknowledge the Federal Highway Administration and Asphalt Research Consortium for the support provided for this study and Prof. Richard Kim from North Carolina State University, USA, for his feedback.

References

[1] Kim, Y.R., Lee, H.J., Little, D.N.: Fatigue Characterization of Asphalt Concrete Using Viscoelaticity and Continuum Damage Theory. Journal of the Association of Asphalt Paving Technologist 66, 520–569 (1997)
[2] Qiu, J., Van De Ven, M.F.C., Wu, S.P., Yu, J.Y., Molenaar, A.A.A.: Investigating self-healing behaviour of pure bitumen using Dynamic Shear Rheometer. Fuel 90, 2710–2720 (2011)
[3] Pronk, A.C.: Partial Healing, A New Approach for the Damage Process During Fatigue Testing of Asphalt Specimen. In: Symposium on Mechanics of Flexible Pavements, Baton Rouge, LA (2006) (ASCE Conference Proceedings)

[4] Bazin, P., Saunier, J.B.: Deformability,Fatigue and Healing Properties of Asphalt Mixes. In: Proc., Second International Conference on the Structural Design of Asphalt Pavement, Ann Arbor, Michigan, pp. 553–569 (1967)
[5] Raithby, K.D., Sterling, A.B.: The effect of rest periods on the fatigue performance of a hot-rolled asphalt under repeated loading. Journal of the Association of Asphalt Paving Technologists 39, 134–152 (1970)
[6] Tayebali, A.A., Deacon, J.A., Coplantz, J.S., Harvey, J.T., Monismith, C.L.: Mix and Mode-of Loading Effects on Fatigue Response of Asphalt-Aggregate Mixes. Proceedings, The Association of Asphalt Paving Technologists 63, 118–151 (1994)
[7] Maillard, S., de La Roche, C., Hammoum, F., Gaillet, L., Such, C.: Experimental Investigation of Fracture and Healing at Pseudo-Contact of Two Aggregates. In: 3rd Euroasphalt and Eurobitume Congress, Vienna (2004)
[8] Park, S.W., Richard Kim, Y., Schapery, R.A.: A viscoelastic continuum damage model and its application to uniaxial behavior of asphalt concrete. Journal of Mechanics of Materials 4, 241–255 (1996)
[9] Lee, H.J., Kim, Y.R.: Viscoelastic Continuum Damage Model of Apshalt Concrete with Healing. Journal of Engineering Mechanics 124, 1224–1232 (1998)
[10] Daniel, J.S., Kim, Y.R.: Development of a Simplified Fatigue Test and Analysis Procedure Using a Viscoelastic, Continuum Damage Model. Journal of the Association of Asphalt Paving Technologist 71 (2002)
[11] Schapery, R.A.: A Theory of Crack Initiation and Growth in Viscoelastic Media. Theoretical Development. International Journal of Fracture 11, 141–159 (1975)
[12] Palvadi, N.S.: Measurement of material properties related to self-healing based on continuum and micromechanics approach. University of Texas, Austin (2011)
[13] Kutay, E.M., Gibson, N., Youtcheff, J.: Conventional and Viscoelastic Continuum Damage (VECD) -Based Fatigue Analysis of Polymer Modified Asphalt Pavements. Journal of the Association of Asphalt Paving Technologist 77, 395–434 (2007)
[14] Bhasin, A., Palvadi, N.S., Little, D.N.: Influence of aging and temperature on intrinsic healing of asphalt binders. Transportation Research Record (2011) (in press)

Evaluation of WMA Healing Properties Using Atomic Force Microscopy

Munir Nazzal[1], Savas Kaya[2], and Lana Abu-Qtaish[3]

[1] Assistant Professor, Civil Engineering Department, Ohio University, Athens
[2] Associate Professor, School of Electrical Engineering and Computer Science, Ohio University, Athens
[3] Graduate Research Assistant, Civil Engineering Department, Ohio University, Athens

Abstract. Warm Mix Asphalt (WMA) technology has received considerable attention in past few years due to its benefits in reducing energy consumption and pollutant emissions during production and placement of asphalt mixtures, widening the paving season, and increasing the pace of the construction process. However, many concerns and questions are still unanswered regarding its long-term performance. One of those questions is the effect of using WMA technology on the healing characteristics of asphalt materials, which has significant impact on their performance.

The fundamental understanding and evaluation of the healing characteristics of WMA requires careful consideration of the micro-mechanisms that influence the adhesive bonds between the asphalt binder and the aggregate, and the cohesive bonds within the asphalt binder. However, all standard laboratory tests that have been used to evaluate the WMA examine their integral, macro-scale behavior only. Therefore, those tests are limited in their ability to validate the healing mechanisms in an asphalt system, as they cannot examine and deconvolve factors contributing to its response at the micro-scale.

In this study, an Atomic Force Microscopy (AFM) based approach was developed and used to evaluate the effect of two types of WMA additives on the healing characteristics of an asphalt binder. The considered WMA additives included Sasobit and Advera. The results of this paper indicated that the use of WMA additives enhances the adhesive intrinsic healing characteristics of the selected asphalt binder. However, the two considered WMA additives showed adverse effects on the cohesive intrinsic healing behavior of that binder. Finally, the Sasobit was found to decrease the rate of crack closure in an asphalt binder.

1 Background

The rising energy costs and the increased awareness of environmental impacts of asphalt mixtures have resulted in an interest in using a new type of asphalt mixtures called Warm Mix Asphalt (WMA). WMA is a generic term for an asphalt mixture placed at lower than conventional temperatures. The use of WMA technologies was developed in Europe with the aim of reducing greenhouse gases produced by manufacturing industries (1). While heat is used to reduce asphalt viscosity and to dry aggregate during mixing of conventional asphalt mixtures, WMA reduces asphalt viscosity by including water or special organic or chemical additives in the mixture. This reduction improves the asphalt mixture workability allowing for its compaction at lower temperatures.

Several studies have been conducted in the last decade to characterize the properties of different types of WMA mixtures (2-4). The results of these studies showed that the emissions were significantly reduced during the production and placement of asphalt mixtures when WMA technologies are used as compared to the conventional Hot Mix Asphalt (HMA). In addition, those studies also demonstrated that WMA additives were able to improve the compactability of asphalt mixtures. Improved compaction was noted at temperatures as low as 190°F (87.8°C) (3). Despite these advantages, results of laboratory tests conducted on WMA mixtures have raised concerns and questions regarding their performance and durability. However, data obtained from WMA field test sections do not support the laboratory test results (3). Some data also suggests that the WMA performance improves with time and may ultimately be equivalent to that of HMA.

The healing characteristics of WMA affect their performance, and might be partially responsible for the differences observed between the laboratory and the field test results. However, these characteristics have not been studied to date. Healing, in this context, can be briefly defined as the process by which the crack growth in asphalt binders or mixtures, which occurs due to repeated loading, is partially or completely reversed.

In recent years, several studies have been conducted on the laboratory investigation of healing behavior of asphalt materials (5-9). The results of these studies clearly indicated the evidence of self-healing of those materials and its significant impact on their performance, particularly fatigue-cracking life. Healing behavior of asphalt binders was evaluated using dynamic shear rheometer (DSR). Although those studies contributed significantly to advancing the understanding the healing behavior of asphalt materials, they have used macro tests that are limited in their ability to validate the healing mechanisms in an asphalt system, as they cannot examine and deconvolve factors contributing to its response at the micro-scale.

The healing phenomena in asphalt materials can be described as a combination of wetting and intrinsic healing processes that occur across a crack interface (9). Wetting is the mechanism in which cracked surfaces come into contact with each other. It depends on the mechanical properties (including viscoelastic properties) and work of cohesion for the asphalt binder. In addition, the intrinsic healing is the strength gained by a wetted crack interface. The intrinsic healing is dictated by the adhesive forces at the asphalt-aggregate interface (adhesive healing) and cohesive forces at asphalt-asphalt interface (cohesive healing). Consequently, the fundamental understanding of the healing behavior of asphalt materials, in general, and WMA in particular requires accurate quantification of the nano and micro-scale properties and mechanisms that influence its response. The use of nanotechnology techniques can be a viable approach for achieving that. Nanotechnology techniques have been used to study biomaterials and geomaterials during the past decade, allowing for the development of disruptive technologies that has started to affect our daily life. In particular, recent research studies were able to develop models that upscale the nanoscale properties of geomaterials to predict with high accuracy the macroscale strength and stiffness of those materials.

One of the nanotechnology techniques that has received increasing attention for examining the behavior of different materials is the Atomic Force Microscopy (AFM) (Figure 1). AFM is a flexible high-resolution scanning probe microscopy technique, which uses a laser-tracked cantilever with a sharp underside tip (probe) to raster over while interacting with the sample. AFM is an ideal tool for measuring nano and micro scale forces within a composite material (10,11). It has been widely used in high-tech materials, polymer, rubber, paint, biomaterials, and paper industries. The forces that can be measured in AFM include, but not limited to, mechanical contact force, friction, van der Waals forces, capillary forces, chemical bonding, electrostatic and magnetic forces. The modern AFM systems can accurately map a particular force in various imaging modes with nano meter resolution or track the dependence of different components as a function of tip-surface distance with sub-nanometer resolution.

During the past decade few research studies have used AFM to study the asphalt materials behavior (i.e. 12,13). In general those studies were limited to examining the asphalt morphology and micro-structure. Recently, Scarpas et al. (14) also developed a finite element model to study the healing phenomenon in asphalt binders, which was based on their behavior observed in AFM images.

In this paper, the recent progress in nanomechanics and material science was utilized to study the healing characteristics of asphalt binders. To this end, a new AFM-based approach was developed and used to examine the influence of two WMA additives on the healing behavior of an asphalt binder. The considered WMA additives included Sasobit and Advera.

2 Development AFM-Based Technique to Evaluate Healing

An AFM-based approach was developed to evaluate the two mechanisms of healing (wetting and intrinsic healing). In this approach, AFM force spectroscopy experiments are employed to measure the nano-scale adhesive and cohesive forces in an asphalt system. Force spectroscopy experiments involves measuring the contact forces between the AFM tip and the sample as the tip approaches, probes, and withdraws from sample surface. In the developed approach, the magnitude of the molecular interactions at the asphalt-aggregate interface (i.e. adhesive forces) is measured by conducting force spectroscopy experiments using silicon nitride tips that resemble the aggregate. In addition, the interaction between asphalt molecules (i.e. cohesive forces) is examined by using tips that are chemically modified by carboxyl (-COOH) group, which is one of the chemical groups found in asphalt binders.

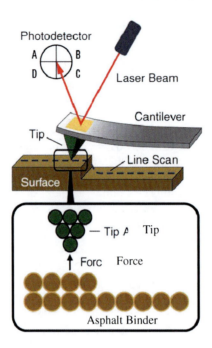

Fig. 1. Schematic diagram of AFM, adapted from (15)

The developed approach also evaluates the wetting mechanism of the healing process, by probing (indenting) the asphalt sample using the AFM tip at a fixed location and indentation depth, to create a nano-crack in the sample. AFM images are then continually taken to record the asphalt crack recovery with time. The AFM imaging is done using tapping (intermittent-contact) mode. The AFM tapping mode imaging is a versatile and powerful tool for scanning the surfaces of soft materials because it was developed to minimize sample deformation and avoid the surface damage found in contact mode AFM. In this imaging mode, the AFM cantilever/tip system is oscillated at its resonant frequency and the piezo-driver is adjusted using feedback control to maintain a constant tip-to-sample distance (set-point) (10). The amplitude of the resultant oscillations changes when the tip scans over the features on the surface. Thus, topographical characteristics of the sample can be obtained. In the developed approach, the topographical images are post-processed and analyzed to measure the closure of the initiated crack with time, which can be used to evaluate the wetting rate for the tested asphalt materials.

An Agilent AFM, 5500 LS, was used for conducting all experiments in this study. In addition, the cantilever used was 125μm long with a drive frequency of 300 kHz. All tests were conducted at room temperature, and the indentation depth was fixed at 750nm.

2.1 Materials and Sample Preparation

A polymer modified asphalt binder meeting specifications for PG 70-22M was used in this study. Different types of WMA technologies were evaluated, this includes: Sasobit, and Advera. Sasobit is paraffin wax produced from coal gasification using the Fischer Tropsch process. It is a fine crystalline, hydrophobic, long-chained aliphatic hydrocarbon. Therefore, the addition of this wax to an asphalt binder causes the binder to become more hydrophobic (16). The addition of Sasobit produces a reduction in the binder's viscosity, allowing production temperatures to be reduced. Advera WMA is an aluminosilicate or hydrated zeolite powder. It contains approximately 18-21% water by mass, which is released in the form of finely dispersed water vapor. This release of water creates a volume expansion of the binder that results in the formation of asphalt foam. This controlled foaming effect allows the asphalt mixture to become fluid at low temperature.

For mixing of WMA additives, the control binder was heated in the oven at 165°C for three hours. The heated binder was then stirred with a lab mixer and heated while the WMA additives were added slowly to it. The Sasobit and Advera loading levels used in this study were 4.5% and 1.5% of the asphalt binder weight, respectively. After mixing, a syringe was used to place about 0.5 ml on the middle of the glass substrate. The glass substrate was then placed in the oven for 15 minutes to allow for the asphalt to spread out. This was found to be the optimum protocol to form uniform and consistent surfaces required for the AFM characterization.

3 Results and Analysis

3.1 Results of Force Spectroscopy Experiments

Force distance curve is the main result obtained from force spectroscopy experiments. This curve presents a plot of the forces acting on the sample as a function of piezo-driver displacement. The forces are calculated based on the cantilever deflection using hooks law:

$$F = -k_c \, d \qquad (1)$$

where F is the acting force on the sample, d is the deflection and k_c is the cantilever spring constant.

Figure 2 presents a typical force distance curves for an asphalt material. This curve can be divided into two regions the approaching region where the tip is brought close to the sample until a contact between the tip and sample occurs and the retracting region where the tip starts to pull away from the sample. In the former region, the tip starts to approach the sample surface by moving towards the sample. Initially, the tip will be far away from the sample and no deflection will happen until it is brought close enough to the surface where it start to deflect due to the repulsive force. The repulsive force increases until reaching to a specified depth of indentation. In the retracting region, initially the repulsive force, hence the deflection, is reduced. However, as the retraction continues the tip sticks to the sample surface due to the attractive forces for a certain time till it finally snaps off the surface and springs back to its original position. The maximum force needed to pull the tip away from the sample is called the pull-off force, which is also the adhesive force ($F_{ad.}$) between the tip and the tested sample.

Force spectroscopy experiments were conducted on at least 32 points for each asphalt sample. The obtained force distance curves were post-processed using a FORTRAN code that was developed to allow appropriate normalization of the raw data, and to obtain the maximum pull-off forces in each force spectroscopy experiment. Figure 3 compares the average adhesive force values of the WMA binders with that of the control asphalt binder, which were obtained in experiments, conducted using silicon nitride tips. It is noted that the inclusion of the WMA additives enhanced the adhesive forces, which are related to the intrinsic healing of asphalt materials. This improvement will induce effects on the corresponding work of adhesion for the asphalt aggregate system.

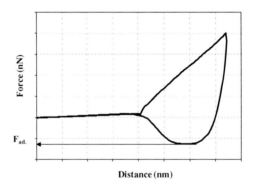

Fig. 2. Force-Distance curve obtained from force spectroscopy experiments on 70-22M binder

Force spectroscopy experiments were also conducted using carboxyl (-COOH) functionalized tips on the control and WMA asphalt binder samples. Figure 4 compares the maximum pull-off force for the WMA binders with that of the control one. It is noted that the WMA additives had decreased the adhesive forces between asphalt binder and -COOH tip, which can be related to the cohesive forces within the asphalt binder itself. The Sasobit resulted in more pronounced reduction in those forces compared to the Advera. This indicates that the effect of the WMA on the adhesive and cohesive forces is different. It is worth noting that the Sasobit is known to increase the hydrophobicity of an asphalt binder (12), which may explain the reduction in the adhesive forces when the hydrophilic – COOH functionalized tips are used.

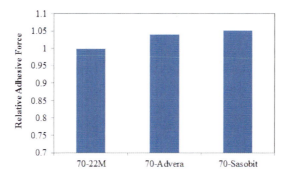

Fig. 3. Adhesive forces obtained using silicon nitride tips

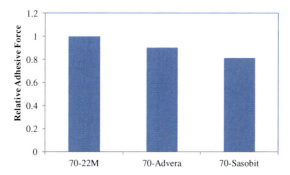

Fig. 4. Adhesive forces obtained using –COOH functionalized tip

3.2 Evaluation of Wetting Mechanism

To evaluate the wetting mechanism, a crack was initiated in the asphalt binder and the closure of the crack with time was monitored by continuously taking AFM images. Figure 5 presents the topographical images that were taken before and after the 70-22 M asphalt sample was probed. As it can be noted, a relatively flat spot was identified and indented; this resulted in a nano-crack as shown in Figure 5b, which mitigated with time, Figures 5 b-d. To evaluate the closure (wetting) rate, the images were post-processed and analyzed to determine how the crack volume changed with time for different points for each sample. Figure 6 presents a typical plot of the crack volume as a function of time for the different binders. It is noted that the volume decreases with time nonlinearly for all asphalt binders evaluated. However, the rate of decrease for each binder is different. To examine that, a power function was fitted through the obtained data. Figure 7 presents the coefficients of the fitted power function for the control and WMA binders. It is noted that the Sasobit had resulted in decreasing the power function coefficient, and thus it reduced the crack closure rate, while the Advera did not have any significant effect. This result suggests that the crack closure rate (i.e. wetting rate) is related to the cohesive forces within the asphalt binder, which is consistent with the healing model proposed by Bhasin et al. (9).

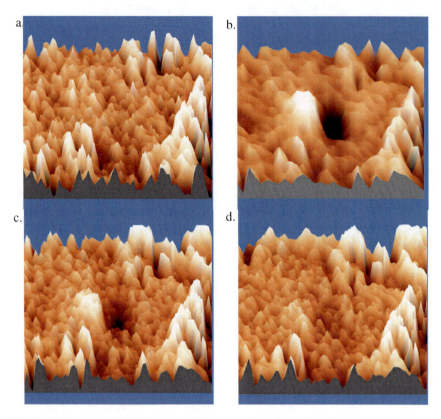

Fig. 5. AFM topographical images: a) Directly before probing b) 163 sec after probing c) 350 sec after probing d) 800 sec after probing

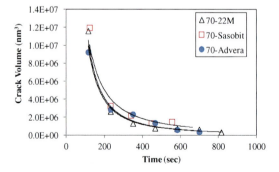

Fig. 6. Crack volume decrease with time

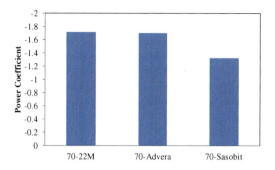

Fig. 7. Rate of decrease of crack volume

4 Conclusions

In this paper, an AFM based approach was developed to evaluate the two main healing mechanism in an asphalt binder, namely wetting and intrinsic healing mechanisms. The developed approach was employed to examine the effect of two WMA additives on the healing characteristics of an asphalt binder. Based on the results of this study the following conclusions can be drawn:

- The Sasobit and Advera WMA additives increased the adhesive forces within the asphalt system, and therefore enhanced its adhesive intrinsic healing characteristics.
- When using -COOH functionalized AFM tips, the results indicated that Sasobit led to a reduction in the adhesive/cohesive forces within the asphalt binder, indicating that it might adversely affect its cohesive intrinsic healing properties.
- The wetting in the asphalt binder was adversely affected by the addition of the Sasobit, but was not influenced by Advera.
- The novel AFM-based technique introduced in this study is a very useful tool to evaluate healing behavior of asphalt binders. However, further tests should be conducted over many crack sizes, using different types of functionalized tips with various geometries to validate results and reduce any uncertainties with the proposed technique.

References

[1] Moulthrop, J., McDaniel, R., McGennis, R., Mohammad, L., Kluttz, R.: Asphalt Mixture Innovations: State of the Practice and Vision for 2020 and Beyond. TR News No. 253 Highway Design and Construction A 2020 Vision 2007, 20–23 (2007)

[2] Nazzal, M., Sargand, S., Al-Rawashdeh, A.: Evaluation of Warm Mix Asphalt Mixtures Containing Rap Using Accelerated Loading Tests. ASTM Journal of Testing and Evaluation 39(3) (2010)
[3] Aschenbrener, T., Schiebel, B., West, R.: Three-Year Evaluation of the Colorado Department of Transportation's Warm-Mix Asphalt Experimental Feature on I-70 in Silverthorne, Colorado. National Center for Asphalt Technology, NCAT Report No. 11-02 (2011)
[4] Ali, A., Abbas, A., Nazzal, M., Powers, D.: Laboratory Evaluation of Foamed Warm Mix Asphalt. International Journal of Pavement Research and Technology (2010) (in print)
[5] Maillard, S., de La Roche, C., Hammoum, F., Gaillet, L., Such, C.: Experimental investigation of fracture and healing atpseudo-contact of two aggregates. In: Proc., 3rd Euroasphalt andEurobitume Congress, European Asphalt Pavement Association and Eurobitume, Vienna, Austria (2004)
[6] Shen, S., Chiu, H.M., Huang, H.: Fatigue and Healing in Asphalt Binders. In: Transportation Research Board 88th Annual Meeting CD (2009)
[7] Bommavaram, R.R., Bhasin, A., Little, D.N.: Use of Dynamic Shear Rheometer to Determine the Intrinsic Healing Properties of Asphalt Binders. In: Transportation Research Board 88th Annual Meeting (2009)
[8] Kim, B., Roque, R.: Evaluation of healing property of asphaltmixture. In: Proc., 85th Annual Meeting of the Transportation Research Board, Washington, DC (2006)
[9] Bhasin, A., Little, D.N., Bommavaram, R., Vasconcelos, K.L.: A framework to quantify the effect of healing in bituminous materialsusing material properties. Int. J. Road Mater. Pavement Des. 8, 219–242 (2008)
[10] Beach, E.R., Tormoen, G.W., Drelich, J.: Pull-off forces measured between hexadecanethiol self-assembled monolayers in air using an atomic force microscope. J. Adhes. Sci. Technol. 167, 845–868 (2002)
[11] Nguyen, T., Gu, X., Fasolka, M., Briggman, K., Hwang, J., Karim, A., Martin, J.: Mapping chemical heterogeneity of polymeric materials with chemical force microscopy. Polym. Mater. Sci. Eng. 90, 141–143 (2005)
[12] Huang, S.C., Turner, T.F., Pauli, A.T., Miknis, F.P., Branthaver, J.F., Robertson, R.E.: Evaluation Of Different Techniques For Adhesive Properties Of Asphalt-Filler Systems At Interfacial Region. J. ASTM Int. 25 (2005)
[13] Schmets, A., Kringos, N., Pauli, T., Redelius, P., Scarpas, A.: Wax Induced Phase Separation in Bitumen. International Journal for Pavement Engineering 11(6) (2010)
[14] Scarpas, A., Robertson, R., Kringos, N., Pauli, T.: A thermodynamic approach to healing in bitumen. Advanced Testing and Characterization of Bituminous Materials (2009)
[15] Sasol Wax Co. More about Sasol wax flex (2008), http://www.sasolwax.com/More_about_Sasolwax_Flex.html
[16] Agilent Technologies, Inc. Agilent 5500 LS AFM/SPM User Manual. Agilent (2009)

Cracking and Healing Modelling of Asphalt Mixtures

Jian Qiu[1,3], Martin F.C. van de Ven[1], Erik Schlangen[2], Shaopeng Wu[3], and André A.A. Molenaar[1]

[1] Delft University of Technology, CiTG, Section of Road and Railway Engineering, P.O. Box 5048, 2600 GA Delft, The Netherlands,
[2] Delft University of Technology, CiTG, Microlab, P.O. Box 5048, 2600 GA Delft, The Netherlands
[3] Wuhan University of Technology, 122, Luoshi Road, Wuhan 430070, People's Republic of China

Abstract. Self healing behaviour of asphalt mixture has been known for many years. This unique behaviour can help asphalt concrete to recover its strength after damage. Healing can also extend the service life of asphalt pavements. A beam on elastic foundation test setup (BOEF) was developed to investigate the self healing behaviour of asphalt mixture in a controlled and effective way. Within this setup, a notched asphalt beam was glued on a low modulus rubber foundation, and a symmetric monotonic load was applied with loading-unloading-healing-reloading cycles. The experimental results indicate the existence of the healing behaviour. Increasing reloading curves are observed for increasing healing time and healing temperatures. In order to further understand the cracking and healing phenomenon, a finite element simulation was carried out with a smeared crack cohesive zone model (CZM). By defining both properties of the bulk asphalt mixture and the cohesive zone, the global load-crack opening displacement (COD) response during cracking and healing process was simulated successfully. The model is also capable to separate the two healing phases including crack closure and strength gain. It is also shown that further implementation of the cracking and healing modelling of asphalt mixtures is important for durable asphalt pavement.

1 Introduction

Self healing behaviour of bituminous materials has been known for many years [1-3]. The significance of the self healing capability is that an asphalt concrete could repair itself under certain conditions such as hot summers and/or rest periods and extend the service life. Hence, the self healing capability is defined as the recovery of mechanical properties like strength, stiffness as well as an increase of the number of load repetitions to failure. According to Phillips, healing is a multi-step process consisting flow, wetting and inter-diffusion [4]. The flow and wetting is related to the crack closure process and the inter-diffusion dominates the gaining of the strength.

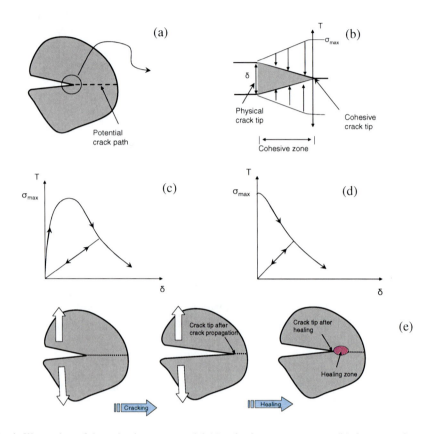

Fig. 1. Illustration of the cohesive zone model (a) cohesive zone concept; (b) the separation displacement and the corresponding traction along a cohesive surface; (c) intrinsic cohesive zone model; (d) extrinsic cohesive zone model; (e) hypothesis of cracking and healing process

In order to quantify and qualify the self healing capability of bituminous materials in a controlled and effective way, a beam on elastic foundation set-up (BOEF) was proposed by the authors of this paper [5, 6]. The experimental results indicate the existence of the healing behaviour. Increasing reloading curves are observed for increasing healing time and healing temperatures. This setup is easy to be implemented into practice for evaluating the self healing behaviours of asphalt mixtures. However, due to the complexity of the setup, there is a need for more understanding of the cracking and healing phenomenon using finite element modelling.

A so-called cohesive zone model (CZM), was developed based on non-linear fracture mechanics [7]. As it is shown in Figure 1a and 1b, a cohesive zone is defined in front of the crack tip with a certain traction-separation behaviour. When the local stress applied is higher than the maximum traction force of the cohesive zone element, the model is active. The post-peak softening is then followed with certain traction-separation relationship until the traction diminishes. This relationship can be either intrinsic or extrinsic (Figure 1c and 1d) [8]. In an intrinsic model, a penalty stiffness is used for the pre-peak stiffness until the maximum traction is reached. After that a descending slope (softening) of the traction-separation relationship is applied. In an extrinsic model, the penalty stiffness is replaced by the stress-strain relationship for the undamaged material until the maximum traction is reached. An extrinsic model is also called a smeared crack model, which is probably more realistic than the intrinsic approach, because it does not assume the pre-existence of the cohesive element during simulation. Figure 1-e illustrates the hypothesis of the cracking and healing process under the CZM concept. The healing process is regarded as a reverse process of cracking. Under certain conditions such as rest period, temperature, compressive stress, etc, the healing process will happen to improve the cohesive strength of the healing zone, hence to bring backwards of the crack tip.

In this paper, the cracking and healing behaviours of asphalt mixtures was modelled using the smeared crack CZM approach under a commercially available FEM code FEMMASSE [9]. The influencing factors and the quality of the results of the modelling were also explored.

2 Experimental

As it is shown in Figure 2, a notched beam was fully glued on a low modulus rubber foundation, and a symmetric monotonic load was applied with loading-unloading-healing-reloading cycles. This setup allows a fully closure of the crack with the help of the rubber foundation, which guarantees an autonomous healing process under different crack phases [5, 6].

The experimental loading-unloading-healing-reloading procedure is given below. First a load was applied at a constant crack opening displacement (COD) speed of 0.001 mm/s until the target COD level was reached. Three target COD levels were selected: 0.2 mm, 0.6 mm and 0.9 mm, respectively. After the target COD level was reached, the specimen was unloaded with a COD closing speed of -0.001 mm/s till the load level had returned to 0. At that moment, the external load was removed, and the recovery of the COD was recorded. A rest period of 1 hour was first applied at 5 °C. Then healing periods of 3 hours and 24 hours and healing temperatures of 5 °C and 40 °C were applied, respectively. After the healing period, the specimens were reloaded at a temperature of 5 °C using a COD speed of 0.001 mm/s till a COD level of 1.5 mm. For specimens healed at a temperature of 40 °C, a conditioning time of at least 2 hours at 5 °C was applied before reloading.

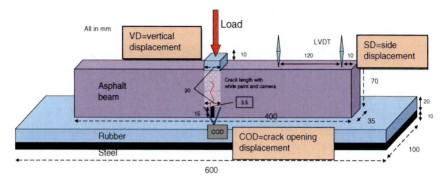

Fig. 2. Schematic of the BOEF setup (all the units are in mm)

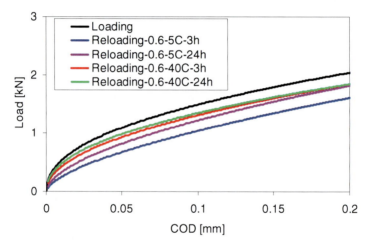

Fig. 3. Typical BOEF test results under different healing conditions for a target COD level of 0.6mm (0.6-5C-3h means healing periods of 3h at healing temperature of 5 °C)

Figure 3 shows the typical load-COD (abbreviation LC) curves for a target COD level of 0.6 mm. It should be noted that, the reloading curves were shifted such that they can be compared directly with the loading curve. Because of the influence of the rubber foundation, the LC curve does not show the abrupt failure like observed in loading tests on simply supported beams. The increase of the reloading curves marks the existence of the self healing phenomenon of asphalt mixtures, and this capability increases with the increasing healing time and healing temperature.

3 FEM Model

Table 1 lists the material parameters that were used for the BOEF analysis. Elastic material properties were assigned to all materials, only the asphalt mixture was modelled visco-elastically. A Generalized Maxwell Model was used to model the visco-elastic properties of asphalt mixtures. The model parameters were calculated from information on the mixture stiffness as a function of loading frequency and temperature which was collected using a dynamic tension-compression test. The visco-elastic material response in frequency domain can be written as Eqn. (1) and Eqn. (2). And the time dependent relaxation modulus can be expressed as Eqn. (3).

$$E^* = E' + iE'' \tag{1}$$

$$E'(\omega) = E_0 \left[1 - \sum_{i=1}^{N} \alpha_i \right] + E_0 \left[\sum_{i=1}^{N} \frac{\alpha_i \tau_i^2 \omega^2}{1 + \tau_i^2 \omega^2}\right] \text{ and } E''(\omega) = E_0 \left[\sum_{i=1}^{N} \frac{\alpha_i \tau_i \omega}{1 + \tau_i^2 \omega^2}\right] \tag{2}$$

$$E(t) = E_0 \cdot \sum_{i=1}^{n} \left(1 - \alpha_i \left(1 - \exp\left(t/t_i\right)\right)\right) \tag{3}$$

Where:
$E'(\omega)$ = storage modulus as a function of frequency, ω [MPa];
$E''(\omega)$ = loss modulus as a function of frequency, ω [MPa];
E^* = complex modulus as a function of frequency, ω [MPa];
$E(t)$ = relaxation modulus as a function of time, t [MPa];
E_0 = stiffness of instantaneous response [MPa];
ω = applied angular frequency [rad/s];
α_i = model parameter, i.e ith Prony E reduction ratio [-];
τ_i = model parameter, i.e. relaxation speed of the ith Prony [s];
n = number of components in the model [-].

Table 1. Material parameters for BOEF analysis

		Elastic material parameters			
		Rubber pad	Rubber foundation	Glue	Steel
Young's Modulus [MPa]		15	6.5	4000	200000
Poisson's ratio		0.49	0.49	0.15	0.15
Visco-elastic material parameters for asphalt mixtures					
E_0 [MPa]	n	Term1	Term2	Term3	Term4	Term5	Term6
19956	α_i	4.52E-01	2.85E-01	2.00E-01	5.71E-02	5.09E-03	6.92E-04
	τ_i [s]	3.05E-03	7.89E-02	9.72E-01	1.15E+01	1.70E+02	1.00E+03

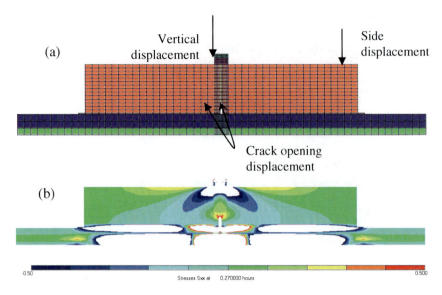

Fig. 4. FEM model assembly of BOEF setup (a) BOEF-2D FEM modelling using FEMMASSE; (b) Horizontal stress distribution Sxx

Figure 4 shows finite element model developed in FEMMASSE to model the BOEF test. A 2D FEM model was developed and plain stress conditions were assumed to occur. The thickness of the model was 35mm. Due to complications of keeping a constant COD by applying a vertical load in the FEM simulation, the simulation was carried out with an experimental vertical load as input. Then, displacements were calculated at three locations, including the COD development (COD), which is measured over the notch with a measuring distance of 20 mm; the vertical displacement (VD); and the side displacement (SD). The SD was measured 10 mm away from the edge of the beam.

A thin layer of elements right above the notch was defined with both visco-elastic and the cohesive zone property. For simplicity, a linear traction-separation relationship was defined to this layer of elements. The traction and maximum separation were chosen as 3 MPa and 0.001 mm, respectively.

4 Results and Discussions

4.1 Modelling of Cracking Behaviour

Figure 5-a compares the simulated COD development with and without damage. The modelling with damage indicates that at a certain displacement, a crack has initiated, and the displacement increases because of crack propagation. Figure 5-b compares the simulated and experimental LC curves. It can be seen that the LC relationship can be simulated successfully if the behaviour of the cohesive zone is modelled appropriately.

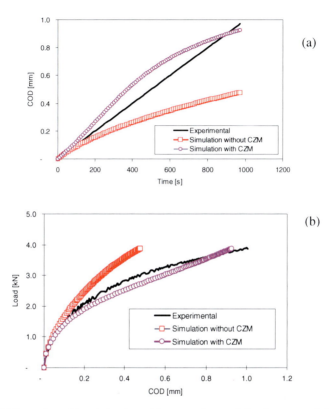

Fig. 5. Modelling of cracking behaviour (a) comparison of development of COD; (b) comparison of Load-COD curves

4.2 Modelling of Unloading Behaviour

Figure 6 shows the simulation results of the immediate reloading curves. It can be seen that the model cannot fully simulate the unloading behaviour. The simulated COD recovers immediately to zero after unloading, which overestimates the experimental COD recovery speed. This is because the experimental crack closure process is a nonlinear process, which is related to the internal compressive stress state due to confining effect of the rubber foundation and relaxation of the asphalt beam. However, the simulation is simplified in such a way that the crack closes linearly with the internal stress state. This results in a rapid recovery of the COD.

Interestingly, the simulated immediate reloading is similar to the experimental reloading after 3 hours of rest at a temperature of 5 °C. It also implies that the recovery in the first hours is related to a crack closure process as shown in the experiment [6].

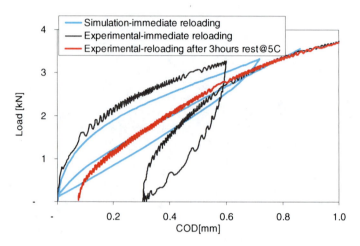

Fig. 6. Simulation of unloading behaviour

4.3 Modelling of Healing Behaviour

It is shown earlier that unloading and a short term healing results in the closure of the crack. So the reloading includes two possible processes, the crack reopens and crack propagates further. It is hypothesised that the crack tip is not only healed totally but also moves backwards due to healing. During simulation, a deduction of the crack length after healing was subjected manually. The same load was applied in all cases in order to see the difference. Figure 7 shows the simulation results of the reloading behaviour.

It can be observed in Figure 7 that the simulated reloading curves are decreasing with increasing crack lengths. However, this effect is less when crack length reaches 20mm and above, then the crack enters the compression zone.

When comparing Figure 7-b with the experimental results shown in Figure 3, the experimental reloading curve with increasing healing time and temperature can be seen as decrease of the crack length or decrease of the crack tip.

As a result, the possible healing can be very important for durable asphalt pavement. It is commonly known that cracking is one of the main causes for early damage of asphalt pavements. Due to healing, the crack can repair itself. This is an autonomous process, which is also believed to be a flow driven process [10, 11]. This process is also very sensitive to the healing conditions such as temperatures, rest periods, compressive stress etc. In the future, optimisation of the healing behaviour should be undertaken to enhance durable asphalt pavement.

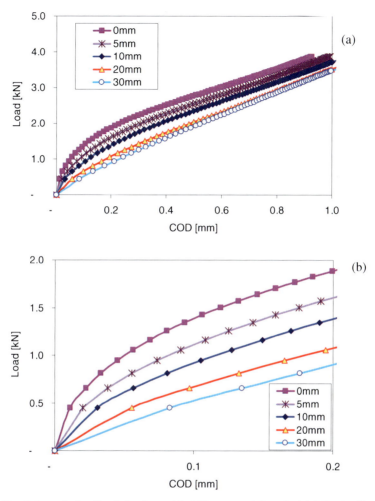

Fig. 7. Simulation of reloading behaviour with different crack lengths (a) full reloading LC curves; (b) detail of reloading LC curves till 0.2mm

5 Conclusions

The cracking and healing modelling was carried out for a special self healing set-up named the Beam on Elastic Foundation test set-up. By using the smeared crack typed cohesive zone model, the cracking and healing behaviour was simulated successfully. Based on the simulation results and discussions, the following conclusions can be made:

o   The load-COD curve is shown to be the most appropriate relationship for experimentally indentifying the cracking and healing behaviour. Upon defining the local cohesive zone parameters, the global load-COD curve was simulated successfully.

- During the unloading process, the experimental crack closes slowly resulting in a non-linear decrease of the COD. Due to the simplicity of the unloading nature of the smeared crack model, the crack surface closes totally after the load was removed. This overestimates the crack closure speed as observed in the tests.
- The reloading behaviour was simulated by using predefined crack lengths. The healing behaviour is then directly related to a decreasing crack length as a crack repairing process.

References

[1] Bazin, P., Saunier, J.B.: In: Proceedings of the Second International Conference on the Structural Design of Asphalt Pavements, Ann Arbor, Michigan, USA, pp. 553–569 (1967)
[2] Raithby, K.D., Sterling, A.B.: Journal of Association of Asphalt Paving Technologists 39, 134–152 (1970)
[3] Van Dijk, W., Moreaud, H., Quedeville, A., Uge, P.: In: Proceedings of the Third International Conference on the Structural Design of Asphalt Pavements, Ann Arbor, Michigan, USA (1972)
[4] Phillips, M.C.: In: Eurobitume Workshop on Performance Related Properties for Bituminous Binders, Luxembourg, p. 115 (1998)
[5] Qiu, J., Molenaar, A.A.A., van de Ven, M.F.C., Wu, S.P., Yu, J.Y.: Materials and Structures (2012), doi:10.1617/s11527-011-9797-7
[6] Qiu, J., Molenaar, A.A.A., van de Ven, M.F.C., Wu, S.P.: In: Proceeding of the Transportation Research Board (TRB) 91st Annual Meeting, Washington, DC, USA (2012)
[7] Hillerborg, A., Modeer, M., Petersson, P.E.: Cement and Concrete Research 6, 773–781 (1976)
[8] Kim, Y.R.: International Journal of Pavement Engineering 12, 343–356 (2011)
[9] FEMMASSE, User manual MLS version 8.5 (2006)
[10] Qiu, J., Van de Ven, M.F.C., Wu, S.P., Yu, J.Y., Molenaar, A.A.A.: Experimental Mechanics (2012), doi:10.1007/s11340-011-9573-1
[11] Qiu, J., Van de Ven, M.F.C., Wu, S.P., Yu, J.Y., Molenaar, A.A.A.: Fuel 90, 2710–2720 (2011)

Effects of Glass Fiber/Grid Reinforcement on the Crack Growth Rate of an Asphalt Mix

C.C. Zheng and A. Najd

College of Highway, Chang'an University, Xi'an, P.R. China

Abstract. This paper presents an application of fracture mechanics to determine crack growth rates of the suggested anti-cracking overlay systems. Two different reinforcing methodologies are applied; 1: addition of chopped glass fibers to the HMA; 2: reinforcing asphalt overlay by glass grids. Asphalt mixture designing tests, three points bending tests and fatigue crack propagation tests were carried out. Fracture toughness K_{IC} is determined for plain and reinforced asphalt concretes. The crack growth rate is determined for each type of anti-cracking systems, the cracking process is analyzed. One of the significant points in this study is the attempt to give better understanding of the crack propagation for multilayer asphalt overlay. The results indicate that the reinforcing materials improve anti-cracking characteristics of the asphalt concrete and composite structure anti cracking overlay gives a good solution for reflective cracking phenomenon over old cracked pavements.

1 Introduction

There is a growing need for more effective rehabilitation methodologies in both developed and developing countries. From a review of the literature, it is apparent that many field methods and analytical techniques have been investigated to minimize reflective cracking in HMA overlays. Many of the early field investigations were based on empirical relationships, which produced results, varied from successful to disastrous. Later research has employed the more favorable mechanistic approach of determining fracture properties of the HMA overlay using fracture mechanics theories. Early research guided by Lytton (1989) was based on identifying fracture properties of geo-synthetic materials using a fracture mechanics based approach. Majidzadeh et al (1985) and Monismith et al. (1980) have also made efforts to predict the fatigue life of an asphalt mixture. Complex geometry and complicated stress systems often necessitate the use of finite element methods (FEM) and extensive computer resources are needed to solve the resulting large systems of equations. RILEM Conference on Reflective Cracking in Pavements in 1989, 1993, 1996, 2000, 2004 and 2008 were organized to point out the main factors and mechanisms involved in the initiation and crack Propagation. Baek and Al-Qadi (2009) employed three-dimensional finite element modeling to investigate the fracture behavior of hot-mix asphalt HMA overlays that resulted in reflective cracking. When steel netting was used, reflective cracking was significantly reduced in the leveling binder due to strong shear deformation support, as well as high tensile strain compensation in the overlay. This also reduces the potential of reflective cracking in the wearing surface. Braham, et al. (2011) examined asphalt mixture laboratory aging protocols from the standpoint of both mixture and binder physical properties that are believed to relate to various forms of pavement cracking.

The main objective of this research is to give more insight into the crack growth and crack resistance characteristics of asphalt concrete mixes in general, and of wearing courses over old cracked asphalt pavement in particular. The influences of reinforcing materials on fatigue crack growth are analyzed. Better understanding of the reinforced composite structure asphalt overlay in retarding crack propagation is achieved.

2 Material Properties

Lime stone aggregate, AH-90 bitumen, chopped glass fibres and glass fibres grid are used in this study. According to asphalt pavements constructions and tests Norm (2001).AC-10I and AK-13BI asphalt mixtures with gradations shown in Figure 1 are used in this paper. Reinforcing materials properties are illustrated in table 1 and table 2.

Table 1. Glass fibres properties

Fibre diameter /mm	Chopped length/ mm	Colour	Tensile strength/ MPa	Elongation %	Modulus of Elasticity /MPa	Density /g.cm^{-3}
0.0058-0.0097	6	Silver White	2000	5	70 000	2.54

Table 2. Glass fibres grid properties

Tensile strength/kN/m		Elongation %	Thermal endure range /°C	Weight g/m^2	Grid size/ mm
Longitudinal direction	Horizontal direction				
35	65	< 5	-100 ~ 280	340	12*12

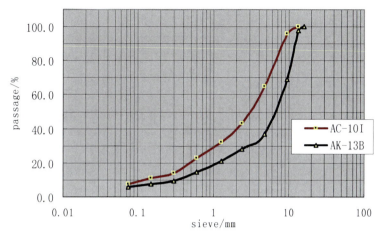

Fig. 1. Gradation of AC-10I and AK-13B

3 Three Points Bending Test

3.1 Specimens Preparation

Testing temperature is 5°C; specimen dimensions are 70mmx 60mm x 250mm; span: 210 MM; For simulating the initiated crack in the overlay, a metal piece with inverted `T` shape was used, fixed on the middle of the metal mould and heated together to the proper temperature before the asphalt mixture spread on it, the mixture should be compacted and rammed by a small metal stick along the two edges of the metal piece to enhance the density in this area. Before testing, bottom surface of the specimen were polished and crack was sharpened by a saw. The initial crack length is 7 mm and crack tip angle is 20 °, as shown in Figure 2.

Fig. 2. Notched asphalt concrete specimen

Four specimens are tested for each of the following material groups:

(1) AK-13B: skid resistance asphalt mixture (gradation 0/16, bitumen 4%)
(2) AC-10I: dense asphalt concrete (gradation 0/13.2, bitumen 5%)

(3) GFRAC: glass fibres reinforced asphalt concrete (AC-10I + glass fibres 1.75‰, bitumen 5%)
(4) GGRAC: glass grid reinforced asphalt concrete (AC-10I + glass grid: The grid is located 10mm above the bottom of the specimen.)

Table 3. Fracture toughness for various asphalt materials

Specimens type	Maximum applied load P_b /N	Critical moment Mc/mm^4	a/h	f(a/h)	Critical stress intensity factor K_{IC} / $MPa*\sqrt{m}$
AK-13B	4636	245459	0.0996	1.8522	**24.3393**
AC-10I	5686	304185	0.0992	1.8524	**29.8602**
FRAC	6535	345030	0.0997	1.8521	**34.3089**
GGRAC	6308	331170	0.0992	1.8524	**32.5634**

3.2 Fracture Toughness K_{IC}

Depending on bending test for notched specimens, fracture toughness can be calculated from the following formula in ZHANG (1998).

$$K_{1c} = \frac{6M_c a^{1/2}}{bh^2} f(a/h) \quad (1)$$

In which: M_c critical moment at failure ; a notch initial length ; b specimen width ; h specimen height.

$$f(a/h) = 1.99 - 2.74(a/h) + 12.97(a/h)^2 - 23.17(a/h)^3 + 24.80(a/h)^4 \quad (2)$$

3.3 Test Results Analysis

From Figure 3, we can notice that even though the crack propagation resistance for both the glass fibers reinforced asphalt concrete (GFRAC), and the glass grid reinforced asphalt concrete (GGRAC) are relatively high.

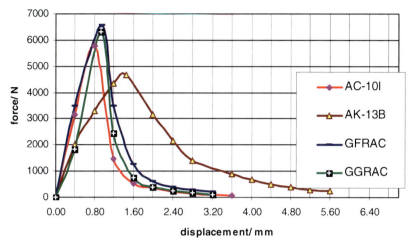

Fig. 3. Force vs. displacement for notched asphalt concrete specimens

Table 3 indicates that the critical stress intensity factor for GFRAC increased only by 14.9% compared with the plain asphalt concrete, and 9.1% for GGRAC compared with the plain asphalt concrete. This result indicates that under high speed rate loading and low temperature, when the crack was initiated, then the reinforcement's efficiency in retarding crack propagation deeply decreased. Because huge stresses concentrate at the crack tip, while the surrounded areas have no sufficient stress dissipating mechanism. The reinforcement has greater efficiency when the crack does not initiate yet, in un-notched specimen the stresses distribute over large area and the reinforcement participation is big.

4 Fatigue Crack Propagation Test

This test is used to measure crack growth rate value (da/dN). this value is used to determine the intrinsic parameters of the material, such as A and m in the fatigue strength law, such as Paris' law. Paris' Law relates the mean crack propagation per cycle da/dN, to the variation of the stress intensity factor ($K_{max} - K_{min}$). In other words, this test makes it possible to accurately relate the values of the measurable load parameters and the values of the mechanical internal parameter, such as the stress intensity factor, so as to measure the efficiency of various anti-cracking processes. In addition to this major goal, this test provides a better understanding to the composite structure asphalt overlay, and the effectiveness of reinforcing material in retarding crack propagation.

4.1 Composite Structure Asphalt Overlay

It is well known that the main function of asphalt wearing course is to provide suitable skid resistance to the traffic loads and good serviceability. In case of asphalt overlay on old cracked asphalt pavement, the overlay will quickly develop cracks due to reflection phenomenon from the old cracked layers. Due to the asphalt wearing course's gradation and structure, one single layer is not sufficient to crack propagation. Therefore, composite structure asphalt overlay is recommended to study the effectiveness of the reinforcements in this system. The specimens were formed in the laboratory in two steps, first, the plain asphalt concrete AC-10I or reinforced asphalt concrete was spread inside a metal mould and compacted to the proper density, then the skid resistance asphalt mixture AK-13B was added above it and then compacted to the proper density. Different reinforcement, gradation and structure have been used to create anti cracking overlay systems. In addition to the main types AC-10I, AK-13B, GFRAC and GGRAC, three types of composite structure asphalt specimens have been tested: Composite AK-13B/AC-10I, Composite AK-13B/GFRAC and Composite AK-13B/GGRAC. The composite structure consists of 35mm AK-13B and 35mm AC-10I or GFRAC or GGRAC. The readers are referred to reference [8] for more details.

4.2 Test Conditions

The formation method of the specimens are typically as it was in the three points bending test. This test was carried out in MTS laboratory of Chang'an University. Loading frequency: 10 Hz; (loading frequency corresponds to a vehicle speed of 70 km/h). Half sine loading wave was applied. The fatigue crack propagation test was performed for the pure opening mode only. The force controlled mode used a maximum load application of 1155 N equivalent to 0.2 K_{IC} for the AC-10I type. Four specimens were tested for each of the seven studied types. Crack detectors (crack propagation gauges) bonded on the two sides of the specimen is used for measuring the crack propagation under the load repetitions, and checked by the dynamic cracking measuring unit.

4.3 Test Results Analysis

4.3.1 For the Basic Types (Non Composite Structure Overlays)

Figure 4 illustrates the cracking progress for the four basic types of the asphalt overlays under load cycles. Obviously the skid resistance type AK-13B took smallest number of load repetitions before failure. This is not surprising, due to the fact that AK-13B has open gradation, high void percentage, and the low bitumen content. The crack takes its way between the grains and the repetitions of load, in this case, leading to the de-bonding mechanism between the aggregates.

The crack propagation of the glass grid reinforced asphalt concrete GGRAC took the same behavior as the plain asphalt concrete at the beginning. This makes it clear that the reinforcing grid does not have a big effect on retarding the concentrated stresses in the crack tip. However, when the crack propagates further upwards and the deformation of the specimen increases, the grid at this stage plays a major role in retarding the crack propagation. The glass grid prevents further opening of the crack due to its low rate of elongation and its high tensile strength. At this point, a transfer of the stresses from the crack tip to the grid occurs and the crack growth rate decreases.

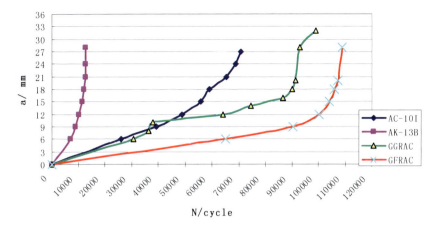

Fig. 4. Crack lengths vs. load cycles

Here a great attention should be paid to the type of grids to be used for retarding crack propagation i.e. only grids with low rate of elongation and high tensile strength should be used. The grid used in this study increases the total applied load cycles value by **40%** compared with the plain asphalt concrete. The crack propagation of the chopped glass fibers reinforced asphalt concrete GFRAC behaves differently from the plain asphalt concrete. Figure 4 shows that cracking occurs in steady progress from the beginning until the failure stage, with low crack propagation velocity along the cracking line, except failure stage. The chopped glass fibers in the asphalt concrete have a bridging effect, which makes the asphalt layer act as a whole. The addition of chopped glass fibers to the asphalt concrete increased the total applied load cycles value by **54%** compared with the plain asphalt concrete.

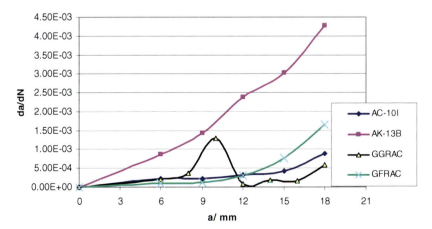

Fig. 5. Crack lengths vs. the velocity of the crack propagation

Figure 5 shows the velocity of the crack propagation in each step during the test, In general, the crack growth rate increases as the crack moves upward, but in glass grid reinforced asphalt concrete, this induction was not typical, possibly because some de-bonding between the grid and the asphalt concrete had happened. The glass grid here may have a negative effect if it is not positioned properly; the grid allows the asphalt layer to deform larger than the plain asphalt concrete can bear, if the de-bonding has happened, then the crack would grow faster than it is expected in plain asphalt layer, due to large deformation which is allowed by the grid. It is noticed that the crack growth rate did not increase in certain steps especially with reinforced asphalt concrete and in some cases the crack propagation stopped and other crack was initiated later.

4.3.2 Determination of the parameters A, n in Paris law

According to the Paris law, we have

$$da/dN = A(\Delta K_1)^n \qquad (3)$$

In which, the stress intensity factor of different crack length can be simulated with finite element method. We then obtain the values of A and n in Table 4 by Paris law from Figure 6.

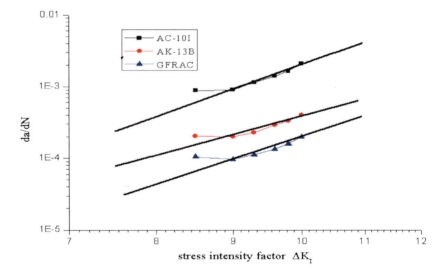

Fig. 6. da/dN vs ΔK_I

The difference propagation behavior between AC-10I and AK-13B is reflected in both n and A.

$$\frac{A_{AK-13B}}{A_{AC\_10I}} = 0.32 \ , \quad \frac{n_{Ak-13B}}{n_{AC-10I}} = 1.49 \qquad (4)$$

Table 4. Values of A and n

TYPE OF SPECIMEN	A	n	R^2
AC-10I	1.6E-06	2.34	0.83
AK-13B	5.1E-07	3.49	0.86
GFRAC	2.5E-07	2.85	0.74

Compared with AC-10I, GFRAC has a lower value A and higher value n, that is to say, the function of glass fiber inside the concrete is to increase value A and decrease value n.

$$\frac{A_{GFRAC}}{A_{AC-10I}} = 0.15 , \quad \frac{n_{GFRAC}}{n_{AC-10I}} = 1.22 \tag{5}$$

4.3.3 For the Composite Structure Anti-cracking Overlays Systems Types

Figure 7 demonstrates the cracking progress for three composite anti cracking asphalt overlays systems. Obviously the plain composite structure AK/AC-10I type took smallest number of load cycles repetition before failure, and the crack growth rate increased rapidly .The composite glass grid reinforced asphalt concrete AK/ GGRAC gave the best result in retarding crack propagation among composite structure asphalt overlay. The grid increased the total applied load cycles value by **238%** compared with the plain composite structure asphalt overlay AK/AC. This result proves that reinforcing grid placed in the lower part of the overlay blocked crack propagation in some extent.

The failure of the composite Glass grid reinforced asphalt concrete AK/ GGRAC was not complete, the ductility was obvious, and the glass grid strands were not broken completely. Also the crack propagation line grew horizontally in some stages, other cracks were initiated beside the main one, but all of them started from the pre-made notch. The addition of chopped glass fibers to the composite overlay AK/GFRAC improved the resistance to the crack propagation, but not as much as the reinforcing glass grid did. The chopped glass fibers were increased the total applied load cycles value by **155%** compared with the plain composite structure asphalt overlay AK/AC-10I. From Figure 7 the composite structure asphalt overlay reinforced with glass grid AK/GGRAC showed high deformation under load cycles compared to the other types at the beginning but without increasing crack growth rate , and the failure of composite AK/ GGRAC happened at displacement value of **30%** higher than it is for the plain asphalt

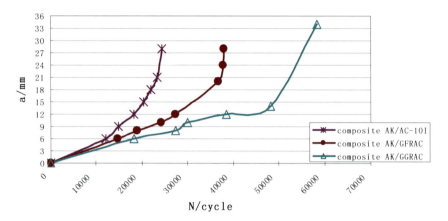

Fig. 7. Crack lengths vs. load cycles for composite anti-cracking overlays systems

concrete, actually, from the curves in Figure 7 we can notice that at the last stage of the crack propagation, the curve becomes horizontal in plain and chopped fibers reinforced composite asphalt overlay, while it continues more vertical in glass grid reinforced composite asphalt overlay, that makes it clear that the composite AK/ GGRAC has higher bearing capacity along the crack propagation process until failure.

5 Conclusions

1) Glass grid reinforcement prevents further opening of the crack due to its low rate of elongation and its high tensile strength.
2) Chopped glass fibers in asphalt concrete have a bridging effect, and the cracking development occurs in steady progress from the beginning until the failure stage, with low crack propagation velocity along the cracking line. Under high speed loading rate and low temperature condition, when the crack was initiated, the reinforcement's efficiency to retard crack propagation is sharply decreased
3) Crack growth rate both in plain and reinforced asphalt concrete grow faster at the beginning then the rate of displacement decreases and again becomes faster until the failure, and it may not increase in certain steps especially with reinforced asphalt concrete.
4) Glass grid composite structure reinforced asphalt concrete AK/GGRAC gives best result in retarding crack propagation among other composite structure asphalt overlay types.
5) The values of A and n in Paris law are obtained for three different asphalt concretes, which is of importance in the prediction of fatigue life of the asphalt mix.

References

[1] Monismith, C.L., et al.: Reflection cracking: analyses, laboratory studies, and design considerations. Proceedings of Association of Asphalted Paving Technologies 49 (1980)
[2] Majidzadeh, K., et al.: Improved methods to eliminate reflection cracking, Report FHWA/RD-86/075. In: Federal Highway Administration, Washington, D.C. (1985)
[3] Lytton, R.L.: Use of geo-textiles for reinforcement and strain relief in asphalt concrete. Geo-textiles and Remembrances 8(3), 26–45 (1989)
[4] Ye, G.Z.: Fatigue shear and fracture behavior of rubber bituminous pavement, Reflective Cracking in Pavements. In: Proceeding of the third international RILEM Conference, Netherlands, Maastricht (1996)
[5] Zhang, D.L.: Asphalt Pavements. China Communication Press, Beijing (1998)
[6] Asphalt Pavements Constructions and Tests Norm (GBJ92-93). China Communication Press, Beijing (2001)
[7] Najd, A.: Evaluation of Asphalt Overlays Anti-cracking Systems over Existing Cracked Asphalt Pavements. A Dissertation of Chang'an university (2005)
[8] Braham, A.F., et al.: The effect of long-term laboratory aging on asphalt concrete fracture energy. Journal of the Association of Asphalt Paving Technologists 78, 416–454 (2009)
[9] Baek, J., Al-Qadi, I.L.: Reflective cracking: modeling fracture behavior of hot-mix asphalt overlays with interlayer systems. Journal of the Association of Asphalt Paving Technologists 78, 789–828 (2009)

Asphalt Rubber Interlayer Benefits in Minimizing Reflective Cracking of Overlays over Rigid Pavements

Shakir Shatnawi[1], Jorge Pais[2], and Manuel Minhoto[3]

[1] Shatec Engineering Consultants, LLC, California, USA
[2] University of Minho, Portugal
[3] Polytechnic Institute of Bragança, Portugal

Abstract. This paper provides an overview of the asphalt rubber interlayer benefits on reflective crack retardation in overlays over rigid pavements. These interlayers are known in California as asphalt rubber absorbing membrane interlayers (SAMI-R) or as asphalt rubber aggregate membrane interlayers (ARAM-I) chip seals. The paper focuses on the performance in terms of field project reviews, laboratory performance tests and finite element analysis. SAMI-R has been given a reflective cracking equivalent thickness of 15 mm of asphalt rubber hot mix overlays or 30 mm of dense graded hot mix overlays. The finite element analysis confirms the quantified reflective cracking benefits of SAMI-R and provides optimum design alternatives to conventional dense grades asphalt concrete overlays. The paper concludes that SAMI-R is effective in minimizing reflective cracking distress and in extending pavement life.

1 Introduction

In the rehabilitation of rigid pavement structures using asphalt concrete overlays, interlayers are often used to minimize reflective cracking. One of these types of interlayers are asphalt rubber chip seals which possess low stiffness and high deformability. These interlayers dissipate the stress and strain energies that accumulate at the crack and joint tips of an rigid pavement which would otherwise get transferred to the underside of the HMA overlay. The dissipation of these high level stresses and strains minimizes the potential of reflective cracking in the HMA overlays. Reflective cracking is considered a major pavement distress which occurs as a result of cracks that reflect through an HMA overlay from cracks or joints of an existing pavement. Another additional benefit of an interlayer is its ability to prevent water intrusion into the lower layers of the pavement structure; thus protecting the structural integrity of the pavement system.

Asphalt rubber chip seals have been used effectively as interlayers over distressed flexible and rigid pavements, and as a surface treatment [1]. In California, these interlayers are known as rubberized stress absorbing membrane interlayers (SAMI-R) or asphalt rubber aggregate membrane interlayers (ARAMI) [2,3], and are often used interchangeably in the pavement technical literature and throughout this white paper. A schematic of a pavement section showing a typical ARAMI is shown in Figure 1. When used as a surface layer such as rubberized open graded friction course, it is called asphalt rubber aggregate membrane (ARAM) or simply a rubberized chip seal.

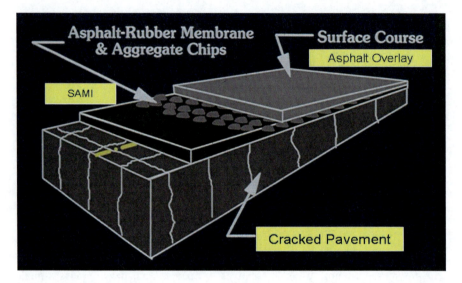

Fig. 1. Schematic of a cracked pavement receiving an ARAMI prior to an HMA overlay

Interlayers can extend the life of preservation and rehabilitation strategies. The magnitude of life extension depends on many factors including existing pavement condition, traffic loading, climatic and environmental conditions, and the type and engineering properties of the interlayer used [1]. The excellent performance of these interlayers is primarily due to the (1) unique elastic properties, and (2) superior aging characteristics of the asphalt rubber binder which can withstand as much as five times more strain than the unmodified asphalt binder [4].

2 Performance of Interlayer Systems

The merits of ARAMI's have been proven both in the field and the laboratory. Additionally, a number of analytical studies using Finite Element Methods (FEM) have demonstrated the efficacy of interlayer systems in minimizing the potential of reflection fatigue cracking in HMA surface courses. In the following, a brief discussion of some related performance studies is presented.

Field performance of many projects in California and Arizona, since the 1970's. has shown the significant benefits of SAMI-R in retarding reflective cracking on HMA overlays [5-12]. These studies concluded the effectiveness of SAMI-R in reflective cracking retardation and the superiority of the pavement systems incorporating SAMI-R's to those without SAMI-R's.

Many laboratory experiments were conducted to investigate the effectiveness of SAMI-R's in retarding reflection cracking in new HMA overlays. Recently, Bin et al. [13] conducted a laboratory simulation study using the Hamburg wheel tracking test to compare the relative reflective cracking performance of various types of interlayers that were placed below a hot mix asphalt overlay over an existing cracked pavement. These interlayers included SBS modified asphalt sand, Asphalt rubber sand, Fiber glass polyester mat and SAMI-R in addition to a control section without an interlayer. Note, the SAMI-R is the interlayer type modeled in this paper. The tests were conducted at a rate of 52 cycles per minute to simulate the development of reflective cracking under a moving load. The test specimens were simply supported beams conditioned at -20 °C for 5 hours prior to load conditioning by the application of 8000 loading cycles to stabilize the deflection. The specimens were then subjected to loading cycles until failure; which was described as the first appearance of a crack at the bottom of the surface layer. It was found that the use of an interlayer would extend pavement life significantly when compared with the option of not using an interlayer. In addition, the SAMI-R interlayer was superior to the other types of interlayers tested in these experiments in retarding reflective cracking and extending the life of the overlay.

Additionally, a number of analytical studies have been conducted using finite element analysis to theoretically investigate the contribution of SAMI-R's to the performance of rigid and flexible pavement systems. Among the early studies are those conducted by Coetzee and Monismith [14] and Chen et al. [15]. In Coetzee and Monismith [14], 48 simulations representing various configurations of cracked concrete pavements overlaid with a rubberized stress absorption membrane followed by an asphalt overlay. In the simulations, the effect of many variables was studied including the asphalt concrete overlay modulus (varied between 100,000 psi to 1,500,000 psi) and thickness (varied between 2 to 4 inches). The modulus of the interlayer was assumed between 1,000 and 20,000 psi with a thickness between 0.125-0.5 inch. The concrete layer modulus was varied between 1,000,000 and 4,000,000 psi and its thickness between 4 and 8 inches. A crack in the concrete layer 0.25-0.5 inch wide was assumed. Finally, the base layer was assumed 12 inch thick and 20,000 psi modulus and the subgrade modulus was varied between 5000-10,000 psi. A general purpose 2-D finite element program was used in the analysis. The analysis confirmed the effectiveness of the low-modulus interlayer in reducing the crack tip effective stress (described by the Von Mises criterion), and the inhibition of reflection cracking resulting from both load (traffic loading) and temperature changes (thermal loading). The study found a significant reduction in crack tip stress with the use of the rubber asphalt interlayer. This was found to be more pronounced in those cases where the overlay modulus is 0.1-0.25 that of the cracked PCC layer. The study has also shown that crack width, interlayer modulus, and overlay thickness have significant effect on the crack tip stress, but that the ratio of overlay modulus to cracked layer modulus appears to be more influential.

Chen et al. [15] analyzed the Arizona three-layer thin-overlay system with the use of a 2-D finite element program in which an asphalt rubber concrete is placed in two lifts each 5/8 inch thick and a low-modulus asphalt rubber interlayer 3/8 inch think placed in between. The bottom lift may be considered as a leveling course. This system, also called SAMI-R in Arizona, was commonly placed on top of cracked Portland cement concrete pavements prior to overlaying with hot mix asphalt. In the analytical studies, a 9 inch PCC layer with a crack 0.3 inch wide was assumed. An HMA overlay of various thicknesses placed over the interlayer system was analyzed under the effect of both moving traffic and thermal loadings. The results of the analysis indicated the significant benefits of using the interlayer system in reducing the critical stresses and strains and in dissipating the stress concentrations at the crack tip for HMA overlays placed over rigid pavements. The study demonstrated a significant reduction in both the effective stress and shear stress above the crack tip upon using an interlayer due to both temperature changes and traffic loading. It was also observed that upon incorporating an interlayer, the effect of overlay thickness becomes less critical leading to more economical designs.

3 Finite Element Analysis

Two dimensional finite element analyses was conducted to study a number of factors and their influence on the performance of rigid and flexible pavement systems incorporating ARAMI's (SAMI-R's) in comparison with systems that did not include these interlayers. Table 1 provides a summary of the parameters used in the analysis. A total of 36 scenarios involving various variables pertaining to concrete pavements overlaid with HMA were analyzed. Cracks 3 mm wide with a 60 cm spacing were assumed. The finite element models studied consisted of an HMA overlay with or without ARAMI layer, with or without leveling course on top of a rigid cracked pavement; thus representing four types of configurations.

Two types of ARAMI's, varying in their stiffness, were used in the analysis. The ARAMI was assumed to be orthotropic with regard to its modulus. The "Soft" ARAMI has a modulus of 7 MPa in the horizontal direction and 100 MPa in the vertical direction. The "Hard" ARAMI was assumed to have a modulus of 35 MPa in the horizontal direction and 100 MPa in the vertical direction. Whenever used, the ARAMI thickness was assumed to be equal to 1.0 cm.

The 2-D finite element model used in the analysis represents an HMA overlay with or without ARAMI layer, with or without leveling course on top of a rigid cracked pavement which, in turn, rest on top of a granular base layer and a subgrade layer. The model was designed considering the existence of full friction between the old and new pavement layers. The materials were modeled assuming a linear elastic behavior.

The mesh of the model was designed as a plain strain problem, by using quadrilateral, two-dimensional structural-solid elements, with eight nodes, with two degrees of freedom at each node. The mesh was designed to apply a load with a dual wheel configuration representing a standard axle wheel of 80 kN, applied on the pavement surface in a representative area of the tire-pavement contact. The finite element model used in the numerical analysis was developed in a general finite elements code, ANSYS(R) Academic Teaching Introductory, V12.1. The 2-D finite element model used in the analysis considers typical values for thickness and stiffness as indicated in Table 1. Many strains and stresses (X, Y, and XY) were determined with the finite element models analyzed at a number of locations; both within the interlayer and in the overlays, as shown in Figure 2.

In cases where a leveling course may be used prior to placement of the interlayer and subsequently the HMA overlay, the Von Mises strain was calculated at the interface between the leveling course and ARAMI and between the ARAMI and the overlay. Also, Von Mises stresses were calculated in the same locations as for strains. Also included for the analysis is the configuration where a rigid pavement receives only a leveling course without any interlayer then an HMA overlay, as shown in Case 19-24 of Figure 2.

In this paper, only the effective stresses and strains defined by Von Mises criterion will be used in evaluating the benefits of the interlayers. The Von Mises stresses and strains have been used by many researchers in evaluating pavement systems [9, 14, 15]. The Von Mises stress is calculated from the principal stresses according to the following equation:

$$\sigma_{VM} = \sqrt{\frac{(\sigma_1 - \sigma_3)^2 + (\sigma_2 - \sigma_3)^2 + (\sigma_1 - \sigma_2)^2}{2}}$$

Table 1. Material properties used in the finite element analysis of rigid pavements

Input parameter	Values
HMA overlay thickness (cm)	2.0, 6.0, 12.0
HMA overlay stiffness (MPa)	2000, 4000
ARAMI thickness (cm)	0 (none), 1.0
ARAMI stiffness (MPa) (Horizontal, Vertical)	Case 1 (Soft interlayer): (7,100) Case 2 (Hard interlayer):(35,100)
Leveling course thickness (cm)	0 (none), 3.0
Leveling course stiffness (MPa)	Equal to that of the HMA overlay
Existing PCC thickness (cm)	20.0
Existing PCC stiffness (MPa)	20,000
Existing PCC crack spacing (cm)	60.0
Existing PCC crack opening (cm)	0.3
Aggregate base layer thickness (cm)	20.0
Aggregate base layer stiffness (MPa)	270
Subgrade stiffness (MPa)	35

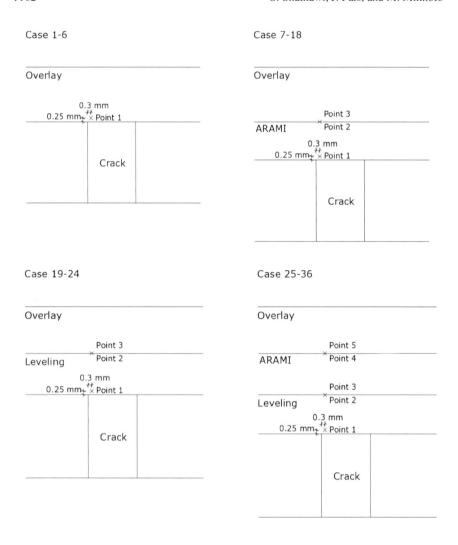

Fig. 2. Location of the points used in the analysis of rigid pavements

where σ_{VM} is the Von Mises stress, and σ_1, σ_2, and σ_3 are the major, intermediate, and minor principal stresses, respectively. A similar equation may be written for the Von Mises strain as follows:

$$\varepsilon_{VM} = \frac{1}{1+\upsilon}\sqrt{\frac{(\varepsilon_1-\varepsilon_3)^2+(\varepsilon_2-\varepsilon_3)^2+(\varepsilon_1-\varepsilon_2)^2}{2}}$$

where ε_{VM} is the Von Mises strain, and ε_1, ε_2, and ε_3 are the major, intermediate, and minor principal strains, respectively. For a 2-D system, the above two equations are reduced by assuming $\varepsilon_3=0$ and $\sigma_3=0$. In order to study the benefits of interlayers in extending the fatigue life of the HMA overlay, the Von Mises stresses and strains were calculated at the underside of the HMA overlay for all the systems with and without interlayers. Once these stresses and strains are calculated, they may be used with appropriate transfer function to calculate fatigue life.

Figure 3 shows the Von Misses stress that develops at the underside of the HMA overlay in the various overlay configurations studied. Similarly, Figure 4 shows the calculated Von Mises strain at the bottom of the HMA overlay.

In order to study the effect of using ARAMI on the fatigue reflective cracking performance of HMA overlays placed over cracked rigid pavements, a transfer function would be required to describe the rate of deterioration as function of the strain at the bottom of the overlay. Since a Von Mises type of strain was used, it would be necessary to have a transfer function with such a strain in its statement. Sousa et al. [9] provided such equations for HMA and gap graded rubberized hot mix asphalt (RHMA-G) as shown in Figure 5. In these equations, the fatigue life represents the number of loading cycles until crack initiation and it does not consider crack propagation. In this study, the overlay is made of HMA (dense graded asphalt concrete) and therefore, the corresponding transfer function shown in Figure 5 will be used in computing the fatigue life of the overlay in terms of repetitions of the 80-kN axle load (1 ESALs). Figure 6 shows the calculated fatigue life for three HMA thicknesses (2, 6, and 12 cm) and the 12 pavement configurations used in the analysis (total 36 cases).

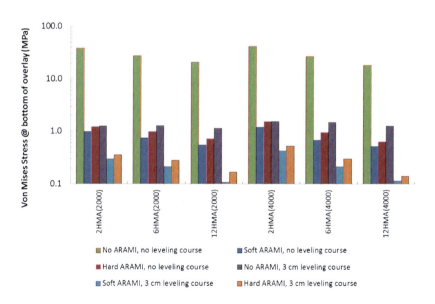

Fig. 3. Von Mises stress at the bottom of the HMA overlay for the rigid pavement configurations

The notation used to describe the configurations in Figure 6 consists of the HMA overlay thickness in cm followed by the HMA stiffness, in parentheses, in MPa. For example, 2HMA(2000) represents the cases where the HMA overlay is 2 cm thick and of 2000 MPa modulus. In each group, a number of cases is considered as described in the legend of Figure 6. The control case is with no ARAMI and no leveling course and represented with the green bar in Figure 6. The soft and hard interlayers are described by their modulus as shown in Table 1. Inspection of Figure 6 reveals the following:

- Strains and stresses are largest for the control cases (i.e., pavement structures without ARAMI or leveling course).
- Using a soft ARAMI results in reduced levels of stress and strain compared to using hard ARAMI.
- The use of a 3 cm leveling course tends to significantly reduce the stresses and strains compared to similar cases without the leveling course. The additional leveling course provides a structural layer that reduces the strain and stress at the bottom of the HMA overlays.
- Increasing the HMA overlay thickness up to 12 cm (when used without ARAMI or leveling course) was not able to reduce the strain and stress to the levels achieved when using ARAMI even with the thinnest HMA of 2 cm. A similar trend was observed in the studies by Coetzee and Monismith [14] and Chen et al. [15]. For example, considering the 2000 MPa modulus HMA, a 12 cm overlay without ARAMI or leveling course would produce a stress of ~20 MPa whereas using an ARAMI with a 2 cm HMA overlay resulted in a stress of only 1 MPa. Similar trends were observed with the Von Mises strains shown in Figure 4. The use of a leveling course tends to diminish the benefits of using thicker overlays by always producing nearly same level of stress and strain regardless of HMA thickness. As an example, for the 2000 modulus HMA, the use of 3 cm leveling course without ARAMI always produced about 1 MPa of stress regardless whether a 2 cm, 6 cm, or 12 cm HMA overlay was used.
- Without ARAMI or leveling course, the HMA overlays exhibit shorter life than when an interlayer or a leveling course was used.
- The use of soft ARAMI resulted in greater extension in overlay life compared to hard ARAMI.
- Without an ARAMI, the HMA overlay tends to fail immediately upon loading due to experiencing "exceptionally high" levels of Von Mises strain in the range of 0.01-0.02 (see Figure 6). It is questionable, however, if the transfer function (Figure 5) used in calculating the fatigue life is valid for this level of strain.

As can be seen, there is significant increase in the life of the overlay with the use of ARAMI. This is due to reduced level of strain upon using these interlayers. It is believed that the interlayers absorb a great amount of the stress and strain and as such only small amount of these stresses and strains arrives at the underside of the overlay.

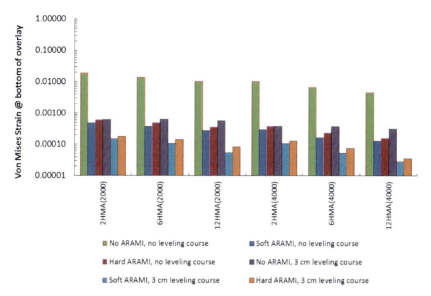

Fig. 4. Von Mises strain at the bottom of the HMA overlay for the rigid pavement configurations

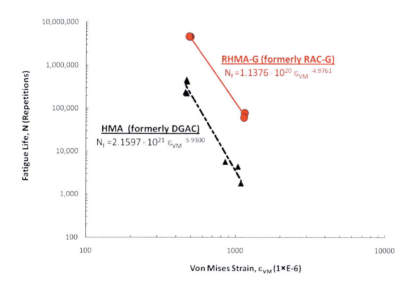

Fig. 5. Fatigue transfer function for HMA and ARHMA mixes (from Sousa et al. [9])

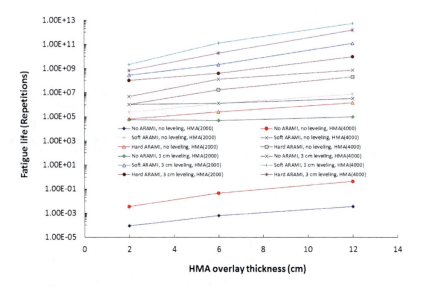

Fig. 6. Fatigue life as function of overlay thickness for the 12 concrete pavement configurations analyzed

4 Conclusions

This white paper has demonstrated the benefits of using overlay systems with asphalt rubber interlayers. The ARAMI (or SAMI-R) has consistently been shown to reduce reflective cracking when used as part of preservation and rehabilitation strategies. Field studies, accelerated wheel tracking experiments, laboratory testing, and analytical studies have all confirmed the significant contribution of these interlayers in extending pavement life and in minimizing reflective cracking in hot mix asphalt overlays. In this paper, 36 rigid pavement configurations representing a variety of cases were modeled and analyzed using the finite element method (FEM) to quantify the benefits of the interlayers in these systems when subjected to loading. The FEM analysis validated the outstanding performance of these composite systems when interlayers were incorporated, and further quantified the benefits of these ARAMI's in terms of critical stress and strain reduction and related pavement life extension. A stress reduction ranging from 92% to 98% was achieved with the use of ARAMI compared to non-ARAMI system. Soft ARAMI's were found to be more effective in reducing stress and strain levels compared to hard ARAMI's. It was also found that the use of leveling course below the interlayers was very beneficial in lowering the strain levels and in increasing pavement life.

References

1. MTAG: Maintenance Technical Advisor Guide (MTAG), California Department of Transportation, Sacramento, CA (2003)
2. Caltrans: Standard Specifications. California Department of Transportation (2006)
3. Greenbook: Greenbook-Standard Specifications for Public Works Construction, 2004th edn., Building News, Anaheim (2004)
4. Green, E.L., Tolonen, W.J.: The Chemical and Physical Properties of Asphalt-Rubber Mixtures, Report No. ADOT-RS-14, vol. (162) (1977)
5. Way, G.B.: Prevention of Reflective Cracking Minnetonka-East, Report Number 1979 GWI. Arizona Department of Transportation (1979)
6. Schnormeier, R.H.: Fifteen Year Pavement Condition History of Asphalt Rubber Membranes in Phoenix. Asphalt Rubber Producers Group, Arizona (1985)
7. de Laubenfels, L.: Effectiveness of Rubberized Asphalt in Stopping Reflection Cracking of Asphalt Concrete. California Department of Transportation, FHWA/CA/TL-85/09 (1988)
8. Predoehl, N.H.: Performance of Asphalt-Rubber Stress Membranes (SAM) and Stress Absorbing Membrane Interlayers (SAMI) in California. California Department of Transportation (1990)
9. Sousa, J., Pais, J., Way, G., Saim, R., Stubstad, R.: A Mechanistic-Empirical Overlay design Method for Reflective Cracking, In: Transportation Research Record, TRB National Research Council, Washington, D.C., pp. 209–217 (2002)
10. Shatnawi, S., Holleran, G.: Asphalt Rubber Maintenance Treatments in California. In: Proceedings of Asphalt Rubber 2003 Conference, Brasilia, Brazil (2003)
11. Van Kirk, J.: Maintenance and Rehabilitation Strategies Using Asphalt Rubber Chip Seals. Proceedings of Asphalt Rubber 2003 Conference, Brasilia, Brazil (2003)
12. Van Kirk, J.: Multi-Layer Pavement Strategies Using Asphalt Rubber Binder. Proceedings of Asphalt Rubber 2006, Palm Springs, California (2006)
13. Bin, Y., Baigang, C., Jun, Y.: Lab Simulation of Reflective Cracking by Load. Proceedings of the AR 2009 Conference, Nanjing, China (2009)
14. Coetzee, N.F., Monismith, C.L.: An Analytical Study of the Applicability of a Rubber Asphalt Membrane to Minimize Reflection Cracking in Asphalt Concrete Pavements. Report to Arizona Refining Company and U.S. Rubber Reclaiming Company, Inc. (1978)
15. Chen, N.J., Divito, J.A., Morris, G.R.: Finite Element Analysis of Arizona's Three-Layer Overlay System of Rigid Pavements to Prevent Reflective Cracking. Association of Asphalt Paving Technologists (1982)

Performance of Anti-cracking Interface Systems on Overlaid Cement Concrete Slabs – Development of Laboratory Test to Simulate Slab Rocking

Katleen Denolf, Joëlle De Visscher, and Ann Vanelstraete

Belgian Road Research Centre (BRRC), Woluwedal 42, B-1200 Brussels

In the past extensive testing was performed at BRRC to determine the laboratory performance of anti-cracking interfaces in case of stresses induced by thermal variations. Several experimental sites were followed during that period to set up correct laying procedures for the different types of products and to give recommendations when applying these products. Recently this study was completed by a four year research project. Important objectives of this project were: the setup of a new test to simulate the vertical movements induced by traffic on overlaid cement concrete slabs, the evaluation of the performance of different anti-cracking interface systems with this new test and the validation of these results by experimental field trials. In this article this new laboratory test to simulate the vertical movement is described in detail and the performances of a reference and four types of anti-cracking interface systems, being a stress absorbing membrane interlayer (SAMI), a geogrid, a geocomposite and a steel reinforcing netting, are compared.

1 Introduction

When an asphalt layer is applied to a pavement of cement concrete slabs, the joints will initiate cracks in the asphalt layer. Such cracks grow from the bottom to the top of the asphalt overlay at an average speed of 2 to 3 cm a year depending on traffic. Experimental sites showed that without an anti-cracking interface system, reflective cracking almost certainly appears at the surface of a 5 cm thick asphalt overlay within 3 years. With an anti-cracking interface system (meaning the complex consisting of an appropriate adhesive layer and an anti-cracking product), reflective cracking can be delayed to 8 years or more, depending on the state of the concrete pavement, traffic and the type of interface system. The main causes for reflective cracking can be related to the movements that occur near the cracks/joints. A first type of movement is the slow horizontal movement by repeated thermal expansion and contraction of the cement concrete as depicted in figure 1.a. A second type of movement is the vertical movement at joints. These movements are caused by the traffic and a loss of carrying capacity of the underlay as shown figure 1.b (shear and deflection). To determine the performance of anti-cracking interface layers in case of thermal (horizontal) movement a laboratory test is available at BRRC since several years. In this research BRRC developed a new test method to simulate slab-rocking (vertical shear movement).

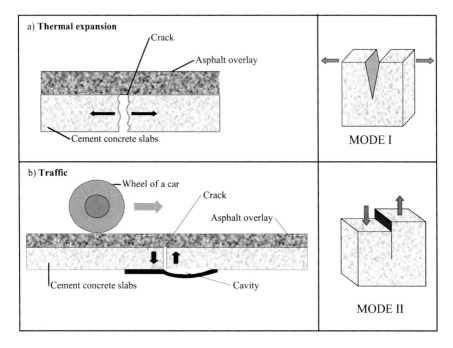

Fig. 1. Schematic overview of two causes of reflective cracking: (a) slow horizontal movement by repeated thermal expansion and contraction of the cement slabs (mode I) and (b) vertical movement at joints and cracks caused by traffic and a loss of carrying capacity of the underlay (mode II).

Development of a new test method to simulate vertical load on cement concrete slabs with an asphalt overlay

A first measurement setup (see figure 2.a) was developed and extensively tested during a first measurement programme. After analysing these results, this setup was modified and optimised (see figure 2.b). A second measurement programme was executed in this optimised setup.

The samples used in both setups have a length of 60 cm and a width of 14 cm. They are composed of a cement concrete support with a thickness of 7 cm covered with an adhesive layer and an anti-cracking product, except for the reference sample where only an adhesive layer was used. Afterwards an asphalt overlay with a thickness of 6 cm was applied. To simulate a joint, a discontinuity of 4 mm wide was implemented in the cement concrete support as shown in figure 2.

The test is executed in a climate chamber at 15°C. Before starting a measurement the test specimens are conditioned at 15°C for 4 hours. To simulate slab rocking a dynamic vertical movement with a frequency of 1Hz is induced by the stamp on one side of the joint: during half a second the stamp exerts a force on the sample, the next half second the sample is not loaded. This cycle is repeated until the sample breaks. On the other side of the joint the sample is clamped and supported by a metal frame as shown in figure 2. During the test, the vertical position of the stamp as a function of time is measured.

To take into account the loss of carrying capacity of the underlay the right hand side of the sample was only supported by a metal roller at its far end in the initial setup (see figure 2.a). After implementing a camera to obtain a detailed image of the crack formation it became clear that in this setup the intended purpose to simulate the pumping of the cement concrete slabs was not entirely achieved because of the total lack of support of the test sample on the right hand side of the joint. To increase the pumping effect we decided to fill the cavity under the sample with a piece of soft foam rubber which offers a limited support to the cement concrete.

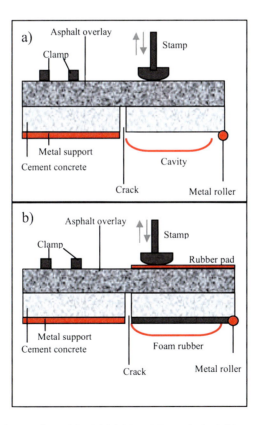

Fig. 2. Schematic overview of the initial (a) and the optimised (b) measurement setup

In the initial setup the stamp exerted a cyclic force of 4 kN on the test specimen. Failure of reference samples was recorded after maximum 1.5 hours of testing. In the optimised setup the reference samples did not fail within 7 hours of measuring when the same load of 4 kN was used. After analysing additional measurements on references with forces of 8 kN, 12 kN and 16 kN, it was decided to use a cyclic force of 12 kN in the optimised setup. A rubber pad was implemented between the stamp and sample to obtain a more homogeneous distribution of the applied force. A picture of the optimised setup can be found in figure 3.

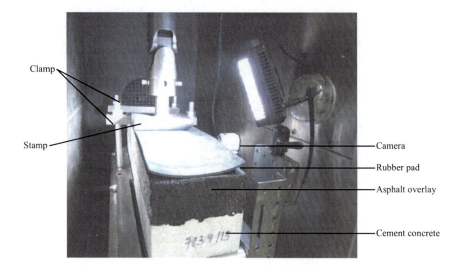

Fig. 3. Picture of the optimized setup

Figure 4.a shows a typical measurement. The slope of the quasi linear part and the testing time corresponding to the inflection point, are two important parameters in this research. The slope of the linear part is a measure for the evolution of micro-cracks or the gradual failure of the test specimen. The inflection point corresponds fairly well with the failure time of the sample and is a measure for the evolution of the macro-cracks. Both parameters are used to evaluate the resistance to reflective cracking of the tested sample.

Figure 4.b zooms in on the linear part of the curve shown in figure 4.a and illustrates clearly the dynamic vertical displacement of the stamp. In figures 4.c, 4.d and 4.e camera images of a sample during a test are depicted. The depicted asphalt overlay was coloured white with chalk to get a better view of the evolution of the cracks. Figure 4.d shows an image of the test specimen at the inflection point and in figure 4.e the evolution of the cracks after failure is depicted.

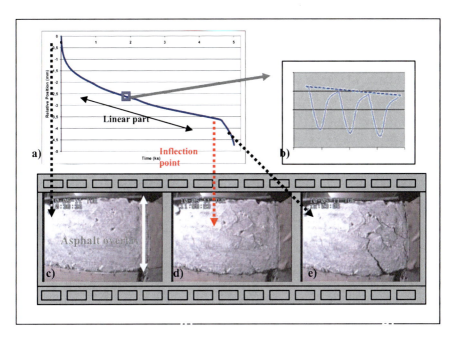

Fig. 4. A typical measurement result (a); a zoom of the linear part of the curve (b) and camera images of a sample during the test: a sample at the beginning of the test (c); at the break point (d) and after the break point (e)

2 Experimental Test Programme

In this research the performances of a reference and four types of anti-cracking interface systems, being a stress absorbing membrane interlayer (SAMI), a geogrid, a geocomposite (composed of a geogrid and a nonwoven geotextile) and a steel reinforcing netting, are compared. As mentioned previously two measurement programmes were carried out. A first one in the initial setup (figure 2.a) and a second one in the optimised setup (figure 2.b).

2.1 Preparation of the Test Samples

As mentioned above the test samples used in this research have a length of 60 cm and a width of 14 cm. The base layer was made of 2 cement concrete blocks, separated by a discontinuity of 4 mm to simulate a crack/joint. The top of the concrete blocks was brushed in order to get a rougher surface to improve adhesion with the upper layers. The top layer was an asphalt concrete wearing course with a thickness of 6 cm. The mixture and the compaction of the asphalt layers were done in the laboratories of BRRC in accordance with the European standards EN 12697-35 and EN 12697-33.

Between the base layer of cement concrete and the asphalt top layer an adhesive layer and an anti-cracking interface product was installed, with exception of the reference samples where only an adhesive layer was applied, as described below:

- Reference test samples
    - An emulsion of 237 g/m² (residual binder content) without polymers was spread on the concrete base
- SAMI test samples
    - A polymer modified bitumen of 2 kg/m² was spread on the concrete base
    - A quantity of 5 kg/m² of coarse aggregate 6.3/10 was spread
- Geogrid test samples
    - An emulsion of minimum 237 g/m² (residual binder content) without polymers was spread on the concrete base
    - The geogrid interface system was placed
    - A polymer modified bitumen of 1.2 kg/m² was applied
    - A quantity of 5 kg/m² of coarse aggregate 6.3/10 was spread
- Geocomposite test samples
    - A polymer modified emulsion of 700 g/m² (residual binder content) was spread on the concrete base
    - A geocomposite interface system was placed
    - A polymer modified emulsion of 500 g/m² (residual binder content) was spread on top of the geocomposite
    - A quantity of 5 kg/m² of coarse aggregate 6.3/10 was spread
- Steel reinforcing netting test samples
    - An emulsion of minimum 237 g/m² (residual binder content) without polymers was spread on the concrete base
    - A steel mesh interface system was placed
    - A slurry surfacing of 18 kg/m² was spread

Note that the application of an emulsion layer with spreading of coarse aggregate (in the case of grids and geocomposites) has two functions:

- it prevents the interface product from cracking when site traffic is passing before overlaying
- it keeps the system flat during overlaying.

It must be noticed that the preparation of the test samples is a crucial phase in this research. Samples that are badly prepared may lead to unreliable measurement results. After the preparation of the samples they should also be handled with care. Wooden supports for the samples were created to avoid damaging of the samples during transport.

2.2 Measurements in the Initial Setup

In the initial setup the following samples were tested:

- 3 references without an anti-cracking product
- 6 SAMI's
- 3 geogrids
- 3 geocomposites
- 4 steel reinforcing nettings

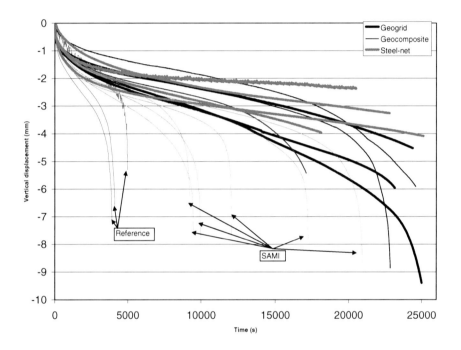

Fig. 5. Measurement results of the initial setup

Figure 5 shows the results of the first measurement programme. The graphs represent the vertical movement of the stamp (in mm) as a function of time (in s). Two interesting parameters can be derived from these graphs:

- the slope of the linear part from each curve as a measure for the micro-cracks or the gradual failure of the samples
- the inflection point in each curve as a measure for the macro-cracks or the time of failure or lifetime of the test specimen

Table 1. Overview of the slope of the linear part and the time of failure for the five different types of test samples measured in the initial setup

Test sample	Slope of the linear part [mm/ks]	Lifetime [s]
Reference	-0.56 ± 0.16	4300 ± 1400
SAMI	-0.19 ± 0.07	13200 ± 5100
Geogrid	-0.14 ± 0.03	> 22000
Geocomposite	-0.11 ± 0.01	21500 ± 9700
Steel reinforcing netting	-0.08 ± 0.03	> 22000

Table 1 leads to the following conclusions and tendencies:

- The samples with an anti-cracking interface layer have a less negative slope than those without one
- The slope of the SAMI samples is more negative than the slope of the samples with a steel reinforcing netting
- The samples with an anti-cracking interface layer have a longer lifetime than those without one
- The lifetime of the SAMI samples is shorter than the lifetime of samples with a geogrid or a steel reinforcing netting
- Within the precision of the test the slopes of the SAMI and geogrid samples can be distinguished
- Although the precision of the test does not allow to distinguish the performances of the SAMI and the geogrid samples, the geogrid tends to perform better
- The steel reinforcing netting samples do not fail within the total duration of the test
- Two out of three geogrid samples do not fail within the total duration of the test

2.3 Measurements in the Optimised Setup

In the optimised setup the following samples were tested:

- 5 references without an anti-cracking interface system
- 4 SAMI's
- 4 geogrids
- 4 geocomposites
- 4 steel reinforcing nettings

The results of these measurements are summarized in table 2.

Performance of Anti-cracking Interface Systems

Table 2. Overview of the slope of the linear part and the lifetime for the five different types of samples measured in the optimised setup

Test sample	Slope of the linear part [mm/ks]	Lifetime [s]	Remarks
Reference 1	-1,91	1180	
Reference 2	-1,88	860	
Reference 3	-1.17	1410	
Reference 4	-1,03	1790	
Reference 5	-0.03	No failure	
SAMI 1	-0.33	4940	
SAMI 2	-0.18	8080	
SAMI 3	-11.81	270	
SAMI 4	-0.15	13450	
Geogrid 1	-2.79	1640	
Geogrid 2	-0.02	No failure	
Geogrid 3	-2.54	730	
Geogrid 4	-0.14	9620	
Geocomposite 1	-0.37	5030	
Geocomposite 2	-3.09	590	
Geocomposite 3	-0.45	2730	
Geocomposite 4	-0.05	No failure	
Steel reinforcing netting 1	-0.02	No failure	
Steel reinforcing netting 2	(-0.22)	(5030)	Difficulties in preparation *
Steel reinforcing netting 3	(-1.95)	(1010)	Difficulties in preparation*
Steel reinforcing netting 4	-0.02	No failure	

*It should be noticed that there were some difficulties in the preparation of the test specimens with the steel reinforcing netting. The steel meshes used for the sample preparation were stored in rolls. Because of the relative small dimensions of our specimen (14 cm by 60 cm) it was rather difficult to completely flatten the steel mesh. For sample 2 and 3 the net curled a bit at the sides which made it impossible to cover the complete net with a slurry surfacing.

Table 2 leads to the following conclusions and tendencies:

- Despite the great care taken in the preparation and handling of the samples there still is a large dispersion in measurement results.
- It is required to test at least four samples per product.
- For many interface layers and in particular the steel reinforcing nettings and geogrids, a correct positioning remains very difficult. The limited size of the specimens makes it hard to place the interface layers in a completely flat and fixed way. The use of nailing could solve this problem. Although this does not correspond with the practice on site, it at least ensures a correct functioning of the product in a laboratory test.

- It is possible to discriminate between good and bad performing systems, but a finer discrimination seems difficult:
    o In general the references performed the worst
    o When prepared correctly the steel reinforcing netting samples performed the best
    o It is difficult to distinguish the performance of SAMI, geogrids and geocomposites
- The ranking obtained with the optimised setup matches the ranking of the initial setup

2.4 Additional Analysis

Because of the rather large dispersion in the measurement results of the optimised setup it was decided to carry out further analyses on the samples to ensure that the dispersion between samples of a given type was not caused by differences in their preparation. The following additional analyses were performed:

- Direct tensile tests to verify the adhesion between the concrete, the anti-cracking interface and the asphalt layer
- Determination of the void content of the asphalt layer of the test samples to verify the compaction of the asphalt layer
- Determination of the binder content and grading of the asphalt layer of the test samples to verify the composition of the asphalt mixture

These analyses were executed on the best and worst performing sample of each type of anti-cracking interface system of the optimised setup.

Direct tensile test. The specimens used in the direct tensile test have a length and a width of 80 mm. The height depends on the used anti-cracking product. Metal plates are attached at both sides with two-component epoxy adhesive. The specimens are conditioned at 10°C (± 1°C) for at least 4 hours before testing. To test the specimens, they are fitted with a clamping system and an appropriate base (figure 6) in a tensile testing machine. A tensile load is applied in displacement controlled mode with a deformation rate of 0.5 mm/min, perpendicularly to the interface until the specimen fails. The strength is calculated from the maximum force and the cross section of the specimen. The adhesive strength is calculated as the average strength of five specimens.

To obtain the required specimens for the direct tensile test the samples tested in the optimised setup were cut in five blocks of 80 mm by 80 mm, three on the left hand side (A, B, C) of the joint and two on the right hand side (D, E) as shown in figure 7. The part of the test sample directly adjacent to the right side of the joint could not be used because of the damage caused by the crack test.

Fig. 6. Cubical specimen clamped in the pulling device - BRRC direct tensile test method

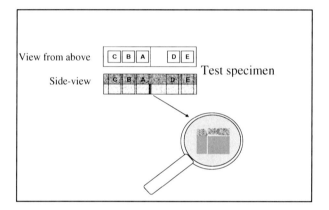

Fig. 7. Schematic overview of the division in blocks of a test sample for the direct tensile test

The direct tensile test led to the following results:

- No difference in bond strength was recorded between the blocks taken on the left and the right side of the joint, except for "geocomposite 4" where a significant difference (larger than the standard deviation) was noted.
- In general two test samples containing the same type of anti-cracking interface system had the same bond strength (no significant difference compared to the standard deviation was observed). Therefore, it can be concluded that the wide dispersion of some results in the crack test is not due to the adhesion in the system concrete – anti-cracking interface system – asphalt.
- The reference test samples had a better bond strength than the SAMI and steel reinforcing netting samples. This is in accordance with their function of "controlled debonding".

- The differences in bond strength of the reference and geogrid samples were negligible compared to the standard deviation. Nevertheless a tendency of higher bond strengths for references compared with geogrids was observed.
- The bond strength of SAMI samples is comparable to that of steel reinforcing nettings.
- The bond strength of geogrid samples was better than that of steel reinforcing nettings.

Analysis of the asphalt layers of the test samples. Analysis on the asphalt layers were done after the direct tensile test, so that the asphalt had been separated from the concrete. Two blocks of asphalt were used: block A and B.

These additional tests showed that the dispersion in the results cannot be attributed to differences in interlayer adhesion or asphalt composition between samples of a given type. The dispersion is most probably related to the inherent rather random character of crack formation in such tests on the one hand, but also on the importance of a flat and fixed installation of the interface product, especially for reinforcing products such as steel reinforcing nettings and grids which have the tendency to curl when making laboratory samples. Nailing is therefore recommended for grids and steel reinforcing nettings. These results also illustrate clearly the consequences on performance of an improper installation on site e.g. when the interface product is not placed flat or is not completely fixed and can still move.

3 Conclusions

– A new laboratory test was developed to simulate the effect of vertical movements (slab rocking) on an asphalt layer placed on different types of anti-cracking interfaces.
– Different test conditions were evaluated. The test conditions had to be selected in such a way that most specimens would fail within a realistic measurement period and a distinction would be possible between good and badly performing systems.
– Two test programmes were done, involving 5 different interface systems. The second programme was done after the optimisation of the setup, which was based on the experience and the results of the first programme.
– The test is capable of making qualitative distinctions between some different systems, but the dispersion on the quantitative data is too large to distinguish products with a similar performance.
– Additional analyses on the test specimens have shown that the dispersion is not due to the quality or the repeatability of the specimen fabrication.

Acknowledgements. The authors would like to thank the Agency for Innovation by Science and Technology (IWT Flanders) for the financial support of this research project (IWT project 060884: "Trillingsgecontroleerd stabiliseren van betonplaten voor duurzame asfaltoverlagingen met scheurremmende lagen").

The Use of Bituminous Membranes and Geosynthetics in the Pavement Construction

Petr Hyzl, Michal Varaus, and Dusan Stehlik

Brno University of Technology, Faculty of Civil Engineering, Department of Roads, Brno, Czech Republic

Abstract. The distress of pavements by cracking is a frequent phenomenon in the road network in the Czech Republic. One of the solutions for this problem is the use of bituminous membranes during construction, repair and reconstruction. Bituminous membranes can be used in the pavement construction for the prevention and postponing of the development of the reflective cracking. When the design of the road structure is unprofessional and an introduction of geofabrics into the pavement structure is intended, there is a danger that the unbonding of the layers will occur. This can cause a significant reduction in the lifetime of the pavement.

The introduction of the new technical regulation TP 147 [1] in the Czech Republic describes the proper use of bituminous membranes and geosynthetics in the pavement structure. The article specifies which particular materials: bituminous emulsions, binders and geofabrics should be used and mentions the regulations dealing with how they are used.

1 Introduction

The formation of cracks in the road network is a problem which occurs not only in Czech Republic, but also in other countries. One of the possibility for eliminating them is the use of bituminous membranes in the pavement structure. Authors of the article were addressed by the Ministry of Transport of Czech Republic to prepare a binding regulation for using bituminous membranes in the pavement structure defining not only the possibilities of their application, but also technical requirements for the materials used, describing the course of construction work and the subsequent inspection of work.

The article aims to get the professional public acquainted with the newly prepared and approved regulation - Technical Conditions No. 147 "Use of bituminous membranes and geosynthetics in the pavement structure" [1], the main ideas of which can certainly be applied in other countries as well.

2 Subject of Technical Conditions – The Regulation in General

The Technical Conditions specify the principles for using bituminous membranes in construction, repair and reconstruction of pavements, airfields and other trafficked areas. Bituminous membranes can be used in the pavement structure for the prevention and postponing of the development of reflective and contracting cracks.

In the new regulation, the Stress Absorbing Membrane Interlayer (SAMI) is defined as a bituminous interlayer for the transfer of horizontal stresses, which is introduced in order to reduce copying cracks into the wearing course. Although the bituminous membrane connects both adjacent courses, it enables, to a certain extent, their independent movement. This causes reduction in the transfer of stress and, in particular, compensation for the horizontal movements of the lower course where the cracks and joints occur.

In design documentation, bituminous membrane is designated as SAMI (Stress Absorbing Membrane Interlayer). Part of the bituminous membrane may be geotextile (GTX) or geocomposite (GCO) with a fibreglass geogrid, when the fibreglass geogrid is designed to absorb horizontal forces in the pavement structure. The binder saturated geotextile forms a bituminous membrane and enables the correct installation – called gluing.

The regulation marginally mentions the use of bituminous membrane as a wearing course with resistance to crack propagation (SAM technology – Stress Absorbing Membrane). Although the use of this SAM technology has not been technicaly verified in the Czech Republic its use is nevertheless recomended since.

3 Use of Bituminous Membranes

In the regulation, the bituminous membrane has been determined as one of the measures used in the construction of the pavement that postpones reflective cracks, particularly in the layers bonded by hydraulic binder. Bituminous membranes can be used also for repairs of concrete pavements (in covering or reinforcing with bituminous layers) or for repairs of flexible pavements (repairs of cracks, change of pavement layers, reinforcement, etc.) which has been disturbed by contracting (i.e. when frost occurs) or reflective cracks.

The use of bituminous membranes for repairs of pavements must be preceded by diagnostics of the pavement with an assessment of the type and frequency of distresses, the type, thickness and quality of layers, binding of layers, assessment of pavement strength, and consultation with specialists.

If the bituminous membrane is used on a new pavement or repaired pavement with the required increase in the pavement strength (by reinforcing), the membrane must be introduced into the calculation and evaluation of the pavement structure.

4 Technical Requirements

4.1 Tack Coat

The bituminous membrane is placed on the tack coat from cationic modified bituminous emulsion. The emulsion for the tack coat must meet the requirements of EN 13808 [8], breaking class 5 and the minimum binder content 38% by weight. The dosed amount of bituminous emulsion under the bituminous membrane is 0.2 kg/m^2 to 0.25 kg/m^2 of residual bitumen (depending on the surface texture and absorbability).

4.2 Bituminous Membrane

The bituminous membrane is a thin layer of modified binder which must comply with the requirements shown in Table 1. If the material used for making the membrane does not meet the parameters in Table 1, it is necessary that the contractor proves the suitability of this material for the given purpose and, at the same time, proves the ability to introduce the material into the functional membrane. The customer will assess the proof and will approve its application.

Table 1. Requirements for binders used for bituminous membrane [1]

Property	Unit	Requirement min.	Requirement max.	Tested acc. to
a) Modified bitumen				
Penetration at 25°C	0.1 mm	40		EN 1426 [2]
Softening point R&B	°C	65[1)		EN 1427 [3]
Fraass breaking point	°C		– 18	EN 12593 [4]
Elastic recovery at 25°C	%	80		EN 13398 [5]
Storage stability	°C		5.0	EN 13399 [6]
Working temperature	°C	170	195	
b) Modified emulsion - made of modified bitumen acc. to EN 14023 and complying with the parameters of EN 13808 [8]				
Binder content	% by weight	63		EN 1428 [7]
Elastic recovery on recovered binder	%	80		EN 13398 [5]
Breaking class		3-5		
Working temperature	°C	60-75		

1) Due to possible vertical movements of concrete pavements slabs, the min. softening point R&B 75°C is required for bituminous membranes.

4.3 Non-woven Geotextile (GTX-N)

The textile used as a protective layer of the membrane must meet the requirements shown in Table 2.

Table 2. Requirements for non-woven geotextile [1]

Property	Unit	Requirement	Tested acc. to
Density	g/m^2	min. 100 - 200	EN ISO 9864 [9]
Thickness at load 2 kN/m^2	mm	min. 1,5	EN ISO 9863-1 [10]
Tensile strength	kN/m	min. 5,0	EN ISO 10319 [11]
The material must be resistant to temperatures up to 160°C for the subsequent laying of the compacted bituminous mixture.			

4.4 Thickness and Position of the Bituminous Membrane in the Pavement Structure

The thickness of the bituminous membrane is achieved by a dosing of the binder. The dosing is dependent on the type of protective layer and the characteristics of the surface. Dosing is shown in Table 3. The range of dosing is given by the macrotexture of the upper layer. A smooth upper layer requires minimum dosage, a coarse-graded mixture or ground surface requires medium dosage and a surface with fine milling is on the maximum dosage. In the case of non-woven textile, dosing depends on the density (thickness) of the textile used. A textile with a higher density will be used for finely milled surfaces.

Table 3. Dosing of modified bitumen in the bituminous membrane [1]

For membrane protection by	Min. dosage [kg/m^2]	Max. dosage [kg/m^2]
Chipping	2.0	3.0
Non-woven textile	1.5	3.0
Slurry seal layer	1.5	2.5

Experience from already implemented constructions is usually utilized in assessing the suitability of the type of bituminous membrane and its minimum and maximum dosage. The bituminous membrane (SAMI) should be placed at such a depth under the surface, that it cannot be damaged by traffic and its function cannot be lost due to low ductility in winter. In the climatic conditions of Czech Republic, the minimum depth is 70 mm. The maximum depth is determined by the thickness of bituminous layers; it is usually max. 200 mm. If the thickness of overlaying by bituminous layers is up to 100 mm, it is not recommended to use the membrane for sections with slow or stopping traffic (crossroads, bus stops, or right lanes for slow traffic) and in the places where there is a centrifugal force action (curves up to 300 m in radius) for heavy traffic roads. Lower thickness of overlaying can be used for layers with a higher stiffness modulus.

4.5 Protective Layer of Bituminous Membrane

Protective layers enable the layening of other structural layers. The following can be, for example, used as protective layers of the membrane:

- introduction of crushed aggregate acc. to EN 13043 [12], the fraction and dosage of which depending on the membrane dosage is shown in Table 4,
- non-woven textile with a density of 100 g/m^2– 200 g/m^2,
- slurry seal layer (in the min. amount of 12 kg/m^2).

Table 4. Dosing of crushed aggregate into the protective layer [1]

Aggregate fraction	Binder content [kg/m^2]	Aggregate dosage [kg/m^2]
2 – 4	2	4 – 5
4 – 8	3	approx. 5
8 – 11	3	approx. 8

4.6 Design Characteristics of Bituminous Membranes

When calculating the stress and deformation of the multi-layered half-space modelling the pavement with bituminous membrane, it is possible – for bitumen types complying with Table 1 and arrangement according to Tables 3 and 4 – to use the values of the modulus of elasticity (at 15°C and load frequency 10 Hz) shown in Table 5. The lateral deformation factor (Poisson's number) of 0.5 will be used in all cases.

Table 5. Design characteristics of the membrane elasticity modulus [1]

Membrane and the type of protective layer	E [MPa]
Membrane with chipping with crushed aggregate	250
Membrane with non-woven textile or slurry seal layer	100

5 Construction Work

5.1 Treatment of the Surface in the Case of Non-rigid Pavement

When constructing a pavement, the surface must meet the requirements and regulations according to which it was made before using the bituminous membrane. The surface must be milled off in case of rehabilitation of the pavement. All distresses of the milled surface are to be repaired by laying and compacting a layer of fine bituminous mixture, or using a jet-patch method or by manually spraying of bituminous emulsion with chipping, eventually in two layers. All cracks are to be cleaned and sealed according to the applicable standards.

5.2 Treatment of Subbase in the Case of Rigid Base Layers

Layers of mixtures bonded by hydraulic binder must harden at least 7 days. The slabs of a concrete pavement must not show mutual vertical movement during loading before the repair by overlaying. If the pavement surface is uneven and disturbed with cracks (irregularities under a 4 m lath greater than 15 mm), the thinnest possible levelling course of bituminous carpet laid on tack coat is to be made. In the case of only local distresses, it is necessary to repair them with a special patching material or a fine-graded bituminous mixture. All cracks and joints are to be cleaned and sealed using the method stated in the applicable standards.

5.3 Making the Bituminous Membrane and Protective Layers

The tack coat is to be applied on a dust free clean surface preferrably dry but perhaps on a surface that is at moust only moist. Modified bituminous cationic emulsions are to be used for the tack coat. Then the bituminous membrane is applied on the tack coat. The bitumen temperature depends on the type of bitumen used. The minimum air temperature at the making of membrane is +10°C. Chipping with coarse crushed aggregate will be sprinkled into hot bitumen. After the aggregate is applied, compaction is not performed, the aggregate is fixed in the membrane by using its dead weight and the adhesion of the bituminous binder to the aggregate. It is forbidden to travel or drive over the chipped membrane with steel rollers. Non-woven textile is to be laid manually into the warm bitumen with a special laying machine that operates parallelly with the pavement axis. When connecting the geotextile by overlapping, it is necessary to carry out manual spraying of the overlap to saturate the geotextile with bitumen. The slurry seal layer is laid with a laying machine on the cold membrane. It is useful to wet the laying machine wheels with water to prevent the bituminous membrane from sticking to the wheels.

5.4 Laying of Other Layers

One must not allow traffic on the bituminous membrane without laying the subsequent layers. Overlaying of the bituminous membrane with protective geotextile layer with an asphalt layer must be performed immediately after laying the geotextile. When the asphalt mixture is laid down on the membrane, the movement of all construction devices must be smooth without sharp curves, full braking and with a reduction of waiting vehicles. In the case of an occurrence of places with disturbed protection, such places must be repaired without delay before laying the subsequent bituminous layer.

The newly approved regulation further contains chapters regarding testing and quality control, climatic limitations and environmental protection during construction work, and also a chapter focused on occupational health and safety.

6 Conclusion

The aim of this article was to acquaint the reader with the main features of the new regulations which deals with the problems of reflective cracks in the pavement structures that use bituminous membranes. Although the application of bituminous membranes has not been so often in the Czech Republic yet, the data of several test sections found on roads of all categories were evaluated before drawing up the requirements. Authors of the regulation (i.e. authors of this paper) believe, that its issue by the Ministry of Transport will help to effect a wider use of this perspective technology on the road network and hope, that the readers can take these suggestions and use them for the construction of roads.

Acknowledgement. This paper has been supported by the research projects TA02030639 „Durable acoustic asphalt pavement courses with utilization of bituminous binders modified by rubber microfiller including innovative technology of rubber milling" and TA02030549 „The most effective utilization of reclaimed asphalt pavement layers for production of new asphalt mixes".

References

[1] TP 147 The use of bituminous membranes and geosynthetics in the pavement construction, Ministry of Transport, Prague (2010)
[2] EN 1426 Bitumen and bituminous binders - Determination of needle penetration, CEN, Brusel (2007)
[3] EN 1427 Bitumen and bituminous binders - Determination of the softening point - Ring and Ball, CEN, Brusel (2007)
[4] EN 12593 Bitumen and bituminous binders - Determination of the Fraass breaking point, CEN, Brusel (2007)
[5] EN 13398 Bitumen and bituminous binders - Determination of the elastic recovery of modified bitumen, CEN, Brusel (2010)
[6] EN 13399 Bitumen and bituminous binders - Determination of storage stability of modified bitumen, CEN, Brusel (2010)
[7] EN 1428 Bitumen and bituminous binders - Determination of water content in bitumen emulsions - Azeotropic distillation Method, CEN, Brusel (1999)
[8] EN 13808 Bitumen and bituminous binders - Framework for specifying cationic bituminous emulsions, CEN, Brusel (2005)

[9] EN ISO 9864 Geosynthetics - Test method for the determination of mass per unit area of geotextiles and geotextile-related products, CEN, Brusel (2005)
[10] EN ISO 9863-1 Geosynthetics - Determination of thickness at specified pressures - Part 1: Single layers, CEN, Brusel (2005)
[11] EN ISO 10319 Geosynthetics - Wide-width tensile test, CEN, Brusel (2008)
[12] EN 13043 Aggregates for bituminous mixtures and surface treatments for roads, airfields and other trafficked areas, CEN, Brusel (2002)

Stress Relief Asphalt Layer and Reinforcing Polyester Grid as Anti-reflective Cracking Composite Interlayer System in Pavement Rehabilitation

Guillermo Montestruque[1], Liedi Bernucci[2], Marcos Fritzen[3], and Laura Goretti da Motta[3]

[1] Universidade do Vale do Paraiba – UNIVAP, Brazil
[2] Escola Politécnica da Universidade de São Paulo – EPUSP, Brazil
[3] Universidade Federal do Rio de Janeiro, COPPE, Brazil

Abstract. An anti-reflective composite interlayer system, composed by a stress relief asphalt layer and a reinforcing polyester grid, was discussed based upon the results of field and laboratory research programs. The stress relief asphalt layers dissipate the stresses in the cracks tip, and reduce vertical and horizontal displacements of the subjacent crack in the new overlay. Polyester grid is a high tensile strength material that slows or even stops the reflective crack in the overlay. The field research program consisted of a highway trial section subjected to an accelerated test by using a transit simulator (HVS – Heavy Vehicle Simulator). A wheel reflective cracking test was carried out in laboratory, using the Displacement Meter CAM (Crack Activity Meter) to measure the horizontal and vertical movements on the crack or joint slab during the wheel load cycles. The results showed that both relief asphalt layer and reinforcing polyester grid work together as composite solution, technically and economically viable, which delay or block the crack reflection.

1 Introduction

Pavement structures are subjected to two types of mechanical effects: thermal and traffic loading. Vertical and horizontal movements are generated during a wheel load passage on a discontinuity in the old pavement surfacing layer. Crack opening and closing movements are due to temperature too. The general term "overlay system" was proposed in order to describe the combination of a bituminous overlay and an interface system, placed on an underlying road structure. The most important component of an overlay system is the asphalt layer itself. Improvements in crack resistance can be obtained in the asphalt layer by modifying its composition or its components. The use of some types of fibers and/or polymer modified binders have proven to be very efficient according to different authors [1-2].

The interface system consists of an interlayer product (stress absorbing membrane, non woven, reinforcement grid, etc...). A great variety of products covering a wide range of tensile stiffness, bond properties and asphalt retention capability are now available on the market, to fulfill different functions in this application. It is necessary to clearly define where and how to use each product in an optimized way. The role of an interlayer system in the road structure depends mainly on its components. They may be:

➢ Taking up the localized stresses in the vicinity of cracks and, then, reducing the stresses in the bituminous overlay on the crack tip. The products in that case act as reinforcement. This is the case of reinforcing polymer grids.
➢ Providing a resilient that can deform horizontally without breaking, in order to absorb large movements taking place in the vicinity of cracks. This is the case for impregnated nonwovens, SAMI (Stress Absorbing Membrane Interlayer) and sand asphalt. Often this function is also described as "controlled debonding".
➢ Providing waterproofing function and keeping the road structure waterproof even after the reappearance of the cracks on the road surface. This is often the case of impregnated nonwovens.

A combination of a stress relief asphalt layer and a reinforcing polyester grid is proposed as an efficient solution of anti-reflective composite interlayer system. In other to study and prove its efficiency, field and laboratory research programs were carried out in Brazil.

2 The Proposed Composite Interlayer System

This composite interlayer system is intended to be a technical and economical alternative to some expensive or time consuming solutions for rehabilitation of severely cracked pavement, such as partial reconstruction, in the case of flexible pavements, and crack & seat or rubblization, in the case of rigid pavements.

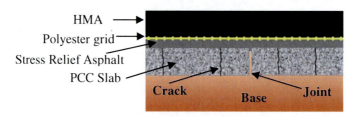

Fig. 1. Layout of the composite system

The proposed composite interlayer system is composed by a stress relief asphalt layer and a reinforcing polyester grid (Figure 1).

2.1 Why Use a Stress Relief Asphalt Interlayer?

The stress relief asphalt interlayer is a highly flexible hot mix asphalt layer (HMA), normally very thin (between 20mm and 30mm), composed by fine aggregates (< 9,5mm) and elastic polymer modified asphalt (PMA), in high contents (between 7% and 7,5% in weight). This interlayer absorbs part of the vertical and horizontal movements of the crack walls, reduces the shear stresses in the interface and delays reflective cracking. Due to the high content of asphalt, it protects the existing pavement structures from water damage.

A laboratory research program was carried out in University of São Paulo (USP) to analyze the behavior of 4 modified asphalt products: 1 rubber and 3 polymer modified asphalt products, from different suppliers [3]. A series of Semi Circular Bending Tests was done. The 3 polymer modified asphalt products performed better than the rubber modified asphalt. Among the polymers, the best result was obtained by Strata, supplied by Betunel.

2.2 Why Use a Reinforcing Polyester Grid?

The reinforcing grid is a bi-directional polymeric mesh, composed by high tensile strength fibers arranged in a grid structure and covered with an asphalt coating or a pressure sensitive adhesive. Due to the tensile and interface adherence properties, the grid absorbs stresses on the crack tip, delaying or even blocking the reflective cracking in the overlay.

The grid can be produced mainly with 3 different raw materials: glass, polyester and polyvinyl alcohol (PVA) fibers. These fibers have different behavior regarding tensile properties and fatigue resistance. As the grid has to guarantee a longer life for the overlay, it should be designed not only to absorb short term tensions, but also to maintain its properties throughout the pavement lifetime. In other words, the fiber strength should not be affected by the traffic dynamic loads during the project lifetime, otherwise the material would lose efficiency and performance.

Technical reports in United States [4-6] showed that glass grids may deteriorate faster under the action of traffic loading, particularly when the differential vertical movements of crack walls (shear mode) is the main factor of reflective cracking. The evaluation of the five road sections of rigid pavements with glass grids showed an early reflection of transverse cracks, which were propagated from the underlying joints of the concrete pavements. In these cases, the glass grid products did not prove to be highly effective in retarding the development of reflection cracks in old jointed concrete pavements. These failures could be explained by the short lifetime of glass fibers when submitted the fatigue, becoming brittle after some load cycles.

A test equipment was developed in the University of São Paulo, with the purpose of evaluating the fatigue behavior of polyester and glass fibers for the use in rigid pavement rehabilitation. This test imposes cyclically alternate vertical displacements, generating fatigue in the fiber in shear mode (Figure 2).

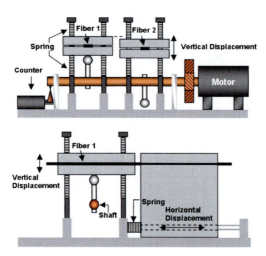

Fig. 2. Layout of the yarn fatigue test equipment

The results showed a superior performance of polyester grid in comparison to glass. The glass broke between 16.000 to 21.000 cycles (Figure 3c) and the polyester grid did not break after 160.000 cycles (Figure 3b), when the test was interrupted [7].

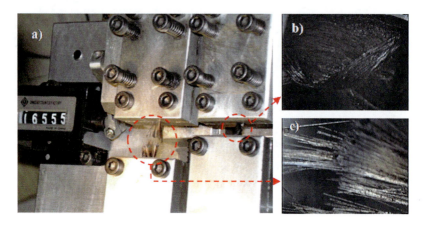

Fig. 3. a) Equipment; b) Polyester fiber (not broken) c) Glass fiber (broken)

Reinforcing grid properties: The grid used in the following field and laboratory researches was Hatelit C 40/17, supplied by Huesker, composed by high tenacity polyester yarns with bituminous coating. Weight (DIN EN 965): 270 g/m². Ultimate tensile strength *(DIN EN ISO 10.319):* ≥ 50 kN/m (longitudinal and transversal). Tensile strength at 3% strain *(DIN EN ISO 10.319):* ≥ 12 kN/m (longitudinal and transversal). Strain at nominal tensile strength *(ISO 10.319):* < 12 % (longitudinal and transversal). Mesh size: 40 mm. Melting point: 250 °C.

3 Field Research: Highway Test with Traffic Simulator

A 100m long test section was constructed on the Rio-Teresopolis highway, under the administration of CRT concessionary. Half section (50m) was rehabilitated with a conventional overlay. Reinforcing polyester grid in combination with 20mm of a stress relief asphalt layer was applied as an interlayer system in the other half section. The total overlay thickness was 70mm (20mm stress relief + polyester grid + 50mm hot mix asphalt). This research was performed by the Federal University of Rio de Janeiro-Brazil [8]. In the conventional section, the cracks came out after 192.000 cycles. The cracks propagated in the entire thickness of the asphalt overlay, from bottom to top. In the section with the composite interlayer system, the test was stopped after 220.500 cycles, because some cracks had appeared on the surface. A specimen was extracted and then it was observed that there was no the crack propagation above the polyester grid (Figure 4). The observed cracks on the surface were top-down cracks (Figure 5a, 5b), which were attributed to a possible oxidation of the hot mix asphalt in the plant. The composite interlayer system was able to block the crack propagation during the period of this test.

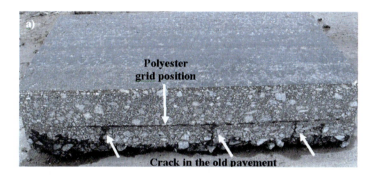

Fig. 4. Crack propagation stopped in the level of the polyester grid

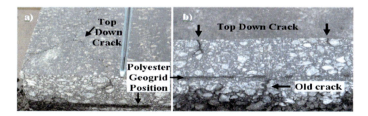

Fig. 5. Detail of the extracted specimen: although top-down cracks were observed on the surface, no crack propagation occurred above polyester grid

4 Laboratory Research: Wheel Reflective Cracking Test

A laboratory research program was carried out in the LTP Laboratory of the São Paulo University, with the objective of studying the pattern of crack incitation and propagation in the proposed interlayer system, as well as measuring its performance in comparison with traditional overlay. The LCPC laboratory simulator, typically used for "Wheel Tracking Rutting Test", was adapted to perform the "Wheel Reflective Cracking Test" in this research. Three overlay systems were tested: the first was the conventional solution (60mm of HMA - Hot Mix Asphalt), the second was the use of 20mm of relief asphalt layer (Strata) plus 40mm of Stone Matrix Asphalt (SMA), and the third was similar to the second, but with the addition of a polyester grid (Hatelit), between Strata and SMA. On the bottom, a concrete slab with a joint in the middle was supported by two types of rubber mats with different densities, in order to impose a differential vertical movement in the joint. Before the placement of the overlay system, vertical and horizontal relative movements of the joint were measured by using the CAM (Figure 6c). The test was developed using three levels of vertical movement or shear displacement (∂_c): low ($\partial_c = 30 \times 10^{-3}$ mm), medium ($\partial_c = 100 \times 10^{-3}$ mm) and high ($\partial_c = 500 \times 10^{-3}$ mm).

The visualization of the crack was enhanced by painting both faces of the specimen in white. A loaded wheel with a pneumatic tire is rolled back and forward on the specimen surface. The test was considered finished when the crack reached the top surface on both faces. Visual observations were made and the number of cycles was recorded. Two advantages were observed: (1) better simulation of field conditions and, (2) both crack initiation and growth can be monitored.

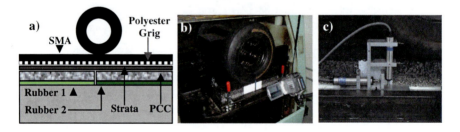

Fig. 6. (a) Test Layout; (b) Crack propagation equipment; (c) Crack Activity Meter installation

4.1 Qualitative Discussion on Crack Initiation and Propagation

First case (conventional - 60mm of HMA): Typically, a dominating reflective crack appeared in the overlay, over the concrete slab joint, and grew vertically upwards (Figure 7a). The cracking initiated earlier the overlay started earlier and its propagation rate was faster. The tests stopped when the crack reached the overlay surface, breaking the specimen in two pieces.

Fig. 7. (a) Typical crack propagation in conventional overlay; (b) Typical crack propagation in Strata/SMA overlay for medium vertical displacement

Second case (20mm of Stress Relief Asphalt + 40mm of SMA): A crack (initiated over the joint) propagated gradually upwards through the stress relief asphalt layer (Strata). At the interface Strata and SMA layers, a horizontal direction change in the crack propagation was observed, due to debonding located between the two layers (Figure 7b). The energy dissipated in the crack horizontal propagation delayed the appearance of the reflective crack in the SMA layer. The horizontal crack propagation stopped when some reflective cracks appeared on SMA layer, propagating vertically towards the surface. The low modulus Strata layer acted as a SAMI ("Stress Absorbing Membrane Interlayer").

Fig. 8. Sequence of the reflective crack propagation in composite interlayer system (strata+ polyester grid + SMA)

Third case (20mm of Stress Relief Asphalt + Polyester grid + 40mm of SMA): A first reflective crack was initiated on one side of concrete slab joint, and its propagation was interrupted when the polyester grid position was reached (Figure 8, crack 1). The same occurred with a second crack initiated on the other side of the joint (Figure 8, crack 2). Another crack appeared in the strata layer with vertical propagation, but also without crossing the geogrid position (Figure 8, crack 3). After a large number of cycles, all cracks in the stress relief layer stopped their propagation in the polyester grid. Then, a new "top-down crack" on layer appeared on the surface of SMA layer (Figure 8, crack 4). The polyester grid acted as reinforcement, interrupting the vertical propagation of reflective cracks.

Comments: It was possible to observe 3 different patterns of reflective cracking. In the first case (conventional), there was a very quickly crack propagation from bottom to top. In the second case (stress relief layer), there was a delay in the crack growth due to dissipation of energy in the horizontal propagation and interface debonding. In the third case, crack growth was first delayed due to the stress relief asphalt layer and then interrupted due to the action of the reinforcing grid. It was possible to observe very well the different behavior of each component the proposed composite interlayer system.

4.2 Quantitative Results and Comparative Performance

The number of cycles of each test was recorded and the results are show in the Table 1. The factor of efficiency of the interlayer system *(FE)* was obtained as follows: $FE = N_{(system)} / N_{(conventional)}$, where N is the number cycles of the wheel load for each case.

Table 1. Fatigue crack propagation results

	Vertical Displacement		
	Low (30×10^{-3}mm)	Medium (100×10^{-3}mm)	High (500×10^{-3}mm)
$N_{Conventional\ HMA}$ (cycles)	10.255	4.385	1.898
$N_{Strata + SMA}$ (cycles)	28.560	8.274	3.240
$FE_{(Strata/\ conventional)}$	*2,8*	*1,9*	*1,7*
$N_{Strata + PET\ grid + SMA}$ (cycles)	194.133	26.760	10.580
$FE_{(System/\ conventional)}$	*18,9*	*6,1*	*5,6*

Regarding reflective cracking, 20mm of a stress relief asphalt layer plus 40mm of SMA overlay showed to be around 2 times more efficient than 60mm of HMA conventional overlay. The inclusion of a polyester reinforcing grid in the interface between the stress relief asphalt layer and the SMA layer increased more than 3 times the number of cycles for medium and high vertical displacements of the bottom joint (100×10^{-3}mm and 500×10^{-3}mm, respectively), and more than 6 times for low vertical displacements (30×10^{-3}mm). The combination of stress relief asphalt layer and a reinforcing polyester grid proved to be a very efficient solution for the simulated situation.

5 Conclusions

A combination of a stress relief asphalt layer and a reinforcing polyester grid was proposed as anti-reflective cracking interlayer system. Field and laboratory research were carried out in order to study and measure the performance of the system. The main conclusions are:

> The LCPC laboratory simulator can be adapted to perform "Wheel Reflective Cracking Test". The advantage of this equipment is to simulate better the field conditions of load and crack displacements.
> The CAM (Crack Activity Meter) plays an essential part in identifying the underlying crack movement.
> The glass fiber grid is not recommended for the proposed system because it becomes brittle after some load cycles and presents a shorter fatigue life in shear mode compared to polyester grid.
> The stress relief asphalt layer (Strata) has a retarding effect on crack initiation and crack propagation.
> The reinforcing polyester grid interrupted the crack propagation in the field and in the laboratory research.
> The stress relief-reinforcement interlayer is a potential composite solution to some reflective cracking problems. In the laboratory tests, this composite system showed an increase in the fatigue life between 5,6 and 18,9 times, in comparison with conventional Hot Mix Asphalt overlay with no interlayer system.

References

[1] Serfass, J.P., De La Mahe, V.B.: Fiber-Modified asphalt overlays. In: Proceedings of the 4th International RILEM Conference: Reflective Cracking in Pavement Research in Practice, Canada, pp. 227–239 (2000)
[2] Mohammad, Yildirim, Y., et al.: Polymer modified asphalt binders Original Research Article. Construction and Building Materials 21(1), 66–72 (2007)
[3] Montestruque, G., Vasconcelos, K., Bernucci, L.: Ensaio de flexão em amostra semi-circular com fenda e análise de imagens para caracterização da resistência à fratura de misturas tipo AAUQ. XXIV ANPET (2010)
[4] FHWA/TX-05/0-4517-1. Performance report on jointed concrete pavement repair strategies in Texas (February 2004)
[5] FHWA/TX-06/0-4517-3. Methods of Reducing Joint Reflection Cracking: Field Performance Studies (2005)
[6] FEP-03-03. Glassgrid pavement reinforcement product evaluation Wisconsin Department of Transportation, Division of Transportation Infrastructure Development (2003)
[7] FAPESP reports. Utilização do crack activity meter na restauração de pavimentos com o sistema anti-reflexão de trincas: Geogrelha – SMA (2008)
[8] Fritsen, M.A.: Avaliação de soluções de reforço de pavimento asfáltico com simulador de tráfego na rodovia Rio Teresópolis. Teses MSc. Rio de Janeiro (2005)

Characterizing the Effects of Geosynthetics in Asphalt Pavements

Stefania Vismara[1], A.A.A. Molenaar[2], Maurizio Crispino[1], and M.R. Poot[2]

[1] Facutly of Civil Engineering DIIAR - Road Section, Politecnico di Milano, Italy
[2] Faculty of Civil Engineering and Geo Sciences- Road and Railway Engineering, Delft University of Technology, The Netherlands

Abstract. Retarding reflective cracking in asphalt overlays is a serious and inevitable challenge to asphalt pavement engineers. Anti-reflective cracking systems using geosynthetic interlayers are believed to be a solution able to improve the pavement performance.

The paper evaluates the effects of the inclusion of different types of geosynthetic and is focusing on the results of monotonic and dynamic tests. Shear tests and indirect tensile strength tests were carried out as interface characterization tools, showing the ability of the geosynthetics to prevent separation.

Non-conventional fatigue tests were performed on reinforced, and unreinforced, notched beams to obtain a better understanding of the behaviour of the geosynthethics in retarding crack propagation. The main parameters, such as the applied load, the crack opening and the vertical displacements were monitored. The void structure of the beams was visualized using X-ray tomography, giving useful information about the interlayer effects on compaction.

Furthermore a finite element model, based on the experimental results, was developed to predict reinforced specimens behaviour.

The findings showed the benefits of using the geosynthetics in asphalt overlays, to prevent crack propagation.

1 Introduction

Reflective cracking is a challenging topic in road engineering, representing one of the major distresses of asphalt overlays. Increasing traffic loads together with thermal loadings and local differential settlements can result in a rapid decrease of the pavement performance [1, 2]. The rehabilitation of damaged roads by simply placing additional layer of asphalt over the existing pavement is rarely a durable solution, because asphalt overlays are not able to take over the tensile stresses [3]. Therefore reinforcing interlayers to delay or retard the crack propagation are needed. Several types of reinforcement have been developed. Geogrids, nets, stress adsorbing membrane interlayers (SAMI) and geocomposites have found increasing application worldwide, showing their benefits to the pavement [4-7]. In particular, geocomposites are intended to fulfill different functions, such as sealing, stress relief and reinforcement and they were demonstrated to prolong the service life of the overlay against cracking [8, 9]. Geosynthetic interlayers are recommended when high horizontal and vertical displacement are likely to occur. The easiness of application together with the advantages in pavement performance makes the geocomposite interlayer a good solution. More findings are however needed to better understand the benefits of the reinforcements to pavement performance.

The present paper is aimed at investigating the effects of geosynthetic interlayers with high modulus glass fibers embedded into asphalt structures. Double-layer systems with and without a geosynthetic interlayer have been prepared, figuring out their effects on asphalt pavements. To gain this aim, conventional monotonic tests to failure have been carried out, such as Leutner shear test and indirect tensile strength tests. Moreover a non conventional fatigue test has been considered to investigate the reinforced structures. Beams on elastic foundation have been subjected to cyclic loadings, while monitoring the crack opening displacements at different locations during the whole test.

Finally, a finite element model has been created to simulate the dynamic stiffness test and to get more information about the stress distribution in both unreinforced and reinforced specimens.

2 Laboratory Activity

A large testing programme was conducted in order to get an insight into the behaviour of asphalt pavements with different interlayer elements. Double-layer systems with and without geosynthetic as reinforcement were then prepared in the laboratory by producing an asphalt mixture having a nominal maximum aggregate size of 11 mm and a 60 penetration grade not-modified bitumen (Table 1). The air voids and the bitumen content were set at 5% by volume and 5% by mass, respectively. The geosynthetic materials used are made up of polypropylene non-woven and glass fibers with a high tensile strength of 50 kN/mm^2 (Type A) and 100 kN/mm^2 (Type B). A total amount of nine double-layer slabs, 500 mm wide and long and 150 or 100 mm thick, were prepared in the laboratory by means of an automatic steel roller compactor (Figure 1), obtaining three slabs per each type of interlayer: (I) with a tack coat interlayer, (II) with geosynthetic interlayer Type A and (III) with geosynthetic interlayer Type B. The recommended amount of emulsified tack coat, consisting of 68% of bitumen was adopted to guarantee adhesion between the asphalt layers in slabs without any reinforcement and between asphalt and geosynthetic in the reinforced slabs.

Table 1. Mixture composition

	AC composition			
	CNR UNI 23/7		UNI EN 10014-64	
mm	%	%passing	Minimum	Maximum
16	0.0	100.0	100	100
11.2	3.8	96.2	75	92
8	20.0	80.0	63	85
5.6	38.6	61.4	55	75
2	67.3	32.7	25	38
0.063	92.3	7.7	6	10

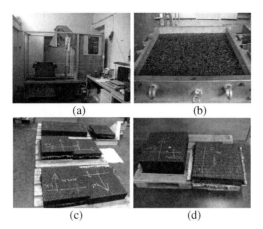

Fig. 1. Slabs preparation: (a) roller compactor, (b) steel mould and (c-d) compacted slabs

A description of the monotonic and cyclic tests to failure performed and their results are reported hereafter in this paper.

2.1 Monotonic Tests

The stiffness, the shear strength of geosynthetic interlayer and its bond between the bottom and the upper layers constitute elements of main importance in reinforced pavements. In order to investigate the interface, traditional shear tests (Leutner test) and indirect tensile strength tests (IDT) were performed on the double-layer specimens drilled from the prepared slabs. The test temperatures were set to 5 and 25°C and the rate of loading was set to 0.85 mm/sec (Marshall speed), according to the European Standards [10, 11]. The specimens had a diameter of 150 mm and a height of 100 mm for the shear test and 100 mm diameter and 84 mm height for the IDT, respectively. In the IDT specimens, the reinforcement was in the middle, perpendicular to the vertical direction, the load was applied along. Three replicates were tested for each condition.

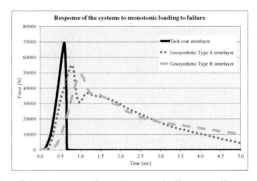

Fig. 2. Example of the response of specimens to indirect tensile strength test at 5°C

The tests exhibited higher values of shear and tensile strength in the specimens with tack coat interlayer compared to those with geosynthetic interlayer.

The shear strength values were, at 5°C (and 25°C): 5.566 (1.159) MPa for the tack coat interlayer, 1.932 (0.302) MPa for the geosynthetic Type A interlayer and 1.669 (0.314) MPa for the geosynthetic Type B interlayer.

The tensile strength values were, at 5° C (and 25°C): 5.167 (1.077) MPa for the tack coat interlayer, 4.197 (0.853) MPa for the geosynthetic Type A interlayer and 3.947 (0.815) MPa for the geosynthetic Type B interlayer.

Despite of that, the graphs obtained show a steep slope in the post-peak regime typical of brittle fracture for the tack coat interlayer and a much slower decay typical of tough fracture for specimens with a geosynthetic interlayer (Figure 2). This softening phase is caused by the bridging effect of the reinforcement.

In detail, the shear tests showed a slippage till failure of the two halves of the specimen in the system without any reinforcement (Figure 3a), while, in presence of geosynthetic type A and type B, the slippage is avoided by the adhesion between the reinforcement and the asphalt layers (Figure 3b and 3c).

The indirect tensile strength test demonstrated the ability of the glass fibers to keep the two halves of the specimens together, as clearly shown in Figure 4.

The interlayer is characterized by good load transmission and adhesion, in particular at 5°C, when the material is stiffer.

Fig. 3. Shear test specimens: (a) tack coat interlayer, (b) geosynthetic interlayer Type A, (c) geosynthetic interlayer Type B

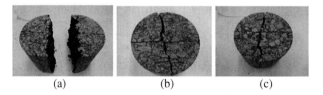

Fig. 4. ITT test specimens: (a) tack coat interlayer, (b) geosynthetic interlayer Type A, (c) geosynthetic interlayer Type B

2.2 Fatigue Tests

The effects of the geosynthetic interlayers studied are mainly connected to the mitigation of reflective cracking. The second part of the paper is then aimed at investigating the effectiveness of the geosynthetic interlayers in reducing crack initiation and propagation.

A non-conventional fatigue test procedure was followed. Double-layer beams 80 mm wide, 500 mm long and 70 mm thick, mechanically notched in the middle of the lower part, were sawn from the slabs and glued to a specifically designed rubber, that was glued to a steel plate (Figure 5). After 4 hours conditioning at the test temperature of 5°C, the beam was subjected to a compressive haversinusoidal load of 5 Hz till it was totally failed. Four peak-peak load levels were chosen: 5, 4, 3 and 2.5 kN, corresponding to 60 to 30% of the tensile strength as measured in the ITT test. A closed circuit of three air tubes, a control unit and a climate chamber, was adopted to maintain the temperature during the test.

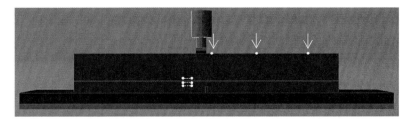

Fig. 5. Fatigue set-up and location of the monitored key-points

The crack propagates along the thickness starting from the corner of the notch, as expected; in fact it generates a stress concentration that causes crack initiation. As shown in Figure 5, each sample was instrumented with three extensometers to measure the crack opening displacement and three vertical deformation transducers to measure the beam curvature, all recorded as a function of the number of cycles.

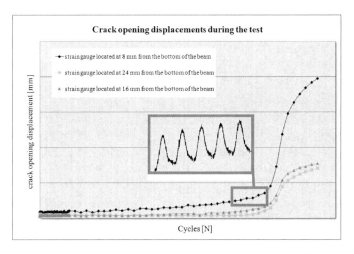

Fig. 6. Typical fatigue test output representing the detail of the cyclic displacements measured

Sructural damage occurs primarly in the form of microcracks, that develop mainly in the brittle matrix and aggregate [4]. This phase corresponds to the first part of the graph reported in Figure 6. After that, debonding occurs at the interface between grain and matrix leading to the crack propagation phase, the beginning of which is represented by the change of slope. The crack develops along the thickness till the top of the beam. After that the third phase occurs, in which the beam is completely broken. A good interlocking guarantees connection and requires additional energy for disconnecting the elements, being a key element to prevent failure of the specimen. Moreover, the geosynthetic interlayer, thanks to its higher stiffness and its good bonding to the asphalt layer, creates a strong bridging zone between the upper and lower layer, delaying the crack propagation.

The crack opening displacements distribution along the thickness during the whole test well explain the behaviour of the different systems analyzed (Figure 7). A single-layer reference beam was also tested.

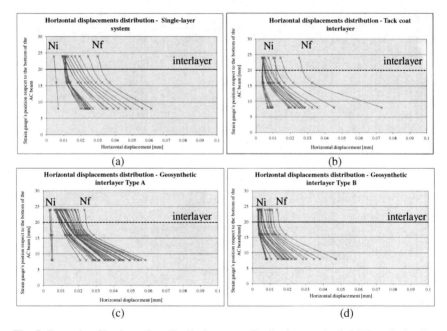

Fig. 7. Example of horizontal cyclic displacement distributions under 2.5 kN cyclic loading from N_i (initial number of cycles) to N_f (failure number of cycle): (a) single-layer system, (b) system with tack coat interlayer, (c) system with geosynthetic interlayer Type A, (d) system with geosynthetic interlayer Type B

The horizontal movements are significantly reduced immediately above and below the geosynthetic; this is clearly shown in Figures 7c and 7d. Once the crack starts to propagate from the notch till the top, the systems without any geosynthetic interlayer show an increasing displacements distribution along the thickness, while the reinforced systems maintain the values below a certain threshold. The fatigue tests conducted at the other three load levels confirmed this behaviour, obtaining similar horizontal displacement distributions.

The characteristic fracture process was monitored by making movies of the tests, highlighting the different path of the crack in the systems studied. A clear picture of the crack path in reinforced system is shown in Figure 8, focusing on the area around the interlayer. The reinforcement remained intact favouring the beam connection. Even after the crack appeared and propagated, the geosynthetic material was still intact, holding the overlay together.

X-ray tomography was also done to get the void structure of the beams around the crack area before and after the tests (Figure 9). Figure 9b showed a non-uniform voids distribution in the structure with the geosynthetic embedded in it. The voids around the interlayer are filled up with the tack coat, constituting a very strong and compacted area. On the other side, the high air void concentration in the upper part constitutes a weak zone, where sometimes the crack starts from.

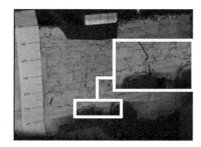

Fig. 8. Crack path in a reinforced beam and detail of the crack around the interlayer

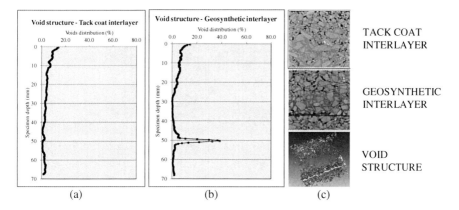

Fig. 9. Example of voids, mortar and rock structure along the thickness, before the fatigue test, for a) tack coat interlayer system and b) geosynthetic interlayer system; c) X-ray tomography of the systems and 3d reconstruction of the voids in a reinforced system

3 Numerical Modelling

As early discussed in this paper, the inclusion of geosynthetic interlayers in asphalt pavements leads to benefits in terms of response to monotonic and cyclic loadings. Thus, stress and strain distributions need to be studied to investigate the behaviour of reinforced structures compared to unreinforced ones. The finite element method (FEM) was adopted to simulate the dynamic indirect tension test (ITT), a simple test used to get an insight into the visco-elastic material response.

A circular specimen clamped between two steel strips is loaded by a dynamic vertical load. In the model the interaction between the specimen and the steel strips, the asphalt layers and the geosynthetic are taken into account by interface elements. The asphalt concrete is assumed to be a visco-elastic material, characterized by Prony series parameters the values of which were obtained experimentally. The reinforced specimens include the geosynthetic, modelled as a composite material made up of textile elements and glass fibers, connected to the asphalt concrete with interface elements representing the tack coat used in the laboratory.

The FE models reported hereafter (Figure 10) compare the stress distributions in unreinforced and reinforced specimens, showing a significant reduction of the maximum principal stress values around the interlayer in the reinforced specimens. Assuming a test temperature of 5°C, a frequency of 0.5 Hz and a target peak deformation of 2 (μm), the unreinforced system presents a central area with high stresses (around 0.50 MPa), while the corresponding reinforced system presents lower stress values (around 0.15 MPa) and a different distribution, in fact immediately above and below the geosynthetic the stresses are close to the minimum values.

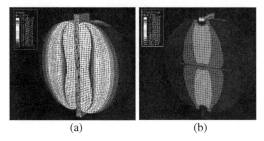

(a)   (b)

Fig. 10. Maximum Principal stress distribution in (a) specimen with tack coat interlayer and (b) specimen with geosynthetic interlayer

4 Conclusion

Preventing crack propagation is the most efficient way to reduce maintenance costs. The main purpose of this study was to provide an insight into the effects of geosynthetic embedded in asphalt pavement.

Based on the findings described, it can be concluded that this type of reinforcement provides benefits to road pavements. The output of monotonic tests to failure showed the ability of the material studied to guarantee a good bond at the interface preventing the two halves of a specimen from separation. The asphalt performs as a brittle material without the geosynthetic interlayer and as a ductile one with the geosynthetic. In fact the cracked geosynthetic specimens can be reloaded again.

By means of a non-conventional fatigue test, it was also proven that such an interlayer reduces the crack opening displacements under cyclic loadings to failure. The horizontal displacements around the interlayer are reduced by about 20% when compared to the unreinforced system. This reduction together with reduction of crack propagation speed is dependent upon compaction, the stones size and characteristics and the degree of bonding, as observed in the fatigue tests. In fact X-ray tomography gave useful information about the void structure, underlining the need of a more homogeneous void distribution. A better compaction technique is then required.

The 3D finite element model of the stiffness test demonstrated that also different stress distributions developed in the systems analysed, highlighting the stress values reduction around the geosynthetic and, consequently, in the whole specimen.

Acknowledgments. The research presented herein was carried out under the collaboration between Politecnico di Milano and TU Delft. The authors gratefully acknowledge the company TenCate Geosynthetics, for supporting the research work and OOMS Nederland Holding bv, for the help in preparing the double-layer slabs.

References

[1] Lytton, R.L.: Journal of Geotextiles and Geomembranes 8(3), 217 (1989)
[2] De Bondt, A.H.: In: Anti-Reflective Cracking Design Of Reinforced Asphaltic Overlays. PhD dissertation. Delft University of Technology (1999)
[3] Tschegg, E.K., Jamek, M., Lugmayr, R.: Engineering Fracture Mechanics 78, 1044 (2011)
[4] Molenaar, A.A.A., Heerkens, J.C.P., Verhoeven, J.H.M.: Effects of Stress Adsorbing Membrane Interlayers. Transportation Research Board, 453 (1986)
[5] Lorenz, V.M.: Transportation Research Record 1117, 94 (1987)
[6] Khodaii, A., Fallah, S.: The Effects of Geogrid on Reduction of Reflection Cracking in Asphalt Overlay. In: Proceedings of 4th National Conference on Civil Engineering. University of Teheran (2008)
[7] Zhongyin, G., Zhang, Q.: Prevention of Cracking Progress of Asphalt Overlayer With Glass Fabrics. In: Proceedings of 2nd International RILEM Conference, Belgium (1993)
[8] Cleveland, G.S., Button, J.W., Lytton, R.L.: Report N. FHWA/TX-02/1777 (2002)
[9] Sobhan, K., Genduso, M., Tandon, V.: Effects of Geosynthetic Reinforcement on Propagation of Reflection Cracking and Accumulation of Permanent Deformation in Asphalt Overlays. In: Larrondo Petrie, M. (ed.) Proceedings of 3rd LACCET, p. 1. Florida Atlantic University, USA (2005)
[10] SN 670-461 (2000). Bestimmung des Schichtenverbunds (nach Leutner). Gültig ab: 1 (August 2000)
[11] UNI-EN 12697-23:2006, Bituminous Mixtures-Test methods for hot mix asphalt-Part 23 (2006)

Geogrid Interlayer Performance in Pavements: Tensile-Bending Test for Crack Propagation

A. Millien[1], M.L. Dragomir[1,2], L. Wendling[3], C. Petit[1], and M. Iliescu[2]

[1] Université de Limoges, GEMH - Génie Civil et Durabilité,
  Centre Universitaire de Génie Civil, Bd. Jaques Derche, 19300 Egletons, France
[2] Université Technique de Cluj-Napoca, 25, rue George Baritiu, 400027, Cluj-Napoca, Roumanie
[3] Département Laboratoire d'Autun - CETE de Lyon, Bd. B .Giberstein, BP141, 71404 Autun cedex, France

Abstract. The roads durability is an objective that gains more and more importance because of the economical – environmental factors. A durable road surface should be provided in the course of repair operations. Geogrids (carbon, glass or steel fibbers) are used for increasing the durability of overlaid asphalt surfaces. They reduce fatigue cracks as well as thermal cracking and prevent structural deformation. This article presents experimental results based on the thermal shrinkage-bending test. The mechanical performances of two double layers complexes (unreinforced or reinforced by grids) are discussed and analysed. In terms of durability, ecology and economical reasons, the three solutions presented in this article, wants to expose the behaviour of a mixture that is reinforced (in three ways) to improve the consumption of eco-elements (like stone and bitumen) and also to improve the life time of the future work. After analyse the three solutions UN, FP and CF we can conclude that a weak mixture (in eco-elements) can be very well improved in using the two reinforcing solution, presented below.

1 Introduction

It's important to say from the beginning that this work is at its first attempt in characterize this reinforcement solutions. Delaying cracks is the purpose of all reinforcement solution, also of the solutions presented in this article. The objective of this article is to present in an international environment like RILEM congress, first achievements after an experimental campaign with the french (exclusive) dispositive. The durability of a pavement is defined by its capacity to resist to traffic and climate effect during conventional lifetime in function of its economic and society importance (LCPC-SETRA, 1994). One of the pavement rehabilitation techniques enable the user to mitigate reflective cracking and rutting with minimum of surface thickness layer. This construction method use grids at the interface between layers. This solution has been studied a lot between 1990 and 2000 mainly for the rigid pavements [1, 2, 3, 4, 5]. Now days, with the environmental aspect, this technique is coming back [6, 7, 8, 9]. Other while we have to better understand the mechanical behavior of reinforced multilayered pavements and interfaces.

This paper is about the reflective cracking performance of two reinforced (carbon fiber and glass fiber grids) systems, compared with a reference one.

Experimental results are from the bending-tensile test. This device has been conceived by Laboratoire Régional des Ponts et Chaussées [10, 11] for traffic effect (bending) and thermal effect (tensile) simulation and reinforcement performances evaluation.

2 Pavement Cracking Backgrounds

It is well known that pavement cracking is due to climate and traffic effects. For semi-rigid pavement or bituminous reinforced pavement, theoretical and experimental studies show that reflective crack initiation is most of the time due to thermal effects [12, 13]. Top-Down cracking can be also observed due to traffic shear effects [14, 15]. Construction defects, such as problems of interlayer bonding [16, 17] and ageing bitumen phenomena [18] are others cracking contributors. Figure 1. [19] give each reflective cracking path in multilayered pavements reinforced or not, in function of interface boundaries.

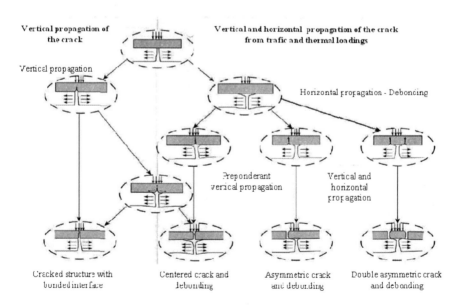

Fig. 1. Path cracking in function of interface conditions (Perez S., 2008)

We can notice debonding and delayed cracking in several cases. For reinforced pavement by grids it is important to get a good contact between HMA and the grid. Most of the time the bonding is realized by the way of tack coat application.

3 The Tensile-Bending Device

The tensile-bending test has been conceived in 1987, the first version [10, 11] was able to compare different systems with a reference one called Stress Absorbing Membrane Interlayer (SAMI) by analysis of reflective cracking velocity. Each sample (560x110x95 mm^3) is submitted to constant temperature (5°C), tensile loading and bending loading in the same time:

- the tensile longitudinal loading is slow and is given by displacement of half part at 0,01 mm/min. This loading is corresponding to thermal effect.
- a cyclic vertical bending is given by pneumatic cylinder (amplitude 0.2 mm, frequency 1 Hz). This cyclic loading is simulating 5 km/h traffic loading.

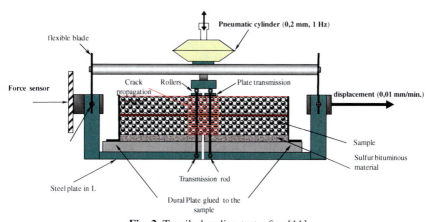

Fig. 2. Tensile-bending test, after [11]

4 Reinforced Systems Tested

The materials are sawed in an experimental pavement constructed in Italy for RILEM Technical Committee (TC ATB/SIB, TG4). The pavement was made of two 50 mm thick layers of AC11 (Standard EN 12697), with pure bitumen 50/70, and calcareous granular material. The interface between each layer is :

- tack coat with 210 g/m² residual bitumen (UN);
- carbon fiber grid with grid size 20x20 mm² (CF) (cf. Table 1);
- glass fiber grid with grid size 33x33 mm² (FP) (cf. Table 1);

Slabs (650x650x100 mm³) have been cut in the pavement in order to perform laboratory tests. The CF and FP slabs have been carved by the two sides, to reach out the fibers. This action has been made to obtain in the final sample a transversal carbon fiber (or glass fiber) and a longitudinal glass fiber, in the middle part of the specimen. Each slab have then been re-cut in the compaction direction to get sample size 560x110x100 mm³ and both layers are reduced to 40 mm thickness each. The densities of each sample are measured according to French Standard NF P 98-250.5

Table 1. Characteristic of Geogrids CF and FP

	Geogrid CF [Longitudinal] Glass fiber	Geogrid CF [Transversal] Carbon fiber	Geogrid FP Glass Polymerized Fiber
Grid size [mm]	20x20		33x33
Tensile modulus of elasticity [N/mm²]]	73,000	240,000	23,000
Elongation at rupture [%]	3 – 4.5	1.5	3
Ultimate tensile force [kN/m]	111 (at 2.7%)	249 (at 1.5%)	211

A sulfur bituminous material 15 mm thick is laid over a tack coat above the lower layer and then, sawed and notched 3 mm in all the thickness of this layer. So the top of the notch is just at the bottom of the HMA bottom layer (cf. Figure 2).

5 Experimental Measurements

Cracking detectors are installed in the front side of samples. On the back side several stain gauges are glued (five - 20mm long) longitudinal strain gauges and four (10 mm long) vertical strain gauges are located such as mentioned in Figure 3 [20].

The collected data are :

- the notch opening displacement, and the vertical displacement due to bending effects. These measurements are done on each side of the sample.
- the tensile horizontal load,
- the vertical crack propagation on the front side,
- the longitudinal and vertical deformations from back side,
- the sample temperature with PT100 thermal sensor.

Datas are stored during time (50 data / minute acquisition rate continuously the first 600 seconds, and then during 5 seconds, every 900 seconds). Each test is repeated three times for the three systems UN, CF and FP.

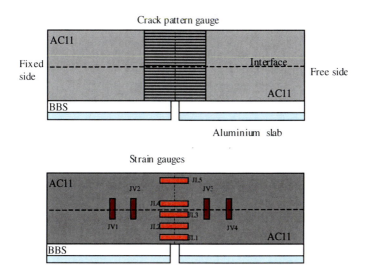

Fig. 3. Instrumentation on both sides of the sample

6 Results

Figures 4-5-6 show the crack growth (measured from the bottom of lower layer) and the mean and amplitude horizontal tensile force against time. After initiation phase, two different macro-crack paths have been observed : a reflective cracking followed by a top down cracking in the upper layer in the same time the reflective crack tip is close to the interface. The crack path is vertical and quite symmetric on both sides. The evolution of mean horizontal tensile force correlate well to the three cracking phases (initiation, reflective cracking and top-down cracking). The FP system response is clearly different from the others (UN, CF), in the same way, this system is the only one that leads to debonding after the crack arrives at the interface (Figures 7-8-9).

The FP grid seems to improve the bearing capacity of the system, even if it`s early debonded.

Figures 7-8-9 present the crack propagation of the three samples tested for each system. Even some dispersion is seen as usual for fatigue tests, these data show that the system failure is obtained first for the UN, second for the CF, third for the FP.

Figures 10 and 11 shows the mean longitudinal strain data from strain gauges during the initiation phase and propagation phase for both kinds of reinforced samples. For both systems, the upper layer is in compression in the beginning of the test and passes in tension after around 5000 seconds. We observe that for CF grid, the strains are greater than for FP grid at the same time that is at the same horizontal displacement and bending number of cycles. But the CF grid system brings a central rigidity about a characteristic length (10 mm) close to the interface. On Figure 11, debonding occurs between 10.255 s and 17.519 s.

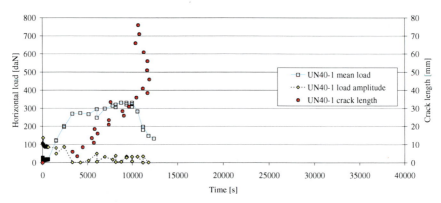

Fig. 4. Crack propagation and horizontal force evolution. (UN 40-1)

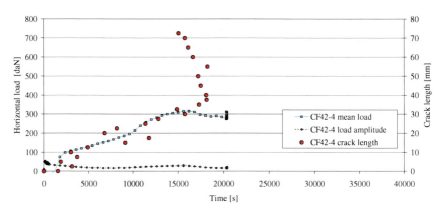

Fig. 5. Crack propagation and horizontal force evolution. (CF 42-4)

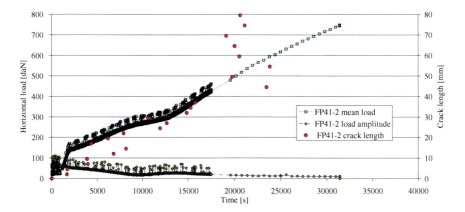

Fig. 6. Crack propagation and horizontal force evolution. (FP 41-2)

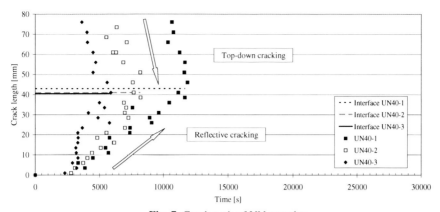

Fig. 7. Crack path of UN samples

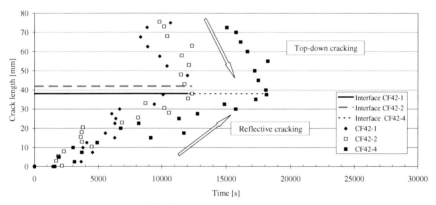

Fig. 8. Crack path of CF samples

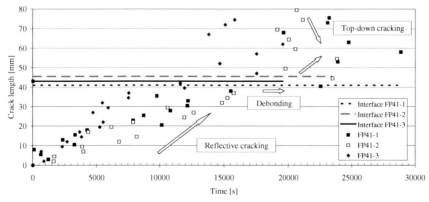

Fig. 9. Crack path of FP samples

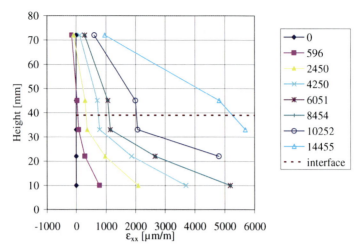

Fig. 10. Longitudinal strains distribution - CF 42-4 (mean values, different curves at different times in second)

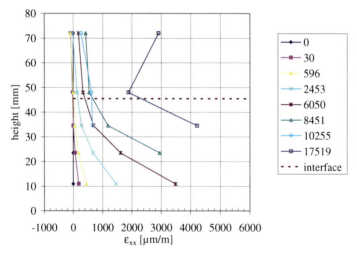

Fig. 11. Longitudinal strains distribution - FP 41-2 (mean values, different curves at different times in second)

7 Conclusions

The present paper shows first results of the TC-RILEM SIB. The tensile-bending test has been chosen in order to compare the reflective cracking performance of each system. We can conclude that the three solutions discussed and analyzed in this article, are performing at interface level (regarding mixture quality). Their capability to resist under the double solicitation bending and traction is in order to their rigidity/flexibility. We have seen that a rigid grid is more efficient under the excitation (traffic and thermal). The FP grid, acts like a metallic reinforcement, but it's more efficient in terms of time performances (it will not corrodes) and also mechanical. The order of efficiency is: FP grid, CF grid and normally the UN witness. The bending components coupled with the thermal traction are the two most common reasons of road cracking. In terms of impact over the environment and economical components, the reinforcement solutions, presented in this article, wants to put in evidence that: the presence at the interface level of a reinforcement element, diminish the mixture quantity. It's very well known that the mixture elements components are in the main part from the environment. Reducing the quantity of bioecological components we contribute to a sustainable road development and also to a sustainable intervention works that must be done periodical. In perspective, other tests will be done without bending component. Monotonic and fatigue shear performances will be evaluated. Scale effects will be studied, because just in this manner we will be able to characterize in a real mode the behavior of these reinforcement solutions.

References

[1] Brown, S.F., Thom, N.H., Sanders, P.J.: A study of grid reinforced asphalt to combat reflection cracking. J. of Association Asphalt Paving Technologist (70), 543–569 (2001)
[2] De Bondt, A.: Effect of reinforcement properties, Reflective Cracking in Pavements Research in Practice. In: Proceedings of the 4th International RILEM Conference, Ottawa, Canada, March 26-30, pp. 13–22 (2000)
[3] Livneh, M., Ishai, I., Kief, O.: Bituminous pre-coated geotextile felts for retarding reflexion cracks. In: Proccedings of the 2nd International RILEM Conference on Reflective Cracking in Pavements, Liège, Belgium (1993)
[4] RILEM Report 18, Prevention of Reflective Cracking in Pavements, State of Art. In: Vanelstraete, A., Francken, L. (eds.) Report of RILEM Technical Committee 157 PRC, BRRC, E&F Spon, Bruxelles (1997)
[5] Vanelstraete, A., Franken, L.: On site behaviour of interface system. In: Proceedings of the 4th International RILEM Conference, Ottawa, Canada, 26-30 Mars, pp. 517–526 (2000)
[6] Bocci, M., Grilli, A., Santagata, F.A., Virgili, A.: Influence of reinforcement geosynthetics on flexion behavior of double-layer bituminous system. In: Advanced Characterization of Pavement and Soil Engineering Materials, Athens, Greece, June 20-22, vol. 2, pp. 1415–1424 (2007)
[7] Florence, C.: Etude expérimentale de la fissuration réflective et modélisation de la résistance de structures cellulaires. In: Thèse de Doctorat de l'ENPC (2005)
[8] Kerzhero, J.P., Michaud, J.P., Hornych, P.: Enrobé armé de fibres de verre - Test sur le manége de fatigre de l'IFSTTAR, RGRA, vol. (890), pp. 48–51 (Décembre 2010-Janvier 2011)
[9] Zielinsky, P.: Fatigue investigation of asphalt concrete beams reinforced with geosynthetics interlayer. In: 6th International RILEM Conference on Cracking in Pavements, Chicago, USA, Juin 16-18, Taylor & Francis (2008)

[10] Vecoven, J.H.: Méthode d'étude des systèmes limitant la remontée des fissures dans les chaussées. In: 1ere Conférence RILEM, Reflective Cracking in Pavements, Liége, Belgique, Mars 8-10, pp. 57–62 (1989)

[11] Dumas, P., Vecoven, J.: Processes reducing reflective cracking, synthesis of laboratory test. In: Proceedings of the 2nd International RILEM Conference on Reflective Cracking in Pavements, Liège, Belgium, March 10-12, pp. 246–253 (1993)

[12] Petit, C., Vergne, A., Zhang, X.: A comparative numerical review of cracked materials. Eng. Fract. Mech. 54(3), 423–439 (1996)

[13] Laveissière, D.: Modélisation de la remontée de fissure en fatigue dans les structures routières par endommagement et macro-fissuration, Thèse de doctorat. Université de Limoges (2002)

[14] Rolt, J.: Top-down cracking: myth or reality. In: The World Bank Regional Seminar on Innovative Road Rehabilitation and Recycling Technologies, Amman, Jordanie, October 24-26 (2000)

[15] Su, K., Sun, L., Hachiya, Y., Maekawa, R.: Analysis of shear stress in asphalt pavements under actual measured tire-pavement contact pressure. In: Proceedings of the 6th ICPT, Sapporo, Japan, pp. 11–18 (July 2008)

[16] Chaignon, F., Roffe, J.C.: Characterisation tests on bond coats: worldwide study, impact, tests, recommendations. International Bitumen Emulsion Federation (IBEF), Bulletin, (9), 12–19 (2001)

[17] Diakhaté, M., Millien, A., Petit, C., Phelipot-Mardelé, A., Pouteau, B.: Experimental investigation of tack coat fatigue performance: towards an improved life time assessment of pavement structure interfaces. Construction and Building Materials 25, 1123–1133 (2010)

[18] Jenner, C.G.J., Uijting, B.G.J.: Asphalt reinforcement for the prevention of cracking in various types of pavements: long term performance and overlay design procedure. In: Proceedings of the 5th International RILEM Conference, Limoges, France, Mai 5-8, pp. 459–465 (2004)

[19] Perez, S.: Approche expérimentale et numérique de la fissuration réflective des chaussées, Thèse de l'Ecole doctorale STS de l'Université de Limoges (2008)

[20] Wendling, L.: Résistance à la fissuration – Résultats des essais de retrait-flexion, Rapport LCPC (littérature grise) - Opération 11P065 (avril 2009)

Theoretical and Computational Analysis of Airport Flexible Pavements Reinforced with Geogrids

Michele Buonsanti, Giovanni Leonardi, and Francesco Scopelliti

Department of Mechanics and Materials MECMAT, Mediterranean University of Reggio Calabria, Italy

Abstract. In recent years the need to increase pavement service life and guarantee high performance has turned a greater attention on the use of pavement reinforcements. In this paper the effectiveness of geogrids as reinforcement of HMA layers in an airport flexible pavement was investigated.

The study proposes a numerical investigation by using the Finite Element Method (FEM) analysing the importance of the geogrids in the pavement behaviour under hard aircraft impact load. The aim of this investigation is to evaluate the stress concentration over the geogrids under impulsive load propagation. The non-homogeneous action is able to develop stress concentration, local damage and fracture with localized weakening. The results show that geogrids can provide a significant contribution to the stress resistance.

1 Introduction

Geosynthetic materials are frequently used to rehabilitate and/or improve pavement mechanical performances [1-7]. Geogrids are the Geosynthetic materials with widespread use for pavement applications, which, depending on the grid constituent material, the mesh shape and size and the stiffness and position in the pavement structure, are able to increase fatigue resistance, reduce rutting and limit reflective cracking. The application of geosynthetics in roads and airfields has become popular in recent years due their high mechanical performances and ability to relieve stresses by reinforcing pavements. Several researches have studied the application of geosynthetics for improving roads and airport pavement performance [5, 6, 8].

Previous studies [2, 5, 6, 9] show that geotextiles provided less resistance against lateral movements than that provided by glass fiber grids. The stiffness of the fabric material reinforcing the hot mix asphalt (HMA) layer needs to be greater than that of the surrounding HMA. High tensile strength and elastic stiffness of glass fiber grids has made them an attractive choice for reinforcing pavement systems. Different studies state that there is a significant benefit in using asphalt geogrids reinforcement: Herbst et al. [10] illustrated an interesting set of data from an experimental site in Austria, where the comparative benefits of geogrids and geotextiles could be directly assessed. Elsing & Sobolewski [11] proposed, in their experience, a factor of 4 on the life of a pavement as a result of the inclusion of a polyester geogrid. The same geogrid was used by Kassner [12] in his experience, he demonstrated the effectiveness of reinforcement at a depth of 100mm over jointed concrete subject to severe temperature variations. Huhnholz [13] presented direct evidence that a polymer geogrid gave a life enhancement factor of at least 3. Penman & Hook [14] described how glass-fiber based geogrids had successfully used as interlayers to extend the design life of asphalt pavements on airport runways, taxiways and aprons. The results obtained by Palacios et al. [15] revealed a partial improvement in reflective cracking resistance due to the incorporation of fiber-reinforced interlayer. It is clear from these studies that many practitioners see significant benefit in using asphalt geogrid reinforcement. Designing a flexible pavement reinforced with glass fiber grid and evaluating the effectiveness of reinforced pavement performance is a complex problem requiring considerable research and study. In this paper, the use of glass-based geogrids as reinforcement materials was analysed by FEM analysis. Use of glass grids in pavement sections is expected to improve pavement performance because of its excellent bonding characteristics with the asphalt and also due to low creep properties.

In the simulation, glass grids were placed within the asphalt layer (HMA). Computer analyses were to investigate the response of the reinforced pavement section under a high impacting load.

2 Mechanical and Constitutive Aspects

Here, we will develop an adequate behaviour mechanical model to perform, subsequently, a computational linear analysis by FEM procedure. For this, we focused our attention over the superficial layer of the pavement section, composed by asphalt mixture (HMA) with the embedded glass grids. The first layer can be modelled as well as a heterogeneous fibres reinforced solid, where the asphalt mixture is the matrix and the geo-grids have the reinforcement role. Under the hypothesis of orthotropic linear elastic behaviour and plane stress, we considered a composite plane solid having strengthening as cross-ply type with, respectively, 0° and 90° fixed way. According to previous theoretical studies [16, 17], let us:

$$D = \begin{bmatrix} E_l & v_{tl} E_l & 0 \\ v_{tl} E_l & E_t & 0 \\ 0 & 0 & G_{lt} \end{bmatrix} \quad (1)$$

The stiffness matrix of the reinforcement mesh where E, v, G respectively are: the elastic modulus (l = longitudinal and t = transversal), the Poisson coefficient and the shear modulus. To make clear, there are two D matrixes, specifically $D°$ and D^{90} since the grids elements can have different thickness. Let us h the total thickness and s the grid element thickness then putting:

$$l_1 = s°/h \qquad l_2 = s^{90}/h \quad (2)$$

The stiffness matrix terms can be split as membrane and flexural types. Under a symmetric geometric composition we found as membrane type the follow:

$$D^m = h \begin{bmatrix} (E_l + E_t)/2 & v_{tl}E_t & 0 \\ v_{tl}E_t & (E_l + E_t)/2 & 0 \\ 0 & 0 & G_{lt} \end{bmatrix} \quad (3)$$

Under a normal biaxial stresses N, the strain field follows as in the form:

$$\begin{bmatrix} \varepsilon_x \\ \varepsilon_y \end{bmatrix} = \frac{1}{h(D_{11}^2 - D_{12}^2)} \begin{bmatrix} D_{11} & -D_{12} \\ -D_{12} & D_{11} \end{bmatrix} \begin{bmatrix} (1-\xi)N_x \\ \xi N_x \end{bmatrix} \quad (4)$$

furnishing the principal stresses in the form:

$$\sigma_l = E_l(\varepsilon_l + v_{tl}\varepsilon_t) \quad (5)$$

$$\sigma_t = E_t(\varepsilon_t + v_{lt}\varepsilon_l)$$

Likewise the flexural matrix in the form:

$$D^f = \frac{h^3}{12} \begin{bmatrix} (E_t - E_l)\beta + E_l & v_{tl}E_t & 0 \\ v_{tl}E_t & (E_l + E_t)\beta + E_t & 0 \\ 0 & 0 & G_{lt} \end{bmatrix} \quad (6)$$

Finally, applying Tsai-Hill criterion for all fibres:

$$\frac{\sigma_l^2}{\sigma_{lR}^2} + \frac{\sigma_t^2}{\sigma_{tR}^2} - \frac{\sigma_l \sigma_t}{\sigma_{lR}^2} = 1 \quad (7)$$

We find that maximum values of the N_x/h relationship depend on the ξ parameter. Rational methods have been developed to analyse the mechanical behaviour of heterogeneous composite solids under various kinds of loading.

Materials properties are derived by methods of micromechanics, whereas structural properties are derived by macro mechanics methods. The mechanical aspects of these heterogeneous solids, has two ways to be analysed. The first one is the micro mechanics approach deals with the resulting materials properties in terms of the constituent materials. Here the most important aspects are local stiffness and basic failure mechanics of the material. Macro mechanics is the latter, deals with the resulting structural properties and structural configuration. Here the most important questions are the stiffness and strength of the entire composite pavement package.

2.1 Micromechanics Approach

There is various methods to treat the constitutive properties of reinforced composite but it's our opinion to apply the more accurate this With the micromechanics approach, the use of the mixture theory [18] allows to find the percentage relations among matrix and reinforcement fibres. We will to use the following notation: $V^{(f)}$ is the volume fraction of the fibres, $V^{(m)}$ the volume fraction of the matrix and $\rho^{(f)}$, $\rho^{(m)}$ the respective mass density. Then the mass density of the complete solids:

$$\rho = \rho^{(f)} V^{(f)} + \rho^{(m)} V^{(m)} \tag{8}$$

Assuming a perfect heterogeneous package, namely no voids, perfect bonding between exactly aligned equally distributed fibres and a homogeneous matrix and considering the strength predictions of mono axial tensile loads. Here should be done two possible characterizations about the failure modes. If the fibres volume fraction is sufficiently large ($\xi > \xi_{min}$) the asphalt matrix will not able to support the entire load after the failure of the fibres which is assumed to take place if the solid is strained to the fibres fracture strain $\varepsilon^{(f)}_{(u)}$. Then the ultimate tensile strength σ^*, when $\xi > \xi_{min}$, assumes the form:

$$\sigma^* = \sigma^{(f)} * \xi + \sigma^{(m)} * (1-\xi) \tag{9}$$

Otherwise, for rather small fibres volume fraction, $\xi < \xi_{min}$, the matrix will be able to support the entire load when the fibres are broken. Then the ultimate strength is:

$$\sigma^* = \sigma^{(m)} * (1-\xi) \tag{10}$$

It's very easy to compare Eqn. (9) to (10) and finding the ξ_{min} value.

2.2 Behaviour under Impact Load Conditions

Focusing the general question that we will to treat us, here we consider the contact conditions among the aircraft wheel and the heterogeneous composite solids (asphalt/fibre-reinforced). Without loss generality we suppose a contact without friction, complete bonding and rigid punch as aircraft wheel impact. So, the contact is modelled as rigid over an orthotropic half-plane ($0 \leq x \leq L$) and, the governing equations relating the loads to the stress fields, follows:

$$p(x) = \frac{\cosh(\pi\eta)}{\sqrt{1-x^2}} \left\{ \frac{2}{\pi} \cos\psi(x) + \sigma^* \kappa [x \sin\psi(x) - 2\eta \cos\psi(x)] \right\} \tag{11}$$

$$q(x) = \frac{\cosh(\pi\eta)}{\sqrt{1-x^2}} \rho \left\{ \frac{2}{\pi} \sin\psi(x) - \sigma^* \kappa [x \cos\psi(x) + 2\eta \sin\psi(x)] \right\} \tag{12}$$

where η, κ and ρ are material parameters. Here $p(x)$ and $q(x)$ represent the contact pressure distribution respectively as contact pressure and contact shear stress. In the equation (11) and (12) the function ψ having the form:

$$\psi(x) = \eta \ln\left(\frac{1+x}{1-x}\right) \tag{13}$$

While for the material parameters the follow relationship appears:

$$\eta = \frac{1}{2\pi} \ln\left(\frac{\pi + \sqrt{vv'}}{\pi - \sqrt{vv'}}\right) \quad k = \frac{1/\sqrt{\alpha\beta}}{\alpha + \beta} \quad \rho = \sqrt{\alpha\beta} \tag{14}$$

Whereas v and v' are the Poisson coefficients in the governing equations of the half-plane depending on the orthotropic properties of the half-plane. Again, α and β are deduced by the elastic constants in the generalised constitutive law for an orthotropic solid. It's easy to see that the contact stress functions approach infinite values and their oscillating behaviour grows stronger as x approaches unity.

3 Computer Analysis

In this study several 3D FEM analysis were performed to analyse non-reinforced and reinforced airport flexible pavement. These simulations were used to investigate the efficiency of glass fiber gird inside asphalt layer on pavement response under a heavy impact caused by aircraft landing gear wheels.

The pavement section is comprised of asphalt concrete and crushed aggregate, as shown in Figure 1. The pavement structure in the application is based on the structure as found for the runway of the Reggio Calabria airport. All pavement layers except the glass grid were modelled by using 3D deformable solid homogeneous elements.

Table 1 shows the elastic properties used in finite element analysis (modulus of elasticity and Poisson's ratio), obtained by conducting laboratory testing on HMA materials and field non-destructive evaluation of granular and subgrade materials.

Glass grid was modelled by using membrane elements.

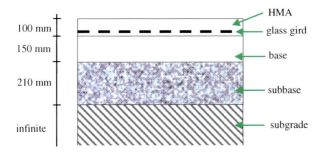

Fig. 1. Pavement section

Table 2 show the glass grid mechanical and dimensional characteristics, it was considered as a linear elastic material since it has very low creep characteristics.

Table 1. Layers thickness and elastic material properties

Layer	Thickness [mm]	Modulus of elasticity [MPa]	Poisson's ratio
Surface	100	7000	0.30
Base	150	2000	0.35
Subbase	210	400	0.35
Subgrade	infinite	70	0.33

Table 2. Typical specification for paving fiber glass grids

Mass	Nominal	[g/m²]	185	
Tensile Strength	Length	[kN/m]	50	
	Elongation at Break	[%]	<5	
	Width	[kN/m]	50	
	Elongation at Break	[%]	<5	
Melting Point	Min.	[°C]	>218	
Grab Strength	Warp	[N]	700	
	Weft	[N]	425	
Dimensions	Grid Size	[mm]	25x25	
	Roll Length	[m]	150	
	Roll Width	[m]	1.5	

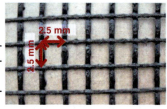

The tyre contact areas considered in the model were Airbus 321 tyre [19]. The most common way of applying wheel loads in a finite element analysis is to apply pressure loads to a circular or rectangular equivalent contact area with uniform tyre pressure [20]. For the finite element model, the contact area, Ac was represented as a rectangle having a length L and a width L' = 0.7·L. To evaluate the pavement load in exceptional condition, the dynamic parameters of a "hard" landing, that caused the broken of some gear components, were considered [21].

Starting from this, considering the damping effect of the gear system, it is possible to calculate the acceleration graph during the hard landing [22]. This value of acceleration was used to calculate the maximum wheel load. Under this load the contact area is:

$$A_c = \frac{397025}{1.36} = 291930 \ (mm^2) \quad (15)$$

Form Eqn. (15) the footprint dimensions are: L = 648 mm and L' = 453 mm. The finite element mesh developed has the following dimensions: 5 m in x and y directions and 2.5 m in the z- direction.

The model presented has 50713 elements and 76586 nodes. Eight-noded linear brick elements C3D8R were used to mesh all the layers of the pavement and four-noded quadrilateral elements M3D4R were used to mesh the glass gird [7].

The loads (vertical and horizontal) were uniformly applied on the surfaces, which were created to be the same size as the wheel imprint of an airbus A321.

Since the boundary conditions have a significant influence in predicting the response of the model, the model was constrained at the bottom (encastre: U1 = U2 = U3 = UR1 = UR2 = UR3 = 0); X-Symm (U1 = UR2 = UR3 = 0) on the sides parallel to y-axis; and Y-Symm (U2 = UR1 = UR3 = 0) on the sides parallel to x-axis. All layers were considered perfectly bonded to one another so that the nodes at the interface of two layers had the same displacements in all three (x, y, z) directions.

Assuming perfect bond at the layer interfaces implies that there will be no slippage at the interface.

This assumption is more applicable to hot mix asphalt layers, since the possibility of slippage is greater at the subbase/subgrade interface [23]. Glass gird was considered embedded in "host" pavement elements using the embedded element technique [24]. Figure 2 shows the deformed shape of the glass fiber gird at impacting instant.

Figure 3 and 4 shows the Mises stress distribution and the deformations in the pavement section.

The results do not shows a significant influence of grid on displacements. The computed displacements under the impacting loads show that the reduction due to grid was about 1%. The reduction of peak stresses in the base layer instead was significant, about the 6% as shown in Figure 5.

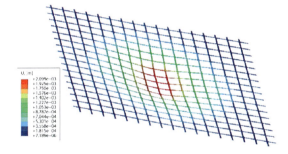

Fig. 2. Deformed glass fiber gird

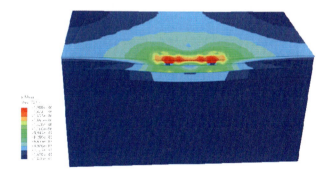

Fig. 3. Mises stress distribution for reinforced pavement

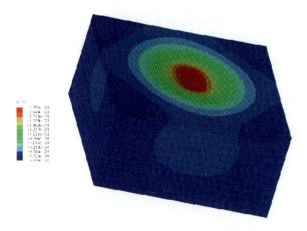

Fig. 4. Deformations in the reinforced pavement section

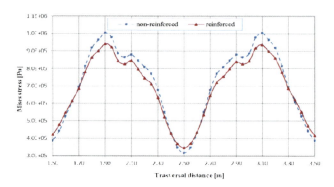

Fig. 5. Mises stress in base layer along the transversal direction of the pavement

4 Discussion and Conclusions

In this paper computational studies were performed to analyse the influence of synthetic fiber glass grids in the HMA layer on the performance of airport pavements. In particular the behaviour of the pavement structure was investigated what it is subjected to the action of a hard landing of an Airbus 321.

The results show how the reinforcement in the asphalt layer causes an interesting reduction in the base vertical stress.

References

[1] Koerner, R.M.: Designing with geosynthetics. Prentice Hall Upper Saddle River, New Jersey (1990)
[2] Lytton, R.L.: Geotext. Geomembranes 8(3), 217–237 (1989)
[3] Barksdale, R.: Transportation Research Board, Washington, DC, USA (1991)

[4] Saad, B., Mitri, H., Poorooshasb, H.: J. Transp. Eng. ASCE 132, 402 (2006)
[5] Cleveland, G.S., Lytton, R.L., Button, J.W.: Transport. Res. Rec. 1849(1), 202–211 (2003)
[6] Kwon, J., Tutumluer, E., Kim, M.: Geosynth. Int. 12(6), 310–320 (2005)
[7] Siriwardane, H., Gondle, R., Kutuk, B.: Geotech. Geol. Eng. 28(3), 287–297 (2010)
[8] Cleveland, G.S., Button, J.W., Lytton, R.L.: Geosynthetics in flexible and rigid pavement overlay systems to reduce reflection cracking. Texas Transportation Institute, Texas A & M University System (2002)
[9] Barksdale, R.D.: Fabrics in asphalt overlays and pavement maintenance. Transportation Research Board (1991)
[10] Herbst, G., Kirchknopf, H., Litzka, J.: Asphalt overlay on crack-sealed concrete pavements using stress distributing media. In: 2nd Int. RILEM Conf. Reflective Cracking in Pavements. Chapman & Hall, Liege (1993)
[11] Elsing, A., Sobolewski, J.: Asphalt-layer polymer reinforcement: long-term experience, new design method, recent developments. In: 5th Int. Conf. Bearing Capacity of Roads and Airfields, Trondheim (1998)
[12] Kassner, J.: Theory and practical experience with polyester reinforcing grids in bituminous pavement courses. In: 1nd Int. RILEM Conf. Reflective Cracking in Pavements, Liege (1989)
[13] Huhnholz, M.: Asphalt reinforcement in practice. In: 3rd Int. RILEM Conf. Reflective Cracking in Pavements, Maastricht (1996)
[14] Hook, K., Penman, J.: The use of geogrids to retard reflective cracking on airport runways, taxiways and aprons. In: Pavement Cracking. CRC Press (2008)
[15] Chehab, G., Chaignon, F., Thompson, M., Palacios, C.: Evaluation of fiber reinforced bituminous interlayers for pavement preservation. In: Pavement Cracking. CRC Press (2008)
[16] Hyer, M.W., White, S.R.: Stress analysis of fiber-reinforced composite materials. McGraw-Hill, New York (1998)
[17] Hult, J.A.H., Rammerstorfer, F.G.: Engineering mechanics of fibre reinforced polymers and composite structures. Springer (1994)
[18] Rajagopal, K.R., Tao, L.: Mechanics of mixtures. World Scientific (1995)
[19] AIRBUS, Airplane Characteristics A321 (1995)
[20] Alkasawneh, W., Pan, E., Green, R.: Road materials and pavement design 9(2), 159–179 (2008)
[21] AAIB, AAIB Bulletin: 6/2009 EW/C2008/07/02. In Accident and Serious Incident Reports (2009)
[22] Buonsanti, M., Leonardi, G., Scopelliti, F., Cirianni, F.: Impact dynamics on granular plate. In: 8th International Conference on Structural Dynamics, EURODYN 2011. Katholieke Universiteit Leuven, Leuven (2011)
[23] Yin, H., Stoffels, S., Solaimanian, M.: Road materials and pavement design 9(2), 345–355 (2008)
[24] Hibbitt, Karlsson, Sorensen: ABAQUS theory manual. Hibbitt, Karlsson & Sorensen (1998)

Optimization of Geocomposites for Double-Layered Bituminous Systems

Francesco Canestrari, Emiliano Pasquini, and Leonello Belogi

Università Politecnica delle Marche, Ancona, Italy

Abstract. In order to improve pavement service life, reinforcement systems can be employed in asphalt layers. In this regard, geocomposites obtained by combining geomembranes with geogrids represent a promising option because they should allow both waterproofing and improved mechanical properties of asphalt pavements. However, the presence of reinforcement may cause an interlayer de-bonding effect that negatively influences overall pavement strength. Given this background, the present research aimed at evaluating the effectiveness of pavement rehabilitation with geocomposites in the laboratory. In particular, the present experimental study intended to implement new products by selecting the optimum combination among different geomembrane compounds, reinforcement types, reinforcement positions and interface conditions. The laboratory investigation was preliminarily organized to perform interface shear tests by means of the ASTRA apparatus. Then, on the basis of the results of the previous phase, the more promising configurations were selected to be further evaluated by means of the three-point bending tests. Specimens were obtained from double-layered slabs compacted in the laboratory. The results presented in this paper enabled the preliminary tuning for the selection of optimized composites to be submitted, in the near future, to performance-related dynamic tests and in situ monitoring of real scale trial sections.

1 Introduction

Road pavements are subjected to high traffic volumes generating accelerated functional and structural distresses which need frequent and expensive maintenance. In this sense, reinforcement systems can be appropriately chosen and employed in asphalt layers in order to improve, when correctly installed, the mechanical properties of pavements against cracking due to repeated loading [1–5] and reflective phenomena [5–8]. However, it is worth noting that the presence of a reinforcement at the interface causes an interlayer de-bonding effect [4, 6, 8, 9] that influences pavement response in terms of stress-strain distribution [10, 11]. Among the wide range of reinforcing products available on the market, geocomposites obtained by combining geomembranes and geogrids can represent a promising option. In fact, this kind of reinforcement system intends to combine improved tensile properties of reinforcements with stress absorbing and waterproofing effects of membranes. Up till now only few studies have documented the extended service life and/or the more cost-effective maintenance process of road structures including membrane interlayer systems reinforced with geotextiles [10, 12, 13] or chopped glass fibres [14]. On the other hand, no literature can be easily found about geogrid-reinforced geomembranes.

Given this background, the research presented in this paper aims at evaluating the effectiveness of asphalt pavement rehabilitation with reinforced geomembranes. In particular, the experimental study intended to implement new products for asphalt interfaces by selecting the optimum combination among different geomembrane compounds, reinforcement types, reinforcement positions and interface conditions. To this purpose, shear and flexural tests were carried out on samples prepared in the laboratory combining two geomembrane compounds, two reinforcement types, two reinforcement positions and four interface conditions. The results presented in this paper enabled the preliminary tuning for the selection of optimized composites to be submitted, in further studies, to performance-related dynamic tests and in situ monitoring of real scale trial sections.

2 Materials and Methods

2.1 Asphalt Concrete

Double-layered slabs were prepared in the laboratory using, for both layers, the same type of asphalt concrete, classified as AC 10 surf 70/100 according to EN 13108-1. This material, prepared with limestone aggregates and plain bitumen, classified as 70/100 according to EN 12591, is characterized by a maximum aggregate size of 10 mm and a bitumen content of 5.6% by the weight of the mix. A preliminary study on the volumetric characteristics of the selected asphalt concrete through 100 gyrations of the Shear Gyratory Compactor showed that an air void content of about 5.5% can be obtained.

2.2 Bonding Materials

During experimental investigation, two types of bituminous materials were applied as bonding agent at the interface of the double-layered slabs: an SBS polymer modified emulsion, hereafter named ME, classified as C 69 BP 3 according to EN 13808, and a water-based elastomeric bituminous primer, hereafter named EP, specifically formulated for the application of the studied geocomposites and characterized by a binder content of 40% by the weight of the material. In both cases, 0.15 kg/m^2 of residual bitumen was spread at the interface before the positioning of the geocomposite.

2.3 Geocomposite Materials

Four geocomposites, obtained combining two membrane compounds and two reinforcing materials, were used as reinforcement for this experimental study.

As far as the bituminous compound is concerned, two polymer modifiers were selected to manufacture the geomembranes: atactic polypropylene plastomeric polymers, hereafter named APP, and styrene-butadiene-styrene synthetic elastomeric copolymers, hereafter named SBS (Table 1).

These polymer modified bituminous membranes were reinforced with two similar fibreglass geogrids characterized by different mesh sizes. In particular, a geogrid having a mesh dimension of 12.5×12.5 mm^2, hereafter named FG12.5, and a geogrid characterized by a mesh dimension of 5.0×5.0 mm^2, hereafter named FG5.0, were selected (Table 1).

Finally, the reinforcement position within the geomembrane was selected as further variable for the optimization of the studied materials: the reinforcing geogrids, in fact, were placed either in the proximity of the upper side or in the proximity of the lower side of the geomembrane (Table 1).

It is worth mentioning that the upper side of all materials was coated with a fine sand ($\Phi < 0.5$ mm) whereas the lower side was characterized by an auto-thermo-adhesive SBS-modified bituminous film (Figure 1).

Table 1. Description of geocomposites

Geocomposite	Geomembrane compound	Reinforcement type	Reinforcement position	Tensile strength [kN/m]
A	APP	FG12.5	Upper side	40×40
B	APP	FG12.5	Lower side	40×40
C	APP	FG5.0	Upper side	40×40
D	SBS	FG5.0	Upper side	40×40

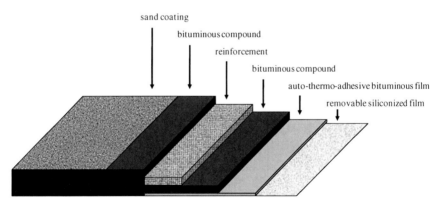

Fig. 1. Typical cross-section of reinforced geomembrane

2.4 Laboratory Specimen Preparation

Double-layered slabs were compacted in the laboratory by means of a roller compactor compliant with EN 12697-33. In the first step, the lower layer was compacted with a thickness of 30 mm, assuming a target of 5.5% air void content. Then, the bituminous bonding material, if planned, was spread at the interface and exposed to the air for breaking. Finally, depending on interface configurations, the appropriate geocomposite was applied and a 45 mm thick upper layer was compacted with the same target of 5.5% air void content. From the laboratory slabs, two types of specimens were obtained: 95 mm diameter cores for Ancona Shear Testing Research and Analysis (ASTRA) tests and 305 mm long, 90 mm wide and 75 mm thick beams for three-point bending (3PB) tests.

2.5 ASTRA Test

The ASTRA device, compliant with the Italian Standard UNI/TS 11214, is a direct shear box, similar to the device usually used in soil mechanics. The specimen is installed in two half-boxes separated by an unconfined interlayer shear zone [11, 15, 16]. During the test, a constant displacement rate of 2.5 mm/min occurs while a constant vertical load, perpendicular to the interface plane, is applied in order to generate a given normal stress. This test returns a data-set where interlayer shear stress (τ), horizontal (ξ) and vertical (η) displacement are reported, allowing the calculation of the maximum interlayer shear stress (τ_{peak}).

2.6 Three-Point Bending Test

The prismatic specimens selected for 3PB tests are placed on supports with a span of 240 mm and subjected to flexural loading at displacement control. Both load and beam deflection in the middle span are measured until failure by means of a load cell and an LVDT, respectively.

Performance of double-layered reinforced systems is indicated by maximum pre-cracking flexural load P_{max}, dissipated energy to failure D, i.e., the area under the load-vertical deformation curve until the maximum pre-cracking load of the system is reached (Figure 2-left), and the total fracture energy or toughness T, i.e., the area under the whole load-vertical deformation curve (Figure 2-right). In particular, the dissipated energy to failure D may account for crack initiation, whereas toughness T could provide an indication of the performance of the reinforcement versus crack propagation [5]. In fact, toughness represents the amount of stress energy required to fracture the system. A system with adequate toughness generally includes a ductile aspect of the fracturing mechanism.

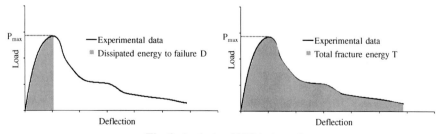

Fig. 2. Analysis of 3PB test results

3 Experimental Program

Ten interface configurations obtained combining different geomembrane compounds, reinforcement types, reinforcement positions and interface bonding materials (Table 2) were considered in order to identify the reinforced solution that would guarantee the best performance in terms of shear and flexural properties. As it can be seen, the curing time (1 hour or 3 hours) of the water-based primer EP was also selected as test variable.

Table 2. Summary of tested configurations

Configuration	1	2	3	4	5	6	7	8	9	10
Geocomposite	A	B	C	C	C	C	D	D	D	D
Bonding material	EP	EP	-	ME	EP	EP	-	ME	EP	EP
Curing time	3h	3h	-	1h	1h	3h	-	1h	1h	3h

The laboratory investigation was preliminarily planned to perform ASTRA interface shear tests on cylindrical specimens. This first phase of the experimental program was subdivided into further two steps. In the first step, geocomposites C and D (configurations 3÷10) were tested with the ASTRA equipment in order to establish the interface condition that assures the best shear properties. Then, the remaining two composite materials (A and B) were tested in combination with the interface condition selected in the previous step. ASTRA tests were carried out at one test temperature (20 °C) and two normal stress conditions (σ = 0.0 and 0.2 MPa) performing three repetitions for each test configurations. This allowed the evaluation of both the standard test condition (σ = 0.2 MPa) and the pure cohesion resistance of the interface (σ = 0.0 MPa) that is the mechanical parameter strongly affected by the presence of the reinforcement. Finally, on the basis of the results of the previous phase, the more promising configurations, characterized by 3 hours cured elastomeric primer at the interface (configurations 1, 2, 6, 10), were selected to be further evaluated by means of three point bending tests. Three-point bending experiments were carried out at 20 °C at a constant rate of 50.8 mm/min on double-layered reinforced systems performing 3 repetitions for each test configuration.

4 Results and Analysis

4.1 ASTRA Test Results

Overall ASTRA test results are presented in Table 3 in terms of mean maximum interlayer shear stress (τ_{peak}) for each configuration obtained applying two normal stresses (σ = 0.0 – 0.2 MPa).

Table 3. Summary of results of ASTRA tests

configuration	1	2	3	4	5	6	7	8	9	10
$\tau_{peak}^{0.0}$	0.17	0.16	0.18	0.15	0.16	0.19	0.30	0.24	0.25	0.35
$\tau_{peak}^{0.2}$	0.23	0.22	0.23	0.20	0.21	0.22	0.42	0.27	0.35	0.40

Influence of geomembrane compound. Results depicted in Figure 3 clearly show that geocomposite D, characterized by elastomeric bituminous compound (SBS), offered higher interface shear resistance than the corresponding material (geocomposite C) prepared with plastomeric modifiers (APP). In fact, for the same interface condition, it can be assumed that the interface shear strength provided by SBS-modified compound was about 60÷80% higher than the corresponding APP-modified membrane. This evident finding, if confirmed by flexural tests, clearly suggests that the production of such reinforcing materials should be addressed towards geocomposites prepared with SBS modified bituminous compound.

Influence of interface condition. Figure 3 also seems to demonstrate that the application of a bonding material at the interface before the laying of the geocomposite, prepared with both SBS and APP compound, inhibits the adhesive properties of the auto-thermo-adhesive film present at the lower side of the tested membranes. In fact, interface shear resistance decreased when the polymer modified emulsion EM or the elastomeric bituminous primer EP were laid at the interface 1 hour before the application of the reinforcing composite material. On the other hand, if the specifically formulated primer EP can cure for at least 3 hours before the application of the geocomposite, the de-bonding effect tends to disappear and the pure cohesion resistance at the interface (obtained with $\sigma = 0.0$ MPa) becomes higher than the one corresponding to configurations without tack coat. On the basis of these results, geocomposites A and B (Table 1) were laid at the interface 3 hours after the application of the elastomeric bituminous primer.

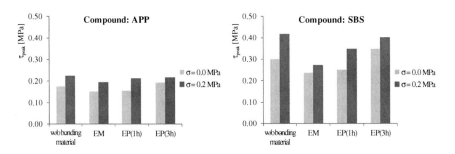

Fig. 3. ASTRA test – influence of compound and interface condition

Influence of reinforcement mesh size. The comparison between configurations 1 and 6 (Table 2) allowed to evaluate the influence of the reinforcement mesh size on interface shear properties of double-layered reinforced bituminous systems. Results summarized in Figure 4-left show very similar performance among the selected materials. This could suggest that the reinforcement placed inside the bituminous geomembrane moderately affects interface shear properties with respect to the influence of the compound. On the other hand, it is worth noting that the geocomposites A and C used for configurations 1 and 6, respectively, were prepared with APP-modified compound that could have negatively levelled maximum interlayer shear stress values τ_{peak}. In this sense, it will be appropriate to study the influence of reinforcement mesh size on shear properties in depth in the following steps of the research project when analogous geocomposites will be prepared with SBS-modified compound.

Influence of reinforcement position. Finally, the comparison between configurations 1 and 2 (Table 2) allowed to evaluate the influence of the reinforcement position on shear performance of reinforced systems. Figure 4-right shows that no difference apparently exists if reinforcement is placed near the lower side or near the upper side of the tested geocomposites. Again, this specific aspect will be further analyzed in the following steps of the research project when analogous geocomposites will be prepared with SBS-modified compound.

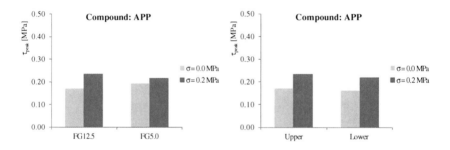

Fig. 4. ASTRA test – influence of reinforcement mesh size and position

4.2 3PB Test Results

Overall 3PB test results are presented in Table 4 in terms of maximum pre-cracking flexural load P_{max} and corresponding deflection δ, dissipated energy to failure D and total fracture energy, i.e., the toughness T. It is worth reminding that all geocomposites, coupled with 3 hours cured elastomeric primer at the interface (configurations 1, 2, 6, 10), were selected to be further evaluated by means of three point bending tests.

Table 4. Summary of 3PB test results

Configuration	Geocomposite	P_{max}	δ	D	T
		kN	mm	N×m	N×m
1	A	4.41	2.51	7.60	22.88
2	B	3.98	2.70	7.40	21.52
6	C	4.10	2.61	7.24	17.14
10	D	4.45	2.63	7.99	20.59

Results listed in Table 4 and shown in Figure 5 denote quite similar behaviours among tested double-layered systems. Nevertheless, it is interesting to note that the best performance in the pre-cracking phase, expressed by P_{max} and D values, was showed by configuration 10 that also exhibited the highest interface properties, i.e., the lowest de-bonding between the asphalt layers. Figure 5 also represents the ductile characteristics conferred to double-layered bituminous systems by introducing a geocomposite reinforcement at the interface. In fact, after the fracture process, the reinforced systems did not rapidly lose their resistance until failure, but they showed a tendency to retain a residual flexural resistance also for high deflection values. This result is due to the fact that the grid inside the geocomposite acts as reinforcement and it is able to absorb part of the applied flexural stress thanks to its tensile properties. This reflects in enhanced post peak energy and thus in a better inhibition of crack propagation which means ductility. Given this background, it clearly appears that the effects of employing reinforcement in a paving system should be properly quantified using toughness as a performance-based parameter.

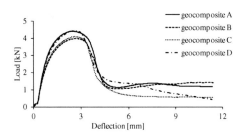

Fig. 5. 3PB mean test results

Influence of geomembrane compound. Figure 6 compares flexural behaviour of geocomposite C, produced with reinforcement FG5.0 and plastomeric bituminous compound (APP), and the corresponding geocomposite D, including elastomeric modifiers (SBS) instead of plastomeric ones. As it can be seen (Table 4), SBS-modified compound showed enhanced performance, both in the pre-cracking phase (P_{max}, D) and in the post-fracture phase (T), with respect to APP-based material, confirming the good interface shear properties evidenced above (Table

3). Such results suggest, for a given reinforcement type, the possibility of predicting flexural behaviour of reinforced bituminous systems starting from their interface shear properties.

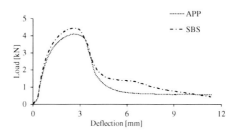

Fig. 6. 3PB test – influence of geomembrane compound

Influence of reinforcement mesh size. The comparison between geocomposite A and C (Table 1) is plotted in the graph of Figure 7-left. It is worth noting (Table 4) that the fibreglass geogrid with a mesh dimension of 5.0 × 5.0 mm^2 (FG5.0) involved a reduction in the flexural properties of reinforced double-layered systems with respect to the fibreglass geogrid having a mesh dimension of 12.5 × 12.5 mm^2 (FG12.5) despite the fact that the two geocomposites were characterized by the same tensile strength (Table 1). This experimental finding could be due to the fact that the thinner mesh wires of reinforcement FG5.0 probably suffered higher damage during the compaction of the upper asphalt layer with respect to geogrid FG12.5.

Influence of reinforcement position. Finally, Figure 7-right gives information about the influence of reinforcement position on flexural performance of reinforced bituminous systems. Experimental data seem to suggest that the reinforcement placed in the proximity of the lower side of the geocomposite negatively affected pre-cracking flexural behaviour of the tested samples with respect to the same reinforcement type placed in the proximity of the upper side of the geocomposite. Similar results, even if less marked, were achieved through ASTRA interface shear tests (Table 3). Thus, it can be asserted that higher interface shear properties lead to enhanced pre-cracking flexural strength.

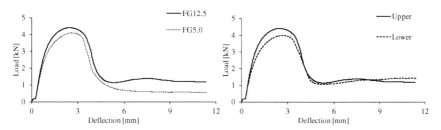

Fig. 7. 3PB test – influence of reinforcement mesh size and position

5 Conclusions

The present research aimed at implementing new geocomposites for pavement rehabilitation by selecting the optimum combination among different geomembrane compounds, reinforcement types, reinforcement positions and interface conditions. To this purpose, ten configurations were analyzed in the laboratory through interface shear tests, by means of the ASTRA apparatus, and three-point bending tests.

On the basis of the experimental results obtained during the present research study, the following general remarks can be drawn:

- geocomposite prepared with SBS-modified bituminous compound showed enhanced interface shear properties and flexural resistance;
- the application of a bonding material at the interface seemed to inhibit the adhesive properties of the auto-thermo-adhesive film present at the lower side of the tested membranes;
- the de-bonding effect due to the application of a bonding material at the interface disappeared when the specifically formulated bituminous primer was exposed to the air for 3 hour for curing;
- selected reinforcement mesh sizes and positions showed a limited influence on interface shear properties of the tested asphalt systems;
- the fibreglass geogrid having a mesh dimension of 12.5×12.5 mm^2 demonstrated higher flexural resistance than the fibreglass geogrid characterized by a mesh dimension of 5.0×5.0 mm^2;
- reinforcement in the proximity of the lower side of the geocomposite seemed to negatively affect the flexural behaviour of the tested double-layered systems with respect to the same reinforcement type placed in the proximity of the upper side of the geocomposite;
- for a given reinforcement, ASTRA shear tests are able to detect the interface conditions that also guarantee valuable flexural properties;
- higher interface shear properties are strictly correlated to enhanced crack initiation resistance, whereas reinforcement characteristics principally affect the crack propagation rate.

The results presented in this paper enabled the preliminary tuning for the selection of optimized composites to be further tested and validated also by means of dynamic four-point bending tests currently in progress.

Acknowledgements. This study was sponsored by INDEX Construction Systems and Products S.p.A. that gave both financial and technical support for the research project.

References

[1] Brown, S.F., Brunton, J.M., Hughes, D.A.B., Brodrick, B.V.: J. Assoc. Asphalt Paving Technol. 54, 18–44 (1985)
[2] Bocci, M., Grilli, A., Santagata, F.A., Virgili, A.: In: Loizos, Scarpas, Al-Qadi (eds.) Proceedings of the International Conference on Advanced Characterisation of Pavement and Soil Engineering Materials, vol. 2, pp. 1415–1424. Taylor & Francis Group, London (2007)

[3] Virgili, A., Canestrari, F., Grilli, A., Santagata, F.A.: Geotext. Geomembr. 27(3), 187–195 (2009)
[4] Ferrotti, G., Canestrari, F., Virgili, A., Grilli, A.: Constr. Build. Mater. 25(5), 2343–2348 (2011)
[5] Lee, S.J.: Can. J. Civ. Eng. 35(10), 1042–1049 (2008)
[6] Caltabiano, M.A., Brunton, J.M.: J. Assoc. Asphalt Paving Technol. 60, 310–330 (1991)
[7] Austin, R.A., Gilchrist, A.J.T.: Geotext. Geomembr 14(3-4), 175–186 (1996)
[8] Brown, S.F., Thom, N.H., Sanders, P.J.: J. Assoc. Asphalt Paving Technol. 70, 543–569 (2001)
[9] Canestrari, F., Grilli, A., Santagata, F.A., Virgili, A.: In: Proceedings of the 10th International Conference on Asphalt Pavements, Quebec City, vol. I, pp. 811–820 (2006)
[10] Shukla, S.K., Yin, J.-H.: In: Shim, J.B., Yoo, C., Jeon, H.-Y. (eds.) Proceedings of the 3rd Asian Regional Conference on Geosynthetics – Now and Future of Geosynthetics in Civil Engineering GeoAsia2004, Seoul, pp. 314–321 (2004)
[11] Canestrari, F., Ferrotti, G., Partl, M.N., Santagata, E.: Transp. Res. Rec. 1929, 69–78 (2005)
[12] Ramberg Steen, E.: In: Nikolaides, A.F. (ed.) Proceedings of the 3rd International Conference on Bituminous Mixtures and Pavements, Thessaloniki, vol. 1, pp. 273–282 (2002)
[13] Vanelstraete, A., De Visscher, J.: In: Petit, C., Al-Qadi, I.L., Millien, A. (eds.) Proceedings of the 5th International RILEM Conference on Reflective Cracking in Pavements, pp. 699–706. RILEM Publications SARL, Bagneux (2004)
[14] Gillespie, R., Roffe, J.-C.: In: Nikolaides, A.F. (ed.) Proceedings of the 3rd International Conference on Bituminous Mixtures and Pavements, Thessaloniki, vol. 1, pp. 159–168 (2002)
[15] Canestrari, F., Santagata, E.: Int. J. Pavement Eng. 6(1), 39–46 (2005)
[16] Santagata, F.A., Partl, M.N., Ferrotti, G., Canestrari, F., Flisch, A.: J. Assoc. Asphalt Paving Technol. 77, 221–256 (2008)

Sand Mix Interlayer Retarding Reflective Cracking in Asphalt Concrete Overlay

Jongeun Baek[1] and Imad L. Al-Qadi[2]

[1] Postdoctroal Researcher, Sejong University, Korea
[2] Professor, University of Illinois at Urbana-Champaign, USA

Abstract. This study evaluated the performance effectivenss of sand mix interlayer in controlling reflective cracking in asphalt concrete (AC) overlaid existing jointed concrete pavement (JCP) using a three-dimentional finite element (FE) model. A cohesive zone model was incorporated into the FE model to characterize the fracture behavior of the AC overlay under transient vehicular loading. A limit state load approach was used to determine the resistance of the AC overlay to reflective cracking in terms of normalized axle load of an overload equivalent to an 80-kN single-axle load. The study concluded that the sand mix interlayer enhanced the fracture resistance of the AC overlay due to its relatively high fracture energy. A macro-crack level of reflective cracking was initiated in the wearing course in the AC overlay earlier than in the leveling binder, so-called crack jumping. The softer the sand mix, the tougher it may be, but it may cause shear rutting in the AC overlay. As the bearing capacity of the JCP becomes lower, more fractured area was developed in the AC overlay and the performance effectiveness of the sand mix interlayer was better.

1 Introduction

Pavement rehabilitation is needed to restore the structural and/or functional capacity of deteriorated pavements. Typical pavement rehabilitations include restoration, recycling, resurfacing, and reconstruction. The proper rehabilitation method is determined based on the type and condition of the existing pavement. For a moderately deteriorated Portland cement concrete (PCC) pavement, resurfacing existing pavement with a relatively thin asphalt concrete (AC) layer, known as an AC overlay, is regarded as an efficient method. AC overlays are designed to support anticipated traffic volume over a specific period of time. When AC overlays are built on a jointed concrete pavement (JCP) or a cracked surface, reflective cracking can develop shortly after the overlay application because of stress intensity at the vicinity of the discontinuities.

Several remedial techniques have been incorporated into AC overlays to control reflective cracking, including placing a thin layer at the interface between an existing pavement and an HMA overlay, rubberizing existing concrete pavement, cracking and sealing existing concrete pavement, and increasing the thickness of the AC overlay. Among these techniques, interlayer systems have been effective in controlling reflective cracking when used appropriately and selected based on their distinct characteristics. Interlayer systems made of softer, stiffer, and tougher materials can absorb excessive stresses, reinforce HMA overlays, and resist crack developments, respectively. The efficiency of these interlayer systems depends on the type and condition of the interlayer systems, installation approach, and characteristics of the existing pavement and AC overlay.

Fracture mechanics based finite element (FE) models were widely used to to examine the fracture behavior of AC overlays. Among them, a cohesive zone model (CZM) has been adapted to facilitate modeling the entire crack process for AC pavements [1–6]. To date, the fracture behavior of AC overlays under more realistic traffic loading has not been investigated. Also, the performance of interlayer systems depends on the circumstances of AC overlay design and installation conditions. Therefore, it needs to understand the mechanism of interlayer systems on controlling reflective cracking due to moving traffic loading in order to (1) evaluate the performance of these interlayer systems and (2) specify their appropriate circumstances relevant to HMA overlay design.

2 Research Objective and Approach

This study evaluated the performance of the sand mix interlayer system in controlling reflective cracking using an FE model. It was built a three-dimensional FE model which consists of the AC overlay and existing JCP. A bilinear cohesive zone model (CZM) was incorporated into the FE model to characterize the fracture behavior of the AC overlay. Using the bilinear CZM, reflective cracking initiation and propagation were simulated. Transient moving vehicular loading was applied across a joint to develop reflective cracking. In order to force reflective cracking development by one pass of load application, various levels of overload were applied. Finally the performance of sand mix interlayer in controlling reflective cracking was examined under various conditions.

3 Finite Element Modelling

3.1 Geometry and Boundary Condition

A three-dimensional FE model was built for a typical AC overlay placed on a JCP. The pavement consists of a 57-mm-thick AC overlay; two 200-mm-thick concrete slabs with a 6.4-mm-wide transverse joint and 6.0 m of joint spacing; a 150-mm-thick base layer; and a 9,000-mm-thick subgrade layer. The AC overlay consists of a 19-mm-thick leveling binder layer and a 28-mm-thick wearing surface. No dowel bars or aggregate interlocking were considered to make the pavement system more voluable to reflective cracking. The dimensions of a concrete slab are 6.0 m in length and 3.6 m in width. Since each concrete slab is geometrically symmetric with respect to the center of the slab, one quarter of the slab was chosen to simplify the pavement model. Symmetric boundary conditions were applied accordingly to the three faces surrounding the two concrete slabs. Infinite elements were placed at a far-field zone to minimize stress wave reflection at the boundary. A vehicular loading with 80 kN of single-axle and dual-assembly tires configuration at a speed of 8 km/h was applied on the AC overlay surface; the travel distance of the loading was 600 mm across the joint. Non-uniform vertical contact pressures measured at approximately 5 km/h were employed for each tread of the tires [7]. The total imprint area of the two tires was 338.8 cm^2. Applied vertical contact pressure was 0.7 MPa on average.

In addition to cotinuum elements, cohesive elements were inserted at the AC overlay directly over the joint where reflective cracking has potential to develop. Figure 1 illustrates the location of the cohesive elements [8]. Actually, the cohesive elements governed by the bilinear CZM connect two parts of HMA overlay sections by means of traction. Since the cohesive elements have zero apparent thickness in a normal direction, the initial geometry of the pavement model is unchanged, despite the insertion of cohesive elements.

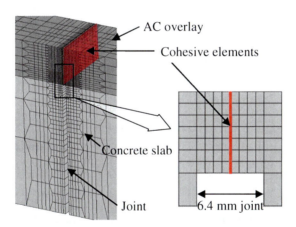

Fig. 1. Geometry of AC overlaid JPCP and the location of cohesive elements

3.2 Material Characterization

Among pavement materials used in the AC overlay model, AC is the key material to govern pavement responses related to reflective cracking. The continuum and fracture properties of the AC were obtained from complex (dynamic) modulus and disk-shape compact tension (DCT) tests conducted at -10°C [9, 10]. Based on complex modulus test results, a linear viscoelastic model was constituted with

Prony series expansion parameters of an instantaneous Young's modulus (E_o), Poisson's ratio (μ), dimensionless shear (g_i) and bulk (k_i) relaxation moduli, and corresponding relaxation time (τ_i). Fracture energy (Γ_c) and tensile strength (T_o) obtained from the DCT test are used to specify the bilinear CZM. Other materials used in sublayers of the AC overlay were characterized simply by using a linear elastic model. Their material properties were selected within typical ranges from the literature. Detailed material properties of the pavement materials are referred in the previous works [9, 10].

3.3 Sand Mix Interlayer

Sand mix interlayer designed to enhance AC's fracture resistance is placed between the wearing course layer and the existing JCP layer as a substitute layer for the leveling binder. Compared to conventional leveling binder, the sand mix interlayer is made of finer graded aggregates and highly polymerized asphalt binder. Corresponding nominal maximum aggregate size (NMAS) of the sand mix and leveling binder are 4.75 mm and 9.5 mm, respectively. The sand mix has 8.6% polymer-modified PG 76-28 asphalt binder; the leveling binder has 5.6% unmodified PG 62-22 asphalt binder. Since the sand mix interlayer has similar compositions to the leveling binder, the bulk and fracture properties were obtained using the same laboratory tests for the AC [9, 10]. The sand mix interlayer has approximately 20% lower relaxation modulus than the leveling binder. However, the fracture energy of the sand mix, 593 J/m^2, is approximately two-fold of that of the leveling binder, 274 J/m^2 at -10°C. In this study, the AC overlay with and without the sand mix interlayer is designated as Design A and Design B, respectively.

4 Effectiveness of the Sand Mix Interlayer

4.1 Quantification of Fractured Area in the AC Overlay

In this study, one pass of an overload was applied to force reflective cracking in the AC overlay instead of considerable number of normal 80-kN loads. A total axle load of the overload is amplified, keeping the same speed, contact area, and normalized vertical contact stress distribution of the normal load. A limit state load is determined when a macro-crack level of reflective cracking occurs in the entire cross section of the AC overlay. Hence, the limit state load can represent the capacity of the AC overlay to withstand reflective cracking. Herein, the fracture area of the AC overlay was quantified with a representative fractured area (RFA) which was proposed in a previous study [5]. The RFA is an average degradation of stiffness of cohesive elements over a specific area, ranging from 0.0 (no crack) to 1.0 (macro-crack development).

The RFA was calculated at several levels of overloads for the AC overlay without the interlayer (Design A). Figure 2 shows RFA variations with respect to a normalized axle load of 80 kN (P_{80}) for Design A. The RFA does not increase notably until $2P_{80}$, then starts to increase rapidly from 0.08 at $3P_{80}$ to 0.85 at $8P_{80}$, and then converges to 1.0. Using a generalized logistic function, RFA could be specified as a function of P_{80}. Using the fitting curve, the RFA that corresponds to a certain overload can be estimated for Design A.

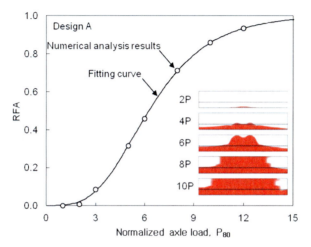

Fig. 2. Fractured area versus overloads (Design A)

4.2 Macro-and-Micro Crack Development

As reflective cracking develops, micro- and macro-cracks exist together in the AC overlay. Among fractured area, macro-cracks exist only in a certain area where the stiffness degradation parameter, D of cohesive elements is equal to 1.0. For Designs A and B, fractured area by micro- and macro-cracks in total overlay, wearing surface, and leveling binder are compared in Figure 3. Macro-cracks were initiated at $8P_{80}$ in Design A and at $10P_{80}$ in Design B. These macro-cracks occurred simultaneously in the wearing surface and leveling binder. In Design A, macro-cracks in the wearing surface and binder layer represent more than 50% of total cracked area; in Design B, macro-cracks in the wearing surface account for more than 50% of total cracked area, but macro-cracks in the binder layer represent less than 50% of total cracked area. For example, at $12P_{80}$, 64.9% and 49.2% of the area in the binder layer is fractured by macro-cracks in Design A and Design B, respectively. This means that the sand mix interlayer system reduced micro-cracks by 15.7%, while the difference in total cracked area is only 1.2%. Hence, the performance of the sand mix interlayer more significantly delays the occurrence of macro-crack-level reflective cracking.

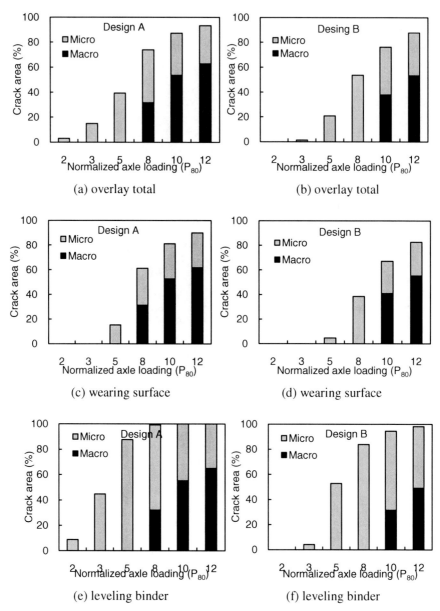

Fig. 3. Percentage of micro- and macro-cracks: (a), (c), and (e) in Design A; (b), (d), and (f) in Design B

4.3 Effect of Fracture Property

The fracture property of sand mix interlayer depends on its component materials. For example, a sand anti-fracture (SAF) interlayer system, a sort of sand mix, possessed a fracture energy of 1800 J/m² [9]. To examine the effect of fracture energy of the sand mix interlayer on controlling reflective cracking, the reflective cracking service life was obtained for three fracture energies of 474 J/m² ($1.0\Gamma_{IC}$), 948 J/m² ($2.0\Gamma_{IC}$), and 1886 J/m² ($4.0\Gamma_{IC}$), with the same cohesive strength of 3.6 MPa ($1.0T°$). As a reference, Design A was added, with a fracture energy of 50% ($0.5\Gamma_{IC}$) and cohesive strength of 70% ($0.7T°$) of the sand mix. For the three Design B cases, the RFA decreased as fracture energy increased. As a result, reflective cracking resistnace factor, Φ_r of Design B which is the ratio of the P_{80} of Design B to that of Design A becomes 1.43, 2.22, and 3.23 for Design B with $1.0\Gamma_{IC}$, $2.0\Gamma_{IC}$, and $4.0\Gamma_{IC}$, respectively. Hence, the sand mix interlayer can delay reflective cracking development; consequently, extends the service life of the HMA overlay regarding reflective cracking.

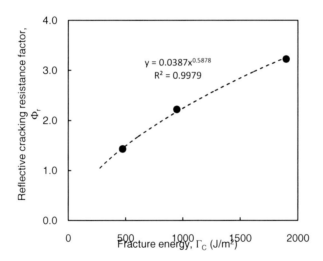

Fig. 4. Reflective cracking control factor versus fracture energy of the sand mix

During the development of reflective cracking, the fracture energy of the material can be degraded progressively. Degradation of fracture energy in the AC overlay was examined in terms of fracture energy damage parameter, D_Γ, which is defined as the ratio of the dissipated energy to fracture energy. D_Γ calculated at the center of the wearing course ($0.5h_{WS}$) was compared with the leveling binder ($0.5h_{LB}$) for the three cases of Design A, Design B with 948 J/m² ($2.0\Gamma_{IC}$). Figure 5 demonstrates D_Γ variations with respect to horizontal distance at a higher level of an overload, $10P_{80}$. For Design A, fracture energy of the AC overlay is fully dissipated, that is, macro-cracks are initiated under the wheel path as well as beyond the wheel path ($0.24W \leq x \leq 0.74W$). In Design B with $2.0\Gamma_{IC}$, macro-cracks occur solely in the wearing surface at $0.32W \leq x \leq 0.62W$, and micro-cracks initiated in the sand mix that replaces the leveling binder in Design A. Hence, macro-crack-level reflective cracking does not develop in the sand mix, but instead jumps to the wearing surface because of the higher crack tolerance of the sand mix. In other words, macro-crack-level reflective cracking is not channelized through the AC overlay. This crack jump phenomenon can play an important role in performance of the AC overlay because it can prevent moisture penetration into underlying pavement layers as well as material loss by pumping.

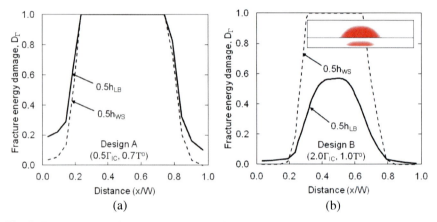

Fig. 5. Fracture energy damage parameter variations at $10P_{80}$ for (a) Design A ($0.5\Gamma_{Ic}$) and (b) Design B ($2.0\Gamma_{Ic}$)

4.4 Effect of Bearing Capacity

The effect of the bearing capacity of the existing JCP on reflective cracking development in Design B was examined. Fractured area in Design B induced by a moderate level of an overload, $5P_{80}$, was obtained for three bearing capacity conditions: 1) E_{BA} of 75 MPa and E_{SB} of 35 MPa, 2) E_{BA} of 300 MPa and E_{SB} of 140 MPa, and 3) E_{BA} of 600 MPa and E_{SB} of 280 MPa. The limit state load, P_{80f} and Φ_r corresponding to each bearing capacity of the JCP for Design A and Design B is compared in Figure 6. As the bearing capacity of the JCP is greater, P_{80f} for Designs A and B increase, but Φ_r decreases because of relatively greater P_{80f} in Design A. It can infer that as the bearing capacity of an existing JCP is lower, the performance of the sand mix interlayer is relatively better; but the enhancement of the reflective cracking service life becomes insignificant.

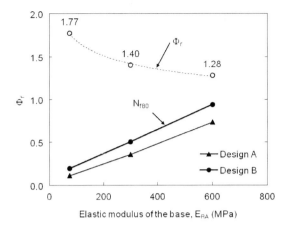

Fig. 6. Reflective cracking resistance factor of Design B versus the elastic modulus of the base

5 Conclusions

In this study, a three-dimensional FE model was built for an AC overlay on an existing JCP with and without sand mix interlayer. The performance of the sand mix interlayer in controlling reflective cracking caused by traffic loading was evaluated under various conditions. The sand mix interlayer is sufficiently effective in controlling reflective cracking. The sand mix interlayer system extends the service life of the AC overlay in terms of reflective cracking. The increase in service life depends on fracture energy of the sand mix. Also, as the bearing capacity of existing JCP increases, the performance effectiveness of the sand mix interlayer gradually decreases. Due to higher fracture tolerance of the sand mix, macro-crack level of reflective cracking is initiated in the wearing course in the AC overlay, so-called crack jumping. In some cases, the crack jump phenomenon can play an important role in the performance of the AC overlay because it can prevent both penetration of moisture into underlying pavement layers as well as material loss by pumping.

Acknowledgements. The authors greatly appreciated the support of their colleagues at the Illinois Center for Transportation at the University of Illinois at Urbana-Champaign. In addition, this work was partially supported by the National Center for Supercomputing Applications (NCSA) under project # TG-ECS090012 and utilized the NCSA Dell Intel 64 Cluster Abe machine.

References

[1] Jenq, Y.-S., Perng, J.-D.: Tran. Res. Rec. 1317, 90–99 (1991)
[2] Soares, J.B., Colares de Freitas, F.A., Allen, D.H.: Crack modeling of asphaltic mixtures considering heterogeneity of the material. In: Proceedings of the 82nd Annual Meeting of the Transportation Research Board (CD-ROM), Washington, D.C., USA (2003)

[3] Paulino, G.H., Song, S.H., Buttlar, W.G.: Cohesive zone modeling of fracture in asphalt concrete. In: Pttit, C., Al-Qadi, I.L., Millien, A. (eds.) Proceedings of the 5th International RILEM Conference–Cracking in Pavements, Limoges, France, pp. 63–70 (2004)
[4] Song, S.H.: Ph.D. Dissertation. University of Illinois at Urbana-Champaign, Urbana, IL, USA (2006)
[5] Baek, J., Al-Qadi, I.L.: Finite element modelling of reflective cracking under moving vehicular loading. In: Roesler, J.R., Bahia, H.U., Al-Qadi, I.L., Murrell, S.D. (eds.) Proceedings of ASCE's 2008 Airport and Highway Pavements Conference, Bellevue, WA, USA, pp. 74–85 (2008)
[6] Kim, H., Wagoner, M.P., Buttlar, W.G.: Const. and Bld. Mat. 23(5), 2112–2120 (2009)
[7] Yoo, P.J., Al-Qadi, I.L.: Trans. Res. Rec. 1990, 129–140 (2006)
[8] Baek, J., Al-Qadi, I.L.: J. of Ass. Asph. Pav. Tech. 78, 638–673 (2009)
[9] Al-Qadi, I.L., Buttlar, W.G., Baek, J., Kim, M.: Report FHWA-ICT-09-44. Illinois Center for Transportation, Urbana (2009)
[10] Jeng, Perng: Eng. Fract. Mech. 85(6), 1234 (1991)
[11] Kim, M., Buttlar, W.G., Baek, J., Al-Qadi, I.L.: Trans. Res. Rec. 2127, 146–151 (2009)

Full Scale Tests on Grid Reinforced Flexible Pavements on the French Fatigue Carrousel

Pierre Hornych[1], Jean Pierre Kerzrého[1], Juliette Sohm[1], Armelle Chabot[1], Stéphane Trichet[1], Jean Luc Joutang[2], and Nicolas Bastard[2]

[1] LUNAM Université, IFSTTAR, CS4 F-44344 Bouguenais, France
pierre.hornych@ifsttar.fr
[2] St Gobain Adfors, Viktoriaallee 3-5, 52066 Aachen, Germany

Abstract. Grids are increasingly used. They have proved their efficiency, but there is presently no widely accepted design method to predict the long term life of grid reinforced pavements. This paper describes a full scale experiment carried out on the large pavement fatigue carrousel of IFSTTAR, to test simultaneously 3 pavement sections with different types of grids, in comparison with an unreinforced pavement structure. The tests are carried out on typical French low traffic pavement structures. Results up to approximately 800 000 loads are presented. The experiment is planed to continue to load the test sections up to at least 1 million loads. During the experiment, the behaviour of the pavement sections has been followed by deflection and rut depth measurements, and surface distress analysis (observation of cracks and other degradations). As observed on the circular APT for low traffic pavements with thin bituminous layers, crack development was following a transversal orientation. This experiment shows the necessity to better understand the grid behaviour by means of modelling, experiments and use of new measurement techniques as planned in the new Rilem TC-SIB and TC-MCD.

1 Introduction

Grids are increasingly used, both for reinforcement of existing pavements, and for improving fatigue and reflective cracking resistance of new pavements. These techniques have proved their efficiency, but require attention to achieve a good bonding between the system and the pavement layers during construction. There is presently no widely accepted design method to predict the long term life of grid reinforced pavements. Therefore, full scale tests are needed to evaluate both the construction procedures and the long term performance of these products [1] [2].

This paper describes a full scale experiment carried out on the large pavement fatigue carrousel of IFSTTAR in Nantes, to test simultaneously 3 pavement sections with different types of grids, in comparison with an unreinforced pavement structure. The paper presents the construction, instrumentation of the pavements, and the first results of the tests.

2 Description of the Full Scale Experiment

The objective of the experiment is to test and compare simultaneously the fatigue behaviour of 4 flexible pavement sections, under typical French axle loading (half axles loaded at 6.5 tons with dual wheels), for a total of 1 million load cycles. The pavement fatigue carrousel of IFSTTAR is a large scale circular outdoor facility, unique in Europe by its size (120 m length, 6 m width) and loading capabilities (maximum loading speed 100 km/h, loading rate 1 million cycles per month, 4 arms equipped possibly with different wheel configurations, lateral wandering of the loads to reproduce real traffic) [3].

2.1 Test Sections and Material Characteristics

The tests are carried out on typical French low traffic pavement structures [4] consisting of an 80 mm thick bituminous wearing course, over a granular subbase (300 mm thick), and a sandy subgrade soil, with a bearing capacity of about 80 MPa. Four structures, each 10 m long, are tested, representing 1/3 of the whole test track. Structures A and B are reinforced with grids incorporating a special film designed to ensured good bonding and replacing the tack coat. Structure C is reinforced with a traditional grid with a tack coat. Structure D is a reference structure without reinforcement. The three different grids are placed in the lower part of the bituminous layer, 2 cm above the interface (Figure 1).

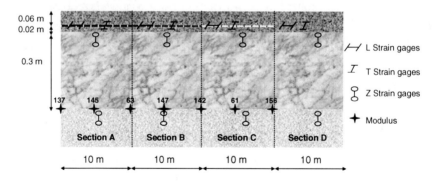

Fig. 1. Pavement test sections and implementation of sensors

The bituminous mix is a standard French 0/10 mm wearing course bituminous material, (BBSG 0/10). This material contains 5.5 % of grade 35/50 bitumen. The mechanical behaviour of this mix was characterised by classical complex modulus and fatigue tests on trapezoidal specimens (EN 12697-31 and EN 12697-24). The specimens had an average void content of 6.2 %. The reference complex modulus of the mix at 15°C and 10Hz is 11320MPa. The Huet Sayegh viscoelastic model parameters obtained for this mix are presented in Table 1. These parameters can be used for viscoelastic pavement structure calculations [5].

Table 1. Material characteristics

E_0 (MPa)	E_{inf} (MPa)	δ	k	h	A_0 (s)	A_1 (s.°C^{-1})	A_2 (s.°C^{-2})
10.0	27180	2.29	0.22	0.65	4.0617	-0.38792	0,0016399

The fatigue law of the mix is approximated by Eqn. (1).

$$\varepsilon/\varepsilon_6 = \left(N/10^6\right)^b \quad (1)$$

With: ε_6, the strain leading to failure for 1 million loads, and b, the slope of the fatigue curve. Experimentally, the fatigue parameters obtained for the mix are $\varepsilon_6 = 116$ μstrains and b = - 0.206.

2.2 Characteristics of the Tested Grids

The three grids tested are all a high-strength open fiberglass geogrid custom knitted in a stable construction and coated with a patent-pending elastomeric polymer and self-adhesive glue (Tensile Strength: 100kN/m×100kN/m). The mesh of grid from test section A is half smaller than those from test section B and C (25×25mm²). Grids of the test section A and B contain a patent-pending, highly engineered film designed to replace the need for a tack coat. These two new grids have shown a better behaviour during specific 3 point bending fatigue tests [6].

3 Construction and Instrumentation of the Test Sections

The pavement structures were built on the existing subgrade of the test track, which is a sand with 10 % fines, sensitive to water. The granular base consisted of 30 cm of 0/31.5 mm unbound granular material (UGM). After construction, this base was covered by a spray seal. A 2 cm thick bituminous layer was first laid and compacted on all 4 sections. This layer cooled rapidly down to about 10 °C (ambient temperature - March 1st, 2011). On sections A and B, the grids were placed without tack coat, due to the adhesive film. On sections C and D (see Figure 2 a-c) a tack coat with 300g/m² of residual bitumen, was applied. Then, longitudinal and transversal strain gauges were put in place (Figure 2.d). The final 6 cm thick bituminous layer was laid on the 4 sections, and compacted successively with a steel drum vibrating roller and a rubber-tyred roller. To ensure melting of the film attached to grids from section A and B, the bituminous mix was put in place at temperatures above 150°C. After compaction, the average in situ void content of the mix was about 7.0 % (0.8 % more than for the specimens tested in the laboratory). The 4 test sections were instrumented with longitudinal and transversal strain gauges placed on the top of the grids (at 6cm depth – see Figure 2.d); temperature sensors and vertical strain gauges at the top of the UGM layer and of the subgrade (Figure 1).

Fig. 2. Construction: a) Grids of section A and B ; b) Grids after compaction; c) Tack coat application; d) Placement of strain gauges before overlay

The circular shape of the test sections required to cut the grids in relatively narrow bands, 5 m long by 1.5 m wide. To cover the test section, one 1.5 m wide band was placed in the centre of the wheelpath, and then two smaller bands on each side (see figure 2 a). Due to this layout, some construction problems occurred on section A, and potentially on section B, during the laying and compaction of the bituminous overlay, and the results obtained on section A will not be presented. Thorough investigations will be made on section B after completion of the testing to verify the state of the grid and of the interface.

4   Initial Measurements and Test Programme

The modulus of the subgrade has been measured with dynamic plate load test and results are shown on Figure 1. These moduli have been measured during construction, during a rainy period, and it is probably the reason of their variability. After construction, drainage has taken place, and the bearing capacity has become more homogeneous. Controls after construction have indicated an average thickness of the bituminous layer of 70 mm, instead of 80 mm. The end of section D is thinner than the other sections, because it is the end of the construction zone, and the transition with another existing pavement structure, which makes the control of the thickness more difficult.

The loading programme started in April 2011. Until September 2011, approximately 800 000 loads have been applied. During the experiment, loading

Full Scale Tests on Grid Reinforced Flexible Pavements 1255

has been stopped approximately every 100 000 cycles to perform various distress measurements. Response of internal transducers has also been recorded regularly. The ambient temperature conditions were practically constant throughout the tests, with daily temperatures in the range 10 to 28°C (mild summer). The rainfall on the test site is presented on Figure 3. It can be noticed that around 600 000 cycles, the rainfall level was 2.5 higher than during the other periods.

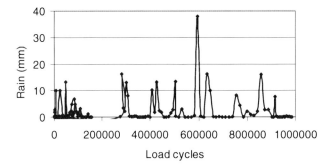

Fig. 3. Rainfall measurements on the test site

The four arms of the fatigue carrousel have been equipped with standard dual wheels, loaded at 65 kN (standard French equivalent axle load). The tyres used are Dunlop 1200 R20 SP321, inflated at 850 kPa (Figure 4). The loading speed was 6 rounds/minute (43 km/h). Its lateral wandering was +/-52.5 cm. Between 50000 and 150000 load cycles, one arm was also equipped with a wide single tyre; loaded at 40 or 50 kN, in order to compare strain distributions under single and dual tyres. Measurements under the wide single tyre will not be discussed here.

Fig. 4. Dual wheel load and its dimensions in mm

5 First Results

5.1 Deflection Measurements

Deflection measurement between the two wheels was performed every 3 meters using a Benkelman beam, under the 65 kN load at about 2 km/h. For temperatures varying between 20 and 28°C, Figure 5 shows that up to 381 000 load cycles, the mean deflection levels were close to 70 mm/100 on the three sections with some scatter which may be due to temperature variations. No significant difference in deflection was observed between the reinforced sections (B, C) and the unreinforced section (D). After 537 000 load cycles, the mean deflection on section B increased to 83 mm/100. On sections C and D, the deflection level remained practically constant up to 813 000 loads. There seems to be some relationship between deflection levels and pavement cracking. On section B, the first cracks appeared after 600 000 cycles, and simultaneously, an increase of the deflection was observed. On sections C and D, where very little cracking was observed, the deflections remained constant.

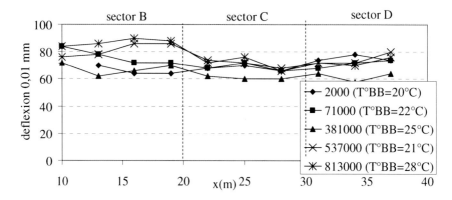

Fig. 5. Results of deflection measurements on the 3 sections

5.2 Rut Depth Measurements

Due to the lateral wandering of the loads, the width of the circulated area is approximately 1.6 m. The transversal profile of the pavement is measured using a 2 meters long ruler, every 3 meters. For each measurement point, the maximum rut depth is determined as the maximum vertical distance between the ruler and the pavement surface. Maximum rut depths measured on the 3 sections, at different load levels are presented on Figure 5. These measurements indicate that, at the beginning, sector B presents a lower rut depth than the other sections, until about 600000 cycles. At this stage, heavy rainfall occurred (Figure 3), and this may explain an increase of the rate of rutting on section B. At, 600000 cycles, section B already presented some cracks (Figure 6), which allowed water to infiltrate in the pavement foundation contrary to the other sections which still presented no damage. On sections C and D, the rut depths are very similar. After 600 000 cycles, the evolution of rutting is the same on all 3 sections, with an average final rut depth of 14.3 mm.

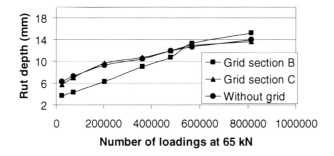

Fig. 6. Rut depth evolution on the 3 sections

5.3 Crack Monitoring

The first cracks were observed on section B, after about 600 000 load cycles, and then on section D after 800 000 load cycles. Section C presents no cracking up to now. Crack patterns were very similar on sections B and D: first, very fine isolated transversal cracks appeared. Then, under traffic, these cracks started to open, and fines started to come out. Other thin transversal cracks developed nearby. The transversal orientation of the cracks is typical of fatigue cracking observed on the carrousel, for pavements with thin bituminous layers [7]. Figure 7 presents the evolution of the extent of cracking, as a function of the level of traffic. It corresponds to the percentage of the length of the pavement affected by cracks (for a transversal crack, the affected length is considered, arbitrarily, to be 500 mm).

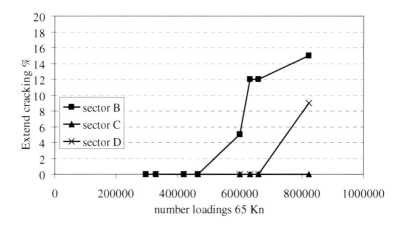

Fig. 7. Extent of cracking, in percent, on the 3 sections

5.4 Strain Measurements

Longitudinal and transversal strain measurements, for a temperature of about 20°C (at 40 mm depth) and a loading speed of 43km/h, are presented on Figure 8. On section C, no measurements are available because the strain gauges were broken during the compaction. The sensors were located in the middle of the wheel path, at position y = 0 mm, and their measurements were recorded for different lateral positions of the dual wheels, varying between -550 and + 550 mm. Positions y = - 210 mm, and + 210 mm correspond to the situation when the centre of one wheel is located on the top of the sensors (Figure 4).

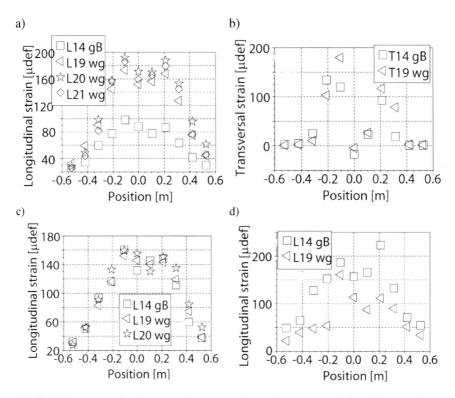

Fig. 8. Maximum longitudinal ε_L and transversal ε_T strains measured: a) and b): ε_L and ε_T at 12000 cycles, 19.2°C; c) ε_L at 152000 cycles, 21.1°C; d) ε_L at 676000 cycles, 21.6°C

At the beginning of the test, at 12000 cycles, (Figure 8 a-b), longitudinal strains measured on section B (gB) (gauge L14) are around half the longitudinal strains measured on section D without grid (wg) (gauges L19, L20, L21). For transversal strains, there is no clear difference between the measurements on section B (gauge T14) and on section D without grid (gauge T19). On section D, longitudinal and transversal strains are of the same level of magnitude. On the reinforced section, longitudinal strains are lower.

At 152000 (Figure 8.c), the longitudinal strains on sections B have increased and they are similar to those on section D without grid. At 676000 cycles (Figure 8 d), the longitudinal strains measured on section B are greater than those measured on section D. The same trend is observed for the deflection (Figure 5), which is higher on section B, after 530 000 cycles, than on section D. This increase of the strains and deflection can be due to the development of cracks as shown on Figure 7. For transversal strains, no significant evolution is observed at the different load levels, and the response of sections B and D remains similar.

Elastic back calculations have been made to estimate the moduli of the unbound granular material and of the soil from deflection measurements. The ALIZE design software used is based on Burmister's multi-layer linear elastic model [8]. Simulations have shown that the soil modulus is 80MPa. According to the French pavement design method, the modulus of the unbound granular materials has been taken equal to 200 MPa (modulus of the soil multiplied by 2.5). The longitudinal strains ε_L and transversal strains ε_T calculated under the centre of the dual wheel axle (position y = 0) are respectively ε_L = 89µdef and ε_T = 15µdef. These results are of the same order of magnitude as the strains measured at the start of the experiment, on section B. After a significant level of traffic, the measured strains increased on section B, indicating a probable deterioration of this section, due to traffic.

6 Conclusion

In this experiment, several different grids were tested as reinforcement of new pavements, with relatively thin bituminous layers (80 mm); the objective was to compare the behaviour of the reinforced sections, in comparison with a reference section without grid.

During construction, there were some difficulties to place the grids on the circular test track, which obliged to cut the grids in relatively narrow bands. Due to this layout, some construction problems, leading possibly to debonding of the reinforced layer, occurred on section B.

Concerning the behaviour of the pavements, it was found that:

- Cracking appeared first on section B after a rainy period, after around 60000 cycles, and section D (without grid) after 800 000 cycles. Section C presents no cracking up to now. The worse performance of section B may be related with the construction problems.
- Rutting was somewhat lower on section B up to 600 000 cycles, and after that, the levels of rutting were very similar on the 3 sections;
- Strain measurements indicated lower longitudinal strains on the reinforced sections at the beginning of the experiment, but this difference disappeared after about 150 000 cycles, leading to similar deformations on all sections.

As the experiment is not finished, it is only possible to conclude that a positive effect of the grids on the resistance to cracking is observed on one section (C). It is planned to continue the loading until at least 1 million loads, in order to attain higher levels of damage, and confirm the differences between the 3 tested sections.

At the end of the experiment, detailed investigations (cores, trenches, FWD) will be made to understand the possible cracking scenarios, and explain the behaviour of section B in particular, which may be due to debonding problems.

These observations, as those reported elsewhere [9-10-11], will also be completed by means of additional modelling, and non destructive testing, in relation with the two new Rilem Technical committees TC-SIB (TG4 – Advanced interface Testing of Geogrids in Asphalt Pavements) and TC MCD (Mechanisms of Cracking and Debonding in asphalt and composite pavements) in TG 3 on "Advanced Measurement Systems for Crack Characterization"

References

[1] Antunes, M.-L., Van Dommelen, A., Sanders, P., Balay, J.-M., Gamiz, E.-L.: Cracking in Pavements. In: Petit, C., Al-Qadi, I.L., Millien, A. (eds.) Proc. of the 5th Int. RILEM Conf., pp. 45–52. Rilem Editions, Paris (2004)
[2] Kerzrého, J.P., Michaut, J.P., Hornych, P.: Revue Générale des Routes et aérodromes, (890), 48–51 (2011)
[3] Autret, P., de Boissoudy, A.B., Gramsammer, J.C.: In: Proc. of the 6th Int. Conf. on Struct. Design of Asphalt Pavements, vol. 1, pp. 550–561 (1987)
[4] Corte, J.F., Goux, M.T.: TRR 1539, 116–124 (1996)
[5] Chabot, A., Chupin, O., Deloffre, L., Duhamel, D.: RMPD. Special Issue on Recent Advances in Num. Simul. of Pavements 11(2), 227–250 (2010)
[6] http://www.sg-adfors.com/Brands/GlasGrid
[7] Hornych, P., Kerzreho, J.P., Chabot, A., Bodin, D., Balay, J.-M., Deloffre, L.: Pavement Cracking. In: Al-Qadi, Scarpas, Loizos (eds.) Proc. of the 6th Int. RILEM Conf., pp. 671–681. CRC Press (2008)
[8] http://www.lcpc.fr/en/produits/alize/index.dml
[9] Florence, C., Foret, G., Tamagny, P., Sener, J.Y., Ehrlacher, A.: Cracking in Pavements. In: Petit, C., Al-Qadi, I., Millien, A. (eds.) Proc. of the 5th Int. RILEM Conf., pp. 605–612. Rilem Editions, Paris (2004)
[10] Perez, S.A., Balay, J.M., Petit, C., Tamagny, P., Chabot, A., Millien, A., Wendling, L.: Pavement Cracking. In: Al-Qadi, Scarpas, Loizos (eds.) Proc. of the 6th Int. RILEM Conf., pp. 55–65. CRC Presse (2008)
[11] Graziani, A., Virgili, A., Belogi, L.: In: Proc. of the 5h Int. Conf. Bituminous Mixtures and Pavements, Thessaloniki, Greece (2011)

Low-Temperature Cracking of Recycled Asphalt Mixtures

N. Tapsoba[1], C. Sauzéat[1], H. Di Benedetto[1], H. Baaj[2], and M. Ech[2]

[1] University of Lyon/ Ecole Nationale des Travaux Publics de l'Etat,
Département Génie Civil et Bâtiment (URA CNRS 1652), Rue Maurice Audin,
69518 Vaulx en Velin Cedex, France
{nouffou.tapsoba,cedric.sauzeat,herve.dibenedetto}@entpe.fr

[2] Lafarge Centre de Recherche, Pôle Formulation et Mise en Œuvre, 95 rue du Montmurier, 38291 Saint-Quentin-Fallavier, France
{hassan.baaj,mohsen.ech}@pole-technologique.lafarge.com

Abstract. The thermo-mechanical behavior of asphalt mixtures, with Recycled Asphalt Pavement (RAP) and other recycled materials was investigated. The cracking behavior at low temperature was studied considering Thermal Stress Restrained Specimen Tests (TSRST). The experimental setup was improved as the radial strains were measured during the performed TSRST. Tri-dimensional behavior could thus be investigated. Mixes made with different RAP content (up to 25%) and manufacturing-waste asphalt roofing shingle content (up to 10%) were studied. The influence of the content of RAP and shingle was analysed. A ranking of the different mixes was proposed based on the classical TSRST outputs, stress and temperature at failure.

Keywords: asphalt mixture, TSRST, radial strain, RAP, manufacturing-waste shingles, failure.

1 Introduction

Cracking is one of the main failure modes of asphalt pavements. Various types of cracks may occur in pavements: fatigue cracking caused by repetition of traffic loading, cracks explained by thermo-mechanical coupling effects resulting from contraction/dilation induced by repeated temperature cycles, cracks caused by frost heaving, etc. In the cold climates, thermal or low-temperature cracking of bituminous pavements is a serious problem. Low temperature properties of asphalt mixtures should be correctly studied in order to ensure adequate structural design.

The Thermal Strain Restrained Specimen Test (TSRST) allows a characterization of this behavior by coupling the thermal and mechanical effects. Different research works were conducted using this test and confirmed its good potential to well assess the thermal cracking resistance of asphalt mixtures ([1- 6]).

In this research, TSRST is used to evaluate the behaviour of asphalt mixtures containing different combinations of recycled bituminous materials. The two types of recycled materials considered in this study are: reclaimed asphalt pavement (RAP) and manufacturing waste asphalt roofing shingles. The use of RAP and shingles in pavement construction increases continuously and this is due to both environmental and economical reasons. However, some studies have shown that the addition of RAP may have a negative influence on low temperature cracking characteristics of the HMA ([7-10]). The impact of the addition of manufacturing-waste asphalt shingles in the HMA on the behaviour of asphalt mixes was also studied ([11, 12]). The results show that the low temperature cracking of mixes with up to 8% of shingles was not affected negatively. The purpose of this study is to determine the effects of RAP and manufacturing-waste shingles, incorporated together in the HMA, on the low temperature cracking performance of asphalt mixture.

In this paper, the experimental test setup and the tested materials are first presented. The second part presents the effect of recycled materials on the low-temperature properties of the tested mixtures. The experimental results are then analysed and compared and the withdrawn conclusions are presented.

2 Experimental Investigation

2.1 Test Equipment and Procedure

TSRST experiments are carried out with a hydraulic press having a maximum capacity of ± 25 kN and a ± 50 mm axial stroke.

A thermal chamber is used for the thermal conditioning of the tested specimens during the test. The temperature is measured using a thermal gauge (PT100 temperature probe) glued at the specimen surface.

Three axial extensometers located at 120° around the specimen are used to measure the axial strain (Figure 1). Axial strain is calculated as the average of the three measurements. To obtain radial strain, non-contact displacement transducers (range 500µm) measure the radial displacement at the mid-height of the specimen. The radial strain is determined from these two measurements.

A general view of the specimen and strain measurement devices is presented in Figure 1.

The principle of the TSRST is to keep the length of the specimen (axial strain) constant while decreasing the temperature inside the thermal chamber at a constant cooling rate of 10°C/h. Cooling incites the specimen to contract, but the servo-hydraulic press prevents it. The thermal stress induced inside the specimen increases until the specimen breaks.

The strain tensor $\underline{\underline{\varepsilon}}$ can defined as the sum of a mechanical strain tensor $\underline{\underline{\varepsilon}}^{mechanical}$ caused by stress field and a thermal strain tensor, $\underline{\underline{\varepsilon}}^{thermal}$ caused by temperature change:

$$\underline{\underline{\varepsilon}} = \underline{\underline{\varepsilon}}^{thermal} + \underline{\underline{\varepsilon}}^{mechanical} \tag{1}$$

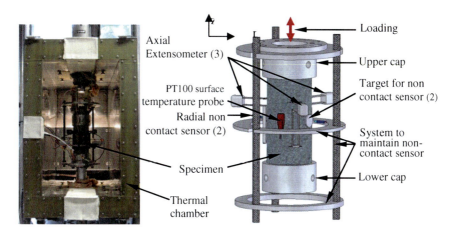

Fig. 1. General view of the experiment (left) and strain measurement devices (right) developed at ENTPE to measure axial strain and radial strain

The thermal strain tensor, assumed isotropic, is equal to:

$$\underline{\underline{\varepsilon}}^{thermal} = \alpha \, \Delta T \, \underline{\underline{\delta}} \qquad (2)$$

with α, the thermal dilation coefficient (supposed independent of the direction), ΔT, the increment of temperature and $\underline{\underline{\delta}}$ isotropic tensor whose determinant is 1.

In our test, with axisymmetric notation it comes:

$$\varepsilon_{axial} = \varepsilon_z = \varepsilon_z^{thermal} + \varepsilon_z^{mechanical} \qquad (3)$$

$$\varepsilon_{radial} = \varepsilon_r = \varepsilon_r^{thermal} + \varepsilon_r^{mechanical} \qquad (4)$$

As the axial strain is restrained, $\varepsilon_z = 0$ for TSRS tests, it comes:

$$\varepsilon_z^{mechanical} = -\varepsilon_z^{thermal} = -\alpha \times \Delta T \qquad (5)$$

$$\varepsilon_r = \alpha \times \Delta T + \varepsilon_r^{mechanical} \qquad (6)$$

Before performing TSRST on bituminous materials, a calibration of the complete device must be carefully carried out. During the tests, the cooling induces some artefact effects on the strain measurement device. The Zerodur®, a material for which the thermal dilation coefficient is perfectly known, was used for calibration. The artefact effects due to cooling are then evaluated and can be corrected during TSRST tests.

2.2 Materials

Seven bituminous mixes with different combinations of RAP and shingles were studied. Table 1 gives RAP and shingles dosage rates and compaction levels for each of the seven mixes tested. The particle size distribution was the same for all mixes (shown in Figure 2). The maximal particle size is 12.5 mm. The binder content of each mix was determined using the Superpave gyratory compactor. A laboratory mixer and a plate compactor with two rubber tyres (French LPC wheel compactor) were used to produce slabs of 600 x 400 x 60 mm. From each slab, five cylindrical specimens were cored and sawn at the DGCB laboratory of ENTPE. These cylindrical specimens were used for the TSRST test. The diameter of tested specimens is 59 ± 0.5 mm and the length 225 ± 3 mm. Three specimens were tested from each mix.

Table 1. RAP & shingle contents and compaction levels of tested materials

	RAM1	RAM2	RAM4	RAM5	RAM6	RAM7	RAM9
RAP (%)	0	15	15	15	20	20	25
Shingles (%)	0	0	3	5	5	7	10
Compaction levels (%)							
specimen 1 (sp1)	94.5	93.6	94.2	95.2	96.0	94.0	94.3
specimen 2 (sp2)	94.6	94.3	93.9	95.0	96.1	95.0	95.2
specimen 3 (sp3)	93.8	93.6	94.0	95.3	95.9	94.5	95.3

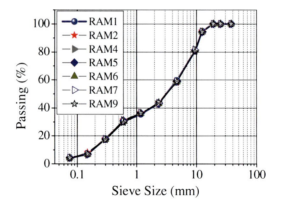

Fig. 2. Aggregate gradation of tested mixes

3 Results and Analysis

3.1 Typical TSRST Results

Main outputs of the TSRST are the variation of the thermally induced stress in function of the temperature and the temperature and the stress values at failure (Figure 3a). Typical TSRST results are shown on Figure 3a.

At failure, stress reaches its highest value which is referred to as the fracture strength. The slope of the stress-temperature curve dσ/dT increases progressively until a certain temperature where it becomes constant (the stress-temperature curve becomes linear). The temperature corresponding to the inflexion point of the curve is designated as the transition temperature. The transition temperature and dσ/dT may play an important role in characterizing the rheological behavior of bituminous mixtures at cold temperatures [13].

The radial strain is measured during the TSRST using a measurement setup designed for this purpose. The variation of radial strain with the temperature is plotted on Figure 3b. When the absolute value of the radial strain increases, it means that the diameter of the sample decreases with the cooling. This diameter decrease is explained by the tensile stresses induced in the sample during cooling.

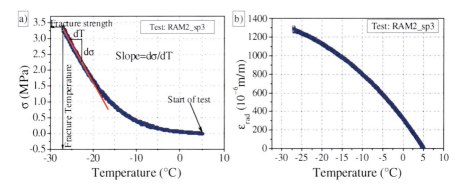

Fig. 3. a) Stress-temperature curve; b) radial strain-temperature curve of TSRST test on RAM2_sp3 specimen

3.2 General Results for the Tested Materials

A total of 21 tests were carried out for 7 different materials. Three specimens were tested of each material. The repeatability of TSRST is first evaluated. Two different characteristics must be studied:

- the general variation of the stress in function of temperature,
- the values of the stress and the temperature at failure.

Concerning the variation of stress in function of temperature, examination of Figure 4 shows that the three curves obtained from three different tests on the mix RAM1 are well superimposed.

In order to quantify the difference between the three tests on a same material, the area under the stress-temperature curve was calculated. This area was limited at the right by the vertical line passing by a chosen reference temperature. Table 2 presents this temperature and the results of calculation. For each mix, a reference test is chosen, which could be considered as an "average" or "representative" test of the considered mix. The calculated area for other tests is compared with the one of "representative" test. The maximum relative difference (in percentage in Table 2) is lower than 15% for all cases. For RAM1 mix, for which the three tests are very similar, the value is close to 3 or 4%. It can be concluded that repeatability of TSRST is rather good, considering the variation of stress in function of temperature, and that the choice of a "representative" test is pertinent to compare the different mixes behavior.

The variation of the stress in function of the temperature is presented in Figure 5 for each mix with the selected "representative" test. Examination of Figure 5 shows different behaviors. Particularly, in the first phase of the tests, the stress relaxation ability of RAM1 mix (without recycled materials) is more important than that of RAM2 mix containing 15% of RAP.

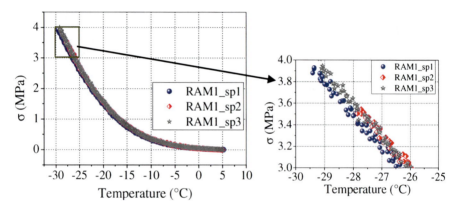

Fig. 4. Stress- temperature for tests on materials RAM1 (Three tests: RAM1_sp1, RAM1_sp2 and RAM1_sp3)

Table 2. Area under stress-temperature curve of tested material and relative difference considering representative test

		RAM1	RAM2	RAM4	RAM5	RAM6	RAM7	RAM9
Area under stress-temperature curve (MPa×°C)	sp1	28.50 (-3.75%)	28.72 (3.71%)	29.60 (ref)	30.40 (-10.0%)	33.68 (3.17%)	31.25 (-4.31%)	29.60 (-2.50%)
	sp2	29.61 (ref)	27.70 (ref)	27.93 (-5.66%)	33.79 (ref)	31.79 (-2.62%)	32.66 (ref)	31.65 (4.28%)
	sp3	29.67 (0.22%)	26.97 (-2.62)	34.08 (-15.1%)	36.08 (6.78%)	32.65 (ref)	33.21 (1.68%)	30.35 (ref)
Limit temperature for area calculation (°C)		-27.9	-27.5	-25.5	-26.2	-24.1	-24.0	-22.2

ref = *representative test.*

Low-Temperature Cracking of Recycled Asphalt Mixtures 1267

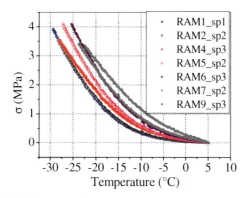

Fig. 5. TSRST response for selected "representative" tested materials

Temperature and stress values obtained at failure are presented in Figure 6 for all performed tests. The average values (temperature and stress) for each mix are also presented. In Table 3, these average values and standard deviation (three tests for each mix) are summarized. For temperature at failure, standard deviation remains below 5 percent for all mixes. For stress value at failure, standard deviation is higher but remains below 10 percent, except on RAM4 material. Even if the repeatability of TSRST is good, for variation of stress in function of temperature, the repeatability for the failure point is not so good.

All these results show that the content of recycled materials influenced the low temperature properties of mixes. As can be seen in Figure 6, the fracture temperature and the variation of stress according to the temperature are affected by the content of recycled material. The fracture temperature of reference material (RAM 1) is better than the others. At high dosage rates shingles and RAP, the low temperature performances decrease. The difference of average temperature between reference material (RAM1) and RAM9 (with 25% of RAP and 10% of shingle) is about 9°C. The values of stress at failure are more scattered and the influence of the recycled materials content is less obvious.

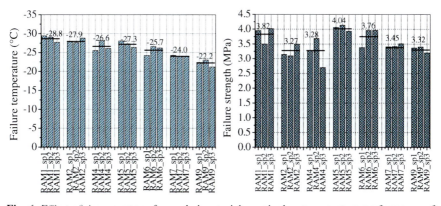

Fig. 6. Effect of dosage rates of recycled materials on the low temperature performance of asphalt mixes

Table 3. Average and standard deviation of tested materials

		RAM1	RAM2	RAM4	RAM5	RAM6	RAM7	RAM9
T_f (°C)	Average of 3 tests	-28.8	-27.9	-26.6	-27.3	-25.7	-24.04	-22.2
	Standard deviation	0.89 (3.1%)	0.73 (2.6%)	1.29 (4.8%)	0.84 (3.1%)	1.28 (5.0%)	0.19 (0.8%)	0.93 (4.2%)
σ_f (MPa)	Average of 3 tests	3.82	3.27	3.28	4.04	4.20	3.45	3.32
	Standard deviation	0.28 (7.3%)	0.18 (5.5%)	0.42 (12.8%)	0.11 (2.7%)	0.35 (8.3%)	0.06 (1.7%)	0.11 (3.3%)

3.3 Influence of RAP and Shingles Content

The values of the failure temperature and failure stress obtained from the TSRST tests for all tested mixes are presented Figure 7 in function of the content of RAP and shingles. The vertical bars at each value represent the standard deviation of the three measures. For asphalt mixes incorporating up to 15% RAP and up to 5% shingles, the low temperature performance seems to be slightly affected as their values of failure temperatures are very close (< 2.2°C) to that of the reference mix (RAM1).

For asphalt mixes incorporating higher substitution rates (>20% RAP and > 7% shingle), the low temperature performance seems to be significantly affected. These mixes cannot be used at the same conditions as the reference mix. Premature transverse cracks may appear on the pavements due to the lack of resistance to thermally induced tension stresses in cold areas requiring bitumen grades to be specified lower than -28°C [12]. However, these mixes may be used in warm regions where the low temperature cracking is not an issue.

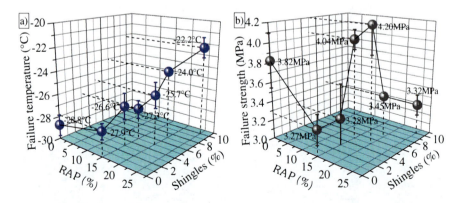

Fig. 7. a) Failure temperature and standard deviations at different dosage rates of RAP and shingles; b) Stress at Failure and standard deviations at different dosage rates of RAP and shingles

4 Conclusion

The objective of the research work presented in this paper was to study the low-temperature cracking of asphalt mixtures containing recycled materials. Seven mixtures with different recycled materials contents were investigated. From the obtained results, the following conclusions could be withdrawn:

- The repeatability of TSRST was evaluated based on the variation of stress in function of temperature and classical outputs, stress and temperature at failure. It is noted that repeatability was very good before failure and reasonable for stress at failure.
- The TSRST set-up was successfully improved which allowed the measurement of the radial strains during the test.
- The impact of the addition of RAP and shingle on the low temperature behavior and cracking characteristics of mixes was investigated. The cracking temperature of asphalt mixes with up to 15% RAP and 5% shingles was quite similar to that of the reference mix (RAM1). Beyond this limit, the low temperature performances may be significantly affected.

References

[1] Monismith, C.L., Secor, G.A., Secor, K.E.: Temperature induced stresses and deformations in asphalt conrete. Proceeding of Association of Asphalt Paving Technologist 34, 248–285 (1965)
[2] Isacsson, U., Zeng, H.: Low-temperature cracking of polymer-modified asphalt. Materials and Structure 31, 58–63 (1998)
[3] Pucci, T., Dumont, A., Di Benedetto, H.: Thermomechanical and mechanical behaviour of asphalt mixture at cold temperature, road and laboratory investigations. International Journal of Road and Pavement Design 5(1), 45–72 (2004)
[4] Di Benedetto, H.: Nouvelle approche du comportement des enrobés bitumineux: résultats expérimentaux et formulation rhéologique. In: Proceedings of the Fourth Rilem Symposium Mechanical Tests for Bituminous Mixes, Characterization, Design and Quality Control, pp. 376–393 (1990)
[5] Olard, F., Di Benedetto, H., Dony, A., Vaniscote, J.-C.: Properties of bituminous mixture at low temperatures and relations with binder characteristics. Materials and Structures 38, 121–126 (2005)
[6] Sauzéat, C., Di Benedetto, H., Chaverot, P., Gauthier, G.: Low temperature behaviour of bituminous mixes: TSRS test and acoustic emission. In: Proceedings of Advanced Characterization of Pavement and Soil Engineering Materials, Athens, Greece, pp. 1263–1272 (2007)
[7] Buttlar, W.G., Rebholz, F.E., Nassar, W.: Detection of Recycled Asphalt Pavement (RAP) in bituminous mixture. Report No ITRC FR 02-2, p. 251. Illinois Department of Transportation (2004)
[8] Behnia, B., Dave, E.V., Ahmed, S., Buttlar, W.G., Reis, H.: Investigation of effect of recycled asphalt pavement (RAP) amounts on low-temperature cracking performance of asphalt mixtures using acoustic emission (AE). Transportation Research Board Annual Meeting, 14 (2011)

[9] You, Z., Mills-Beale, J., Fini, E., Goh, S.W., Baron, C.: Evaluation of Low-Temperature Binder Properties of Warm-Mix Asphalt, Extracted and Recovered RAP and RAS, and Bioasphalt. Journal of materials in Civil Engineering 23(11), 1569–1574 (2011)

[10] Ma, T., Bahia, H.U., Mahamoud, E., Hajj, E.Y.: Estimating Allowable RAP in Asphalt Mixes to Meet Target Low Temperature PG Requirements. Journal of the Association of Asphalt Paving Technologists 79, 473–496 (2010)

[11] Baaj, H., Paradis, M.: Use of Post-Fabrication Asphalt Shingles in Stone Matrix Asphalt Mix (SMA- 10): Laboratory Characterization and Field Experiment on Autoroute 20 (Québec). Proceedings, Canadian Technical Asphalt Association 20, 365–384 (2008)

[12] Baaj, H., Ech, M., Lum, P., Forfylow, R.W.: Behaviour of asphalt mixes modified with ASM & RAP. CTAA. Proceedings, Canadian Technical Asphalt Association 25, 461–481 (2011)

[13] Jung, D.H., Vinson, T.S.: Low temperature cracking: Test selection. In: Strategic Highway Research Program SHRP-A-400 contract A-033A, Washington DC, p. 109 (1994)

Thermal Cracking Potential in Asphalt Mixtures with High RAP Contents

Qazi Aurangzeb[1], Imad L. Al-Qadi[2], William J. Pine[3], James S. Trepanier[4], and Ibrahim M. Abuawad[1]

[1] Graduate Research Assistant, Department of Civil and Environmental Engineering, University of Illinois at Urbana-Champaign, Urbana, IL 61801
{aurangz1,abuawad1}@illinois.edu

[2] Founder Professor of Engineering, Director of Illinois Center for Transportation, The Department of Civil and Environmental Engineering, University of Illinois at Urbana-Champaign, Urbana, IL 61801
alqadi@illinois.edu

[3] Research Engineer, Heritage Research Group, Indianapolis, IN, 46268
Bill.Pine@heritage-enviro.com

[4] HMA Operations Engineer, Bureau of Materials and Physical Research, IDOT, Springfield, IL, 62704
james.trepanier@illinois.gov

Abstract. Asphalt recycling is a key component of the sustainable practices in the pavement industry. Use of reclaimed asphalt pavement (RAP) minimizes the construction cost as well as consumption of natural resources. Adding RAP, though, is believed to make the asphalt mixtures prone to thermal cracking by increasing their stiffness. A semi-circular bend (SCB) test was conducted to evaluate the low temperature cracking potential of asphalt mixtures with RAP, whereas, a flow number test was conducted to provide some insight into the gained stiffness. Eight asphalt mixtures with a high amount (up to 50%) of RAP were designed using two material sources. Two additional softer binders were used to prepare testing samples in order to evaluate the effect of binder grade bumping. Flow number test data showed significant improvement in potential rutting resistance of asphalt mixtures when RAP was added. However, thermal cracking potential may increase when RAP is added to asphalt mixtures. This effect could be reduced when binder double-pumped grade is used.

1 Introduction

Reclaimed asphalt pavement (RAP), acquired by milling or ripping and breaking the asphalt pavement, is primarily used in asphalt pavement recycling. Minimizing the requirement of new aggregates and asphalt binder in mixtures with RAP has twofold benefits; cost savings and conservation of natural resources. Addition of RAP introduces significant complexities in asphalt concrete mix design due to RAP's aged asphalt binder and degraded aggregates. The asphalt binder becomes stiffer and brittle in the field due to many factors including oxidation, volatilization, polymerization, thixotropy, syneresis, and separation [1]. Asphalt pavements made with recycled material without accounting for the effects of RAP are bound to fail prematurely. The most common issue with designing asphalt mixtures with high RAP contents is controlling the mix volumetrics including voids in mineral aggregate (VMA).

Researchers have reported contradictory results regarding VMA of asphalt mixtures with RAP; a few studies reported a decrease in VMA [2, 3, 4, & 5], while others reported an opposite trend [6, 7]. It is important to highlight that unless the volumetrics of the asphalt mixtures with and without RAP are comparable, conclusions as to the effectiveness of including RAP in asphalt mixture can be misleading. Moreover, use of corresponding aggregate bulk specific gravity, G_{sb}, values for the RAP is crucial in assuring comparable volumetric properties between the various blends. Research work that uses, aggregate effective specific gravity, G_{se}, for RAP can result in blends that are not truly comparable in VMA, even though the calculated VMA values suggest they are.

Thermal cracking caused by high thermal stresses developed at low temperatures is one of the predominant asphalt pavement failure modes in cold regions, such as the northern part of the US. Propensity of asphalt pavements to crack under thermal and traffic loading is believed to be amplified when RAP is used due to the stiffened aged RAP asphalt binder. Li et al. [8] investigated the effect of RAP percentage and sources on the thermal cracking of asphalt mixtures by performing semi-circular beam (SCB) test. The results indicated that asphalt mixtures with 20% RAP exhibited similar fracture resistance abilities to that of control asphalt mixtures. The addition of 40% RAP to asphalt mixture significantly decreased the low-temperature fracture resistance. Tam et al. [9] looked into the thermal cracking of recycled asphalt mix and showed that recycled asphalt mixtures are less resistant to thermal cracking than control asphalt mixes. The thermal cracking potentials of laboratory and field mixes were analysed using the McLeod's limiting stiffness criteria and pavement fracture temperature (FT) method. Gardiner and Wagner [10] found that adding RAP to asphalt mixtures decreased the rutting potential and increased the potential for low temperature cracking.

This paper presents the effect of incorporating high RAP contents in asphalt mixture on its low temperature cracking and rutting potentials. The flow number test was performed to estimate the changes in potential rutting due to RAP addition into the asphalt mixtures, whereas, SCB test was conducted to measure the fracture energy of the asphalt mixtures at low temperature. Glass transition temperatures for all binders used in this this study were measured to better understand the asphalt mixture behaviour at low temperature.

2 Material and Mix Designs

In this study, virgin aggregates and RAP materials were procured from Illinois Department of Transportation (IDOT) District 1 and District 5. District 1 and District 5 virgin aggregate materials were mainly crushed dolomitic limestone. RAP was obtained in two sizes, i.e. +9.5 mm and -9.5 mm from both the sources.

In addition to the use of the base binder (PG 64-22), two softer binders, i.e. PG 58-22 and PG 58-28, were used to evaluate the effect of binder grade bumping on the fracture potential of the asphalt mixtures with RAP. The performance grades of extracted RAP binders were found to be PG 82-10 and PG 88-10, respectively, for District 1 and District 5 RAP materials. Table 1 shows the true and PG binder grades of the acquired asphalt binders. Although at the borderline, PG 58-28 turned out to be PG 58-22, which may have implications on fracture behavior of the asphalt mixtures.

For each material source, a control (0% RAP) mix and three asphalt mixes with 30%, 40%, and 50% RAP were designed in accordance with the IDOT specifications utilizing the Bailey method [11] of aggregate packing. Control gradation was effectively achieved by fractionating both virgin aggregates and RAP materials in different sieve sizes. Details of the mix design process are available elsewhere [12]. Similar VMA of the asphalt mixtures, as shown in Table 2, would ensure that the difference in performance of the four asphalt mixes is not driven by volumetric changes.

Table 1. True and PG grades for virgin and RAP binders

Binder Type	True Grades	PG Grades
District 1 PG 64-22	66.7-24.2	64-22
District 5 PG 64-22	67.0-22.9	64-22
PG 58-22	62.3-22.4	58-22
PG 58-28*	61.4-27.4	58-22
District 1 RAP	82.4-13.7	82-10
District 5 RAP	89.3-14.9	88-10

* Not a true PG 58-28

3 Experimental Program

This paper is part of a study [12] that encompassed a detailed experimental plan to evaluate the laboratory performance of asphalt mixtures with high percentages of RAP (up to 50%). While the results from SCB test are discussed in detail, results from flow number (FN) test are described briefly to explain the effect of RAP on asphalt mixture potential rutting.

A flow number test is used as a performance indicator for permanent deformation resistance of asphalt mixtures. The flow number is usually considered at the cycle number where the strain rate starts increasing with loading cycle (as shown in Figure 1).

A higher flow number indicates higher resistance to permanent deformation (rutting). The flow number test was performed at 58 °C, a total deviator stress of 200 kPa, and a frequency of 10 Hz. The test was conducted until the completion of 10,000 cycles or 5% permanent strain, whichever occurred first.

Table 2. Asphalt mixtures' volumetrics

Mix Type	Asphalt Content (%)	Air Voids (%)	VMA (%)	VFA[1] (%)
D1[2]-Control	4.9	4.0	13.7	70.8
D1-30% RAP	4.9	4.0	13.6	70.6
D1-40% RAP	5.1	4.0	13.7	70.8
D1-50% RAP	5.0	4.0	13.7	70.8
D5[2]-Control	5.2	4.0	13.8	71.0
D5-30% RAP	5.2	4.0	13.8	71.0
D5-40% RAP	5.2	4.0	13.6	70.8
D5-50% RAP	5.2	4.0	13.5	70.4

[1] Voids filled with Asphalt
[2] D1 = District 1, D5 = District 5

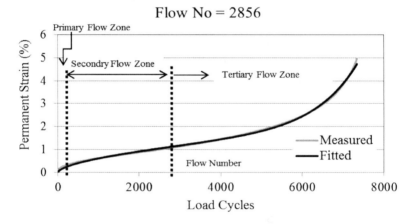

Fig. 1. Typical flow number data

The SCB test is used to evaluate low temperature cracking potential of asphalt mixtures. The test output includes fracture toughness and fracture energy which are used to assess the asphalt mixture's potential for low temperature cracking. A 50-mm-thick specimen was used instead of 25-mm-thick, typical test specimen thickness, due to the relatively large nominal maximum aggregate size (NMAS), 19 mm, of the asphalt mixtures. The test was conducted at 10 °C above the lower base binder performance grade, i.e. -12 °C for PG 64-22. A sitting load of 0.1 kN was applied before actual test loading. The test was controlled using the crack mouth opening displacement (CMOD) rate at 0.1 mm/min. The test was stopped when the load level dropped to 0.1 kN. The test setup is shown in Figure 2. Three replicates of asphalt mixtures were tested for each combination of RAP content and binder type as presented in Table 3. Control mixture was prepared with base binder only, i.e. PG 64-22. Asphalt mixtures with various RAP contents were prepared using all three binders (PG 64-22, PG 58-22, and PG 58-28) to study the effect of binder grade bumping on the asphalt concrete fracture behavior.

Fig. 2. Setup for semi-circular bend test

Table 3. Test matrix for SCB test

Binder Grade	RAP Content (%)			
	0	30	40	50
PG 64-22	3	3	3	3
PG 58-22	0	3	3	3
PG 58-28	0	3	3	3

4 Results and Discussion

4.1 Flow Number Test Results

Figure 3 shows the average flow number for three replicates for each mix and binder type combination for all asphalt mixtures. A consistent trend of flow number increase with increasing RAP content was observed in District 1 asphalt mixtures. Since higher flow number implies higher resistance to permanent deformation, asphalt mixtures with 50% RAP showed the highest resistance to rutting followed by the mixtures with 40% and 30% RAP, and the control mix.

Although the effect of softer binder i.e. binder grade bumping is obvious and quite consistent on the flow number of the asphalt mixtures with RAP, the effect diminishes with an increase in RAP content used in the mix. While the flow number of asphalt mixture with 30% RAP using PG 58-28 is 57.5% less than that with 30% RAP using PG 64-22, only 11% reduction in the flow number of asphalt mixture with 50% RAP resulted when double-bumped grade binder was used. This diminished effect can be attributed to the increasing effect of the RAP binder when a higher RAP content is used in the mixtures.

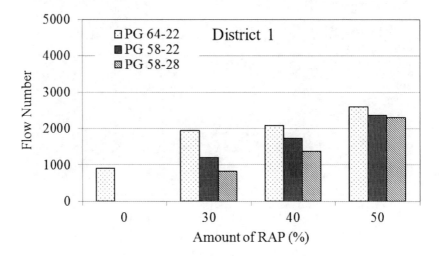

Fig. 3. Flow number for District 1 mixtures

Flow number results for District 5 asphalt mixtures are shown in Figure 4. The District 5 control mixture showed the highest rutting potential (least flow number) compared to other mixtures. The effect of increasing RAP content is quite obvious with base binder grade, i.e. PG 64-22. Using soft binders, grade bumping, resulted in reducing the flow number.

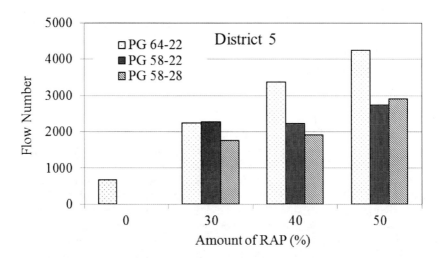

Fig. 4. Flow number for District 5 mixtures

4.2 Semi-circular Bend Fracture Test

Fracture energy is determined from the fracture work obtained over the area under the load-CMOD curve. Since the plane stress conditions might have been violated when a 50-mm-thick specimen was used instead of the 25-mm-thick specimen, fracture toughness of the mixtures was not determined in this study. Higher fracture energy suggests more energy is required to create a unit surface area of a crack. Therefore, the lower the fracture energy, the greater the potential for thermal cracking.

Glassy transition temperature (T_g) was measured for all binders, i.e. virgin binders and extracted RAP binders for both District 1 and District 5. A Differential Scanning Calorimeter (DSC) was used to determine T_g of the asphalt binders. DSC measures the amount of energy (heat) absorbed or released by a sample as it is heated, cooled or held at a constant (isothermal) temperature. The T_g for the binders is shown in Table 4.

Table 4. Glassy transition temperatures

S. No.	Binder Type	T_g (°C)	
		Onset[1]	Peak
1	District 1-PG 64-22	-9.9	-16.3
2	District 1- RAP extracted RAP Binder	-12.1	-14.5
3	District 5-PG 64-22	-11.8	-16.1
4	District 5- extracted RAP binder	-10.2	-14.7
5	PG 58-22	-16.1	-18.3
6	PG 58-28	-17.2	-18.2

[1] the temperature at the onset of the spike

Figure 5 shows the effect of RAP content and binder grade on the fracture energies of District 1 asphalt mixtures. The control mixture showed the highest fracture energy, which sharply plummeted for the asphalt mixture with 30% RAP. Further addition of RAP content in asphlat mixtures didn't result in significant change in the fracture energy.

The SCB test measures global fracture energy which consists of localized fracture behavior plus the energy dissipated due to creep. At -12 °C, the material is still at the verge of viscoelastic range as shown by the T_g values. Hence, asphalt concrete creep contributes to the total measured fracture energy. Therefore, greater energy is required to initiate and propagate a crack above T_g. In addition, at relatively low temperature, the crack tends to propagate in a straight path irrespective of the presence of aggregate and mastic, whereas, at a temperature above T_g, cracks tend to circumnavigate the aggregate particles and propagate through the softer mastic. This longer meandering path may, in turn, cause an increase in the fracture energy.

The effect of binder grade bumping was also evaluated on the fracture properties. An increase in the fracture energy was observed when single-bumped grade binder (PG 58-22) was used with District 1 asphalt mixtures with 30% RAP, whereas, binder grade double bumping (PG 58-28) did not show any further improvement in the fracture energies when compared to the mixtures prepared with single-bumped grade binder PG 58-22. For asphalt mixtures with 40% RAP, the fracture energy slightly increased as the binder becomes softer. This increase in fracture energy is expected as the mixture becomes more ductile and its resistant to cracking improves. For asphalt mixtures with 50% RAP content, binder grade double bumping improved fracture energy. The behavior of asphalt mixtures with 50% RAP using PG 58-22 could not be explained; and appeared as an outlier. It is important to note that the low-temperature true grade of PG 58-28 was determined to be -27.4 which may explain the insignificant difference in the fracture energies of mixtures prepared with PG 58-28 and PG 58-22 binders.

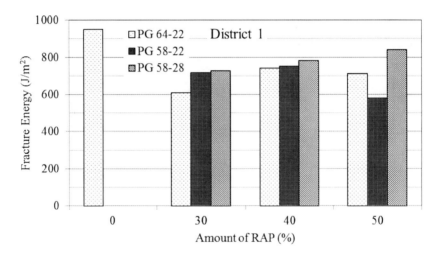

Fig. 5. Fracture energies for District 1 mixtures

For District 5, fracture energy results, presented in Figure 6, do not manifest effect of RAP on fracture behavior. A decrease in fracture energy was observed when 30% RAP was added in asphalt mixture. No further significant reduction in fracture energy was noticed when RAP content was increased beyond 30% in the asphalt mixtures. In addition, it was evident that softer binder, especially double-bumping grade, improved the fracture energy.

Although the effect of RAP content and the use of softer binder on the fracture energy are evident, the asphalt mixture's relatively large NMAS and testing at a temperature greater than the binder's T_g influenced the total fracture energy results. Hence, testing at a temperature below T_g is recommended.

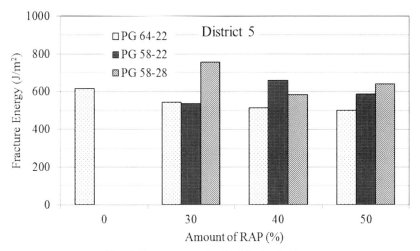

Fig. 6. Fracture energies for District 5 mixtures

5 Summary and Conclusion

Thermal cracking potential of asphalt mixtures with high RAP contents was investigated. Two material sources were utilized to design asphalt mixtures with 0%, 30%, 40%, and 50% RAP. Flow number test data showed significant improvement in potential rutting resistance of asphalt mixtures when RAP was added. However, thermal cracking potential may increase when RAP is added to asphalt mixtures. Using softer binder proved to be helpful in reducing the effect of RAP aged binder on mixture stiffening at low temperature. Fracture test results should be carefully evaluated when asphalt mixture is tested above binder's T_g value because the resulting global fracture energy may include creep effect.

Acknowledgement. This publication is based on the preliminary results of ICT-R27-37, *Impact of High RAP Contents on Pavement Structural Performance*. ICT-R27-37 was conducted in cooperation with the Illinois Center for Transportation; the Illinois Department of Transportation, Division of Highways; and the U.S. Department of Transportation, Federal Highway Administration.

The contents of this paper reflect the view of the authors, who are responsible for the facts and the accuracy of the data presented herein. The contents do not necessarily reflect the official views or policies of the Illinois Center for Transportation, the Illinois Department of Transportation, or the Federal Highway Administration. This paper does not constitute a standard, specification, or regulation. The significant help provided by Jim Meister, Illinois Center for Transportation Senior Research Engineer, is also appreciated.

References

[1] Roberts, F.L., Kandhal, P.S., Brown, E.R., Lee, D., Kennedy, T.W.: Hot Mix Asphalt Materials, Mixture Design, and Construction, 2nd edn. Napa Education Foundation, Lanham (1996)

[2] Al-Qadi, I.L., Carpenter, S.H., Roberts, G.L., Ozer, H., Aurangzeb, Q.: Paper No. 09-1262, CD-ROM. Transportation Research Board of National Academies, Washington, DC (2009)
[3] West, R., Kvasnak, A., Tran, N., Powell, B., Turner, P.: TRR: Journal of the Transportation Research Board (2126), 100–108 (2009)
[4] Kim, S., Sholar, G.A., Byron, T., Kim, J.: TRR: Journal of the Transportation Research Board (2756), 109–114 (2009)
[5] Mogawer, W.S., Austerman, A.J., Engstrom, B., Bonaquist, R.: Paper No. 09-1275, CD-ROM. Transportation Research Board of National Academies, Washington, DC (2009)
[6] Daniel, J.S., Lachance, A.: TRR: Journal of the Transportation Research Board (1929), 28–36 (2005)
[7] Hajj, E.Y., Sebaaly, P.E., Kandiah, P.: Use of Reclaimed Asphalt Pavements (RAP) in Airfields HMA Pavements. Final Report AAPTP Project No. 05-06, Airfield Asphalt Pavement Technology Program (2008)
[8] Li, X., Marasteanu, M.O., Williams, R.C., Clyne, T.R.: TRR: Journal of the Transportation Research Board (2051), 90–97 (2008)
[9] Tam, K.K., Joseph, P., Lynch, D.F.: TRR: Journal of the Transportation Research Board (1362), 56–65 (1992)
[10] Gardiner, M.S., Wagner, C.: TRR: Journal of the Transportation Research Board (1681), 1–9 (1999)
[11] Vavrik, W.R., Huber, G., Pine, W.J., Carpenter, S.H., Bailey, R.: Bailey Method for Gradation Selection in HMA Mixture Design. Transportation Research E-Circular, E-C 044. Transportation Research Board of National Academies, Washington, DC (2002)
[12] Aurangzeb, Q., Al-Qadi, I.L., Abuawad, I.M., Pine, W.J., Trepanier, J.S.: Achieving Desired Volumetrics and Performance for High RAP Mixtures. Accepted in TRR: Journal of the Transportation Research Board. Paper No. 12-2232, CD-ROM. Transportation Research Board of National Academies, Washington DC (2012)

Micro-mechanical Investigation of Low Temperature Fatigue Cracking Behaviour of Bitumen

Prabir Kumar Das, Denis Jelagin, Björn Birgisson, and Niki Kringos

Division of Highway and Railway Engineering, Transport Science Department, KTH Royal Institute of Technology, Sweden

Abstract. In an effort to understand the effect of low temperature fatigue cracking, atomic force microscopy (AFM) was used to characterize the morphology of bitumen. In addition, thermal analysis and chemical characterization was done using differential scanning calorimetry (DSC) and thin-layer chromatography / flame ionization detection (TLC/FID), respectively. The AFM topographic and phase contrast image confirmed the existence of bee-shaped microstructure and different phases. The bitumen samples were subjected to both environmental and mechanical loading and after loading, micro-cracks appeared in the interfaces of the bitumen surface, confirming bitumen itself may also crack. It was also found that the presence of wax and wax crystallization plays a vital role in low temperature cracking performance of bitumen.

1 Introduction

Low temperature cracking is known to be one of the major distresses that result in both structural and functional problems for asphalt pavements. When the temperature decreases, asphalt mixtures tend to contract which results in an increase in tensile stresses inside the pavement. If these tensile stresses exceed the tensile strength of the asphalt mixture, cracks may start propagating inside the material and this often results in visible transverse cracks on the pavement surface [1]. These cracks may then lead to other types of pavement deteriorations, since water can now freely access the structure and traffic loading may give added local impact forces. Cracking is therefore a significant contributor to the overall reduction of the pavement service life and an increase in maintenance cost.

There are, of course, many different causes for cracks to initiate and propagate in an asphaltic pavement. Even on top of a rut, which is caused at higher temperatures, often a crack appears. Nevertheless, cracking may develop due to a critical low ambient air temperature drop which is known as low temperature cracking or due to several thermal cycles which is known as thermal fatigue cracking. This paper is mainly focused on thermal fatigue cracking. Asphalt mixtures consist of a matrix of bitumen and mineral aggregates where each component has its own stiffness, strength and contraction-expansion rates. Because of this change in properties, cracks may develop in the interfaces between the aggregate and bitumen, as shown in Figure 1 (a). The existence of microstructures in bitumen matrix found in several studies proves the heterogeneity of bitumen [2-4]. This inhomogeneity is creating internal interfaces within the bitumen matrix. Bitumen have therefore zones that, just like mineral-bitumen interfaces, have a tendency to act as natural stress inducers. Thus, at low temperatures when the bitumen becomes stiff, this induced stress may cause cracking which could propagate through bitumen, as shown in Figure 1 (b).

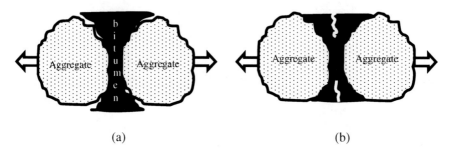

Fig. 1. Two possible ways of crack initiation due to thermal cracking: a) interface and b) through bitumen

Over many years, a great number of efforts have been made to investigate the complexity of the thermal fatigue problem of asphalt pavements. Models for predicting low temperature cracking have also been proposed. Nevertheless, until now a test method is not available to investigate mastic (i.e., bitumen and fillers) and bitumen resistance to thermal fatigue. Also, most of the existing specifications and theories do not address cracking of bitumen itself due to thermal fatigue. Yet, according to many researchers, bitumen is an important component that needs to be considered when investing low temperature or thermal fatigue cracking of asphalt mixtures [5-6].

For this reason, the present study is focusing on a micro-mechanical investigation of the thermal fatigue cracking behaviour of the bitumen itself, where the hypothesis is that bitumen itself may also crack due to its inherent the heterogeneities under environmental and traffic loading.

1.1 Low Temperature Properties of Bitumen

Bitumen is a thermoplastic material that at high temperatures has a very low and at low temperatures has a very high viscosity. Initially when a pavement cools, the bitumen is viscous enough to shrink due to its stress relaxation behaviour. The bitumen stiffens upon further cooling and reaches a certain temperature at which the thermal stress can no longer relax through this viscous mechanism. This phenomenon also occurs with bitumen when subjected to several thermal cycles. From a chemical point of view at lower temperatures the fluid non-polar molecules begin to organize themselves into a structured form. Since bitumen contains both polar and non-polar components, at low temperatures these non-polar molecules combine with the already-structured polar molecules and make asphalt more rigid and likely to fracture when exposed to stresses.

In addition to this phenomenon, waxes are also often an important component to consider in the thermal fatigue behaviour of bitumen, even though its behaviour is not fully understood to date. Natural waxes may be present at all times at various concentration in bitumen, depending on the crude source type and refinery process, and sometimes commercial waxes are added into bitumen or asphalt mixtures to allow for sufficient flow when reducing the mixing and compaction temperature [7]. Wax modification may also have a positive healing effect on bitumen, thus improving the long-term performance of asphalt pavements [8-9]. Crystallizing wax in bitumen may have negative effect on bitumen properties and may increase the sensitivity to plastic deformation or cracking in asphalt pavement. Thus wax crystallization and melting in bitumen is considered an important issue when it comes to quality and performance [10].

1.2 The Importance of Understanding the Fundamental Behaviour

Thermorheological behaviour of bitumen is an important factor for understanding the performance-based optimization of asphalt mixture. The temperature dependent rheological behaviour of bitumen depends on the chemical structure and intermolecular associations (structuring) [4]. This structuring is largely responsible for the physical properties of bitumen thus the prediction of the performance of asphalt pavements should also directly be related to this [2, 11]. Structuring may occur at various ranges from molecular to macroscopic but most of the asphalt researchers have concerned themselves with the microstructures of the bitumen itself. Bitumen is a complex mixture of molecules of different size and polarity, for which microstructural knowledge is still rather incomplete. Once we have knowledge at that level, by upscaling material parameters from microscale observations, a multiscale model can be proposed which will allow us to predict cracks in the pavement, based on the fundamental knowledge on smaller sales.

From extensive atomic force microscopy (AFM) investigations shown in earlier papers [2-4, 11-14], it has been shown that bitumen has the tendency to phase separate under certain kinetic conditions, leading to a predominant clustering of two types of phases, illustrated in Figure 2. To ensure that the found phenomena of phase separation in bitumen is not a side effect of sample preparation or a surface effect in the AFM, several other experiments were also performed, among which an extensive neutron scattering study [12]. From mechanical considerations, it is known that the interfaces between two materials with different stiffness properties serve as natural stress inducers. This means that when the material is exposed to mechanical and or environmental loading, these interfaces will attract high stresses and are prone to cracking.

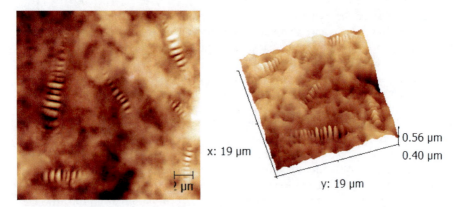

Fig. 2. Topographic 2D (left) and 3D right AFM image (19μm × 19 μm) of bitumen indicating evidence of microstructures

Thus, it would be of great interest to understanding the conditions under which this phase behaviour occurs and the speed, or mobility, of this microstructure appearance. For this reason, this paper is presenting preliminary results of an on-going project to characterize bitumen behaviour on multiple scales.

3 Experimental Study

3.1 Material

An unaged bitumen was used in this study provided by Nynan Bitumen, Sweden. The physical-chemical properties of this bitumen are presented in Table 1. The penetration and softening point were measured in accordance with the European standards EN 1426 and EN 1427, respectively. The chemical characterization (SARA analysis) of bitumen was done by using the thin-layer chromatography with flame ionization detection (TLC-FID) of which the detail procedure can be found in a previous study [15].

Table 1. Physical chemical properties of bitumen

	Characteristics	Result
Physical properties	Penetration (25°C, 0.1 mm)	86
	Softening point (°C)	46.4
Chemical components	Saturates (%)	11
	Aromatics (%)	55
	Resins (%)	19
	Asphaltenes (%)	15

3.2 Atomic Force Microscopy (AFM)

As described in earlier section, a detailed knowledge of microstructure is needed to understand the physico-chemistry of bitumen, which can serve as the direct link between the molecular structure and the rheological behaviour. Optical microscopy techniques have been employed to have a better understanding and visualization of bitumen microstructures [10]. However, because of the opacity and adhesive properties of bitumen, optical microscopy has not received much attention from the asphalt industry. To overcome some of the limitations of optical microscopy, researcher in the asphalt field have chosen to use scanning probe microscopy such as the AFM. AFM is capable of measuring topographic features at atomic and molecular resolutions as compared to the resolution limit of optical microscopy of about 200nm. Moreover, AFM has the advantage of imaging almost any type of surface which opens the window for investigating micro-cracks due to thermal fatigue.

(a)   *(b)*

Fig. 3. (a) Silicon bar of 60mm × 20mm × 7mm (b) bitumen film on top of it

Sample preparation: In this study, AFM (using a VECO Dimension 3100 device) was used for the microstructural characterization of the investigated bitumen. Approximately 30mg of hot bitumen was carefully placed on a rectangular silicon bar (60mm × 20mm × 7mm) and it was spread out with a blade to form a relatively thin film, as shown in Figure 3. Then the bitumen film was cooled to room temperature and covered to prevent dust pick-up. After that the sample was annealed at 25°C for a minimum of 24 h before imaging [13]. The scanning was carried out in tapping mode at room temperature (25°C) and scan rate was 1Hz. With Veeco Nanoprobe TM cantilevers with spring constant 40N/m and resonant frequency 300 kHz. AFM images were acquired at several locations on the sample surface to have complete information of the bitumen sample.

3.3 Thermal Fatigue Procedure

The bitumen film over the silicon bar was then subjected to thermal fatigue with heating and cooling cycles. A freezer was used to regulate the low temperature at -20°C and a room with controlled temperature at 25°C was used for thawing. The experimental method of thermal fatigue cycles (each cycle took 40 minutes) in this study is explained in the Figure 4. In between each thermal cycle the sample was subjected to additional tensile stress. This was done by bending at the mid-point of the silicon bar at a controlled angle of 3 degrees. The sample underwent 15 cycles of thermal loading; after the last freezing, the sample was placed under the AFM to investigate change in micro-structure due to the thermal fatigue.

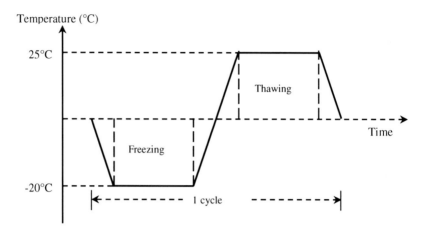

Fig. 4. Thermal loading cycle

3.4 Differential Scanning Calorimetry (DSC)

DSC analysis was conducted to obtain the heat flow characteristics of bitumen. The tests were performed using a TA instrument (model TA Q2000) equipped with a refrigerated cooling system. Approximately 15 mg of bitumen was scraped from the sample container and hermetically sealed into DSC sample pan. The sample pan was then placed horizontally on a hot plate at 80°C for 15 minutes to let the bitumen touch the pan bottom and also level out the surface. After that the sample was annealed at 25°C for a minimum of 24 h. The sample was cooled from 25°C to -80°C and kept at this temperature for 5min. Data was recorded during heating from -80°C to +90°C and cooling from +90°C to -80°C. Finally, a second heating was done from -80°C to +90°C and data was recorded. In total, for each sample one cooling and two heating scans data was recorded. The heating and cooling rate was 10°C/min. Wax content was determined from endothermic pick during the heating scan and while calculation a constant melting enthalpy of 121 J/g was used as reference [10].

4 Results and Discussions

4.1 Basic Morphology

The basic morphology of the bitumen was investigated in a clean room environment with a controlled temperature. Typical topographic and phase contrast image of the bitumen surface obtained at 25°C are presented in Figure 5, where one can easily observe the existing of phase separation in the bitumen matrix as reported earlier by several researchers [2-4, 11-14]. The rippled microstructures are also observed which are often referred to as bee-structures. The pale and dark lines indicate rise and drop of the topographic profile against the background, which are also known as peaks and valleys, respectively. A phase shift less than 10° can be detected between the bright and darker network. The shades in the phase contrast images indicate the relative stiffness of the phases as obtained from sample-tip interaction. The greater sample-tip interaction results softer phase and dark shade in the phase image. Thus, the evidence of different stiffness in the bitumen matrix can also be observed.

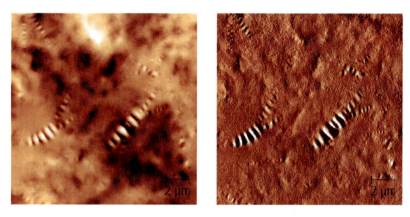

Fig. 5. AFM topographic (left) and phase contrast (right) images of bitumen at 25°C (15μm × 15μm)

All of the previous research [2-4, 11-14] on bitumen under AFM mentioned the so called bee-shaped structures, thus it would be a great interest to zoom into bees to have a clear idea. Figure 6 shows a line profile along a bee which illustrates the earlier mentioned higher and lower parts of the bee-shaped structures. The mean topographical change and the distance between higher parts of bees are recorded as 68 nm and 567 nm, respectively. These measured values are consistent to those reported by Loeber et al. (1996), Jäger et al. (2004) and Masson et al. (2007) [2,3,14].

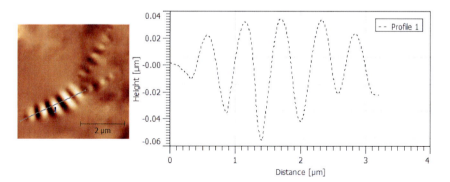

Fig. 6. Topographic image (5μm × 5μm) and a line profile along bee-shaped structure

4.2 DSC Characterisation

DSC has widely applied to determine glass transition temperature, wax content, wax crystallisation starting and melting out temperature of bituminous materials. In the case of bitumen, the interpretation of DSC curve is rather complicated since there are several overlapping phenomena to consider. Bitumen is a complex mixture of different molecules thus the glass transition phenomenon of bitumen occurs over a large temperature range. It can be seen from heat flow diagram (cf. Figure 7), the glass transition stated at around -40°C and followed by a weak exothermic effect around -10°C caused by cold crystallization of wax which could not crystallize through the cooling cycle due to limited mobility. In the cooling cycle, an exothermic transition occurs at around +50°C which usually is interpreted as starting of crystallisation. It can be also found that the crystallization takes place during a range of temperatures, indicating the presence of molecules with different crystallization points. At around +40°C to +60°C the wax completely melted. The wax content according to DSC was found 6.2%.

According to Pauli et al. (2011) [4], this wax crystallization is influencing for much of the development and appearance of microstructures, including the well-known bee-structures (cf. Figure 5). In addition, paraffin waxes could exhibit crystalline forms at ambient temperature, which could also responsible for microstructuring. Thus, the appearance of surface structuring is related to the wax types and concentrations. It is important to consider wax crystallization of bitumen used in asphalt pavement because it occurs in a temperature range that coincides with the pavement service temperature. Moreover, it is highly probable that wax crystallization affect low temperature properties of bitumen.

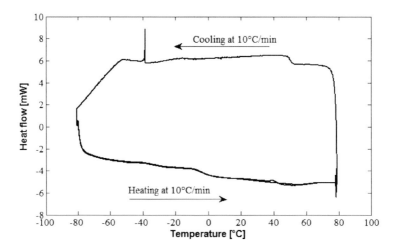

Fig. 7. Heat-flow diagram by DSC

4.3 AFM after Thermal Fatigue Cycles

From the above discussion, it can be concluded that bitumen has the tendency to phase separate out structures with predominantly wax. The interface between different phases with different stiffness could generate high stresses due to environmental or mechanical loading. In this study, the sample was exposed to thermal cycles and controlled tensile stresses thus the degraded material properties results into micro-cracks (crazing pattern), as depicted in Figure 8. If this process would continue, these micro-cracks would continue developing and finally form macro-crack.

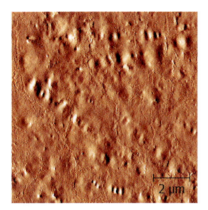 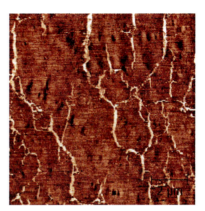

Fig. 8. Evidence of micro-crack through AFM scanning: topographic (left) and phase contrast image (right)

5 Conclusions

Atomic force microscopy was used to reveal the microstructures of the bitumen surface which was subjected to both thermal fatigue and controlled stress. As expected, the micro-cracks were found at the interface of the different phases due to stress concentration. This phenomenon proofs bitumen itself may also crack under certain kinetic conditions. In addition, wax crystallization is a very important factor to understand the physico-mechanical behaviour of bitumen as micro structuring is associated with it.

References

[1] Al-Qadi, I.L., Marwa, M., Elseifi, M.A.: J. Trans. Res. Record 1919, 87–95 (2005)
[2] Loeber, L., Sutton, O., Morel, J., Valleton, J.-M., Muller, G.: J. of Microscopy 1, 32–39 (1996)
[3] Masson, J.-F., Leblond, V., Margeson, J., Bundalo-Perc, S.: J. of Microscopy 3, 191–202 (2007)

4. Pauli, A.T., Grimes, R.W., Beemer, A.G., Turner, T.F., Branthaver, J.F.: Int. J. of Pavement Engineering 12(4), 291–309 (2011)
5. Soenen, H., Vanelstraete, A.: Performance indicators for low temperature cracking. In: Sixth International RILEM Symposium on Performance Testing and Evaluation of Bituminous Materials, pp. 458–464 (2003)
6. Soenen, H., Ekbland, J., Lu, X., Redelius, P.: Isothermal hardening in bitumen and in asphalt mix, Euroasphalt & Eurobitumen Congress, report no. In: Euroasphalt & Eurobitumen Congress, report no, Vienna, vol. 50, pp. 1351–1363 (2004)
7. Edwards, Y.: Influence of waxes on bitumen and asphalt concrete mixture performance, Ph.D. thesis. KTH Royal Institute of Technology, Sweden (2005)
8. Das, P.K., Tasdemir, Y., Birgisson, B.: Construction and Building Materials 30, 643–649 (2012)
9. Das, P.K., Tasdemir, Y., Birgisson, B.: Road Materials and Pavement Design 13(1) (2012) (in press)
10. Lu, X., Langton, M., Olofsson, P., Redelius, P.: J. of Materials Science 40, 1893–1900 (2005)
11. Lesueur, D., Gerard, J.-F., Claudy, P., Létoffé, J.-M., Planche, J.-P., Martin, D.: Journal of Rheology 40(5), 813–836 (1996)
12. Schmets, A., Kringos, N., Pauli, T., Redelius, P., Scarpas, T.: Int. J. of Pavement Engineering 11(6), 555–563 (2010)
13. Pauli, A.T., Branthaver, J.F., Robertson, R.E., Grimes, W., Eggleston, C.M.: Atomic force microscopy investigation of SHRP asphalts. ACS Division of Fuel Chemistry Preprints 46(2), 104–110 (2001)
14. Jäger, A., Lackner, R., Eisenmenger-Sittner, C., Blab, R.: Road Materials and Pavement Design 5, 9–24 (2004)
15. Lu, X., Isacsson, U.: Construction and Building Materials 16(1), 15–22 (2002)

The Study on Evaluation Methods of Asphalt Mixture Low Temperature Performance

Tan Yiqiu, Zhang Lei[*], Shan Liyan, and Ji Lun

Harbin Institute of Technology, Harbin 150090, China

Abstract. In this paper, beam bending test, thermal stress restrained specimen test (TSRST) and direct tensile relaxation test were adopted to evaluate two categories (AC and SMA), six kinds of asphalt mixtures low temperature performance.

It can be concluded from test results that low temperature cracking resistances performance of asphalt mixture evaluated by the index of bending strength is not consistent with the failure strain obtained from beam bending test, it is because the low temperature performance of asphalt mixture determined by many factors, not only determined by strength characteristics or deformation capacity. By the same token, relaxation time obtained from direct tensile relaxation test cannot be used to evaluate the low temperature performance of asphalt mixture;

The curve of strain versus stress can be obtained through the beam bending test of asphalt mixture. The greater bending the strain energy density, the better low temperature performance. From the regression analysis it can be found that there is a good correlation between the bending strain energy density and fracture temperature. This correlation shows that the critical values of bending strain energy density can be used to determine the low temperature performance of asphalt mixture in the absence of fracture temperature.

1 Background

Low temperature cracking induced by seasonal and daily thermal cyclic loading is one of the main critical distresses in asphalt pavements which is a significant cause of premature pavement deterioration and quite harmful to the service life and quality of the road [1, 2]. Therefore, how to improve the anti-cracking performance of asphalt pavement is a hot research filed, and how to evaluate low temperature anti-cracking performance accurately is the most important premise [3-5].To understand the behavior of asphalt mixture at low service temperatures and to predict their field performance, a constitutive stress-strain relationship must be described. For this purpose, many models were established [6-8]. To evaluate low temperature performance of asphalt mixture, various methods are developed and many researches are carried out [9-11].

From the literature reviewed it can be found that lots of work focuses on the properties of asphalt mixture and the influence factors at the low temperature, few of them compared the direct tensile relaxation test results with other test results.

---

[*] Corresponding author.

2 Objective

Through comparison of the test results and analysis of the relationship between indices obtained from three kinds of tests, an index which can reflect low temperature performance of asphalt mixture accurately will be proposed.

3 Materials and Methods

3.1 Materials Used

Asphalt. Three types of asphalt binders namely AH-90, diatomite modified asphalt and SBS modified asphalt were used in this study. The basic properties of the studied binders as per Chinese specifications (JTJ052-2000) are shown in Table 1.

Table 1. The Basic Properties of Asphalts

Evaluation index		AH-90	Diatomite modified asphalt	SBS modified asphalt
Penetration, 100g, 5s (0.1mm)	25℃	91	79	95
Ductility 5cm/min (cm)	10℃	>100	36.5	>100
	15℃	>140	54.8	>100
Softening Point $T_{R\&B}$ ℃	--	44.8	47.6	74

Aggregate and grading. Figure 1 shows the recommended gradation limits by Chinese specifications (JTGF40-2004) for dense graded asphalt mixture (AC-16) and SMA mixture (SMA-13) and the selected gradation in this research was in the middle of the limits.

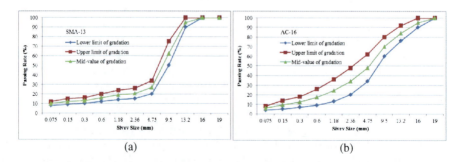

Fig. 1. The grading curve used in test: (a) the gradation of AC-16 and (b) the gradation of SMA-13

The properties of the aggregates are presented in Table 2 and Table 3.

Table 2. Mechanical Indexes of Coarse Aggregate

Index	Test results		Technical requirements
	10-20mm	5-10mm	
Crushing value %	11.2	——	≯ 28
Los Angeles abrasion value %	16.3	——	≯ 30
Water Absorption %	0.48	0.51	≯ 2.0
Apparent density (g/cm^3)	2.896	2.885	≮ 2.60
Asphalt Adhesion	5		≮ 4
Elongated particle contents %	8.5	8.6	≯ 15

Table 3. Density of Different Size Aggregate

Mesh Size (mm)	16	13.2	9.5	4.75	2.36
Density (g/cm^3)	2.815	2.770	2.783	2.768	2.753
Mesh Size (mm)	1.18	0.6	0.3	0.15	0.075
Density (g/cm^3)	2.715	2.765	2.703	2.727	2.625

Mixture design. The mix design procedures for AC-16 and SMA-13 in this paper were determined as per the Chinese specification JTJ052-2000 (T0703-1993). Locally available materials that meet the normal AC-16 and SMA-13 specifications were used to produce the mixes. Six kinds of asphalt mixture (A, B, C, D, E and F) are made under the condition of optimal asphalt content to maintain consistency through the study. The results are shown in the following Table 4.

Table 4. The Category of Asphalt Mixtures

Grading type	Asphalt type	OAC (%)	Mixture ID
AC-16	AH-90	4.7	A
	Diatomite modified asphalt	5.4	B
	SBS modified asphalt	5.3	C
SMA-13	AH-90	6.0	D
	Diatomite modified asphalt	6.5	E
	SBS modified asphalt	6.2	F

It should be mentioned that diatomite as a modifier used in asphalt can improve the performance of asphalt mixture.

3.2 Performance Tests

Three performance tests in the laboratory were adopted. The tests performed were beam bending test (three points bending test), Thermal Stress Restrained Specimen Test (TSRST) and direct tensile relaxation test.

Beam bending test was carried out as per Chinese specification with Materials Testing System (MTS-810). Through this test bending strength and bending strain can be obtained, its calculation equations are shown as follow.

$$R_B = 3LP_B / 2bh^2, \varepsilon_B = 6hd / L^2 \quad (1)$$

In which, R_B is bending strength (MPa); ε_B means bending strain; P_B is peak of load (N); d is the deflection when specimen destroyed (mm).

Thermal Stress Restrained Specimen Test (TSRST) and direct tensile relaxation test are carried out using the developed equipment by highway lab in Harbin Institute of Technology (HIT) and as shown in Figure 2. This equipment consists temperature control cabinet (-60 °C to 20 °C with ±0.5 °C accuracy). TSRST and direct tensile relaxation test were done on small beams of the dimension 220 × 35 ×35mm.

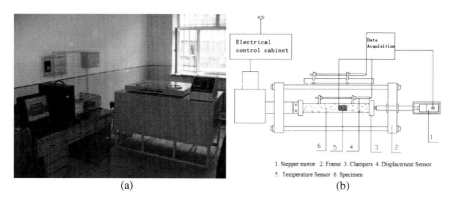

(a)   (b)

Fig. 2. TSRST Equipment: (a) TSRST Equipment and (b) schematic diagram of TSRST equipment

From the curve of Time-Stress, relaxation time and relaxation modulus can be obtained. Relaxation module were calculated as follow:

$$G(t) = \frac{\sigma(t)}{\varepsilon} \quad (2)$$

The definition of relaxation time in this paper is that the total time of the stress reduces to 50% initial stress. Three of tests were to monitor the low temperature behavior of mixes.

4 Results and Discussion

4.1 Thermal Stress Restrained Specimen Test Results and Discussion

Thermal Stress Restrained Specimen Test results are shown in Table 5.

Table 5. The Summary of TSRST Results

Evaluation index	Mixture ID					
	A	B	C	D	E	F
Fracture temperature (°C)	-21.4	-24.5	-26.1	-23.5	-25.3	-30.1
Coefficient of variation (%)	6.58	4.16	5.24	2.13	2.20	2.79
Fracture strength (MPa)	3.96	4.93	4.96	3.37	4.52	4.56
Coefficient of variation (%)	13.0	7.0	10.8	12.2	15.2	18.9
Transition temperature (°C)	-16.2	-17.7	-17.8	-18.3	-20.1	-20
Coefficient of variation (%)	4.05	2.04	2.25	8.84	5.12	2.88
Slope (MPa/°C)	-0.444	-0.39	-0.361	-0.305	-0.436	-0.287
Coefficient of variation (%)	22.9	10.8	28.8	34.5	30.4	17

Because TSRST can simulate the actual material properties with the temperature changes as well as the fracture temperature evaluation index has explicit meaning, it was selected as reference in this paper.

It can be inferred from Table 5 that different kinds of asphalt mixtures have different growth slope, and it may result in different fracture temperature. Table5 also shows that the low temperature performance of the six asphalt mixtures results ordered from F as higher value to A as lower one (F> C> E> B> D> A). From test results, an interesting phenomenon can be found in both AC asphalt mixture (A, B and C) and SMA asphalt mixture (D, E and F), the asphalt mixture mixed with neat asphalt has the worst low temperature performance and the asphalt mixture mixed with SBS modified asphalt has the best low temperature performance. It illustrates that the asphalt type has a great effect on the low temperature performance of asphalt mixes, and the modified asphalt can increase the low temperature performance of asphalt mixes.

4.2 Direct tensile Relaxation Test Results and Discussion

Direct tensile relaxation test results are shown in Figure 3.

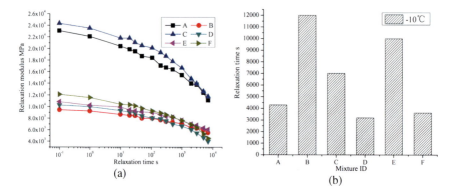

Fig. 3. Direct tensile relaxation test results (-10°C) (a) the relaxation curves of asphalt mixtures and (b) the relaxation time of different asphalt mixtures

From Figure 3 (a) it can be found that the six mixes have different relaxation abilities at the same tensile stress level. The neat asphalt mixes (A and D) and SBS modified asphalt mixes (C and F), the SMA asphalt mixes have a lower relaxation modulus than AC asphalt mixes and it also can be inferred from Figure 3 (b) that the SMA asphalt mixes have a less relaxation time than AC asphalt mixes. It may be because the OAC of SMA asphalt mixes are higher than AC asphalt mixes', that is the SMA asphalt mixes are softer than AC asphalt mixes at the same low temperature, which results in the lower relaxation modulus and less relaxation time.

Relaxation time can reflect the asphalt mixture relaxation ability. The shorter relaxation time means the better ability of stress relaxation. Under the low temperature condition, the better relaxation ability will reduce the temperature stress faster, which may lessen the low temperature cracking. From Figure 3 it can be inferred that the relaxation time of the six asphalt mixtures ordered from B as higher value to D as lower one (B> E> C> A> F> D). Analyzing the results can be found that the diatomite modified asphalt mixtures have the longest relaxation time. From the definition of relaxation time, the B and E mixes have the worst low temperature property, and the A and D mixes have the best low temperature property. But considering the results of TSRST, there are some conflicts. In the TSRST, the F and C mixes are proved to be the best and the A and D mixes are the worst. That is because the relaxation time only reflect one aspect low temperature property of mixes, however, as known, the low temperature property of mixes are determined by many factors. So the results from direct tensile relaxation cannot be used to evaluate the low temperature performance of asphalt mixture but can be a reference with other index.

4.3 Beam Bending Test Results and Discussion

Beam bending test results are shown in Figure 4.

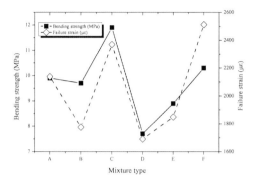

Fig. 4. Beam Bending Test Results

It can be concluded from Figure 4 that low temperature cracking resistances performance of asphalt mixture, evaluated by the indexes of bending strength and failure strain obtained from beam bending test, are not consistent. It is because the low temperature performance of asphalt mixture determined by many factors, not only determined by strength characteristics but also deformation capacity, any one of them is inappropriate to evaluate asphalt low temperature performance.

According to the material damage principle, the crack is due to the energy accumulated greater than the critical value, so the strain energy density of asphalt mixture is much better than other and adopted to evaluate the low temperature performance in this study.

The strain energy density of asphalt mixture obtained from beam bending test, and the area under the strain – stress curve when the beam is failure is the bending strain energy density of asphalt mixture. The greater bending strain energy density, the better low temperature performance. The critical values of bending strain energy density of the above six kinds of asphalt mixture are shown in Table 6.

Table 6. The Critical Value of Bending Strain Energy Density of Asphalt Mixtures

Mixture type	A	B	C	D	E	F
Critical value of bending strain energy density (KJ/m^3)	7.1	8.3	12.3	7.9	11.0	13.4

The order of bending strain energy density results is the same to the TSRST results. In order to further study the relevance between fracture temperature and bending strain energy, the linear regression method is adopted. The regression results are shown in Figure 5.

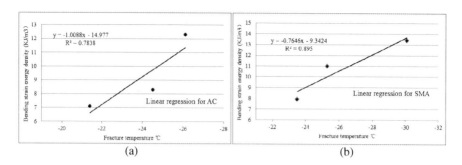

Fig. 5. The linear regression results between fracture temperature and bending strain energy for AC and SMA: (a) the results of AC-16 and (b) the results of SMA-13

From Figure 5 it can be found that the relevance between fracture temperature and bending strain energy, both AC asphalt mixes and SMA asphalt mixes, is good, respectively 0.7838 and 0.895. This results illustrate bending strain energy can be used to evaluate the low temperature performance of asphalt in the absences of TSRST.

5 Conclusions and Further Work

From the above discussion, some conclusions can be made as to the materials used in this research. These conclusions are:

1. From the TSRST results, it can be found that the materials used in this study that SBS modified asphalt can improve the low temperature performance of asphalt mixture significantly, the fracture temperature is 7 °C lower than the neat asphalt mixture; The diatomite modified asphalt has the similar effect.
2. From the direct tensile relaxation test results, it can be found that the six mixes have different relaxation abilities at the same tensile stress level. The SMA asphalt mixes have a lower relaxation modulus and a less relaxation time than AC asphalt mixes.
3. From the beam bending test, it can be concluded that the low temperature property evaluated by the indexes of bending strength and failure strain are not consistent, it is because the low temperature performance of asphalt mixture is determined by many factors, not only determined by strength characteristics or deformation capacity. By the same token, relaxation time obtained from direct tensile relaxation test cannot be used to evaluate the low temperature performance of asphalt mixture, but it can be a reference with other index.
4. From the relevance analysis between fracture temperature and bending strain energy, it can be found there is a good relevance between them. So when the TSRST is in the absence, bending strain energy can be used to evaluate the low temperature performance of asphalt mixtures.

Acknowledgement. The authors are grateful to the financial support by National Natural Science Foundation of China (50808058 and 51108138), and many thanks should be given to Li Xiaolin for her help.

References

[1] Rajbongshi, P., Das, A.: Estimation of temperature stress and low-temperature crack spacing in asphalt pavements. Journal of Transportation Engineering 135(10), 745–752 (2009)
[2] Raad, L., Saboundjian, S., Sebaaly, P., Epps, J.: Thermal cracking models for AC and modified AC mixes in Alaska. Transportation Research Record (1629), 117–126 (1998)
[3] Apeagyei, A.K., Buttlar, W.G., Reis, H.: Assessment of low-temperature embrittlement of asphalt binders using an acoustic emission approach. Insight: Non-Destructive Testing and Condition Monitoring 51(3), 129–136 (2009)
[4] Ma, T., Mahmoud, E., Bahia, H.U.: Estimation of reclaimed asphalt pavement binder low-temperature properties without extraction: Development of testing procedure. Transportation Research Record (2179), 58–65 (2010)
[5] Al Qadi, I.L., Dessouky, S., Yang, S.H.: Linear viscoelastic modeling for hot-poured crack sealants at low temperature. Journal of Materials in Civil Engineering 22(10), 996–1004 (2010)
[6] Selvadurai, A., Au, M.C., Phang, W.A.: Modeling of Low-Temperature Behavior of Cracks in Asphalt Pavement Structures. Canadian Journal of Civil Engineering 17(5), 844–858 (1990)
[7] Kim, H., Buttlar, W.G.: Finite element cohesive fracture modeling of airport pavements at low temperatures. Cold Regions Science And Technology 57(2-3), 123–130 (2009)
[8] Li, X., Marasteanu, M.O.: Cohesive modeling of fracture in asphalt mixtures at low temperatures. International Journal of Fracture 136(1-4), 285–308 (2005)
[9] Khattak, M.J., Baladi, G.Y., Drzal, L.T.: Low temperature binder-aggregate adhesion and mechanistic characteristics of polymer modified asphalt mixtures. Journal of Materials in Civil Engineering 19(5), 411–422 (2007)
[10] Sui, C., Farrar, M.J., Tuminello, W.H., Turner, T.F.: New technique for measuring low-temperature properties of asphalt binders with small amounts of material. Transportation Research Record (2179), 23–28 (2010)
[11] Li, X.J., Marasteanu, M.O.: Investigation of low temperature cracking in asphalt mixtures by acoustic emission. Road Materials and Pavement Design 7(4), 491–512 (2006)

Permanent Deformations of WMAs Related to the Bituminous Binder Temperature Susceptibility

F. Petretto[1], M. Pettinari[1], C. Sangiorgi[2], and Andrea Simone[3]

[1] PhD Student, DICAM, University of Bologna, Bologna
[2] Researcher, DICAM, University of Bologna, Bologna
[3] Associate Professor, DICAM, University of Bologna, Bologna

Abstract. The purpose of this study is to evaluate the growth of permanent deformations in Warm Mix Asphalts related to their binder temperature susceptibility. Rutting builds up in the pavement under the repeated loading generated by the traffic along the wheels path. The distress develops in two different ways, the pre-consolidation rutting, deformation at variable volume, and the instability rutting, deformation at constant volume. These phenomena are directly related to the loading time, the shape of aggregates, the content of air voids, the type of binder and its temperature susceptibility. This study focuses on the binder contribution to rutting when subjected to high temperature variation. An accurate laboratory investigation was completed to assess how this material property influences the different propagation of permanent deformations. The EN 12697 - 22 Wheel Tracking test was performed at three different temperatures (40 - 50 - 60°C) on two gap graded asphalt mixtures. In order to evaluate how, at these temperatures, the properties of the bitumen act on the asphalt concrete tendency to exhibit permanent deformations, a series of DSR rheological tests were also conducted.

1 Introduction

Permanent deformation (rutting) is a common form of pavement distress. Since today researchers have used different fundamental, empirical and simulative tests to evaluate the rutting potential of asphalt concretes. The main aim of this research is to provide a deeper insight into the contribution to the permanent deformation resistance coming from the different components of asphalt mixtures, related to their temperature susceptibility.

A large portion of the rut resistance of a mix is attributable to the inter-granular friction of the aggregate particles. The binder holds the aggregates together, but as the temperature increases, its stress dependency changes and the binder properties become prevalent in defining the rut resistance. To confirm how the asphalt components influence the rutting behavior in relation to their temperature dependence, a series of laboratory tests were conducted on a Warm Mix Asphalt (WMA) and on a PmB Hot Mix Asphalt (PmB HMA).

Two gap graded mixtures, with a maximum nominal size of 11 mm, have been chosen to conduct this research, these mixes are based on the volumetrics of a typical SMA used in Italy for road construction. The binder content and type were varied to observe the different behavior of a traditional PmB and a Warm Binder. Resistance to permanent deformations was evaluated using both the Wheel Tracking tests at different temperatures on compacted slabs and a series of DSR tests on the sole bituminous binders.

2 Background on Permanent Deformations

The permanent deformation of asphalt concrete has a major impact on the performance of a pavement throughout its life. Rutting not only reduces the service life of pavements, but it may also affect basic vehicle handling, which can be hazardous to all road users. Rutting develops gradually as the number of load applications increases and it appears as longitudinal depressions in the wheel paths. It is caused by a combination of densification and shear deformation inside the asphalt mixture [1].

The densification is the further compaction of asphalt mixtures generated by traffic after construction. When compaction is poor, the channelized traffic provides a repeated loading action in the wheel path and completes the consolidation. A considerable amount of rutting can occur if thick layers of asphalt are consolidated by the traffic especially when the initial air voids content is higher. The lateral plastic flow of the asphalt mixtures due to wheel tracks results in rutting. The use of excessive binder in the mix causes the loss of internal friction and cohesion between the aggregates, what provokes that traffic loads are supported by the asphalt cement rather than by the aggregate structure. Plastic flow can be minimized using large size aggregates, angular and rough textured coarse and fine aggregates, stiffer binders, as well as by providing suitable compaction during construction [1].

Under hot conditions or under sustained loads, asphalt binder behaves like a viscous liquid and flows. Viscous liquids are sometimes called "plastic" because, once they start flowing, they do not return to their original position. This is why, in hot weather, some asphalt pavements flow under repeated wheel loads and wheel path ruts appear (*Figure 1*). However, the rheological behavior of the binder phase has a great importance on the mobility of aggregates. It is known that bitumen is thermo-sensitive. At low temperatures, visco-plastic deformation is less important and is even blocked due to stiffening of the binder. Nevertheless, when increasing the temperature, the lubricating effect of bitumen increases and promotes visco-plastic deformations [1].

The bitumen temperature susceptibility may be evaluated by measurements of various viscous and elastic parameters (e.g., storage and loss modulus, dynamic, and complex viscosities) at different temperatures and frequencies [2]. Asphalt mixtures containing binders with lower temperature susceptibility should be more resistant to cracking and rutting at low and high temperatures, respectively. Temperature susceptibility is usually defined as the change in binder properties as a function of temperature. Since binder properties may be characterized by means of various parameters, different approaches have been proposed to evaluate the temperature susceptibility [3].

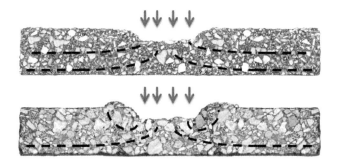

Fig. 1. Pre – consolidation Rutting (above), Instability Rutting (below)

2 Experimental Program

In the present study two mixtures with the same gradation and different binders were compared. The permanent deformations resistance was evaluated by means of a series of WT tests conducted on compacted slabs. The binder influence on the rutting resistance was defined via rheological tests.

2.1 Mixtures Characterization

The aggregates gradation used for both mixes is a typical gap graded curve (*Table 1*). This kind of mix is widespread in Italy and known as Splitt Mastix Asphalt (SMA). The aggregates sources are calcareous with a maximum nominal size of 11 mm.

Table 1. Mixes gradation

Sieve [mm]	16	11	8	5.6	4	2	1	0.5	0.25	0.125	0.0063
Passing [%]	100	95.4	80.5	65.3	52.8	37.9	25.1	17.4	12.7	8.9	6.2

2.2 Binder

The tested binders are commonly used in Italy to produce wearing course mixtures, their properties are summarized in table 2. Both asphalt cements were obtained from a single base bitumen modified by a SBS polymer (5% in weight). Furthermore, one bitumen was additivated with a wax in order to evaluate its temperature susceptibility in comparison to the unadditivated binder. The wax used in this study is the Sasobit®, produced by Sasol Wax (Germany). It is classified as a modifier or asphalt flow improver. It has a long chain aliphatic hydrocarbon and it is obtained from coal gasification using the Fischer-Tropsch process. It melts after mixing at 85° to 115°C causing a marked reduction of the binder viscosity. The manufacturer reports a reduction in mixing and compaction temperatures of 30 to 50°C [4].

Table 2. Binders properties

Property	Binder A	Binder B	Eur. Stand.
Penetration [dmm]	35	32	EN 1426
Softening Point [°C]	78	80	EN 1427
Dyn. Visc. @ 160°C	0.75 Pa s	0.45 Pa s	EN 13701 – 2
Breaking Point [°C]	-16	-15	EN 12593
Pol. Mod. SBS [%]	5.00	5.00	-
Wax [%]	-	2.00	-

2.3 Mix Design

In this study the mixes are compacted in accordance to EN 12697 – 33, with a standard roller compactor. The optimum binder content was evaluated to achieve the target air voids of 6.00%. The bitumen amount varied of about 0.4 % between Mix A and Mix B, the binder content is referred to the mass of aggregates. The 6.00% target air voids was chosen as it is a common requirement in many Italian specifications for wearing courses (*Table 3*).

Table 3. Mixture properties after compaction

Property	Mix A	Mix B	Eur. Stand.
Binder content [%]	6.80	6.40	EN 12697 – 1
Max. density [g/cm^3]	2.405	2.405	EN 12697 – 5
Air voids [%]	6.03	6.00	EN 12697 – 8

3 Analysis of Results

3.1 Wheel Tracking Tests

The Wheel Tracking tests were conducted in accordance to EN 12697 – 22. Each mix was tested at three different temperatures: 40, 50 and 60°C. Six slabs were compacted and tested for each mixture.

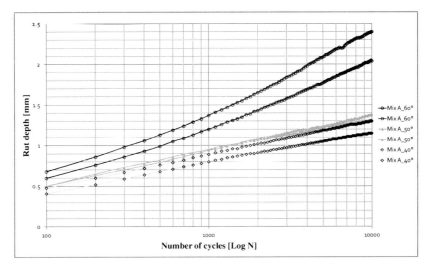

Fig. 2. Mix A_WT @ 40 - 50 - 60°C

In figure 2 and 3, the rutting curves illustrate the different behaviors of the mixes. The data display a linear trend for both mixtures at 40 and 50° C. It may occur that, at these temperatures, the materials tend to compact under repeated loading with a change in volume. The performance drastically changes with the increase in temperature from 50 to 60°C. Both mixes, darker lines, exhibit a non-linear trend that reveals a higher deformation rate, generating deeper ruts.

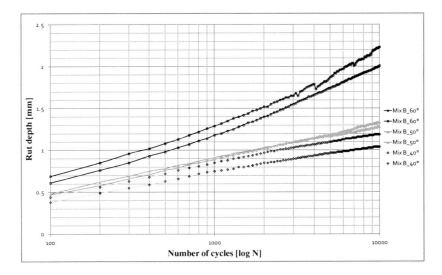

Fig. 3. Mix B_WT @ 40 - 50 - 60°C

Analyzing the data, the final rut depth accumulated after the test at 40 and 50°C is comparable in the two mixes, even though Mix B has a slightly better performance (*Table 4*). Comparing the average rut depth accumulated at those test temperatures by Mix A and Mix B, the difference ranges from 2 to 9% of the highest value (Mix A).

Table 4. Mix A & Mix B_WT results

Sample	Testing temperature [°C]	Final rut depth [mm]	Average rut depth [mm]
Mix A_1a – 1b	40	1.32 – 1.15	**1.24**
Mix B_1a – 1b	40	1.04 – 1.19	**1.12**
Mix A_2a – 2b	50	1.38 – 1.30	**1.34**
Mix B_2a – 2b	50	1.28 – 1.33	**1.31**
Mix A_3a – 3b	60	2.40 – 2.04	**2.22**
Mix B_3a – 3b	60	2.23 – 2.01	**2.12**

The accumulated deformation at 60°C is the highest: each mix significantly reduces its performance. In both mixes at this temperature, the rut depth increases more than 90% with respect to the correspondent values obtained at 40°C (*Table 4*). The average rut depths at 60°C are similar and their difference is less than 5%.

The WT tests show that also at the highest temperature, the WMA performs better than the HMA, but to confirm the effective increase in performance, it would be necessary to carry out a series of test using different wax contents.

The 2% of Sasobit® was chosen since the optimum percentage of wax addition ranges between 2 and 3% by weight, considering the effectiveness of using such an additive and the overall economics [4].

The different mechanical behavior shown by the mixes at the lowest temperatures (40-50°C), if compared to the highest (60°C), suggests an exhaustive volumetric study. In fact, a mixture rutting resistance is influenced by its binder rheology related to the variation in temperature, as the aggregates interlocking phenomena terminates. The analysis was conducted on the slabs tested with the Wheel Tracking (*Table 5*). The results demonstrate that, at the lowest temperatures, both mixes reduce their air voids content following a re-arrangement of the aggregates skeleton under the repeated loading. Differently, at 60°C, a reduced volume change is recorded, indicating that instability rutting occurs. In this case, the bitumens behavior influences the materials response more than the aggregates arrangement. Consequently the cyclic loading generates an accumulated deformation with the displacement of material and with a reduced volume change [5].

Table 5. Volumetric analysis after testing (Initial values, Table 3)

Testing Temp. [°C]	Mix A – Average Air voids [%]	Mix B – Average Air voids [%]
40	5.35	5.53
50	4.83	5.19
60	5.23	5.46

Better to understanding the deformation process at higher temperatures, a series of rheological test were conducted aiming to assess how the bitumens consistency characterizes the mixtures resistance.

3.2 Rheology Tests

Rutting is more prevalent at high temperatures than at intermediate or low ones; the properties related to rutting should therefore be measured in the upper range of service temperatures. The binder has the leading role on rutting resistance, for this reason a series of DSR tests were performed (two replicates). The 25-mm plate was used and measurements were taken in frequency sweep to cover the range from 0.01 to 1.47 Hz at five different temperatures (5, 10, 25, 40 and 55 °C) and all data were referred to 25°C. Master curves show a comparable materials behaviour even though Binder A has generally a lower G* than Binder B within all the frequency range. The same binder plots a phase angle curve typical of

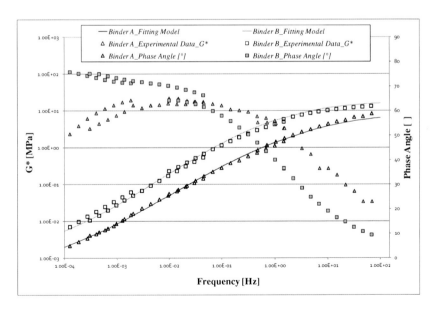

Fig. 4. Experimental data and Fitted Master Curves, $T_{rif} = 25°C$

PmBs. On the contrary, Binder B has a discontinuous phase angle trend at frequencies below 0.1 Hz, despite it carries an equivalent amount of modifier. This is potentially due to the percentage of wax used to additivate and it could be explained analyzing the wax rheology: at lower frequencies, i.e. higher temperatures, could correspond the wax melting point (*Figure 4*). Successively the experimental data were fitted using a sigmoidal model developed by Medani-Huurman for asphalt concretes [6].

This model has a typical S-shape and to better fit the data, a lower G^*_{min} value was chosen. This variable is fitted to better minimize the model root mean square using the solver function in the Excel spreadsheet (*Figure 4*).

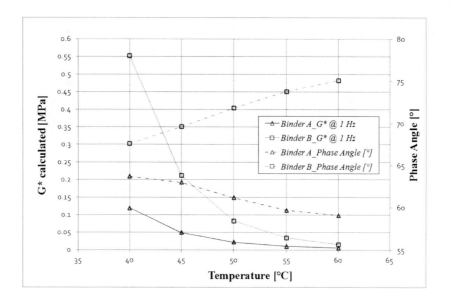

Fig. 5. G^* Calculated Temperature Sensitivity Curves @ 1 Hz and Phase Angle

The model permits to obtain the complex modulus G^* at one reference frequency in correspondence of different temperatures; this representation is also known as Temperature Sensitivity Curve. The chosen frequency is 1 Hz as closest DSR frequency to WT testing. Figure 5 shows that Binder A and Binder B have a different response under temperature variation. The wax bitumen is more susceptible to the increase of temperature, even if the additive generates a raise in shear performance, like shown in the master curves [4].

Analyzing the phase angle data, the trends are different; in Binder A, the value decreases with temperature increase, while opposite is Binder B response. This aspect was met during the Phase Angle-Frequency curves interpretation and was observed during WT testing. In fact, WT results have shown an higher increase of accumulated deformations passing from 50 to 60°C tests.

Better to understanding the mixes behavior, in Figure 6 are shown the trend of the loss modulus G'' and the storage modulus G' for the tested binders. The G'', associated with viscous effects, dominates in all the higher temperatures range for both asphalt binders. The wax bitumen G' decreases of 9% with the temperature variation if compared to the value obtained at 40°C. In the same range, the G'' of Binder A increases of 13%: this reveals an higher viscous behavior and a loss in performance at higher temperatures (*Figure 6*).

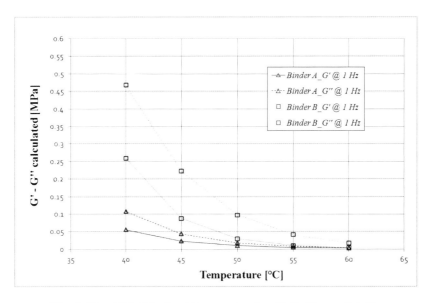

Fig. 6. G' and G'' Calculated Temperature Sensitivity Curves @ 1 Hz

Although the WMA bitumen has an higher phase angle than the HMA binder, it allows the accumulation of less permanent deformations, as it has been proven with the WT tests. This is mainly due to the higher stiffness of the wax bitumen that are shown in the master curve graphs.

4 Conclusions

The results have shown that adding a 2% wax to the studied PmB bitumen improves its mechanical performances. This was confirmed by the results obtained from the Wheel Tracking tests carried out on two different mixes and by the rheological tests on binders.

The WT results have shown that there is a significant change in performance for both mixes when tested at 40-50°C and 60°C. The WMA performs slightly better at all temperatures and tests. The volumetric properties evaluated on the tested slabs, confirm that both mixes change their behavior in accordance with the test temperature. At 40 and 50°C the mixtures are compacted under the repeated load, the air voids content decreases and a deformation with volume change occurs. At 60°C instead, the rut depth increases for both mixes with a reduced decrease of air voids content. These results suggest that at 40 and 50°C the pre-consolidation rutting is prevalent while at the highest temperature instability rutting develops.

These findings advise to investigate the influence of the bitumen properties on rut resistance. Rheological data show how higher stiffness magnitudes (Binder B) ensure better rutting performances. Although the phase angle of Binder A is lower than that of Binder B and displays an opposite trend (*Figure 5*), the wax affects the binder G* Modulus reducing the tendency to accumulate permanent deformations.

It is foreseen a rheological analysis on the wax in order to assess how its amount and melting point are related to the bitumen phase angle at high temperatures.

References

[1] Di Benedetto, H., Perraton, D., Sauzéat, C., De La Roche, C., Bankowski, W., Partl, M., Grenfell, J.: Rutting of bituminous mixtures: Wheel Tracking tests campaign analysis. Materials and Structures 44, 969–986 (2011)
[2] Lu, X., Isacsson, U.: Rheological characterization of styrenebutadiene-styrene copolymer modified bitumens. Construction & Building Materials 11, 23–32 (1997)
[3] Kumar, P., Mehndiratta, H.C., Singh, L.: Comparative study of rheological behavior of modified binders for high-temperature areas. Journal of Materials in Civil Engineering, ASCE, 978–984 (October 2010)
[4] Liu, Z., Wen, J., Wu, S.: Influence of Warm Mix Asphalt Additive on Temperature Susceptibility of Asphalt Binders. In: Proceedings of 4th International Conference on Bioinformatics and Biomedical Engineering, iCBBE (2010)
[5] Petretto, F., Pettinari, M., Sangiorgi, C., Dondi, G.: The mix gradation influence on the permanent deformations resistance of compacted wma. In: Proceedings of 5th International Conference on Bituminous Mixtures and Pavements, Thessaloniki, Greece, pp. 1403–1411 (2010)
[6] Medani, T.O., Huurman, M., Molenaar, A.A.A.: On the computation of master curves for bituminous mixes. In: Proceedings of 3rd Eurasphalt & Eurobitume, EAPA (2004)

Cracking Resistance of Recycled Asphalt Mixtures in Relation to Blending of RA and Virgin Binder

M. Mohajeri, A.A.A. Molenaar, and M.F.C. Van de Ven

Delft University of Technology, Road and Railway Engineering Section, 2628CN Delft, The Netherlands

Abstract. The degree of blending between the reclaimed asphalt (RA) binder and the virgin bitumen during the asphalt recycling mixing process is presumed to greatly influence the performance properties of recycled asphalt mixtures. Studies on the effect of different mixing methods using different quantities of RA in the laboratory, showed that tensile strength of RA mixtures and their fatigue characteristics are affected to some extent by the preheating conditions and moisture content of the RA The effects however are not very significant. These somewhat unexpected results are believed to be due to the low void content of the mixtures tested but they also might be influenced by the amount of blending that has taken place between the RA binder and the virgin binder.

This paper will report on how well the virgin binder blends with the old binder which is part of the RA. It is believed that by knowing precisely the degree of blending of RA binders, one can develop much more realistic mix designs and modify or select better mixing processes. This will result in more durable pavements.

Nano-indentation is employed to measure the mechanical properties of the binder layer and to determine the degree of blending between the hard RAbinder and the soft virgin binder. Also the characteristics of the interface zone between binder and aggregate were measured in this way. For this purpose mono sized sea sand particles and glass beads up to 4mm were mixed with a virgin soft and hard bitumen separately to produce different mixtures in order to simulate different levels of blending and interface properties. RA was simulated by mixing aggregate with hard bitumen. Nano-indentation was employed to determine the resilient modulus and hardness at each location and these results were used to determine the degree of blending.

Although current reported nanoindentation results in this paper didn't allow to determine blending degree between two binders; further nanoindentations will be performed on differnt types of samples to analyze the degree of blending.

1 Introduction

Using high amounts of reclaimed asphalt in producing new asphalt concrete mixtures has become common practice in the Netherlands as well as many other countries. Environmental issues including strict regulations on dumping old asphalt, has resulted in a situation in the Netherlands whereby producing asphalt mixtures for base layers with at least 50% RA has become inevitable. Since RA binder is already hardened due to aging in the pavement during its service life, one need to add 50% soft bitumen to compensate its hardness in order to meet the stiffness requirements defined for hot mix asphalt. Figure 1 shows how RA binder which is still covering the RA aggregates might be blended with the virgin binder that is covering the virgin aggregates.

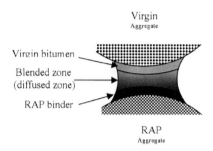

Fig. 1. Blended zone between RA and virgin aggregate

The question however is whether this hypothetical model is correct. In order to check this a full scale trial was made in which virgin aggregate was preheated in the inner drum of a double drum mixer and then mixed with RA in the outer drum of the double drum mixer. The results showed that the virgin aggregate was coated by the binder of the RA. Figure 2 shows how the mixture looked like when leaving the drum. The figure shows that the RA binder is not sticking only to the RA aggregates but also sticks to the virgin aggregate. This simple test showed that the model as proposed in figure 1 might be too simple to simulate reality.

Fig. 2. Mixture of 50%RA with virgin aggregate without adding virgin bitumen

However when producing recycled mixtures with a double barrel drum, the hot virgin aggregates first come in contact with the RA before virgin bitumen is added to the mixture. This would imply that in this mixing process both the RA and virgin aggregates are coated with RA binder and afterwards with virgin binder. If this is the case then the principles outlined in figure 1 are still valid.

The grade of added bitumen in the recycled asphalt mixture is usually determined by blending charts or equations such as the log-pen rule assuming that 100% of RA and soft virgin binder will thoroughly and homogenously blend.

Several laboratory and field studies have however shown that this latter assumption is seldom completely valid [1-5]. Even in some cases designers don't count on any blending in the mixture and they consider the RA aggregates as black rocks. However the experimental results shown in figure 2 indicate that this assumption is not completely true when mixing is done with a double barrel mixer.

Non homogenous and partial blending of RA with soft virgin bitumen could decrease the fatigue life [6]. These results indicates the necessity of investigating the blending degree between RA binder and virgin bitumen. This includes a study on the effect of the mixing procedures used since each research has followed its own method of mixing and this has strong effect on the extent in which blending happens.

Several researches have been done to determine the blending degree by means of two different approaches being a *Mechanistic* and an *Interface Detection* approach. In the *Mechanistic approach* , they have tried to indirectly determine the amount of blending by comparing rheological properties of RA mixture and ideal mixtures in which extracted RA is already blended with virgin bitumen prior to mixing [7,8,9].

In the *Interface Detection approaches*, blending measurements are mainly done based on studying the diffusion between two binders or between binder and rejuvenator. Stage extraction [4,10], DSR [11] and FTIR-AR tests [12] have often been used to determine the degree of blending degree by measuring the degree of diffusion.

Difference in viscosity, rheology and chemical composition were considered in previous studies to differentiate two different binders. In this study the authors have tried to detect and observe the interface between two binders by differentiating them by means of *nanoindentation* .Also fatigue properties of RA mixtures prepared by three mixing methods, are presented in this Paper. Initially the goal of these tests was to study the effect of blending on fatigue properties due to RA preheating and moisture conditions.

3 Materials

3.1 Materials for Fatigue Testing

30% and 60% crushed RA were use to produce 8 mixtures with three different mixing methods. The first method is the so called "Standard Method" (SM) which is currently used and prescribed in the Netherlands for the design of mixtures containing RA. The disadvantage of this method is that the RA is preheated to a high temperature which is not used in practice. Therefore a second method is used in which the RA is preheated to $130°$ C as is done in practice by using a parallel drum. This is called the "Partial Warming" (PW) method. The third method is the so called "Upgraded Method" (UPG) and was meant to simulate to some extent the mixing procedure of the double drum mixer. The mixing conditions are shown in table 1. Preheating temperatures in PW and UPG methods are in the range of real plants preheating temperatures to achieve 170°C final mixture.

Table 1. Mixing conditions in sample preparation for fatigue test

Lab mixing method	code	Virgin aggregate Preheating temperatures (°C)		RA conditions		Produced sample identification
		Virgin	RA	Moisture	Content	
Standard Method	SM	170	170	0%	0, 30, 60	SM0, SM30 SM60
Partial Warming	PW	240,330	130	0%	30, 60	PW30,PW60
Upgraded method	UPG	290,345 430,515	23	0%, 4%	30, 60	UPG0-30,UPG0-60 UPG4-30,UPG4-60

The aggregate gradation, bitumen and filler content were kept constant in all mixtures. 70/100 pen graded virgin bitumen was used as soft binder to add to the mixture. The reference SM0 mixture (without RA) was prepared with 40/60 pen bitumen. All mixtures were compacted at 170° C and aged in oven for 30 minutes. The diameter of fabricated cylindrical samples was 100 mm.

3.2 Materials for Nanoindentation

Standard mono-size glass beads and mono-size river sand was used in the mixtures. Pen 20/30 bitumen was used to simulate the RA binder. The soft binder was a pen 160/220 bitumen.

Table 2. Aggregae sizes in each mixture

Specimen ID	Type and size of aggregate coated with soft bitumen (160/220)	Type and size of aggregate coated with hard bitumen (20/30)
BL1	GB 2mm	GB 2mm
BL2	GB 1mm	GB 2mm
BL3	GB 1mm	GB 1mm
BL4	S 4-2mm	S 1-2mm
BL5	S 1-2mm	S 4-2mm
BL6	S 0.5-1mm	S 0.5-1mm
BL7	S 0.5mm	S 0.5-1mm

S: Sand, GB: Glass Bead.

50% of the aggregates were coated by layer of 20μm soft binder and the rest were coated with the hard one prior to mixing. The required amount of bitumen to coat the aggregate was determined by calculating the specific area of the aggregates.

Different sizes of glass beads and sieved river sand were used to make 7 specimens (BL1~BL7). BL3 was used for nanoindentation and the rest of the specimens were used to make thin sections for optical microscopy. The properties of each mixture are listed in table 2.

Since the mixtures were consist of mono-size aggregtes and no filler was added, they were not strong enough for sawing and polishing. Therefore they were molded in a paper cup without compaction and then they were impregnated with epoxy to stabilize them (figure 3). Another sample (BL-0) was made of granite stone slices. One side of each stone was covered by a thin layer of hard and the other with soft bitumen. Then they were put on top of each other in such a way that soft bitumen was exposed to the hard bitumen. Since as the first step we wanted to see a clear zone between the two binders, we just kept them at 50°C for an hour in oven. Although this temperature might not allow full blending between two binder, it was hoped that a limited blending area in the interface zone between the two binders could be detected.

All samples were cut to proper sizes. Then they were polished consequently with several rough and soft sand papers to achieve smooth surfaces. All samples were constantly cooled by ice water at 0°C. Other samples were polished with sand paper and rollers to make thin slices (30μm) for optical microscopy. Figure 3 illustrates a sections of BL3 and a cut sample of BL0 which were used for nanoindentations in this research.

Fig. 3. BL3 (left) and BL0 (right)

4 Experiments and Results

4.1 Indirect Tensile Strength and Fatigue Results

ITS test was conducted on each mix design in order to determine the stress level at which the stress controlled fatigue test was supposed to be done. All ITS tests were done at 5°C and the constant displacement speed of 5mm/min. The results are shown in figure 4. However Increasing RA in SM and PW has improved the ITS in SM, PW and UPG0 but ITS in UPG method with 4% moisture and 60%RA shows less ITS comparing with 30% RA. It was most probably caused by less RA binder blending in this mixture due to high amount of RA with 4% moisture.

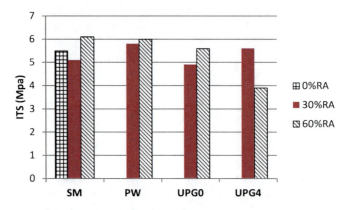

Fig. 4. Indirect tensile strength results

Indirect tension fatigue testing was done on cylindrical samples. All fatigue tests were conducted at the same stress level of 220kPa at constant temperature of 20°C and at 8 Hz frequency. Also a non-standard test was performed because it was hypothesized that the mixing of the super-heated aggregates with the moist and cool RA could have a negative effect of the adhesion of the binder to the aggregates and doing fatigue tests on submerged specimens might reveal this effect. Results of both fatigue test under dry standard conditions and saturated conditions are presented in figure 5.

According to figure 4, increasing RA content in PW and UPG method didn't have negative effect on fatigue life while it is the other way around in SM. Furthermore fatigue results obtained for the UPG samples are not less than those obtained from the other mixing methods (SM and PW). Also adding moisture to

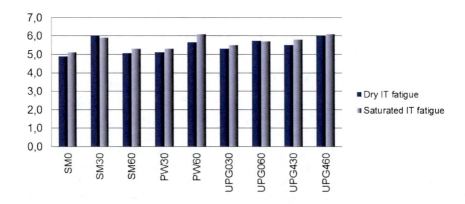

Fig. 5. Fatigue reults obtained under dry and saturated conditions at 20°C, 8Hz and a stress level of 220kPa

RA prior to mixing has not deteriorated the fatigue properties. Comparing results of two fatigue test method shows saturation conditioning also didn't have negative effect on fatigue life probably because of the low air voids content in the compacted samples.

Because the UPG and PW mixing methods using different amounts of RA did not result in lower fatigue results when compared to the reference mixture SM0, it was hypothesized that full blending most probably had occurred. It is admitted however that no proof for this statement exists.

4.3 Nanoindentation Results

All indentation tests were done with a G200-Nanoindentor instrument using a Berkovich tip. An extra setup was attached beneath the sample holder to maintain the surface temperature at -10°C during the indentation. Figure 6 is illustrating a load vs. penetrating displacement curve.

First indentation was done on a selected line on BL3 (see figure 6). Indentation started from a point on the glass bead and ended on the binder. This was done to observe properties of the aggregate and binder interfaces. 100 indentations were done every 3μm. The maximum penetration was 200 nm; the effect of two indenting impression on two neighboring points was not taken into account.

In analyzing the data attention had to be paid to the fact that the binder behavior is viscoelastic and the stone can be taken as an elastic material. The software default calculation method was used to analyze the indentations made on both the aggregate and the binder. The software calculates the modulus at the maximum load.

Values for hardness and modulus at max load for each point are plotted in figure 7.

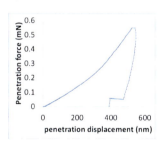

Fig. 6. An example of load vs. penetrating displacement in Nanoindentation test

Another indentation test was done on a stone-binder-stone interface in BL-0. Nano indentions were again done at -10°C with a maximum penetration depth of 500nm. 150 points were indented every 3μm. The modulus results are shown in figure 8. In this figure the modulus of binder zone is shown separately below the main chart in order to magnify the modulus difference in this zone.

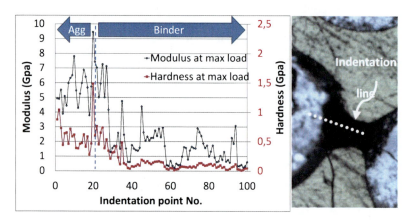

Fig. 7. Modulus and hardness on a glass-bitumen interface (BL-3)

4.4 BL-3 Results

According to figure 7, modulus values are almost in the same range until point 20. They are between 4 and 8 GPa. Then between point 20 and 35, the modulus gradually starts to decrease to less than 2 GPa. This zone, which is as wide as 45 micron, is the interface between binder and the glass. The modulus decrease in this zone could have two reasons;

(1) The binder zone close to the aggregate might be stiffer because of losing a part of its light fraction due to absorbance by or polarity of the aggregate. This phenomenon has been studied earlier by means of bitumen stage extraction [10].
(2) Figure 9 shows how binder film thickness look likes around the aggregate on the surface of the mixture. Thus measured modulus values could vary due to thickness of the binder at each point.

In the zone between point 35 to 100, one can observe repetitively increase and decrease in the modulus within almost every 20 points (every 60 μm). repetitive rise and fall in the modulus curve could suggest the presence of higher molecular weight micelles which are dissolved in a softer and lower molecular weight medium as is hypothesized in [13]. Peak values on the micelles are between 2 and 3 GPa. Away from the peak values (on the micelles) the modulus decreases gradually to less than 0.5 GPa. The peak values are decreasing in each repetition most probably because of moving from the hard to the soft binder. Knowing the properties of each binder separately enable us to recognize the interface or border between the two bitumen.

4.5 BL-0 Results

Nano-indentation on BL-0 was done in a different sequence. As described earlier, BL-0 consists of parallel granite stone sheets. It was done on a line starting from one stone and ending on the other stone by passing the bituminous "bridge" consisting of soft and hard binder.

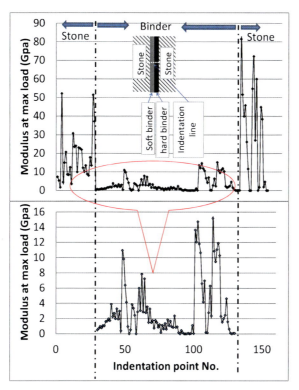

Fig. 8. Modulus results obtained from nanoindentation on a stone-binder-stone interface in BL-0 (below: data obtained on the binder zone)

Three different zones are detectable in figure 8. First of all two stone zones can be recognized which consist of points 1-30 and points 130-150. The third zone is the binder zone having a thickness of 300µm. This zone consists of points 31-129. The modulus values of the stone zone are between 25 and 70 GPa. In the binder zone again one can see periodic rise and fall at every 20 points. Also there is a region (points no. 100-125) in which the modulus is less than the stone modulus and is much higher than the values for the bitumen obtained in the area from points 30-99.

This could indicate the hard bitumen but this is not sure because the values in the 100-125 area are very high for bitumen even for a hard bitumen. Because of this it is not clear whether the "jump" in modulus at position 100 indicates the interface between the two types of bitumen.

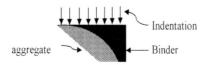

Fig. 9. Binder film thickness around the aggregate while surveying the surface of a cut sample.

5 Conclusions

Based on the results reported in this paper, the following conclusions have been drawn.

- The amount of RA as well as the moisture content of the RA does not have negative effects on the fatigue properties of the investigated recycled mixtures when compared to those of a reference mixture made of virgin materials.
- The degree of blending between the RA binder and virgin bitumen seems to be a key factor controlling the mechanical properties of recycled mixtures.
- The results of the indentation tests successfully showed the interface between binder and aggregates thanks to their huge difference in their elasticity modulus values. Detecting the interface between two binders interface was however difficult.
- Modulus values in the binder zones as determined by means of nanoindentation showed a continuous increase and decrease at every 60 µm. The peak values could represent the denser and harder functional group of bitumen molecules such as asphaltenes and low modulus values could be due to the softer phase of molecules.
- Modulus measurement by determining the slope of the curve in unloading part is very sensitive to the shape of the curve which sometimes ended up with very scattered results during continuous indentation.

References

[1] McDaniel, R.S., Soleymani, H., Anderson, R.M., Turner, P., Peterson, R.: NCHRP Final Report (9-12), TRB, Washington D.C. (2000)
[2] Oliver, J.W.H.: Int. Road. Mat. Pav. Design 2(3), 311–325 (2001)
[3] Stephens, J.E., Mahoney, J., Dippold, C.: Connecticut Department of Transportation, No. JHR 00-278 (2001)
[4] Huang, B., Li, G., Vukosavljevic, D., Shu, X., Egan, B.: Transportation Research Report (1929), p. 37–45 (2005)
[5] Carpenter, S.H., Wolosick, J.R.: J. Transportation Research Record 777, pp. 15–22. TRB, Washington, D.C (1980)
[6] Huang, B., Zhang, Z., Kingery, W., Zuo, G.: Cracking in Pavements. In: Proceedings of the 5th International RILEM, pp. 631–650 (2004)

7. Al-Qadi, I.L., Carpenter, S.H., Roberts, G., Ozer, H., Aurangzeb, Q.: Research report of Illinois Center for Transportation. No. ICT-R27-11 (2009)
8. Bennert, T., Dongre, R.: J. Tran. Res. Board, 1, 75–84 (2010)
9. Parashant, S., Mehta, Y.A., Nolan, A.: In: Proceeding of Transportation Research Board, 89th Annual Meeting, n.10- 0917 (2010)
10. Noureldin, A.S., Wood, L.E.: J. Tran. Research Record 1115, 51–61 (1987)
11. Karlsson, R., Isacsson, U.: In: Proceedings of 72nd AAPT Annual Meeting, vol. 72, pp. 563–606 (2003)
12. Karlsson, R., Isacsson, U.: J. Material Science 38, 2835–2844 (2003)
13. Read, J., Whiteoak, D.: The shell Bitumen Handbook, pp. 29–37. Shell bitumen, London (2003)

Warm Mix Asphalt Performance Modeling Using the Mechanistic-Empirical Pavement Design Guide

Ashley Buss and R. Christopher Williams

Iowa State University, Ames, IA

Abstract. Warm mix asphalt (WMA) is a cost effective means for reducing the mixing and compaction temperature of hot mix asphalt (HMA). These additives, which come in several forms, reduce the mixing and compaction temperatures by approximately 30°C. The challenge for researchers has been determining how the additives impact or change traditional HMA mixes. The purpose of this research is to use collected laboratory data as inputs into the Mechanistic-Empirical Pavement Design Guide (MEPDG) to determine if there are statistically significant differences in the amount of pavement cracking when comparing HMA and WMA. The MEDPG is a model that has been developed into software for the purpose to provide state-of-the-practice pavement design for both new and rehabilitated pavement structures. The pavement design includes site condition parameters such as traffic, climate, subgrade, and existing pavement conditions in the case of rehabilitation. The MEPDG evaluates a proposed design for various distresses such as rutting, fatigue cracking, longitudinal cracking, transverse cracking, and roughness. Samples from field produced WMA mixes have been tested and material information has been collected and statistically analyzed over a period of several years. Data analysis showed some statistically significant differences in laboratory dynamic modulus (E*) data when comparing HMA control mix with the WMA mix. This study will use the laboratory measured E* as input into the MEPDG model to predict the pavement cracking in order to determine if there are differences in the pavement distresses. Within the design inputs, various factors can be manipulated such as various traffic loading and pavement thickness. Changing these inputs show which site condition variables have the most impact on pavement performance, how pavement distresses may differ when comparing HMA and WMA and how this translates to differences in pavement life.

Keywords: dynamic modulus-warm mix asphalt-pavement performance.

1 Introduction

WMA is produced at approximately 30°C lower than traditional HMA mixes. The benefits of using WMA additives includes reduced fuel consumption, reduced emissions, cooler temperature paving, longer haul distances and less compaction effort. One hypothesized disadvantage is that the lower temperatures lead to incomplete drying of aggregate which may lead to moisture susceptibility issues. The long term effects of using WMA additives in traditional HMA mixes are being discussed in many owner agencies across the nation. The purpose of this study is to begin investigating the impact of laboratory performance test results using the Mechanistic-Empirical Pavement Design Guide (MEPDG) as a model. In the dynamic modulus data, there are statistical differences that appear among various treatment effects. The MEPDG uses dynamic modulus data as a parameter for predicting pavement distress over a 20 year design life. This study looks at how the dynamic modulus data impacts the predicted pavement performance and if there are differences between the various factors.

Four field WMA mixes and four HMA control mixes were used in this research project. Each mix was produced for a different project at different plant locations. The WMA was produced first and the HMA control mixture was produced on the following day unless weather delayed paving. The corresponding control mixes to each WMA mix differed only by the WMA additive. For each project, loose HMA and WMA mix was collected at the time of production and binder from the tank was collected for each mix. The WMA additives were terminally blended and no laboratory binder blending was performed. The field sampled binder and mix was taken to the Iowa State University asphalt laboratory for subsequent asphalt binder testing and mix performance testing.

1.1 MEPDG Background

The MEPDG is a software program that utilizes both mechanistic and empirical design methods. The AASHO road test, performed in the 1950's, is what many of the empirical pavement design principles are currently based on. Since the 1950's the typical traffic loads have increased and design of pavement material has improved, e.g. polymer-modified asphalts. The MEPDG provides a framework in which the engineer determines design inputs for traffic, desired reliability, climate, and pavement structure [1]. The MEPDG also allows for engineers to assign a "level of reliability" to their pavement designs. The higher the level of reliability, the more conservative the pavement design will be to account for variability. There are also different levels of input depending on how much data was collected for this particular pavement design. Level 1 is the most detailed data and Level 3 is general design inputs. The various input levels impact the reliability because it is assumed there is more uncertainty in Level 3 inputs; therefore, the program accounts for the higher degree of variability in the different levels. The MEPDG also allows for design of rehabilitated pavements. The ability of the engineer to input detailed material information, in this case E^* and $G^*/\sin(\delta)$, allows for the engineer to see how differences in the pavement materials will impact the pavement design.

2 Objectives

The objective of this paper is to begin investigating how adding warm mix asphalt may potentially impact pavement performance using the MEPDG. This research focuses this effort by looking at three main objectives. First, determine how changes in E* will impact pavement performance and determine which types of pavement cracking change the most. Second, show how changes in E* are sensitive to various traffic loading and pavement designs. Finally, investigate if the variables studied impact long term pavement performance when E* values prove to be statistically different.

3 Experimental Plan

Prior studies have shown that the MEPDG is sensitive to the E* values of the AC layer and that reasonable pavement performance prediction can be obtained using the software which gives reasonable pavement performance results [2,3]. The research presented in this paper is part of a two phase study. Phase I of the experimental plan gives the background information of the project and explains the variables studied. Phase II explains how the data gathered in phase I is used in predicting pavement performance in order to compare the variables important in this study.

The Phase I experimental plan involves field produced mixes that incorporate both an HMA control and a WMA experimental mix. This was repeated for four different control/experimental mix pairs, testing a total of three different WMA additives. Each of the mixes were produced at a different locations within the state of Iowa. The experimental (WMA) and control (HMA) mix design was the same except for the addition of a WMA technology and day to day plant and stockpile variability. Mix was compacted at each job site and collected in order to reheat samples and compact at a later date. This was done to investigate whether reheating WMA changed any mix properties which could impact an agency's current quality control/quality assurance program. Half of all dynamic modulus samples were moisture conditioned according to AASHTO T-283 [4]. The moisture conditioning investigated how the E* changes due to the moisture induced damage and portrays the moisture susceptibility of each mix. Dynamic modulus tests were performed on all samples and the E* results were statistically analyzed.

Phase II experimental plan uses the laboratory data collected in Phase I as inputs into the MEPDG software in order to investigate the long term impact of the variables studied in Phase I. Phase I statistical analysis of the E* data shows that the three main comparisons are statistically significant. Phase II will help show whether the statistical differences in the E* laboratory data correspond with predicted pavement performance using the MEPDG as a model. Model runs will be performed by varying only the asphalt binder and the E* properties. Three different pavement designs and traffic loads were investigated to detect sensitivity to layer thickness in conjunction with reduced traffic loads. The model will help make comparisons between the following variables: HMA vs. WMA, field compacted vs. lab compacted (reheated) samples, moisture conditioned vs. not moisture conditioned and a high, medium and low traffic pavement design. Figure 1 shows the variables tested for each field mix.

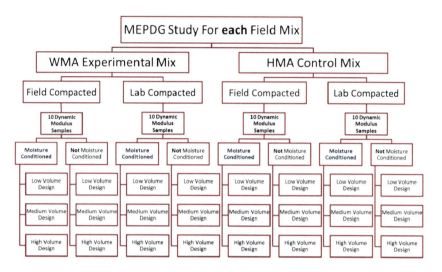

Fig. 1. Model simulation variables for each mix

3.1 Materials and Performance Tests

The projects for this study are shown in Table 1 with the exception of Field Mix 1 (FM1) which was a county project that used Revix®/Evotherm 3G® as the WMA technology. Other technologies studied include Sasobit® and Astec Double Barrel Green® foaming. Sasobit® is similar to paraffin waxes that are found in crude oil, except that it has a higher molecular weight. Sasobit is a crystalline, long chain aliphatic polymethylene hydrocarbon produced from natural gas using the Fischer-Tropsch (F-T) process [5]. Evotherm is a chemical modifier produced by Meadwestvaco. Evotherm is added to the asphalt typically at a dosage rate of 0.25 to 0.75 percent by weight of binder [6]. The Astec Double Barrel Green® is a plant modification which uses a multi-nozzle manifold to make warm mix using only water injection as the foaming agent [7].

Table 1. Project summary for Iowa DOT WMA projects

Code	Year	Project Number	WMA Technology	Design	Binder Grade	RAP
FM2	2009	NHSX-218-9(129)--3H-34	Revix/Evotherm	12.5mm/10M ESALS	PG 64-28	17%
FM3	2009	STP-143-1(4)--2C-18	Sasobit	12.5mm/3M ESALS	PG 64-22	20%
FM4	2009	STP-065-3(57)--2C-91	Water Injection	12.5mm/3M ESALS	PG 64-22	19%

Sample preparation was performed at the mixing plant in the field and in the laboratory. To produce the field samples, loose mix was collected and the Superpave Gyratory Compactor (SGC) was used to compact samples at the plant the day of production without reheating the mix. The laboratory samples were compacted in the same manner but compaction occurred at a later date and mix was reheated.

Dynamic modulus tests were performed according to NCHRP 547. The dynamic modulus samples, compacted to the precise size, are 100 mm diameter and 150 mm in height. Each field produced mix has ten field compacted dynamic modulus samples as well as ten laboratory compacted dynamic modulus samples. Half of the lab compacted samples and half of the field compacted samples were moisture conditioned and represent the experimental samples whereas the unconditioned samples are the control samples. The purpose of dynamic modulus testing is to define the materials stress to strain relationship under continuous sinusoidal loading [8]. Dynamic modulus testing measures the stiffness of the asphalt under dynamic loading at various temperatures and frequencies thus it is used to determine which mixes may be more susceptible to performance issues including rutting, fatigue cracking and thermal cracking. The test was performed under strain controlled conditions. The target strain was 80 microstrain which is considered to be well within the elastic region of the material. The dynamic modulus test is considered to be a non-destructive test at low levels of strain in theory. The strain response of the material was measured using 3 LVDTs that were positioned on mounted brackets at the beginning of each test. The test is performed at three temperatures (4, 21, 37°C) and nine frequencies (25, 15, 10, 5, 3, 1, 0.5, 0.3, 0.1 Hz) for each sample and yields 27 test results per sample. The dynamic modulus values (E*) are used to construct master curves which can be used to compare the various categories [8].

3.2 MEPDG Input Design Parameters

Three pavement designs were used to see how the pavement distresses varied from different thicknesses and traffic loading. The pavement structures, Figure 2, represent low, medium and high traffic level designs with average annual daily truck traffic (AADTT) of 100, 700 and 2000, respectively. The traffic distributions utilized the default values regardless of traffic level. The pavement structures are based on typical Iowa roadway thicknesses that use standard Iowa aggregates, for each of the given AADTT traffic levels. The climate file remained the same for all model runs and was generated by interpolating several Iowa stations. A typical Iowa subgrade classification of A-7-6 was used. All MEPDG inputs were a level three design with the exception of the material properties of the asphalt layers. All data inputs remained the same except for the pavement designs, traffic levels and asphalt material properties.

The MEPDG requires dynamic modulus inputs for 5 temperatures and 6 frequencies. The dynamic modulus testing was performed at 3 temperatures and 9 frequencies. The E* data can be shifted based on the theory of time-temperature superposition [8]. If an asphalt sample is loaded at a high frequency at a lower temperature, the material response can be correlated to a lower frequency at a higher temperature using shift factors. The relationship between temperature and shift factor is linear. A linear equation can be used to determine the shift factor at a higher or lower temperatures which can then be used to shift the E* values to give the E* value that corresponds to material responses at -10°C and 54°C.

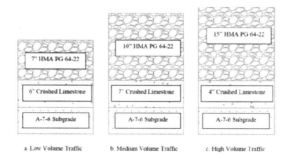

a. Low Volume Traffic b. Medium Volume Traffic c. High Volume Traffic

Fig. 2. Pavement designs for low, medium and high traffic levels

4 Results

A statistical analysis was conducted in detail during Phase I of the experiment. The ANOVA table, Table 2, shows the important factors studied and their interactions. A significance level of α < 0.05 was used. For the analysis, each mix was analyzed separately. The assumptions needed for the ANOVA analysis are: independence, normally distributed data and equal variance. The assumptions of normal distribution and independence were met but the variance was not constant. In order for the constant variance assumption to be satisfied, a square root transformation was performed on the E* values.

Table 2. ANOVA tables for $\sqrt{E^*}$ statistical analysis separated by mix

	Source	DF	Type I SS	Mean Square	F Value	Pr > F (α)	Higher Average E* Treatment
Field Mix #1	mix	1	6076743.8	6076743.8	492.57	<.0001	HMA
	comp	1	2399557.4	2399557.4	194.5	<.0001	Lab
	mix*comp	1	76266.3	76266.3	6.18	0.0132	
	mcond	1	6062662.8	6062662.8	491.43	<.0001	No Moisture Conditioning
	mix*mcond	1	23343.8	23343.8	1.89	0.1694	
	comp*mcond	1	61417.7	61417.7	4.98	0.026	
	mix*comp*mcond	1	825077.8	825077.8	66.88	<.0001	
	Source	DF	Type I SS	Mean Square	F Value	Pr > F (α)	Higher Average E* Treatment
Field Mix #2	mix	1	1015250.7	1015250.7	90.31	<.0001	HMA
	comp	1	197722	197722	17.59	<.0001	Field
	mix*comp	1	8961.4	8961.4	0.8	0.3722	
	mcond	1	925236.4	925236.4	82.3	<.0001	No Moisture Conditioning
	mix*mcond	1	1051377.6	1051377.6	93.53	<.0001	
	comp*mcond	1	680	680	0.06	0.8058	
	mix*comp*mcond	1	597982	597982	53.19	<.0001	
	Source	DF	Type I SS	Mean Square	F Value	Pr > F (α)	Higher Average E* Treatment
Field Mix #3	mix	1	4891633	4891633	364.96	<.0001	HMA
	comp	1	1550413	1550413	115.68	<.0001	Lab
	mix*comp	1	922370	922370	68.82	<.0001	
	mcond	1	3612270	3612270	269.51	<.0001	No Moisture Conditioning
	mix*mcond	1	289625	289625	21.61	<.0001	
	comp*mcond	1	108601	108601	8.1	0.0045	
	mix*comp*mcond	1	532800	532800	39.75	<.0001	
	Source	DF	Type I SS	Mean Square	F Value	Pr > F (α)	Higher Average E* Treatment
Field Mix #4	mix	1	3249873	3249873	319.58	<.0001	WMA
	comp	1	4709	4709	0.46	0.4964	
	mix*comp	1	1017906	1017906	100.1	<.0001	No Moisture Conditioning
	mcond	1	3356027	3356027	330.02	<.0001	
	mix*mcond	1	140236	140236	13.79	0.0002	
	comp*mcond	1	194105	194105	19.09	<.0001	
	mix*comp*mcond	1	133330	133330	13.11	0.0003	

Phase II results are shown in Figures 3, 4, and 5 [9]. The figures present alligator cracking, total rutting and IRI, respectively, as calculated by the MEPDG. The data is categorized by all of the variables studied. There are two data points in each category, one field compacted and the other is the reheated laboratory response. The differences between field and lab compacted can be observed by noting how far apart the data points in each category are from each other. All pavement distresses appear to follow the same trend between the various pavement distresses. The medium level pavement design consistently had higher pavement distresses with a few exceptions. The interactions of "mix" (HMA vs. WMA), "moisture conditioning" or "mcond" (conditioned vs. not conditioned), and "compaction" or "comp" (field vs. laboratory compaction) were evaluated in any combination. For this study, the MEPDG model used averages so only two way interactions of the factors listed were evaluated. These interactions can be compared with the laboratory data to determine if there are trends in both the laboratory data and the pavement performance model.

For FM1, there is a large difference between field and laboratory compacted HMA samples as shown by the large separation of the black dots in each category. Moisture conditioned samples appear to have slightly higher average but it does not appear to be significant even though moisture conditioning was a statistically significant factor for the E* laboratory data as shown in Table 2. The differences between average pavement distresses for HMA and WMA don't appear to be significant except in the case of IRI. The HMA has a higher average roughness compared to the WMA values. There are differences between the pavement performance data and the E* data. This may be due to averaging E* for the model runs, in order to reduce the number of runs and also the ANOVA analysis looks at overall trends but doesn't specifically break each E* value into its specific category. Interaction plots were plotted using averages to see if there may be interactions that showed up in the E* data. The interaction plots showed an interaction between mix and moisture conditioning which was not evident in laboratory E* data.

Field mix 2 shows the pavement performance for HMA and WMA are similar with the exception of several categories showing WMA with a slightly higher average pavement distress for the moisture conditioned samples. There doesn't appear to be a difference in the pavement distresses when comparing whether the samples were moisture conditioned or not. The data points with each category are spaced close together which indicates that there is no noticeable difference in the modeled pavement distresses when comparing field or laboratory (reheated) compaction. The interaction plots developed using the distress data reflects the mix and moisture conditioned interaction shown in the E* ANOVA table.

Field mix 3 shows similar trends to field mix 2. There are little differences in the pavement performance data for all variables. The only noticeable differences is the HMA average distress appears to be slightly lower than WMA for the moisture conditioned samples, alligator cracking and total rutting. The interaction of mix and compaction is the only detectable interaction in the pavement distresses.

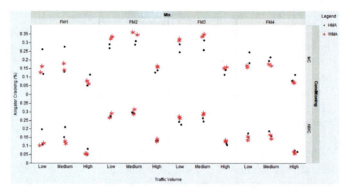

Fig. 3. MEPDG predicted alligator cracking

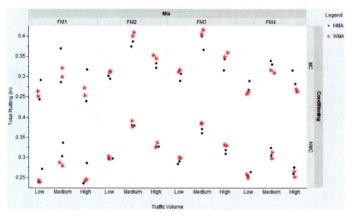

Fig. 4. MEPDG predicted total rutting

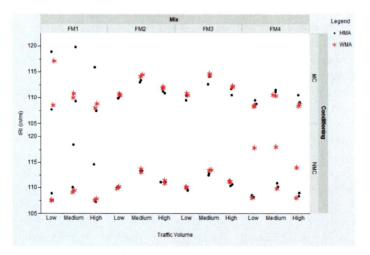

Fig. 5. MEPDG predicted IRI

Field mix 4 doesn't show differences in the variables for the alligator cracking and the total rutting but there is a large difference in the category of WMA/NMC/Field compacted for the IRI values. This is interesting because the other two pavement performance distresses did not indicate this difference. Mix*compaction appeared to be an interaction that also appear in the pavement performance data.

5 Conclusions

The MEPDG can be used as a tool to help designers reasonably choose the pavement design that best fits their needs based upon pavement performance predictions. The MEPDG predicted pavement responses show that, in most cases, there was little to no difference when comparing HMA and WMA over a long period of time. The data shows some differences between the various treatment conditions and some distress responses that reflect the phase I laboratory data analysis but specific trends were not seen in every mix variable studied. This may be due general field variability. Total rutting and alligator cracking followed similar trends but the IRI would, at times, display a result that wouldn't match with the rutting and alligator cracking trends. The pavement designs showed similar trends in most cases, with the medium level pavement design having the highest distress levels. The ANOVA table in Phase I appeared to show more differences than the MEPDG pavement performance data. In this study, average E* values were used for the model runs. Each mix had 24 categories for a total of 96 runs. In order to study the distribution of all mix samples, 960 runs will need to be performed. Doing this will help to show statistically what the differences are and further strengthen the conclusions. Generating an MEPDG run for each sample will give a distribution and variance for each sample set within each treatment category. This will allow a more detailed statistical analysis of the MEPDG pavement performance data. The MEPDG is a powerful tool for pavement design and material engineers; however, further model validation and calibration is necessary but continuing these efforts will provide for faster pavement material evaluation and pavement designs which result in longer pavement life.

References

[1] NCHRP, Guide for Mechanistic-Empirical Design of New and Rehabilitated Pavement Structures.National Cooperative Highway Research Program, NCHRP 1-37A (2004)
[2] Mohamed, M.E., Witczak, M.W.: Verification for the Calibrated Fatigue Cracking Model for the 2002 Design Guide. In: Annual Meeting of the Association of Asphalt Paving Technologists, Long Beach, CA (2005)
[3] Mohamed, M.E., Witczak, M.W.: Verification for the Calibrated Permanent Deformation Model for the 2002 Design Guide. In: Annual Meeting of the Association of Asphalt Paving Technologists, Long Beach, CA (2005)

[4] AASHTO: Designation T283-07 Resistance of Compacted Hot Mix Asphalt (HMA) to Moisture Induced Damage. In: Standard Specifications for Transportation Materials and Methods of Sampling and Testing, 27th edn. American Association of State Highway and Transportation Officials, Washington, D.C (2007)
[5] Sasol, *Sasobit*® Asphalt Technology (2002), http://www.sasolwax.us.com/sasobit.html (retrieved April 6, 2011)
[6] Meadwestvaco, *Evotherm*® Warm Mix Asphalt (2011), http://www.meadwestvaco.com/Products/MWV002106 (retrieved on September 22, 2011)
[7] Astec, Inc., Double Barrel *Green*® (2011), http://www.astecinc.com/index.php?option=com_content&view=article&id=117&Itemid=188 (retrieved on September 27, 2011)
[8] Witczak, M.: NCHRP Report 547: Simple Performance Tests: Summary of Recommended Methods and Database. In: National Highway Research Council, Transportation Research Board, Washington, D.C. (2005)
[9] SAS Institute Inc., *JMP*® 9.0.0, Cary, NC. (2010)

Shrinkage and Creep Performance of Recycled Aggregate Concrete

Jacob Henschen[1], Atsushi Teramoto[2], and David A. Lange[3]

[1] Graduate student, University of Illinois at Urbana-Champaign
[2] Graduate student, Nagoya University, Japan
[3] Professor, University of Illinois at Urbana-Champaign

Abstract. With the growing emphasis on sustainability in the concrete industry, there has been a renewed interest in the use of recycled concrete as aggregate in new concrete. Recycled aggregate concrete is not a new concept, but it is normally met with resistance due to reduction in strength and an increase in drying shrinkage. The organizations that allow for recycled concrete aggregates in concrete give few guidelines. One of the common requirements is that the recycled aggregates should be soaked prior to use. The goal of this study was to investigate how the initial moisture state of the aggregates affect the creep and shrinkage properties. Tests included free shrinkage of prisms, restrained shrinkage ring tests and an actively restrained tensile creep test. By using the recycled concrete aggregate at or near SSD conditions, the mixtures were workable and had lower free shrinkage. The recycled aggregate mixes did have higher tensile creep but this is not detrimental in paving applications where it would work to prevent shrinkage cracking.

1 Introduction

The O'Hare Modernization Program (OMP) is a major public works program to improve the O'Hare International Airport in Chicago. Construction at O'Hare generates a large volume of concrete waste every year as in-place pavements are removed even as new aprons, taxiways, and runways are constructed. The OMP and the City of Chicago has placed high value on sustainability goals at the airport, and has aspired to be seen as an international sustainability leader. Toward that end, OMP is today using recycled concrete aggregate (RCA) as an aggregate source for new concrete. RCA in concrete is an established concept dating back to the 1940s, but nevertheless the use of recycled concrete has struggled to gain wide acceptance [1]. Most research and field trials have demonstrated that the use of RCA in concrete has negative effects on both fresh and hardened properties. The creep and shrinkage of RCA concrete tends to be adversely affected, and both of these properties are associated with premature degradation of concrete pavement.

In recent years, more research has been pursued to determine if these negative effects can be overcome. One step that has become commonplace in manuals of practice is to presoak the RCA to control the effect of high water demand on the fresh concrete [2,3]. Other research has considered alterations to the mixing procedure in order to improve the bond between the paste and RCA to improve the hardened properties and improve workability [4, 5]. Other researchers have developed methods to remove the old mortar from RCA, essentially reclaiming the virgin aggregate from the old concrete [6]. Many organizations and researchers advocate soaking RCA prior to using it in concrete, but there is a lack of detailed understanding regarding the connection between initial moisture state and its effect on hardened properties [2, 3]. The initial moisture of the aggregate affects the internal moisture of hardened concrete and therefore can be expected to affect the hardened properties of the concrete. Since the absorption capacity of the RCA is higher than virgin aggregate, a significant amount of free water must be added to the concrete mixture to compensate for air dry RCA. Additionally, the rate of adsorption is generally too slow to allow RCA to achieve SSD condition prior to the time of cement setting. Thus, this approach leaves the concrete with a higher w/c than anticipated, leading to problems associated with high w/c concrete mixtures.

This study is part of larger research program studying recycled concrete aggregates. In the concrete mixtures, the initial moisture content of the RCA was varied from oven dry, to 80% of SSD and SSD. These mixtures used RCA for the entire coarse fraction and were compared to a mixture using only virgin coarse aggregate. In concrete pavements, drying shrinkage and tensile creep have a significant effect whether or not the pavement will crack due to environmental factors. The tests presented here were selected to assess the resistance of RCA pavement to cracking.

2 Materials

For all of these tests, a consistent source of RCA was used supplied by a contractor working with the OMP. Old airport pavements were broken up, crushed, washed and graded to a CA7 gradation. The RCA did contain up to 3% bitumen, the content was small enought that it would not adversely affect the perfomrance of resulting concrete. The RCA was free from steel and other contaminants.

The mixture proportions for the virgin and RCA concrete are found in Table 2.

The virgin aggregate was used at an air dry state. The coase aggregate in the virgin aggregate concrete was a blend of two gradations in order to closely replicate the RCA gradation. The two-stage mixing procedure proposed by Tam [4, 5] was used for all mixtures because it was shown to improve fresh and hardened concrete properties. All of the specimens were prepared in accordance to ASTM C192 [7]. Prior to casting the specimens, slump and air content were measured and the results are recorded in Table 1. Cylinders were cast and tested in splitting-tension according to ASTM C496 [8].

Table 1. Virgin and recycled concrete mixture compositions

	SG	VAC (kg/m³)	RCA (0, 80, 100) (kg/m³)
Water	1.00	129	129
Cement	3.15	246	246
Fly Ash	2.20	61	61
CA 7	2.60	848	0
CA 16	2.70	237	0
RCA	2.41	0	977
Sand	2.57	759	853
Slump	cm	2.5	5.1
Air	%	2.7	2.5
	w/c	0.42	0.42

3 Experimental Procedure

The test program was designed to assess the performance of the RCA concrete in a paving application. The tests included free shrinkage of prisms, restrained ring shrinkage, actively restrained tensile creep, and internal moisture. The free shrinkage tests followed a modified ASTM C490 [9] procedure. The prisms were covered with foil tape on two opposite sides to allow for one directional drying which replicates the drying conditions that pavements are subjected to.

The shrinkage ring tests followed the AASHTO PP34 [10] standard which is a 76.2 mm ring of concrete cast on the outside of a 12.7 mm steel ring. The steel ring passively restrains the concrete from drying shrinkage and by doing so tensile stresses develop in the ring. The strain measurements taken on the inside of the steel ring can be used to determine the stress developed in the concrete. In addition, the time to cracking is noted for the tests which can be used to compare the cracking resistance of different mixtures. This test does not allow creep or drying shrinkage to be directly measured, but It is usefull to assess relative early age cracking resistance between different concrete mixtures.

Tensile creep properties were directly measured using an actively restrained tensile creep test. This test uses a hydraulic actuator to actively restrain a dogbone shaped specimen. Using an second dogbone that is unrestrained, the free shinkage of the material can also be determined. By using the measurements from the two specimens, tensile creep properties can be calculated [11].

4 Results and Discussion

The prism free shrinkage results, located in Figure 1, show that all of the mixtures were within 100 microstrain of each other after 45 days. In this trial all of the RCA mixtures had lower shrinkage than the virgin aggregate mixture.

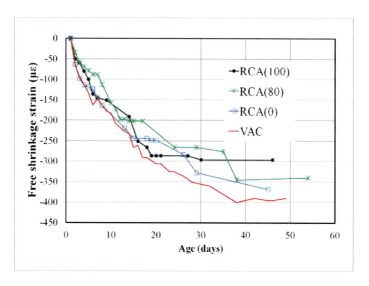

Fig. 1. Free shrinkage of concrete prisms

This result disagrees with what some other researchers have found [12, 13], but some research has shown that presoaking the RCA leads to a significant reduction in drying shrinkage [14]. Comparing the RCA mixtures, free shrinkage increased with decreasing initial moisture at later ages (past 40 days). An explanation for this could be that since the RCA is more porous, the mixtures started with higher water contents (mix water and water stored in the aggregate). The RCA would have provided a stable volume in which to store the additional water allowing the internal moisture of the specimens to remain higher over time when compared to the virgin aggregate mixture. The higher moisutre levels would result in less drying shrinkage. Even though all of the RCA mixtures performed quite well, it is apparent that keeping the RCA at or near SSD resulted in lower drying shrinkage. This behavior is due to the amount of free water in the cement paste prior to set. In the SSD RCA there is no correction for initial moisture taken and so the apparent and the actual w/c are the same. For the oven dry RCA there is a large correction in the water and therefore the apparent w/c is higher than the actual. The oven dry RCA does absorb some of the free water from the cement paste but from measuring absorption of the aggregates the time to full saturation is much greater than the time it takes for the cement to set. This results in the real w/c being somewhere between the actual and the apparent. Since the real w/c is higher than the actual, the resulting concrete would be expected to have lower strength and higher free shrinkage.

The strain measurements from the ring test are shown in Figure 2. Each of the RCA mixtures has lower strain than the virgin aggregate mixture until about 28 days in age. After this point the strain in the RAC(80) and RAC(100) mixtures exceeds that of the virgin aggregate mixture.

Shrinkage and Creep Performance of Recycled Aggregate Concrete   1337

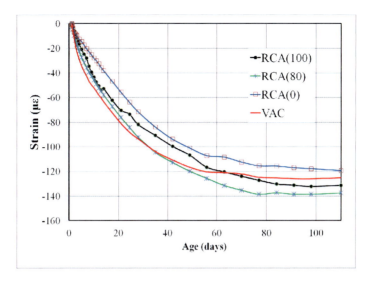

Fig. 2. Strain developed in steel ring

None of the speciments cracked during the test which demonstrated that all of them had high cracking resistance. After, 80 days the strains began to level off which meant that the tensile creep in the specimens were equalizing with the shrinkage. Beyond this point the specimens would probably never crack. The ring test deals with both the drying shrinkage and tensile creep behavior of a material. These competing properties make the results difficult to interpret directly, but it does show reasonably well that none of the mixtures have an affinity to cracking.

The free shrinkage results from the restrained tensile creep test in Figure 3 show the same trend that the free shrinkage prisms did when comparing the RCA mixtures to the virgin aggregate mixtures. The trend within the RCA mixtures for the dogbone differs in that the SSD and oven dried RCA mixtures are nearly identical and the RCA(80) mixtures falls in between the VAC and other RCA mxitures. Though the dogbone results deviate from the prisms, this behavior is somewhat mirrored in the ring test where at early ages the RCA(80) mixture had a higher strain value than the other two RCA mixtures. The total creep results from the dogbone are shown in Figure 4. Again this does not agree with some previous research that shows creep to be higher in RCA mixtures [Gomez, Domingo]. The total creep behavior is similar to the dogbone free shrinkage where the RCA(100) and RCA(0) mixtures have lower values than the VAC. The RCA(80) mixture had the highest total creep of all of the mixtures. Figure 5 shows specific creep which normalizes the creep strain with the applied restraint force. Since this is a test for early age creep behavior, it is better to have higher specific creep. Having higher tensile creep for given loading will help to prevent restrained concrete from cracking due to drying shrinkage. This is the reason that the RCA(100) was the first to crack even though it had low values of shrinkage. The RCA(80) does not follow the behavior of the other RCA mixtures. In the dogbone test, the free shrinkage and creep values are higher than the other mixtures. This behavior is unexpected and further testing will will be needed to understand it. For the purposes of this testing program, it was more important that the RCA mixtures performed as well as the VAC mixtures.

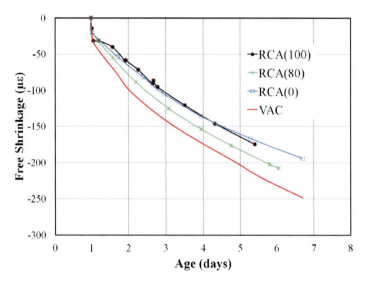

Fig. 3. Free shrinkage measured in restrained creep test

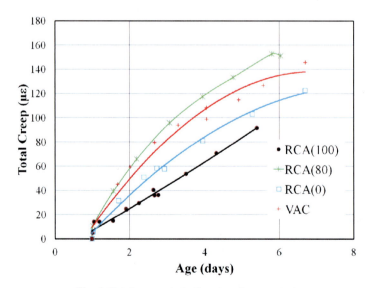

Fig. 4. Total creep strain from tensile creep test

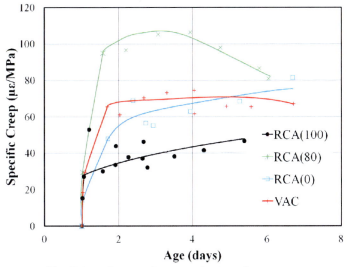

Fig. 5. Specific creep from restrained tensile creep test

Table 2 shows the comparison between the splitting tesile strength and stress at fracture for the restrained creep tests. The ratios between stress and strength give indication of the cumulative damage caused by tensile creep. The RCA(80) had the highest total creep and this may be explanation for why it exhibited the lowest stress to strength ratio. This result suggests that creep and or internal microcracking of RCA(80) may have served to dissipate shrinkage strain and initiate fracture at a relatively low load.

Table 2. Comparison of strength results

		RCA(100)	RCA(80)	RCA(0)	VAC
Split Tensile Strength (7day)	Mpa	2.45	2.74	1.83	3.08
Fracture Stress (Tensile creep test)		1.97	1.69	1.50	2.18
Fracture Stress/ Split Tensile		0.80	0.62	0.82	0.71

Considering the free shrinkage results, the RCA mixtures performed better than the virgin aggregate mixture, and by increasing the moisture in the aggregates decreased the shrinkage. This relation did not hold true in the restrained creep test which showed that the RCA(80) had the highest early creep and the RCA(100) had the lowest of all the mixtures. As discussed above, higher early creep can help to prevent the concrete from cracking. This effect can be seen in the age at which the tensile creep tests cracked. The RCA(80) cracked 1.5 days after the RCA(100) cracked even though the free shrinkage of the RCA(80) was higher throughout the test. The RCA(0) also performed very well in these tests. The problem with using the RCA at oven dry conditions is that workability was quickly lost and the concrete became very difficult to place. Loss of workability (i.e. slump loss) has been seen in literature related to applications of RCA concrete as one of the main reasons most researchers favor presoaking the RCA prior to its use in concrete.

5 Conclusions

The free shrinkage, restrained ring, and restrained tensile creep tests all demonstrated that concrete produced with RCA can perform as well or better than concrete with virgin aggregate. None of the mixtures produced for this study exhibited abnormally high free shrinkage or high susceptibility to cracking. The key to effectively using RCA is to presoak the aggregate, and use it when it is at or just below SSD condition. By doing this, the concrete will exhibit reduced shrinkage and similar cracking resistance to virgin aggregate concrete. Using RCA in dry state will decrease the workability of the fresh concrete as well as increase shrinkage and possibly reduce early age creep. Using the RCA at just below SSD results in a balance between free shrinkage and early creep.

Acknowledgements. The authors gratefully acknowledge support for this research research from the O'Hare Modernization Program through the Center of Excellence for Airport Technology.

References

[1] Hansen, T.: RILEM Tech Com.-37-DRC 19(111), 201–246 (1986)
[2] ACI 555R-01, ACI Man of Con Practice (2001)
[3] AASHTO MP16, Std Spec for RCA as Coar Agg in PCC (2009)
[4] Tam, W., et al.: Cem. and Con. Res. 35, 1195–1203 (2005)
[5] Tam, W., et al.: Const. and Build. Mat. 22, 2068–2077 (2008)
[6] Tam, W., et al.: Res. Conserv. and Recy. 50, 82–101 (2007)
[7] ASTM C192, Making and Cur Lab Spec (2000)
[8] ASTM C496, Splitting Tens Str of Cylin (1996)
[9] ASTM C490, Use of App for Determining Length Change (2000)
[10] AASHTO PP34, Std Prac for Est Crack Tendency (1998)
[11] Altoubat, S., Lange, D.: ACI Materials Journal (98), 323–331 (2001)
[12] Gomez-Soberon, A., et al.: ACI Spec. Pub. 209, 461–474 (2002)
[13] Domingo-Cabo, A., et al.: Const. and Build. Mater. 23, 2545–2553 (2009)
[14] Corinaldesi, V., Morinconi, G.: Waste Management 30, 655–659 (2010)

Effect of Reheating Plant Warm SMA on Its Fracture Potential

Zhen Leng[1], Imad L. Al-Qadi[2], Jongeun Baek[3], Matthew Doyen[4], Hao Wang[5], and Steven Gillen[6]

[1] Postdoctoral Research Associate, Department of Civil and Environmental Engineering, University of Illinois at Urbana-Champaign, USA
[2] Founder Professor of Engineering, Department of Civil and Environmental Engineering, University of Illinois at Urbana-Champaign; Director, Illinois Center for Transportation, USA
[3] Research Professor, Sejong University, Korea
[4] Graduate Research Assistant, Department of Civil and Environmental Engineering, University of Illinois at Urbana-Champaign, USA
[5] Assistant Professor, Department of Civil and Environmental Engineering, Rutgers University, USA
[6] Material Manager, Illinois State Toll Highway Authority, USA

Abstract. The primary objective of this study is to evaluate the fracture characteristics of stone mastic asphalt (SMA) with warm mix additives during its curing process. To this end, three SMA mixtures, prepared using different warm mix technologies (i.e. Evotherm additive, Sasobit additive, and foaming process) and one conventional SMA, were evaluated using the semi-circular bending (SCB) test. To investigate the aging effect due to mixture reheating, specimens tested in this study were compacted using both fresh and reheated plant mixtures. Specimens were tested at $-12°C$ at 1 day, 3 days, 7 days, 3 weeks, 6 weeks and 12 weeks after compaction. The fracture energies of tested specimens were determined by calculating the areas underneath the loading-crack mouth opening distance (CMOD) curves obtained from the SCB test. This study revealed that the effect of curing time on the fracture potential of warm SMA is not statistically significant. In addition, reheating mixture for testing specimen preparation increases the fracture potential of both warm SMA and control SMA mixtures. The Evotherm SMA provided greater fracture potential than the control SMA when fresh plant mixtures were used to prepare the SCB test specimens. However, this relation was reversed when the SCB test was performed on specimens prepared using reheated plant mixtures, suggesting the importance of laboratory aging on laboratory test results.

1 Introduction

The use of warm-mix asphalt (WMA) in the U.S. has increased rapidly in recent years due to its significant environmental benefits. Compared to traditional hot-mix asphalt (HMA), WMA production consumes less energy, emits less greenhouse gas, and allows for an extended construction season. Three WMA techniques have been commonly used to facilitate the mixing of binder and aggregates: organic additives, foaming techniques, and chemical additives. Using organic additives and foaming techniques reduce the effective viscosity of binder while chemical additives serve as a surface-active agent to reduce friction at bitumen-aggregate interfaces. However, due to different WMA preparation techniques, the physical and chemical properties of the mixture can be altered, which can result in different mechanical behavior of the mixture. Therefore, many studies have been conducted to evaluate the short-term and long-term performance of mixtures produced with various WMA techniques [1-6]. However, most of these studies have focused on the WMA rutting potential at high temperature, while the WMA low-temperature fracture potential is still not well studied.

2 Research Objective and Scope

The objective of this study is to experimentally characterize the fracture resistance of warm mix with various additives. Specifically, three stone mastic asphalt (SMA) mixtures, prepared using various warm mix technologies (i.e., Evotherm additive, Sasobit additive, and foaming process), and one conventional control SMA, were evaluated using the semi-circular bending (SCB) test. The following research tasks were conducted:

1) Investigated the curing time effect on the fracture potential of the warm and control SMA mixtures.
2) Evaluated the aging effect due to mixture reheating on the fracture potential of the warm and control SMA mixtures.

3 Testing Material and Experimental Plan

3.1 Testing Material

Three WMA techniques were used in this study to produce the warm SMA: Evotherm additive (a chemical additive), Sasobit additive (an organic additive), and foaming process. A typical SMA binder mix that has been used by Chicago area contractors on many large-scale expressway overlay projects was selected as the control mixture. As Table 1 shows, the control SMA and the Evotherm SMA had the same mixture design, which includes PG 64-22 binder with 12% ground tire rubber (GTR) and 8% recycled asphalt pavement (RAP). PG 64-22 binder with 12% GTR was used in the foamed SMA; but it contained 13% RAP. SBS-modified PG 70-22 binder, 5% RAP, and 5% recycle asphalt shingle (RAS) were used in the SMA with Sasobit additive. The compaction temperatures of the three warm SMA's were 15-25 °C lower than that of the control SMA.

Table 1. Composition of Asphalt Mixtures with Various Warm Mix Additives

Mix	N_{des}[a]	NMAS[b] (mm)	Binder	RAP	RAS	Compaction Temp. (°C)	WMA Additive
Control SMA	80	12.5	6.2% PG 64-22 with 12% GTR	8%	NA	152	NA
Evotherm SMA			6.2% PG 64-22 with 12% GTR		NA	127	0.5% of binder
Foamed SMA			6.2% PG 64-22 with 12% GTR	13%	NA	127	1.0% of binder[c]
Sasobit SMA			6.2% PG 70-22 SBS modified	5%	5%	127-137	1.5% of binder

[a] N_{des} = Design number of gyrations
[b] NMAS = Nominal maximum aggregate size
[c] Water was considered as the WMA additive.

3.2 Semi-circular Bending Test and Specimen Preparation

Fracture potential of asphalt concrete (AC) mixtures was characterized based on the fracture energy obtained from the SCB fracture test (Figure 1) [7]. 150-mm-diameter cylindrical specimens were sliced into 50-mm thick cylinders and cut in half along the diameter. A 15-mm notch was cut into each half of the specimen. The test was performed at a temperature of -12°C, which is 10°C warmer than the low-temperature binder grade. The test was conducted at a constant crack mouth opening displacement (CMOD) rate (0.7mm/min), and the load, displacement, and CMOD were recorded. The work of fracture was obtained from the SCB test by calculating the area under the load-CMOD curve using Eqn. (1).

$$W_f = \int P du \tag{1}$$

where W_f is work of fracture, P is applied load and u is CMOD. The fracture energy was then obtained by dividing the work of fracture by the area of the ligament.

Fig. 1. Semi-circular beam fracture test setup

3.3 Testing Plan

One key issue for any laboratory performance test is that the measured mixture property may be affected by the type of mixtures used for preparing the specimen. Using reheated plant mixture may cause additional aging to the mixture, while using lab-mixed materials may not completely represent the field mixtures either. The most appropriate way to represent the curing that occurs in the field is to compact the specimens in the field using fresh plant mixtures without reheating. Another issue which may affect the fracture potential of the warm mix is the testing time after the specimen is compacted. Due to the use of warm mix additives, a time-dependent hardening, called curing, may occur in WMA as asphalt binder in the WMA regains its original viscosity and/or a certain amount of entrapped moisture is evaporated from the WMA.

In this study, the testing plan was designed to consider the effects of both mixture reheating and curing time on the WMA fracture potential. To evaluate the curing time effect on the mixtures' fracture property, SCB specimens were compacted in the plant using fresh mixtures right after they were produced. These specimens were then transported to the Advanced Transportation Research and Engineering Laboratory (ATREL) of the University of Illinois, and tested at 1 day, 3 days, 7 days, 3 weeks, 6 weeks, and 12 weeks after compaction. To investigate the aging effect due to mixture reheating on its fracture potential, loose mixture samples were collected from the plant. These loose samples were reheated and then compacted at ATREL. The SCB specimens prepared using the reheated mixtures were tested at the same curing times as those of the specimens prepared using fresh plant mixtures.

4 Test Results and Discussion

4.1 Air Void Contents of Prepared Specimens

The air void contents of the gyratory-compacted specimens were measured before each test. Table 2 summarizes the measured air void contents of the specimens using fresh and reheated plant mixtures. The data show that the average air void contents are within the range of 6.0±0.5%.

Table 2. Air Voids for Testing Specimens

Mix	Fresh Mix		Reheated Mix	
	Average Air Void (%)	COV[a]	Average Air Void (%)	COV
Control SMA	6.5	5%	6.4	3%
Evotherm SMA	5.9	2%	5.9	2%
Foamed SMA	5.5	2%	5.5	6%
Sasobit SMA	6.0	3%	5.9	4%

[a]Coefficient of Variance.

4.2 Effect of Curing Time on Fracture Potential

Figure 2 shows the determined fracture energy at -12°C for the control SMA mixture and the mixtures containing various warm mix additives using specimens prepared with fresh plant mixtures. Note that the testing results for 3-week and 7-week Evotherm SMA mixtures are missing, because these specimens were damaged during the specimen preparation. Table 3 summarizes the variation of fracture energy due to curing time for various SMA mixtures.

Table 3. Variation of Fracture Energy Due to Curing Time (Fresh Plant Mix)

Fracture Energy (J/m^2)	SMA Mixture			
	Control	Evotherm	Foam	Sasobit
Average	1275.9	1137.9	1206.9	896.6
COV	6%	3%	12%	6%

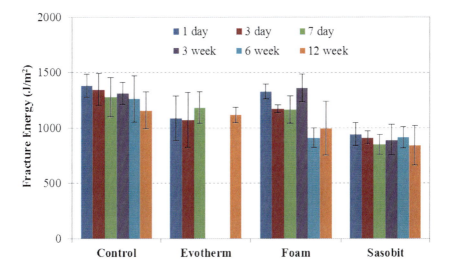

Fig. 2. Fracture energy of various fresh SMA mixtures

The test results were statistically analyzed using the Statistical Analysis System (SAS) program. A Fisher LSD (Least Significant Difference) test was performed at a significant level of 0.05. The statistical significance of the changes in the fracture energy as a function of curing time was analyzed. The test results were ranked using letters and the letter was changed when the mean was statistically different from others. The letter A was assigned to the best performer followed by the other letters in alphabetic order. A double letter, such as A/B, indicates that the difference in the means is not statistically significant and that the results could fall in either group.

Table 4 presents the effect of curing time on the fracture energy when the data are clustered for all mixture types. The results show that there is relatively small or no significant difference in fracture potential as a function of curing time. As the curing time increases, the fracture potential slightly increases. Except for 6 weeks, the fracture potential at the other five curing periods can be considered statistically same.

Table 4. Fisher LSD Test Results for the Effect of Curing Time

Curing time					
1day	3day	7day	3week	6week	12week
A	A/B	A/B	A/B	B	A/B

4.3 Effect of Aging on Fracture Potential due to Mixture Reheating

In this study, the loose mixtures collected from the asphalt plant were reheated in the laboratory to investigate the influence of sample reheating on fracture potential. The reheating process can artificially age the mixture because chemical reactions may take place in the reheating process regardless of the duration of heating. The reheated specimens were tested at the same curing times as those of the specimens without reheating. Figure 3 presents the fracture energy for the control SMA mixture and the mixtures containing various warm mix additives using specimens prepared with reheated plant mixtures. Table 5 summarizes the variation of mixture properties due to curing time for various SMA mixtures after reheating. It is clear that fracture energy has been reduced for all mixes due to the lab heating.

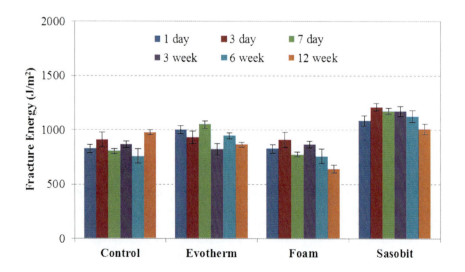

Fig. 3. Fracture energy of various reheated SMA mixtures

Table 5. Variation of Fracture Energy Due to Curing Time (Reheated Plant Mix)

Fracture Energy (J/m²)	SMA Mixture			
	Control	Evotherm	Foam	Sasobit
Average	862.1	931.0	1137.9	793.1
COV	9%	9%	7%	12%

An aging ratio was used to quantify the extent of binder hardening effect on mixtures' fracture potential. The aging ratio was calculated as the ratio of the fracture energy of the reheated mixtures with respect to that of the fresh mixtures. The average fracture energy at different curing times for each mixture with and without reheating was used to calculate the aging ratio. Figure 4 compares the aging ratios due to sample reheating for various SMA mixtures. The results show that the reheating process reduces the fracture resistance of all four mixtures. This is expected because the viscosity of binder could increase significantly during the reheating process and the binder becomes stiffer and more brittle. It was also found that the effect of reheating is more significant for the control mixture, compared to the mixtures containing warm mix additives. This is probably because the reheating temperature for the control mixture (152°C) is higher than those for the mixtures containing warm mix additives (127 to 137°C), which are corresponding to their compaction temperatures.

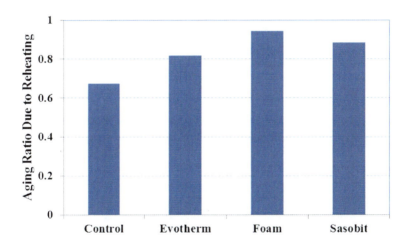

Fig. 4. Aging ratios of mixture properties due to reheating

4.4 Performance Comparison between Mixtures

Since the foamed SMA and Sasobit SMA had different mixture components than the control SMA, their performances were not directly compared. Instead, this paper focuses on the performance comparison between the control SMA and Evotherm SMA, which had the same job mix formula.

Fisher LSD test was conducted to evaluate the effect of fresh mixture type on the fracture potential for the control SMA and Evotherm SMA. The result shows that compared to control mixture, the mixture containing Evotherm had statistically higher fracture potential. This could be due to the effect of the less aged binder in the asphalt mixture containing Evotherm, because these two mixtures shared the same mixture components except for the Evotherm additive.

On the other hand, Fisher LSD results show that the order of the control SMA and Evotherm SMA was reversed after the mixture reheating. As indicated earlier, this could be related to the fact that reheating temperature for the control SMA mixture was greater than that of the Evotherm SMA mixture. Therefore, different aging effects due to variation in mixture reheating temperatures must be taken into account when comparing mixture fracture potentials for hot and warm mixtures using reheated samples. Otherwise, the performance of the warm and hot mixtures could be erroneously presented.

5 Summary

This study characterized the fracture potential of warm SMA at various curing times using both fresh and reheated plant mixtures. The following summarizes the findings of this study:

- The effect of curing time on the fracture potential of warm SMA is not statistically significant.
- Reheating mixture for testing specimen preparation increases the fracture potential of both warm SMA and control SMA mixtures. Due to the relatively higher reheating temperature, the reheating effect is more pronounced for the control SMA mixture compared to the warm SMA mixture.
- For the two mixtures with the same job mix formula, the Evotherm SMA provided greater fracture potential than the control SMA when fresh plant mixtures were used to prepare the SCB test specimens. However, this relation was reversed when the SCB test was performed on specimens prepared using reheated plant mixtures. This suggests that lab reheating temperature is important. Higher temperature used for control SMA compared to warm SMA caused significant greater aging to the mixture and hence lower fracture energy.
- A further study using lab-mixed mixture to prepare SCB test specimens is currently underway.

Acknowledgement. This publication is based on the results of the Illinois Tollway Project, Short-Term Performance of Modified Stone Matrix Asphalt (SMA) Mixes Produced with Warm Mix Additives. This project was conducted in cooperation with the Illinois Center for Transportation and the Illinois Tollway. The contents of this study reflect the views of the authors, who are responsible for the facts and the accuracy of the data presented herein. The contents do not necessarily reflect the official views or policies of the Illinois Tollway. The assistance of the research engineers, Jeffrey Kern and James Meister, and graduate students, Ibrahim Abuawad, Sarfraz Ahmed, Qazi Aurangzeb, Seonghwan Cho, Khaled Hasiba, Jaime Hernandez-Urrea, Brian Hill, Hasan Ozer, Alejandro Salinas, Pengcheng Shangguan, and Songsu Son, at the Advanced Transportation Research and Engineering Laboratory of University of Illinois, in the laboratory testing, is greatly appreciated.

References

[1] Al-Qadi, I.L., Leng, Z., Baek, J., Wang, H., Doyen, M., Gillen, S.L.: Early-age Performance Characterization of Evotherm SMA: Laboratory Testing Using Plant-compacted Specimens and On-site Stiffness Measurement. Transportation Research Record: Journal of the Transportation Research Board, TRB (accepted 2012)

[2] Diefenderfer, S., Hearon, A.: Laboratory Evaluation of a Warm Asphalt Technology for Use in Virginia, FHWA/VTRC 09-R11. Virginia Transportation Research Council (2008)

[3] Hurley, G., Prowell, B.D.: Evaluation of *Sasobit*® for Use in Warm Mix Asphalt. NCAT Report 05-06. National Center for Asphalt Technology, Auburn University, Auburn (2005)

[4] Hurley, G., Prowell, B.D.: Evaluation of Evotherm® for Use in Warm Mix Asphalt, NCAT Report 06-02. National Center for Asphalt Technology. Auburn University, Auburn, Alabama (2006)

[5] Prowell, B.D., Hurley, G.C., Crews, E.: Field Performance of Warm-Mix Asphalt at National Center for Asphalt Technology Test Track. Transportation Research Record: Journal of the Transportation Research Board, No.1998, pp. 96–102 (2007)

[6] Xiao, F., Amirkhanian, S.N., Putman, B.: Evaluation of Rutting Resistance in Warm-Mix Asphalts Containing Moist Aggregate. Transportation Research Record: Journal of the Transportation Research Board, No.2180, pp. 75–84 (2010)

[7] Li, X., Marasteanu, M.: Evaluation of the Low Temperature Fracture Resistance of Asphalt Mixtures Using the Semi-Circular Bend Test. Journal of the Association of Asphalt Paving Technologists 74, 401–426 (2004)

Fatigue Cracking Characteristics of Cold In-Place Recycled Pavements

Andreas Loizos, Vasilis Papavasiliou, and Christina Plati

National Technical University of Athens (NTUA), Greece

Abstract. For the rehabilitation of heavily damaged pavements, Cold In-Place Recycling (CIR) offers an attractive alternative, in comparison to other pavement rehabilitation options. One issue, however, that still has limited research information available is the fatigue cracking characteristics of CIR pavements, especially in regards to the in situ behavior of heavy-duty pavements. For this reason, a field experiment was conducted on a CIR heavy-duty highway pavement that utilized the foamed asphalt technique for rehabilitation. In order to achieve this goal, a systematic monitoring of pavement performance and a data analysis research study was performed by the Laboratory of Highway Engineering of the National Technical University of Athens (NTUA) for approximately one year on the in service pavement. The present study focuses on the in situ estimation of the fatigue cracking characteristics / tendencies of CIR pavements, using in situ Non Destructive Tests (NDTs) and facilitated with advanced analysis tools. The data analysis results, indicate improved fatigue cracking characteristics of the cured foamed asphalt recycled material and a low fatigue cracking tendency of the asphalt concrete overlay.

1 Introduction

As the cost of hot mixed asphalt mixtures continuously increases and the availability of good materials is limited, Cold In-Place Recycling (CIR) has gained ground as an attractive alternative to other pavement rehabilitation options. CIR is an advantageous rehabilitation technique, as it is ideally suited for the reworking of the upper layers of distressed pavements [1]. Among the multiple cold recycling systems available, the foamed asphalt (FA) technique [2] has gained popularity in recent years for its efficient use of salvaged construction materials.

The majority of research related to the specific technique, until recently, has concentrated on material characterization and mix designs performed in the laboratory. In recent years however, research has also began to focus on the characterization of the in situ behavior of FA mixtures and CIR pavements, including strain response analysis based on field collected data.

The research done by [3] on the behavior of FA treated materials based mainly on Heavy Vehicle Simulator (HVS) tests showed that the resilient modulus of the FA treated base layer starts at a relatively high value and then decreases under the action of traffic until a constant resilient modulus is reached. In contrast to the above mentioned research, more recent analysis results of Non Destructive Tests (NDT) on a heavy-duty CIR recycled pavement using the foamed asphalt technique [4], showed high modulus values of the cured FA material and no obvious evidence of reduction with time and traffic. Taking into account that the recycled layer plays the role of the base / subbase, an adequate structural condition (expressed by the modulus of the recycled material) is needed, in order to reduce the induced critical stains. Moreover, due to the fact that in heavy-duty CIR pavements the thickness of the asphalt concrete (AC) overlay is lower in comparison with conventional flexible pavements, it is very important to investigate the in situ fatigue cracking tendency of the AC overlay for the performance of the CIR pavement.

For this reason a respective field experiment was undertaken by the NTUA Laboratory of Highway Engineering on sections of a heavily used Greek highway that was rehabilitated with CIR utilizing foamed asphalt as a stabilization agent. A comprehensive research study was performed involving monitoring the performance approximately one year of the in service pavement using NDTs and related data analysis tools. The fatigue cracking tendency of the AC overlay mix for the design period was investigated by calculating the in situ critical strains after the curing of the recycled material and using the appropriate fatigue law. The main findings of the data analysis concerning the conducted field experiment are presented and discussed in the present research work.

2 Test Sections and Mix Design Characteristics

The experiment was performed along three test sections hereafter referred to as S1, S2 and S3, respectively. Due to the fact that the base and the subbase of the pavement were initially constructed from a cement bound material (CBM) with a nominal thickness of 38 cm, the FA recycled layer is based on a CBM remaining layer.

Prior to the CIR implementation, foamed asphalt mix designs were undertaken on several blends of material recovered from test pits. These blends were treated with foamed bitumen using the appropriate laboratory unit and several briquettes were manufactured for testing purposes to determine the indirect tensile strength (ITS), the unconfined compressive strength (UCS), the cohesion (c) and the angle of internal friction (Φ), as well as the determination of the indirect tensile stiffness modulus (ITSM). The aim of the mix design was to establish the application rates for foamed asphalt and active filler (cement), in order to achieve optimal strengths and to determine the strength characteristics of the mix.

According to the mix design, 3% foamed bitumen (from 80/100 Pen grade bitumen) and 1% ordinary Portland cement as an active filler was used for the composition of the FA mix. The decision to introduce 1% cement was based on the improvement in the achieved soaked strengths. Since milling machines produce few fines, it was decided to introduce 30% (by volume) natural fine sandy material to blend with the recovered material. For analytical pavement design purposes the maximum stiffness modulus was considered to be 3000 MPa, which is equal to the maximum expected (at the design stage) moduli of the FA layer, as estimated from ITSM laboratory [5] results on the FA specimens.

The AC overlay was comprised of two layers, a 50 mm binder course (a dense AC mix) and a 40 mm semi-open graded surface course. For analytical pavement design purposes the stiffness modulus of the AC overlay (composite modulus of both binder and surface courses) was considered to be 3000 MPa. The profile of the investigated pavement is presented in Figure 1.

3 Field Data Collection and Analysis

Non Destructive Tests (NDT) were conducted approximately one year after the sections were opened to traffic, in order to evaluate the pavement after the curing of the FA material [4].

In situ NDT using the Falling Weight Deflectometer (FWD) were performed along the outer wheel path of the heavy traffic lane at nine, twelve and thirteen specific test points of the test sections S1, S2 and S3 respectively.

The GPR system of the NTUA Laboratory of Highway Engineering [6] was utilized to gather the data required for the analysis to estimate the thickness of the recycled pavements. This data is useful for the backanalysis procedures. The system used, is appropriate for the evaluation of the upper part of the pavement structure since it produces reliable information to an approximately 0.70 m penetration depth [7]. The system follows the principles of [8] and is supported by the appropriate software [9].

A detailed backanalysis was performed using the Elmod software [10]. Considering the level of the subgrade at the bottom of the CBM layer, the backanalysis model consisted of four layers. For backanalysis purposes, the thicknesses of all layers were estimated using the GPR analysis results. Figure 1 shows the pavement modelling for the backanalysis, including the loading condition parameters (R, P) and the layers characterization through the parameters of the layer properties i.e. moduli (E_i), thickness (h_i) and Poisson ratio (v_i).

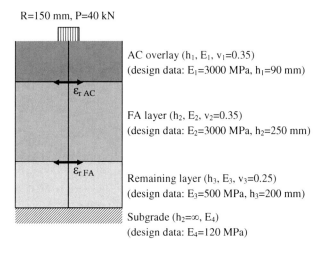

Fig. 1. Pavement modelling for structural pavement analysis (moduli, strains)

The horizontal (tensile) strains at the bottom of the FA layer (ε_r FA), as well as at the top of the FA layer, i.e. bottom of the AC layers (ε_r AC) (see Figure 1), were calculated using a multi-layer linear elastic analysis [11]. These locations were selected, taking into account the results of the strain response analysis based on back-calculated moduli at characteristic locations within the body of the recycled layer in a similar pavement structure, which is further detailed in [12]. Tensile strains at these locations are considered to be critical in regards to possible fatigue failure and are directly related to the performance of the pavement structure. The load used for the calculations was a 40 kN single wheel with a 150 mm radius. The backcalculated moduli as well as the GPR-determined thicknesses of the different pavement layers were also used for the analysis. For analysis purposes, the modulus of the AC overlay in terms of the composite modulus of both binder and surface courses (as determined from the backanalysis) was considered. Strains were also calculated using data from the analytical pavement design following the related pavement model (Figure 1).

A laboratory fatigue test (ITFT) was performed in the laboratory on characteristic dense AC mixes used in Greece, similar to the one used in the AC overlay. The fatigue characteristics of the AC material, expressed by the relevant fatigue law, were further used in the analysis, in order to estimate the fatigue cracking tendency of the AC overlay.

4 Analysis Results

4.1 GPR-Determined Layer Thicknesses and Backcalculated Moduli

The GPR analysis confirmed, more or less, the design thickness data. More specifically, according to the GPR analysis results, the AC overlay of the three test sections were on average 19 mm to 35 mm thicker than the design value (90 mm). The GPR-determined FA layer thicknesses averages ranged from 247 mm to 275 mm (design value: 250 mm). The thicknesses of the remaining CBM layer averages ranged from 115 mm to 200 mm (design value: 200 mm). The GPR-determined average thicknesses of the test sections S1, S2 and S3 and the design data are presented graphically in Figure 2.

The temperature measured at the mid depth of the bituminous overlay during the related monitoring ranged between 29 and 31°C. Taking this fact into account, no adjustment of the backcalculated AC and FA moduli was conducted. Figure 3 illustrates the average backcalculated moduli of the AC, FA and remaining CBM layers for the test sections S1, S2 and S3. It must be noted that limited outliers (very high moduli) were discarded from the analysis, in order to eliminate some inaccurate modulus estimations. The coefficient of variation (CoV) of the moduli values ranged for the AC from 24% to 29%, for the FA from 42% to 47% and for the CBM from 28% to 50%. From the analysis results, the following conclusions were drawn.

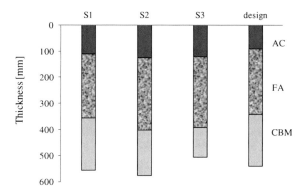

Fig. 2. GPR-determined average thicknesses and design data

During monitoring approximately one year after construction, the average backcalculated moduli were much greater than the related pavement design values. This is probably an indication of completion of the curing procedure of the recycled material, also referred to in [13].

The very high average moduli values of the FA layer and the remaining CBM layer (especially for test section S3) is an indication of a stiff recycled pavement structure.

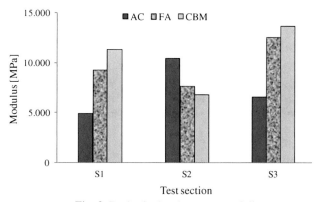

Fig. 3. Backcalculated average moduli

As shown in Figures 2 and 3, the "soil support" of the FA layer (i.e. the remaining CBM layer) of the three test sections differs in thickness and stiffness. In addition, the backcalculated AC and FA moduli were also varied. The variation of the moduli for the three test sections may result in a different behaviour, concerning the tendency of fatigue cracking of the AC overlay and FA layers. This was investigated, taking into account the strains induced in the body of the CIR pavement.

4.2 Calculated Tensile Strains

The average tensile strains were calculated at the bottom of the AC overlay (ε_r AC) and at the bottom the FA layer (ε_r FA), (see Figure 1) of the test sections S1, S2 and S3, respectively. The tensile strains at the same locations in the body of the CIR pavement were also calculated using the pavement design data. The results are presented graphically in Figure 4.

According to the analysis results, at the bottom of the AC overlay the average tensile stains were lower than the calculated one using data from the pavement design. At the bottom of the FA layer the average tensile stains were much lower than the calculated strain using data from the pavement design.

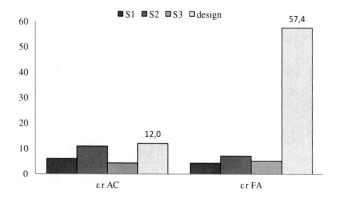

Fig. 4. Calculated average tensile strains

The very low values of the tensile strain at the bottom of the FA layer (ε_r FA), is an indication that fatigue cracking is not critical for the recycled material. However, the stiff pavement structure could result in an increased tendency for reflection of possible cracks coming from the surface of the remaining CBM layer. The latest comment may be supported through the results of a research work done by [14]. The strains at the bottom of the AC overlay (ε_r AC) were also lower than the relative one taking into account the pavement design data. More precise results and conclusions concerning the "damage" and consequently the tendency of fatigue cracking of the AC overlay were drawn, taking into account the analysis results using the in situ fatigue characteristics of the AC overlay.

4.3 In Situ Fatigue Characteristics of the AC Overlay

The fatigue characteristics of the AC material can be expressed by an equation with the general form (1).

$$N_{AC} = a * \varepsilon_{r\,max}^{b} \tag{1}$$

Where
- N_{AC}: Cycles to failure of the AC material
- $\varepsilon_{r\,max}$: Maximum (critical) tensile strain
- a and b: Coefficients determined from laboratory tests

Each traffic load causes a certain amount of damage to the pavement structure that accumulates over time and eventually leads to pavement failure. The damage per pass, or the relative damage (d_{rel}) of a pavement material represents the damage caused by a standard axle load and it is expressed by the following equation (2).

$$d_{rel} = \frac{1}{N} \qquad (2)$$

Where
- N: Cycles to failure of the material

The in situ fatigue cracking tendency of the AC overlay was estimated using the damage ratio (d_{ratio}), i.e. comparing the relative damage (d_{rel}) using both the design data and the in situ collected data after the curing of the recycled material (1 year). The d_{ratio} was calculated using the following equation (3).

$$d_{ratio} = \frac{d_{rel}(design)}{d_{rel}(field)} = \frac{N_{field}}{N_{design}} = \left(\frac{\varepsilon_{r\,field}}{\varepsilon_{r\,design}}\right)^b \qquad (3)$$

Where:
- d_{ratio}: Damage determined from recycled pavement design data, in comparison with the damage determined from in situ collected data (1 year)
- d_{rel}(field): Relative damage of the AC overlay determined from in situ collected data (1 year)
- d_{rel}(design): Relative damage of the AC overlay determined from pavement design data
- N_{field}: Cycles to fatigue failure of the AC overlay determined from in situ collected data (1 year)
- N_{design}: Cycles to fatigue failure of the AC overlay taking into account the recycled pavement design data
- $\varepsilon_{r\,field}$: Maximum (critical) tensile strain determined from in situ collected data (1 year)
- $\varepsilon_{r\,field}$: Maximum (critical) tensile strain determined from pavement design data
- b: Coefficient defined from laboratory fatigue tests (b=-4.78)

The damage ratio (d_{ratio}) was calculated at each test section S1, S2 and S3, respectively, using equations (1), (2) and (3). The average tensile strains at the bottom of the AC overlay (ε_r AC) were used for the analysis (see Figure 4). The results (considered as average d_{ratio} values) are presented in Table 1.

Table 1. Damage ratio of the AC overlay determined from recycled pavement design data, in comparison with damage determined from in situ collected data

	d_{ratio}
Test section S1	28
Test section S2	1.6
Test section S3	179

According to the results presented in Table 1 and considering equation (3), it can be concluded that the relative damage of the AC overlay is much lower than the relative damage using the design data. Consequently, severe fatigue cracking problems are not expected for the AC overlay. Moreover, the variation of the d_{ratio} values is indicative for the different fatigue cracking behaviour of the AC overlay for the three test sections.

5 Conclusions

The present research study has attempted to contribute towards providing increased information regarding the fatigue cracking characteristics of cold in-place recycled (CIR) pavements using the foamed asphalt (FA) technique, with a focus on the fatigue cracking tendencies of the AC overlay. The main findings and discussion points are the following:

The in situ critical strains in the body of the FA layer were much lower than the maximum calculated tensile stain using data from the pavement analytical design. This fact is an indication of improved fatigue cracking characteristics of the recycled material.

The relative damage of the AC overlay was much lower than the relative one using the design data. Consequently, low fatigue cracking tendency of the AC overlay is expected in terms of the performance of the CIR pavement.

Acknowledgments. The authors would like to thank the Greek Ministry of Public Works and the involved national and international bodies for supporting the research work of this research study.

References

[1] Wirtgen, Wirtgen cold recycling manual, Wirtgen GmbH, Windhagen, Germany (2004)
[2] Asphalt Academy: The Design and Use of Foamed Bitumen Treated Materials, Interim Technical Guideline 2, 1st edn., Pretoria (2002)
[3] Long, F., Theyse, H.: Mechanistic-Empirical Structural Design Models for Foamed and Emulsified Bitumen treated Materials. In: Proceedings of the 8th Conference on Asphalt Pavements for Southern Africa (CAPSA 2004), Sun City, pp. 553–567 (2004)

[4] Papavasiliou, V., Loizos, A.: Field Behavior of Foamed Bitumen Pavement Material. In: Proceedings of the 7th International RILEM Symposium on Advanced Testing and Characterization of Bituminous Materials, Rhodes (2009)
[5] ASTM, Standard Test Method for Indirect Tension for Resilient Modulus of Bituminous Mixtures, D4123-82, Pennsylvania (2006)
[6] Geophysical Survey Systems Inc.: RADAN for Windows NT, Version 4.0, User's Manual, North Salem, New Hampshire (2002)
[7] Saarakento, T.: Use of Ground Penetrating Radar in relation with FWD. In: FWD / Backanalysis Workshop, Proceedings of the 6th International Conference of the Bearing Capacity of Roads, Railways and Airfields (BCRA), Lisbon, pp. 24–26 (2002)
[8] ASTM, Standard Test Method for Determining the Thickness of Bound Pavement Layers Using Short-Pulse Radar Non-Destructive Testing of Pavement Structures, D4748, Pennsylvania (2006)
[9] RoadScanners: Road Doctor Software. Version1.1, User's Guide, Rovaniemi, Finland (2001)
[10] Dynatest, ELMOD: Pavement Evaluation Manual (2001)
[11] BISAR, User Manual. Bitumen Business Group (1998)
[12] Loizos, A., Papavasiliou, V., Plati, C.: Early Life Performance of Cold-In Place Pavement Recycling Using the Foamed Asphalt Technique. In: Transportation Research Record: Journal of the Transportation Research Board, No. 2005, pp. 36–43 (2007)
[13] Loizos, A., Papavasiliou, V., Plati, C.: Aspects Concerning Field Curing Criteria for Cold-In Place Asphalt Pavement Recycling, CD Compedium. In: 87th Annual Meeting, Transportation Research Board, Washington, D.C. (2008)
[14] Molenaar, A.A.A., Leewis, M.: Design of Sand-Cement Base Courses Using Fracture Mechanics Principles. In: Proceedings of the 6th International Symposium on Concrete Roads, Madrid, pp. 103–113 (1990)

Author Index

Abuawad, I.M. 1271
Abu-Qtaish, L. 1125
Ahmed, S. 409
Airey, G.D. 347, 643, 783, 793
Albayati, A.H. 921
Albert, J.R. 307
Aliha, M.R.M. 359
Allen, B. 953, 975
Allou, F. 675
Almeida, A. 379
Al-Qadi, I.L. 1241, 1271, 1341
Al-Rub, R.A. 399
Ambassa, Z. 675
Ameri, M. 359
Amorim, E.F. 115
Anderson, M. 453
Angellier, N. 771
Arsenie, I.M. 653
Artamendi, I. 953, 975
Artières, O. 201
Aurangzeb, Q. 1271
Ayatollahi, M.R. 359

Baaj, H. 1261
Bacchi, M. 201
Baek, C. 465
Baek, J. 1241, 1341
Bahia, H. 147
Bakhshi, M. 635
Balay, J.M. 697
Baldo, N. 719, 729
Bali, A. 21
Banadaki, AD. 487
Bańkowski, W. 1029

Bastard, N. 1251
Baumgardner, G.L. 901
Belogi, L. 1229
Bernucci, L. 1189
Bhasin, A. 1115
Bianchini, P. 201
Biligiri, K.P. 751
Birgisson, B. 103, 299, 1281
Blankenship, P. 453
Bodin, D. 697
Botella, R. 61
Boulanouar, A. 21
Braham, A.F. 997
Brake, N.A. 571
Breitenbücher, R. 581
Breysse, D. 697
Brill, D.R. 337
Buannic, M. 805
Buehler, M.J. 911
Buonsanti, M. 1219
Buss, A. 1323
Butt, A.A. 299
Buttlar, W.G. 287, 409

Canestrari, F. 1229
Casey, D.B. 347
Castaneda, D.I. 191
Chabot, A. 51, 1017, 1251
Chatti, K. 571
Chazallon, C. 653
Chen, J. 561
Chen, L. 255
Chen, Y. 879
Cheung, L.W. 277

Cocurullo, A. 783, 793
Collop, A.C. 93, 347, 643
Crispino, M. 1199

da Motta, L.G. 1189
Das, A. 1039
Das, P.K. 1281
Dave, E.V. 409
De Bondt, A.H. 327, 941
de Lara Fortes, A.C. 115
Denolf, K. 1169
De Visscher, J. 1169
de Vos, E.R. 1059
Diakhaté, M. 771
Di Benedetto, H. 223, 665, 805, 1261
Doligez, D. 653
Dondi, G. 389
Doré, G. 1017
Dortland, G. 201
Doyen, M. 1341
Dragomir, M.L. 1209
Duchez, J.L. 653

Ech, M. 1261
Eko, R.M. 675
Elseifi, M. 1
Erkens, S. 889, 1091
Eslaminia, M. 497

Falchetto, A.C. 11
Farrar, M.J. 233
Fauchard, C. 179
Feng, D. 561
Fernandes, A. 607
Ferreira, A. 429
Filho, P.L. de O. 245
Fini, E.H. 911
Fontana, P. 591
Fritzen, M. 1189

Gallet, T. 805
Gao, Y. 527
Gelpke, R. 475
Ghatak, A. 1039
Gideon, C.S. 1069
Gillen, S. 1341
Gomes, A.M. 625
Grellet, D. 1017
Grenfell, J.R. 347, 643, 783, 793

Grzybowski, K. 125
Guddati, M.N. 465, 487, 497
Guilbert, V. 179
Guo, E.H. 337

Habib, N.Z. 859
Hagos, E.T. 849
Hakim, H. 751
Hakimzadeh, S. 287
Hammoum, F. 51
Hanson, D. 453
Harnsberger, P.M. 233
Harvey, J. 537
Hasan, M.R.M. 71
Hébert, J.F. 157
Henschen, J. 1333
Hesami, E. 103
Homsi, F. 697
Hopman, P.C. 941
Hornych, P. 179, 201, 1017, 1251
Houben, L. 369
Hribar, D. 1049
Huang, Y. 419
Hun, M. 51
Huurman, M. 475
Hyzl, P. 1181

Iliescu, M. 1209
Ishihara, S. 317

Jacobs, M.M.J. 475, 941
Jean-Michel, P. 1017
Jelagin, D. 103, 299, 1281
Jenkins, K. 1007
Jiang, C. 169
Joutang, J.-L. 1251
Judycki, J. 41

Kaganovich, E. 211
Kaloush, K. 751
Kamaruddin, I. 859
Karadelis, J.N. 549
Kaya, S. 1125
Kebede, N.A. 287
Kerzrého, J.-P. 179, 1017, 1251
Khedoe, R. 941
Kiani, B. 635
Kim, M. 1

Kim, Y.R. 465, 487, 497
King, G. 453
Kluttz, R. 687
Komiyama, M. 859
Kong, P.K. 277
Kringos, N. 103, 299, 1281
Krishnan, J.M. 1069
Krysiński, L. 1029

Lange, D.A. 191, 1333
Larcher, N. 771
Laurent, J. 157
Leegwater, G.A. 889, 1091
Lefebvre, D. 157
Lei, Z. 1291
Leng, Z. 1341
Leonardi, G. 1219
Leung, G.L.M. 277
Li, M. 527
Li, N. 827
Li, X. 561
Little, D.N. 399, 487, 1115
Liu, Q. 1081
Livneh, M. 761
Liyan, S. 1291
Loizos, A. 1351
Lopp, G. 879
Luding, S. 1103
Lun, J. 1291
Lura, P. 591

MacRae, G. 815
Maggiore, C. 643
Magnanimo, V. 1103
Malárics, V. 507
Manganelli, G. 389
Mansourian, A. 359
Marasteanu, M.O. 11
Marques, M.J. 601, 607
Martínez, A. 61
Masad, E. 399
Matintupa, A. 137
McCarthy, L.M. 307
Medina, V.A. 625
Mensching, D.J. 307
Merine, G.M. 849
Micaelo, R. 429
Millien, A. 1209
Minelli, F. 615

Minhoto, M. 441, 1157
Miró, R. 61
Mitiche-Kettab, R. 21
Mohajeri, M. 1311
Mohammad, L.N. 1
Molenaar, A.A.A. 687, 739, 827, 1135, 1199, 1311
Mondal, S. 1039
Monteiro, L.V. de A. 245
Montestruque, G. 1189
Motamed, A. 1115
Mudford, C.J. 997
Müller, H.S. 507
Muraya, P.M. 83

Najd, A. 1145
Napiah, M. 859
Nazzal, M. 1125
Nguyen, Q.T. 665

Ogundipe, O.M. 93
Olard, F. 223, 837

Pais, J. 441, 1157
Palvadi, S. 1115
Papavasiliou, V. 1351
Pasetto, M. 719, 729
Pasquini, E. 1229
Pauli, A.T. 233
Pérez-Jiménez, F. 61
Petho, L. 267
Petit, C. 675, 771, 1209
Petretto, F. 31, 1301
Pettinari, M. 31, 1301
Phillips, P. 953, 975
Picariello, F. 31
Pine, W.J. 1271
Pinheiro-Alves, M.T. 601, 607
Pirskawetz, S. 591
Plati, C. 1351
Plizzari, G.A. 615
Poot, M.R. 1199
Pouget, S. 223
Pradena, M. 369
Pramesti, F.P. 739
Prince, S. 125
Pronk, A.C. 827
Pszczoła, M. 41

Qian, Z.   255
Qiu, J.   1135

Rahmani, T.   635
Reggia, A.   615
Reinke, G.H.   901
Ribeiro, A.B.   601, 607, 625
Ribeiro, L.F.M.   115
Richardson, J.   93
Rolim, A.L.   245
Roque, R.   879
Rowe, G.M.   125, 901
Ruot, C.   805

Saarenketo, T.   137
Said, S.   751
Saleh, M.F.   517, 815
Sangiorgi, C.   31, 1301
Santos, L.P.   379
Sauzéat, C.   223, 665, 805, 1261
Savard, Y.   157
Scarpas, T.   687
Schlangen, E.   1081, 1135
Scholten, E.   687
Scopelliti, F.   1219
Scott, A.   815
Sequeira, A.R.   601
Sha, A.   985
Shafiee, M.H.   869
Shatnawi, S.   441, 1157
Shekarchizadeh, M.   635
Shinohara, Y.   317
Siekmeier, J.   707
Sievering, C.   581
Simard, D.   837
Simone, A.   389, 1301
Simonin, J.-M.   179
Sohm, J.   1251
Souliman, M.   751
Souza, H.N.C.   245
Souza, R.   429
Stehlik, D.   1181
Stempihar, J.   751
Sudyka, J.   1029
Sybilski, D.   1029

Tabatabaee, H.   147
Tabatabaee, N.   869
Takarli, M.   771

Tan, I.M.   859
Tanaka, K.   317
Tapsoba, N.   1261
Tebaldi, G.   11, 879
Teltayev, B.   211
Teramoto, A.   1333
ter Huerne, H.L.   1103
Themeli, A.   653
Thirunavukkarasu, S.   465, 497
Thodesen, C.   83
Thom, N.H.   93
Toth, C.   267
Trepanier, J.S.   1271
Trichet, S.   179, 1251
Tsai, Y.   169, 419
Tu, S.   985
Turos, M.I.   11
Tutumluer, E.   707

Underwood, B.S.   465

Valdés, G.   61
van Bochove, G.   1081
van de Ven, M.F.C.   71, 739, 827, 849, 1081, 1135, 1311
van Dommelen, A.   1091
Vanelstraete, A.   1169
van Montfort, J.   1081
van Vliet, D.   889, 1091
Varaus, M.   1181
Varin, P.   137
Velasquez, R.   147
Veloso, L.A.C.M.   245
Vignali, V.   389
Vismara, S.   1199
Voskuilen, J.L.M.   71

Walther, A.   963
Wang, H   1341
Wang, Z.   169
Ward, C.   953, 975
Wendling, L.   1209
Williams, R.C.   1323
Willis, J.R.   687
Wistuba, M.   963
Wong, W.G.   277
Wu, R.   537
Wu, S.   827, 1135

Author Index

Xiao, Y.   707
Xu, Y.   549

Yeow, T.   815
Yin, H.   337
Yiqiu, T.   1291

Yotte, S.   697
Yusoff, N.I.M.   783, 793

Zeiada, W.   751
Zheng, C.C.   1145
Zhong, Y.   527

RILEM Publications

The following list is presenting our global offer, sorted by series.

RILEM PROCEEDINGS

PRO 1: Durability of High Performance Concrete (1994) 266 pp., ISBN: 2-91214-303-9; e-ISBN: 2-35158-012-5; *Ed. H. Sommer*

PRO 2: Chloride Penetration into Concrete (1995) 496 pp., ISBN: 2-912143-00-4; e-ISBN: 2-912143-45-4; *Eds. L.-O. Nilsson and J.-P. Ollivier*

PRO 3: Evaluation and Strengthening of Existing Masonry Structures (1995) 234 pp., ISBN: 2-912143-02-0; e-ISBN: 2-351580-14-1; *Eds. L. Binda and C. Modena*

PRO 4: Concrete: From Material to Structure (1996) 360 pp., ISBN: 2-912143-04-7; e-ISBN: 2-35158-020-6; *Eds. J.-P. Bournazel and Y. Malier*

PRO 5: The Role of Admixtures in High Performance Concrete (1999) 520 pp., ISBN: 2-912143-05-5; e-ISBN: 2-35158-021-4; *Eds. J.G. Cabrera and R. Rivera-Villarreal*

PRO 6: High Performance Fiber Reinforced Cement Composites (HPFRCC 3) (1999) 686 pp., ISBN: 2-912143-06-3; e-ISBN: 2-35158-022-2; *Eds. H.W. Reinhardt and A.E. Naaman*

PRO 7: 1st International RILEM Symposium on Self-Compacting Concrete (1999) 804 pp., ISBN: 2-912143-09-8; e-ISBN: 2-912143-72-1; *Eds. Å. Skarendahl and Ö. Petersson*

PRO 8: International RILEM Symposium on Timber Engineering (1999) 860 pp., ISBN: 2-912143-10-1; e-ISBN: 2-35158-023-0; *Ed. L. Boström*

PRO 9: 2nd International RILEM Symposium on Adhesion between Polymers and Concrete ISAP '99 (1999) 600 pp., ISBN: 2-912143-11-X; e-ISBN: 2-35158-024-9; *Eds. Y. Ohama and M. Puterman*

PRO 10: 3rd International RILEM Symposium on Durability of Building and Construction Sealants (2000) 360 pp., ISBN: 2-912143-13-6; e-ISBN: 2-351580-25-7;
Eds. A.T. Wolf

PRO 11: 4th International RILEM Conference on Reflective Cracking in Pavements (2000) 549 pp., ISBN: 2-912143-14-4; e-ISBN: 2-35158-026-5; *Eds. A.O. Abd El Halim, D.A. Taylor and El H.H. Mohamed*

PRO 12: International RILEM Workshop on Historic Mortars: Characteristics and Tests (1999) 460 pp., ISBN: 2-912143-15-2; e-ISBN: 2-351580-27-3;
Eds. P. Bartos, C. Groot and J.J. Hughes

PRO 13: 2nd International RILEM Symposium on Hydration and Setting (1997) 438 pp., ISBN: 2-912143-16-0; e-ISBN: 2-35158-028-1; *Ed. A. Nonat*

PRO 14: Integrated Life-Cycle Design of Materials and Structures (ILCDES 2000) (2000) 550 pp., ISBN: 951-758-408-3; e-ISBN: 2-351580-29-X, ISSN: 0356-9403;
Ed. S. Sarja

PRO 15: Fifth RILEM Symposium on Fibre-Reinforced Concretes (FRC) – BEFIB'2000 (2000) 810 pp., ISBN: 2-912143-18-7; e-ISBN: 2-912143-73-X;
Eds. P. Rossi and G. Chanvillard

PRO 16: Life Prediction and Management of Concrete Structures (2000) 242 pp., ISBN: 2-912143-19-5; e-ISBN: 2-351580-30-3; *Ed. D. Naus*

PRO 17: Shrinkage of Concrete – Shrinkage 2000 (2000) 586 pp.,
ISBN: 2-912143-20-9; e-ISBN: 2-351580-31-1; *Eds. V. Baroghel-Bouny and P.-C. Aïtcin*

PRO 18: Measurement and Interpretation of the On-Site Corrosion Rate (1999) 238 pp., ISBN: 2-912143-21-7; e-ISBN: 2-351580-32-X; *Eds. C. Andrade, C. Alonso, J. Fullea, J. Polimon and J. Rodriguez*

PRO 19: Testing and Modelling the Chloride Ingress into Concrete (2000) 516 pp., ISBN: 2-912143-22-5; e-ISBN: 2-351580-33-8; Soft cover, *Eds. C. Andrade and J. Kropp*

PRO 20: 1st International RILEM Workshop on Microbial Impacts on Building Materials (2000) 74 pp., e-ISBN: 2-35158-013-3; *Ed. M. Ribas Silva (CD 02)*

PRO 21: International RILEM Symposium on Connections between Steel and Concrete (2001) 1448 pp., ISBN: 2-912143-25-X; e-ISBN: 2-351580-34-6;
Ed. R. Eligehausen

PRO 22: International RILEM Symposium on Joints in Timber Structures (2001) 672 pp., ISBN: 2-912143-28-4; e-ISBN: 2-351580-35-4; *Eds. S. Aicher and H.-W. Reinhardt*

PRO 23: International RILEM Conference on Early Age Cracking in Cementitious Systems (2003) 398 pp., ISBN: 2-912143-29-2; e-ISBN: 2-351580-36-2;
Eds. K. Kovler and A. Bentur

PRO 24: 2nd International RILEM Workshop on Frost Resistance of Concrete (2002) 400 pp., ISBN: 2-912143-30-6; e-ISBN: 2-351580-37-0, Hard back;
Eds. M.J. Setzer, R. Auberg and H.-J. Keck

PRO 25: International RILEM Workshop on Frost Damage in Concrete (1999) 312 pp., ISBN: 2-912143-31-4; e-ISBN: 2-351580-38-9, Soft cover;
Eds. D.J. Janssen, M.J. Setzer and M.B. Snyder

PRO 26: International RILEM Workshop on On-Site Control and Evaluation of Masonry Structures (2003) 386 pp., ISBN: 2-912143-34-9;e-ISBN: 2-351580-14-1, Soft cover; *Eds. L. Binda and R.C. de Vekey*

PRO 27: International RILEM Symposium on Building Joint Sealants (1988) 240 pp., e-ISBN: 2-351580-15-X; *Ed. A.T. Wolf, (CD03)*

PRO 28: 6th International RILEM Symposium on Performance Testing and Evaluation of Bituminous Materials, PTEBM'03, Zurich, Switzerland (2003) 652 pp., ISBN: 2-912143-35-7; e-ISBN: 2-912143-77-2, Soft cover; *Ed. M.N. Partl (CD06)*

PRO 29: 2nd International RILEM Workshop on Life Prediction and Ageing Management of Concrete Structures, Paris, France (2003) 402 pp.,
ISBN: 2-912143-36-5; e-ISBN: 2-912143-78-0, Soft cover; *Ed. D.J. Naus*

PRO 30: 4th International RILEM Workshop on High Performance Fiber Reinforced Cement Composites – HPFRCC 4, University of Michigan, Ann Arbor, USA (2003) 562 pp., ISBN: 2-912143-37-3; e-ISBN: 2-912143-79-9, Hard back;
Eds. A.E. Naaman and H.W. Reinhardt

PRO 31: International RILEM Workshop on Test and Design Methods for Steel Fibre Reinforced Concrete: Background and Experiences (2003) 230 pp.,
ISBN: 2-912143-38-1; e-ISBN: 2-351580-16-8, Soft cover; *Eds. B. Schnütgen and L. Vandewalle*

PRO 32: International Conference on Advances in Concrete and Structures, 2 volumes (2003) 1592 pp., ISBN (set): 2-912143-41-1; e-ISBN: 2-351580-17-6, Soft cover; *Eds. Ying-shu Yuan, Surendra P. Shah and Heng-lin Lü*

PRO 33: 3rd International Symposium on Self-Compacting Concrete (2003) 1048 pp., ISBN: 2-912143-42-X; e-ISBN: 2-912143-71-3, Soft cover;
Eds. Ó. Wallevik and I. Níelsson

PRO 34: International RILEM Conference on Microbial Impact on Building Materials (2003) 108 pp., ISBN: 2-912143-43-8; e-ISBN: 2-351580-18-4;
Ed. M. Ribas Silva

PRO 35: International RILEM TC 186-ISA on Internal Sulfate Attack and Delayed Ettringite Formation (2002) 316 pp.,ISBN: 2-912143-44-6;e-ISBN: 2-912143-80-2, Soft cover; *Eds. K. Scrivener and J. Skalny*

PRO 36: International RILEM Symposium on Concrete Science and Engineering – A Tribute to Arnon Bentur (2004) 264 pp., ISBN: 2-912143-46-2;
e-ISBN: 2-912143-58-6, Hard back; *Eds. K. Kovler, J. Marchand, S. Mindess and J. Weiss*

PRO 37: 5th International RILEM Conference on Cracking in Pavements – Mitigation, Risk Assessment and Prevention (2004) 740 pp., ISBN: 2-912143-47-0; e-ISBN: 2-912143-76-4, Hard back; *Eds. C. Petit, I. Al-Qadi and A. Millien*

PRO 38: 3rd International RILEM Workshop on Testing and Modelling the Chloride Ingress into Concrete (2002) 462 pp., ISBN: 2-912143-48-9; e-ISBN: 2-912143-57-8, Soft cover; *Eds. C. Andrade and J. Kropp*

PRO 39: 6th International RILEM Symposium on Fibre-Reinforced Concretes (BEFIB 2004), 2 volumes, (2004) 1536 pp., ISBN: 2-912143-51-9 (set); e-ISBN: 2-912143-74-8, Hard back; *Eds. M. Di Prisco, R. Felicetti and G.A. Plizzari*

PRO 40: International RILEM Conference on the Use of Recycled Materials in Buildings and Structures (2004) 1154 pp., ISBN: 2-912143-52-7 (set); e-ISBN: 2-912143-75-6, Soft cover; *Eds. E. Vázquez, Ch. F. Hendriks and G.M.T. Janssen*

PRO 41: RILEM International Symposium on Environment-Conscious Materials and Systems for Sustainable Development (2005) 450 pp., ISBN: 2-912143-55-1; e-ISBN: 2-912143-64-0, Soft cover; *Eds. N. Kashino and Y. Ohama*

PRO 42: SCC'2005 – China: 1st International Symposium on Design, Performance and Use of Self-Consolidating Concrete (2005) 726 pp., ISBN: 2-912143-61-6; e-ISBN: 2-912143-62-4, Hard back; *Eds. Zhiwu Yu, Caijun Shi, Kamal Henri Khayat and Youjun Xie*

PRO 43: International RILEM Workshop on Bonded Concrete Overlays (2004) 114 pp., e-ISBN: 2-912143-83-7; *Eds. J.L. Granju and J. Silfwerbrand*

PRO 44: 2nd International RILEM Workshop on Microbial Impacts on Building Materials (Brazil 2004) (CD11) 90 pp., e-ISBN: 2-912143-84-5; *Ed. M. Ribas Silva*

PRO 45: 2nd International Symposium on Nanotechnology in Construction, Bilbao, Spain (2005) 414 pp., ISBN: 2-912143-87-X; e-ISBN: 2-912143-88-8, Soft cover; *Eds. Peter J.M. Bartos, Yolanda de Miguel and Antonio Porro*

PRO 46: ConcreteLife'06 – International RILEM-JCI Seminar on Concrete Durability and Service Life Planning: Curing, Crack Control, Performance in Harsh Environments (2006) 526 pp., ISBN: 2-912143-89-6; e-ISBN: 2-912143-90-X, Hard back; *Ed. K. Kovler*

PRO 47: International RILEM Workshop on Performance Based Evaluation and Indicators for Concrete Durability (2007) 385 pp., ISBN: 978-2-912143-95-2; e-ISBN: 978-2-912143-96-9, Soft cover; *Eds. V. Baroghel-Bouny, C. Andrade, R. Torrent and K. Scrivener*

PRO 48: 1st International RILEM Symposium on Advances in Concrete through Science and Engineering (2004) 1616 pp., e-ISBN: 2-912143-92-6; *Eds. J. Weiss, K. Kovler, J. Marchand, and S. Mindess*

PRO 49: International RILEM Workshop on High Performance Fiber Reinforced Cementitious Composites in Structural Applications (2006) 598 pp., ISBN: 2-912143-93-4; e-ISBN: 2-912143-94-2, Soft cover; *Eds. G. Fischer and V.C. Li*

PRO 50: 1st International RILEM Symposium on Textile Reinforced Concrete (2006) 418 pp., ISBN: 2-912143-97-7; e-ISBN: 2-351580-08-7, Soft cover; *Eds. Josef Hegger, Wolfgang Brameshuber and Norbert Will*

PRO 51: 2nd International Symposium on Advances in Concrete through Science and Engineering (2006) 462 pp., ISBN: 2-35158-003-6; e-ISBN: 2-35158-002-8, Hard back; *Eds. J. Marchand, B. Bissonnette, R. Gagné, M. Jolin and F. Paradis*

PRO 52: Volume Changes of Hardening Concrete: Testing and Mitigation (2006) 428 pp., ISBN: 2-35158-004-4; e-ISBN: 2-35158-005-2, Soft cover; *Eds. O.M. Jensen, P. Lura and K. Kovler*

PRO 53: High Performance Fiber Reinforced Cement Composites HPFRCC5 (2007) 542 pp., ISBN: 978-2-35158-046-2; e-ISBN: 978-2-35158-089-9, Hard back; *Eds. H.W. Reinhardt and A.E. Naaman*

PRO 54: 5th International RILEM Symposium on Self-Compacting Concrete, 3 Volumes (2007) 1198 pp., ISBN: 978-2-35158-047-9; e-ISBN: 978-2-35158-088-2, Soft cover; *Eds. G. De Schutter and V. Boel*

PRO 55: International RILEM Symposium Photocatalysis, Environment and Construction Materials (2007) 350 pp., ISBN: 978-2-35158-056-1; e-ISBN: 978-2-35158-057-8, Soft cover; *Eds. P. Baglioni and L. Cassar*

PRO 56: International RILEM Workshop on Integral Service Life Modelling of Concrete Structures (2007) 458 pp., ISBN 978-2-35158-058-5; e-ISBN: 978-2-35158-090-5, Hard back; *Eds. R.M. Ferreira, J. Gulikers and C. Andrade*

PRO 57: RILEM Workshop on Performance of cement-based materials in aggressive aqueous environments (2008) 132 pp., e-ISBN: 978-2-35158-059-2; *Ed. N. De Belie*

PRO 58: International RILEM Symposium on Concrete Modelling CONMOD'08 (2008) 847 pp., ISBN: 978-2-35158-060-8; e-ISBN: 978-2-35158-076-9, Soft cover; *Eds. E. Schlangen and G. De Schutter*

PRO 59: International RILEM Conference on On Site Assessment of Concrete, Masonry and Timber Structures SACoMaTiS 2008, 2 volumes (2008) 1232 pp., ISBN: 978-2-35158-061-5 (set); e-ISBN: 978-2-35158-075-2, Hard back; *Eds. L. Binda, M. di Prisco and R. Felicetti*

PRO 60: Seventh RILEM International Symposium (BEFIB 2008) on Fibre Reinforced Concrete: Design and Applications (2008) 1181 pp, ISBN: 978-2-35158-064-6; e-ISBN: 978-2-35158-086-8, Hard back; *Ed. R. Gettu*

PRO 61: 1st International Conference on Microstructure Related Durability of Cementitious Composites (Nanjing), 2 volumes, (2008) 1524 pp., ISBN: 978-2-35158-065-3; e-ISBN: 978-2-35158-084-4; *Eds. W. Sun, K. van Breugel, C. Miao, G. Ye and H. Chen*

PRO 62: NSF/ RILEM Workshop: In-situ Evaluation of Historic Wood and Masonry Structures (2008) 130 pp., e-ISBN: 978-2-35158-068-4; *Eds. B. Kasal, R. Anthony and M. Drdácký*

PRO 63: Concrete in Aggressive Aqueous Environments: Performance, Testing and Modelling, 2 volumes, (2009) 631 pp., ISBN: 978-2-35158-071-4; e-ISBN: 978-2-35158-082-0, Soft cover; *Eds. M.G. Alexander and A. Bertron*

PRO 64: Long Term Performance of Cementitious Barriers and Reinforced Concrete in Nuclear Power Plants and Waste Management – NUCPERF 2009 (2009) 359 pp., ISBN: 978-2-35158-072-1; e-ISBN: 978-2-35158-087-5; *Eds. V. L'Hostis, R. Gens, C. Gallé*

PRO 65: Design Performance and Use of Self-consolidating Concrete, SCC'2009, (2009) 913 pp., ISBN: 978-2-35158-073-8; e-ISBN: 978-2-35158-093-6; *Eds. C. Shi, Z. Yu, K.H. Khayat and P. Yan*

PRO 66: Concrete Durability and Service Life Planning, 2nd International RILEM Workshop, ConcreteLife'09, (2009) 626 pp., ISBN: 978-2-35158-074-5; e-ISBN: 978-2-35158-085-1; *Ed. K. Kovler*

PRO 67: Repairs Mortars for Historic Masonry (2009) 397 pp., e-ISBN: 978-2-35158-083-7; *Ed. C. Groot*

PRO 68: Proceedings of the 3rd International RILEM Symposium on 'Rheology of Cement Suspensions such as Fresh Concrete' (2009) 372 pp., ISBN: 978-2-35158-091-2; e-ISBN: 978-2-35158-092-9; *Eds. O.H. Wallevik, S. Kubens and S. Oesterheld*

PRO 69: 3rd International PhD Student Workshop on 'Modelling the Durability of Reinforced Concrete' (2009) 122 pp., ISBN: 978-2-35158-095-0; e-ISBN: 978-2-35158-094-3; *Eds. R. M. Ferreira, J. Gulikers and C. Andrade*

PRO 71: Advances in Civil Engineering Materials, Proceedings of the 'The 50-year Teaching Anniversary of Prof. Sun Wei', (2010) 307 pp.,ISBN: 978-2-35158-098-1; e-ISBN: 978-2-35158-099-8; *Eds. C. Miao, G. Ye, and H. Chen*

PRO 74: International RILEM Conference on 'Use of Superabsorsorbent Polymers and Other New Additives in Concrete' (2010) 374 pp., ISBN: 978-2-35158-104-9; e-ISBN: 978-2-35158-105-6; *Eds. O.M. Jensen, M.T. Hasholt, and S. Laustsen*

PRO 75: International Conference on 'Material Science - 2nd ICTRC - Textile Reinforced Concrete - Theme 1' (2010) 436 pp., ISBN: 978-2-35158-106-3; e-ISBN: 978-2-35158-107-0; *Ed. W. Brameshuber*

PRO 76: International Conference on 'Material Science - HetMat - Modelling of Heterogeneous Materials - Theme 2' (2010) 255 pp., ISBN: 978-2-35158-108-7; e-ISBN: 978-2-35158-109-4; *Ed. W. Brameshuber*

PRO 77: International Conference on 'Material Science - AdIPoC - Additions Improving Properties of Concrete - Theme 3' (2010) 459 pp., ISBN: 978-2-35158-110-0; e-ISBN: 978-2-35158-111-7; *Ed. W. Brameshuber*

PRO 78: 2nd Historic Mortars Conference and RILEM TC 203-RHM Final Workshop – HMC2010 (2010) 1416 pp., e-ISBN: 978-2-35158-112-4; *Eds J. Válek, C. Groot, and J.J. Hughes*

PRO 79: International RILEM Conference on Advances in Construction Materials Through Science and Engineering (2011) 213 pp., e-ISBN: 978-2-35158-117-9; *Eds Christopher Leung and K.T. Wan*

PRO 80: 2nd International RILEM Conference on Concrete Spalling due to Fire Exposure (2011) 453 pp., ISBN: 978-2-35158-118-6, e-ISBN: 978-2-35158-119-3; *Eds E.A.B. Koenders and F. Dehn*

PRO 81: 2nd International RILEM Conference on Strain Hardening Cementitious Composites (SHCC2-Rio) (2011) 451 pp., ISBN: 978-2-35158-120-9, e-ISBN: 978-2-35158-121-6; *Eds R.D. Toledo Filho, F.A. Silva, E.A.B. Koenders and E.M.R. Fairbairn*

PRO 82: 2nd International RILEM Conference on Progress of Recycling in the Built Environment (2011) 507 pp., e-ISBN: 978-2-35158-122-3; *Eds V.M. John, E. Vazquez, S.C. Angulo and C. Ulsen*

PRO 85: RILEM-JCI International Workshop on Crack Control of Mass Concrete and Related Issues concerning Early-Age of Concrete Structures – ConCrack 3 (2012) 237 pp., ISBN: 978-2-35158-125-4, e-ISBN: 978-2-35158-126-1; *Eds F. Toutlemonde and J.M. Torrenti*

RILEM REPORTS

Report 19: Considerations for Use in Managing the Aging of Nuclear Power Plant Concrete Structures (1999) 224 pp., ISBN: 2-912143-07-1; e-ISBN: 2-35158-039-7; *Ed. D.J. Naus*

Report 20: Engineering and Transport Properties of the Interfacial Transition Zone in Cementitious Composites (1999) 396 pp., ISBN: 2-912143-08-X; e-ISBN: 2-35158-040-0; *Eds. M.G. Alexander, G. Arliguie, G. Ballivy, A. Bentur and J. Marchand*

Report 21: Durability of Building Sealants (1999) 450 pp., ISBN: 2-912143-12-8; e-ISBN: 2-35158-041-9; *Ed. A.T. Wolf*

Report 22: Sustainable Raw Materials – Construction and Demolition Waste (2000) 202 pp., ISBN: 2-912143-17-9; e-ISBN: 2-35158-042-7; *Eds. C.F. Hendriks and H.S. Pietersen*

Report 23: Self-Compacting Concrete state-of-the-art report (2001) 166 pp., ISBN: 2-912143-23-3; e-ISBN: 2-912143-59-4, Soft cover; *Eds. Å. Skarendahl and Ö. Petersson*

Report 24: Workability and Rheology of Fresh Concrete: Compendium of Tests (2002) 154 pp., ISBN: 2-912143-32-2; e-ISBN: 2-35158-043-5, Soft cover; *Eds. P.J.M. Bartos, M. Sonebi and A.K. Tamimi*

Report 25: Early Age Cracking in Cementitious Systems (2003) 350 pp., ISBN: 2-912143-33-0; e-ISBN: 2-912143-63-2, Soft cover; *Ed. A. Bentur*

Report 26: Towards Sustainable Roofing (Joint Committee CIB/RILEM) (CD 07), (2001) 28 pp., e-ISBN: 2-912143-65-9; *Eds. Thomas W. Hutchinson and Keith Roberts*

Report 27: Condition Assessment of Roofs (Joint Committee CIB/RILEM) (CD 08), (2003) 12 pp., e-ISBN: 2-912143-66-7

Report 28: Final report of RILEM TC 167-COM 'Characterisation of Old Mortars with Respect to Their Repair' (2007) 192 pp., ISBN: 978-2-912143-56-3; e-ISBN: 978-2-912143-67-9, Soft cover; *Eds. C. Groot, G. Ashall and J. Hughes*

Report 29: Pavement Performance Prediction and Evaluation (PPPE): Interlaboratory Tests (2005) 194 pp., e-ISBN: 2-912143-68-3; *Eds. M. Partl and H. Piber*

Report 30: Final Report of RILEM TC 198-URM 'Use of Recycled Materials' (2005) 74 pp., ISBN: 2-912143-82-9; e-ISBN: 2-912143-69-1 – Soft cover; *Eds. Ch. F. Hendriks, G.M.T. Janssen and E. Vázquez*

Report 31: Final Report of RILEM TC 185-ATC 'Advanced testing of cement-based materials during setting and hardening' (2005) 362pp.,ISBN: 2-912143-81-0; e-ISBN: 2-912143-70-5 – Soft cover; *Eds. H.W. Reinhardt and C.U. Grosse*

Report 32: Probabilistic Assessment of Existing Structures. A JCSS publication (2001) 176 pp., ISBN 2-912143-24-1; e-ISBN: 2-912143-60-8 – Hard back; *Ed. D. Diamantidis*

Report 33: State-of-the-Art Report of RILEM Technical Committee TC 184-IFE 'Industrial Floors' (2006) 158 pp., ISBN 2-35158-006-0; e-ISBN: 2-35158-007-9, Soft cover; *Ed. P. Seidler*

Report 34: Report of RILEM Technical Committee TC 147-FMB 'Fracture mechanics applications to anchorage and bond' Tension of Reinforced Concrete Prisms – Round Robin Analysis and Tests on Bond (2001) 248 pp., e-ISBN 2-912143-91-8; *Eds. L. Elfgren and K. Noghabai*

Report 35: Final Report of RILEM Technical Committee TC 188-CSC 'Casting of Self Compacting Concrete' (2006) 40 pp., ISBN 2-35158-001-X; e-ISBN: 2-912143-98-5 – Soft cover; *Eds. Å. Skarendahl and P. Billberg*

Report 36: State-of-the-Art Report of RILEM Technical Committee TC 201-TRC 'Textile Reinforced Concrete' (2006) 292 pp., ISBN 2-912143-99-3; e-ISBN: 2-35158-000-1, Soft cover; *Ed. W. Brameshuber*

Report 37: State-of-the-Art Report of RILEM Technical Committee TC 192-ECM 'Environment-conscious construction materials and systems' (2007) 88 pp., ISBN: 978-2-35158-053-0; e-ISBN: 2-35158-079-0, Soft cover; *Eds. N. Kashino, D. Van Gemert and K. Imamoto*

Report 38: State-of-the-Art Report of RILEM Technical Committee TC 205-DSC 'Durability of Self-Compacting Concrete' (2007) 204 pp., ISBN: 978-2-35158-048-6; e-ISBN: 2-35158-077-6, Soft cover; *Eds. G. De Schutter and K. Audenaert*

Report 39: Final Report of RILEM Technical Committee TC 187-SOC 'Experimental determination of the stress-crack opening curve for concrete in tension' (2007) 54 pp., ISBN 978-2-35158-049-3; e-ISBN: 978-2-35158-078-3, Soft cover; *Ed. J. Planas*

Report 40: State-of-the-Art Report of RILEM Technical Committee TC 189-NEC 'Non-Destructive Evaluation of the Penetrability and Thickness of the Concrete Cover' (2007) 246 pp., ISBN 978-2-35158-054-7; e-ISBN: 978-2-35158-080-6, Soft cover; *Eds. R. Torrent and L. Fernández Luco*

Report 41: State-of-the-Art Report of RILEM Technical Committee TC 196-ICC 'Internal Curing of Concrete' (2007) 164 pp., ISBN: 978-2-35158-009-7; e-ISBN: 978-2-35158-082-0, Soft cover; *Eds. K. Kovler and O.M. Jensen*

Report 42: 'Acoustic Emission and Related Non-destructive Evaluation Techniques for Crack Detection and Damage Evaluation in Concrete' – Final Report of RILEM Technical Committee 212-ACD (2010) 12 pp., e-ISBN: 978-2-35158-100-1; *Ed. M. Ohtsu*

RILEM Publications Published by Springer

RILEM BOOKSERIES (Proceedings)

VOL. 1: Design, Production and Placement of Self-Consolidating Concrete (2010) 466 pp., ISBN: 978-90-481-9663-0; e-ISBN: 978-90-481-9664-7, Hardcover; *Eds. K. Khayat and D. Feyes*

VOL. 2: High Performance Fiber Reinforced Cement Composites 6 – HPFRCC6 (2011) 584 pp., ISBN: 978-94-007-2435-8; e-ISBN: 978-94-007-2436-5, Hardcover; *Eds. G.J. Parra-Montesinos, H.W. Reinhardt and A.E. Naaman*

VOL. 3: Advances in Modeling Concrete Service Life – Proceedings of 4th International RILEM PhD Workshop held in Madrid, Spain, November 19, 2010 (2012) 170 pp., ISBN: 978-94-007-2702-1; e-ISBN: 978-94-007-2703-8, Hardcover; *Eds. C. Andrade and Joost Gulikers*

VOL. 5: Joint fib-RILEM Workshop on Modelling of Corroding Concrete Structures (2011) 290 pp., ISBN: 978-94-007-0676-7; e-ISBN: 978-94-007-0677-4, Hardcover; *Eds. C. Andrade and G. Mancini*

For the latest publications in the RILEM Bookseries, please visit http://www.springer.com/series/8781

RILEM STATE-OF-THE-ART REPORTS

VOL. 1: State-of-the-Art Report of RILEM Technical Committee TC 207-INR 'Non-Destructive Assessment of Concrete Structures: Reliability and Limits of Single and Combined Techniques' (2012) 390 pp., ISBN: 978-94-007-2735-9; e-ISBN: 978-94-007-2736-6, Hardcover; *Ed. D. Breysse*

VOL. 2: State-of-the-Art Report of RILEM Technical Committee TC 225-SAP 'Application of Super Absorbent Polymers (SAP) in Concrete Construction' (2012) 165 pp., ISBN: 978-94-007-2732-8; e-ISBN: 978-94-007-2733-5, Hardcover; *Eds. V. Mechtcherine and H-W. Reinhardt*

VOL. 3: State-of-the-Art Report of RILEM Technical Committee TC 193-RLS 'Bonded Cement-Based Material Overlays for the Repair, the Lining or the Strengthening of Slabs or Pavements' (2011) 198 pp., ISBN: 978-94-007-1238-6; e-ISBN: 978-94-007-1239-3, Hardcover; *Eds. B. Bissonnette, L. Courard, D.W. Fowler and J-L. Granju*

VOL. 4: State-of-the-Art Report prepared by Subcommittee 2 of RILEM Technical Committee TC 208-HFC 'Durability of Strain-Hardening Fibre-Reinforced Cement-Based Composites' (SHCC) (2011) 151 pp., ISBN: 978-94-007-0337-7; e-ISBN: 978-94-007-0338-4, Hardcover; *Eds. G.P.A.G. van Zijl and F.H. Wittmann*

VOL. 5: State-of-the-Art Report of RILEM Technical Committee TC 194-TDP 'Application of Titanium Dioxide Photocatalysis to Construction Materials' (2011) 60 pp., ISBN: 978-94-007-1296-6; e-ISBN: 978-94-007-1297-3, Hardcover; *Eds. Yoshihiko Ohama and Dionys Van Gemert*

VOL. 7: State-of-the-Art Report of RILEM Technical Committee TC 215-AST 'In Situ Assessment of Structural Timber' (2010) 152 pp., ISBN: 978-94-007-0559-3; e-ISBN: 978-94-007-0560-9, Hardcover; *Eds. B. Kasal and T. Tannert*

For the latest publications in the RILEM State-of-the-Art Reports, please visit http://www.springer.com/series/8780

Printed by Publishers' Graphics LLC